Instrument Rating Manual

International Standard Book Number 0-88487-069-3

JS314299A

INTRODUCTION

I. SUBJECT CONTENT OF THE COURSE

The main feature of the *Jeppesen Sanderson Instrument Rating Manual* is its completeness. With this book you will receive the total training (ground phase) required for an instrument rating. In addition to the knowledge areas required by FAR Parts 61 and 141, subjects such as gaining instrument flight experience, advanced aircraft performance, up-to-date instrument flying techniques, medical facts for pilots, and much more are included. Through this total training concept you will be better prepared for the FAA written exam and the required flight instruction.

The general knowledge areas presented in this manual are as follows:

(1) Instruments, Systems, and Attitude Flying
(2) Meteorology
(3) Weather Forecasts and Reports
(4) National Airspace System
(5) Radio Navigation Systems
(6) Instrument Navigation Charts
(7) Instrument Approaches
(8) ATC Procedures
(9) Airman's Information Publications
(10) Flight Computer
(11) Flight Planning Considerations
(12) Emergency IFR Procedures
(13) Safety of Flight
(14) Job Performance

II. HOW TO USE THE COURSE

You can achieve the best results from this course if you follow the instructions and guidance specifically as suggested.

A. Start the course immediately! Schedule your time for a program of regular study. If you are not yet enrolled in an instrument flight training program, visit your local flight training school for a brief orientation to instrument flying.

B. Start at the beginning and proceed through the material in the order presented, adhering to the instructions in the following steps.

1. *Study* the text material thoroughly.
2. Satisfy yourself that you understand each part before continuing to the next part, and follow this procedure throughout the course. After you have completed the subject coverage, complete the "Pilot Job Performance" exercises in section 14, then take the final exam.
3. For best results, complete this course as soon as possible, preferably within 60 days.
4. A final examination and exam answer key are included as an integral part of this book. Also included are all the necessary IFR charts. The instrument flight charts are intended for instructional purposes only and should not be used for actual flight activities.

If you have any questions or need explanations about any part of the course, we are prepared to offer assistance at any time.

Just write: **Marketing Manager, Aviation Products**
Jeppesen Sanderson, Inc.
55 Inverness Drive East
Englewood, Colorado 80112

TABLE OF CONTENTS

1 INSTRUMENTS, SYSTEMS, AND ATTITUDE FLYING

ENGINE INSTRUMENTS AND SYSTEMS

Engine instruments indicate power, temperatures, and general working conditions of the engine. In addition, they provide valuable information regarding the engine support systems such as oil, ignition, fuel, induction, and electrical.

POWER OUTPUT INDICATORS

TACHOMETER

On training aircraft with fixed-pitch propellers, the tachometer is the only power output indicator. Since there is a direct relationship between engine speed and power output, engine power is adjusted by reference to the tachometer. Many tachometers have markings which indicate maximum cruise power at various altitudes. The instrument shown in figure 1-1 has maximum cruise power markings at sea level, 5,000 feet, and 10,000 feet MSL.

Fig. 1-1. Tachometer

MANIFOLD PRESSURE

As shown in figure 1-2, the manifold pressure gauge is basically a barometer which measures absolute pressure in the intake manifold. By adjusting throttle position, the pilot controls the pressure and, therefore, volume of fuel-air mixture to the engine. The amount of fuel-air mixture is then indicated by the manifold pressure gauge.

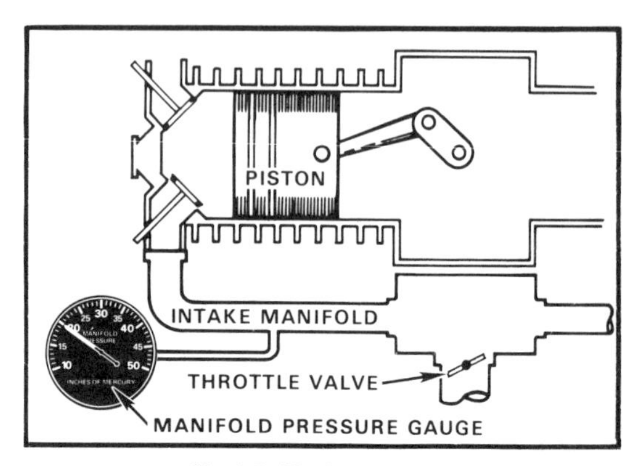

Fig. 1-2. Manifold Pressure

On an aircraft with a constant-speed propeller, the tachometer and manifold pressure gauges provide an accurate indication of power output. A specific, known power setting can be established by adjusting the propeller and throttle to predetermined engine speed and manifold pressure values.

TEMPERATURE GAUGES

To provide adequate service life and reliability, the aircraft engine must be operated within specified temperature limits. Engine temperature normally is measured at the hottest cylinder and displayed on a cylinder-head temperature gauge. However, on aircraft not equipped with such a gauge, the pilot must rely on an oil temperature gauge for this information. This source is reliable since engine oil will maintain approximately the same temperature as the engine in which it is being used.

As shown in figure 1-3, both engine temperature gauges are color coded. The green colored area is the normal operating range while the red line denotes the maximum allowable temperature.

Since the temperature of engine exhaust gases is a reliable indication of the fuel-air mixture, many aircraft are equipped with an *exhaust gas temperature* (EGT) gauge. If the engine is leaned until the EGT reaches maximum, then enriched for approximately a 100° Fahrenheit drop in temperature, the *best power* fuel-air mixture is utilized. The best economical mixture results when the engine is leaned to peak EGT, then enriched until the temperature decreases 25° to 50° (depending on the engine). A typical EGT gauge is shown in figure 1-4.

OIL SYSTEM

Oil for the engine usually is stored in an oil sump at the bottom of the crankcase. Oil is

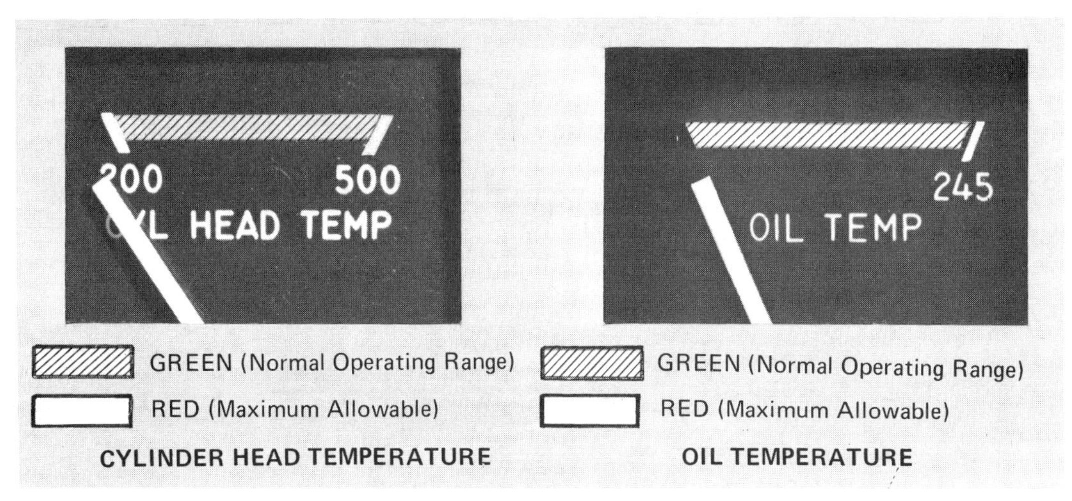

Fig. 1-3. Engine Temperature Gauges

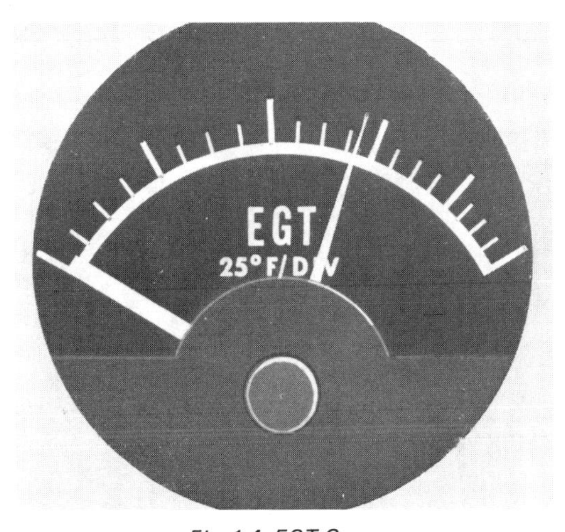

Fig. 1-4. EGT Gauge

pumped under pressure from the sump, to the bearings in the engine, then it drains back into the oil sump for recirculation. A typical oil system is shown in figure 1-5.

The two indicators provided to monitor the engine oil system are the oil temperature and pressure gauges. Both are located on the instrument panel in easy view of the pilot.

Although the main purpose of engine oil is to lubricate the moving parts of the powerplant, another important function is to cool the engine by radiating away heat collected as oil circulates through the engine. The consequences of starting a flight with an insufficient supply of oil can be quite serious, because normal oil consumption could reduce the supply to a dangerously low level.

IGNITION SYSTEM

The electrical spark across the points of the spark plug, which ignites the fuel-air mixture and begins the power stroke, does not come from the electrical system of the airplane. Since the ignition system is completely separate from the electrical system, the engine will continue to run if the battery or generator fails, or if the master electrical switch is turned off in flight. (See Fig. 1-6).

The two spark plugs in each cylinder are fired by the dual ignition system. The current for the ignition is generated by magnetos attached to the rear of the engine and driven by gears connected to the crankshaft, as shown in figure 1-7. If one of the gear drives, magnetos, some of the wiring, or spark plugs in one system fails, the engine will continue to run normally with only a slight loss of power.

FUEL SYSTEM

Figure 1-8 shows the fuel system of a typical small aircraft. In high-wing aircraft, gravity forces fuel from the tanks to the engine carburetor. In low-wing aircraft, an engine-driven pump delivers fuel under pressure to the carburetor. In addition, an electric pump is installed in many aircraft to provide extra pressure for starting and to insure a flow of fuel if the engine-driven pump fails.

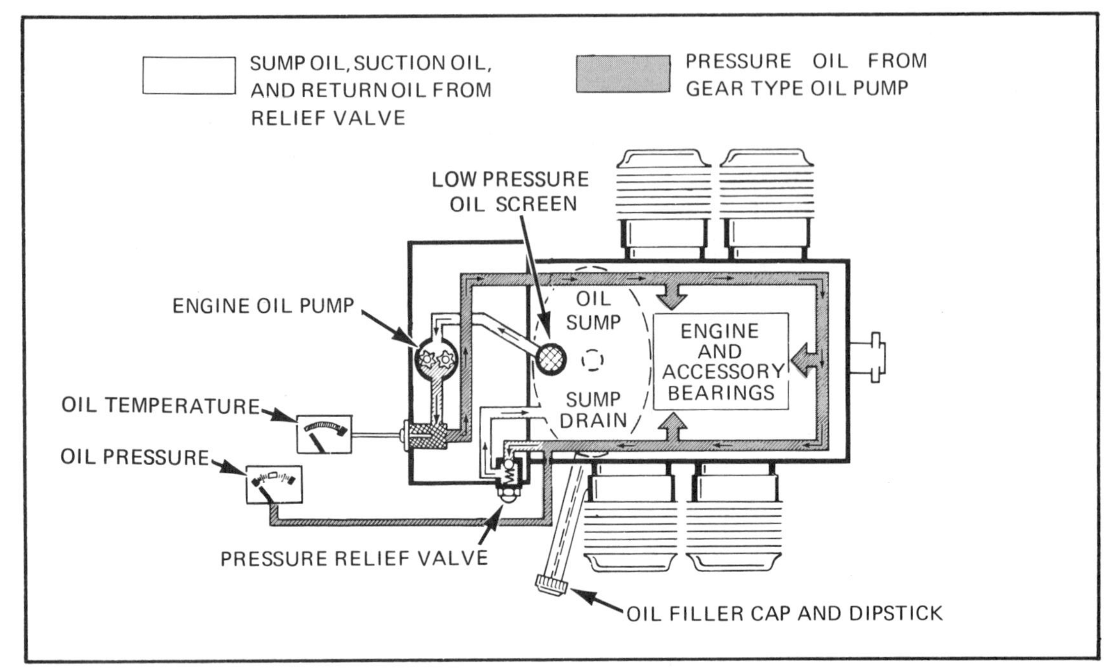

Fig. 1-5. Engine Oil System

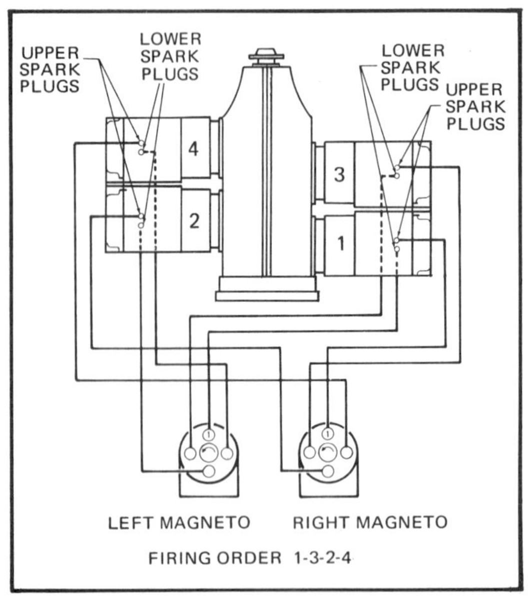

Fig. 1-6. Ignition System

The preflight inspection should include a comparison between the fuel gauge readings and observed quantities in the tanks. At the same time, the fuel color should be checked to be sure it is the recommended grade. (Red-tinted fuel is grade 80, green is 100, and blue is 100 low lead.) Next, the area beneath the tanks should be inspected for colored stains indicating fuel leaks.

Fig. 1-7. Magnetos

Each sump drain or fuel strainer should be actuated to eliminate water and contaminated fuel. Impurities resulting from condensation, fuel truck contamination, or vandalism might otherwise remain in the system. The pilot's operating handbook or flight manual must be studied for tank capacities, consumption rates, and emergency procedures.

A pilot must be familiar with the operation of all fuel controls (tank selector, boost pumps, mixture controls, and primers) to preclude error or delay during emergencies or night operations.

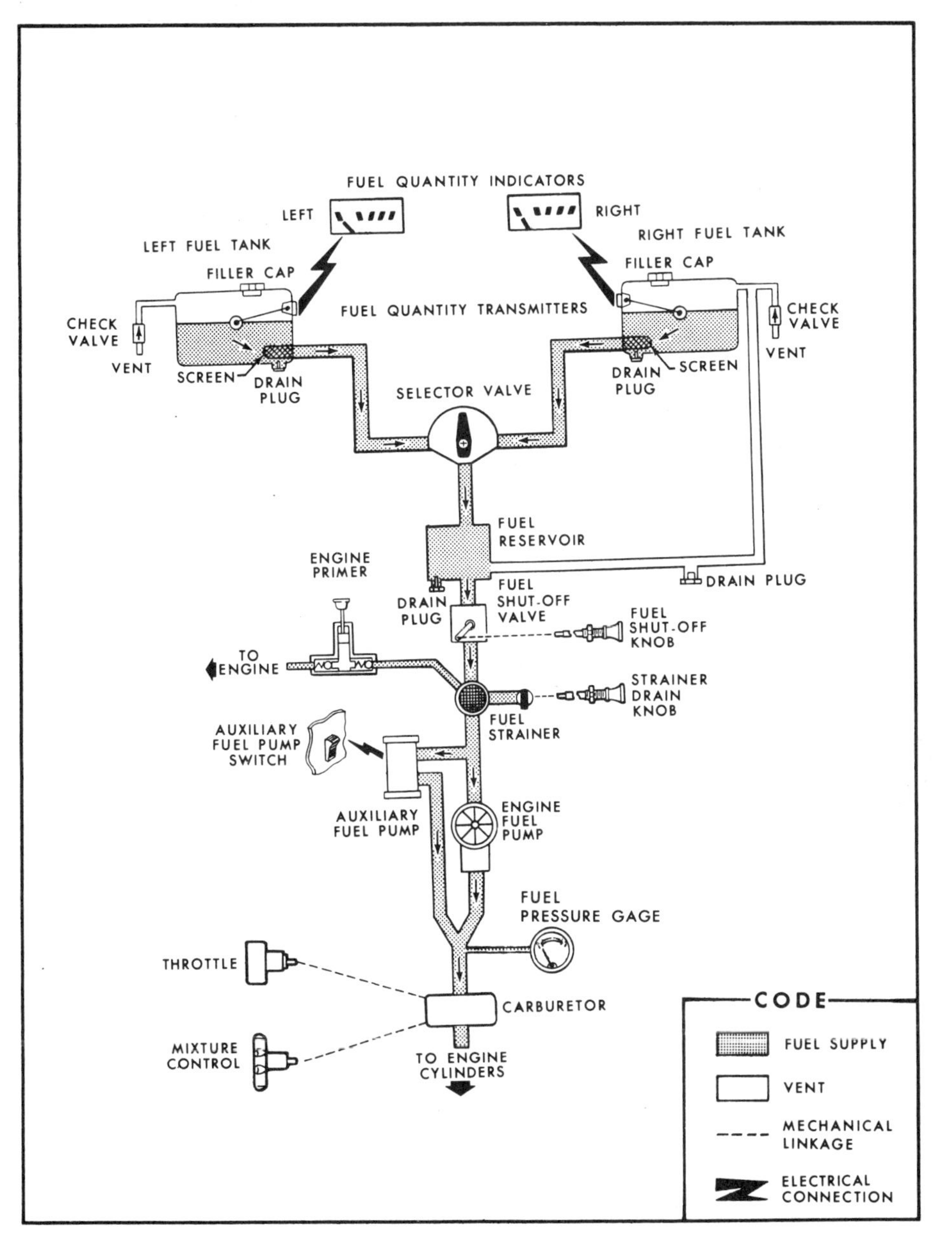

Fig. 1-8. Fuel-Pump Type Fuel System

INDUCTION SYSTEMS

CARBURETOR

The carburetor air intake and filter are located in the front portion of the engine cowling. The filter should be checked periodically for cleanliness since operation on gravel or dirt runways and taxiways eventually may cause reduced air flow. After the air is filtered, it passes through a venturi (narrowed neck) in the carburetor, as shown in figure 1-9.

When air flows rapidly through the venturi, a low pressure area in the throat of the carburetor is created. Because the float chamber is vented to the atmosphere and a low-pressure area exists in the venturi, *atmospheric pressure forces the fuel to flow from the float chamber into the carburetor throat.* The aircraft carburetor is designed to meter the correct quantity of fuel, mix it with air in the proper proportion, and atomize or vaporize the fuel-air mixture before it enters the combustion chamber.

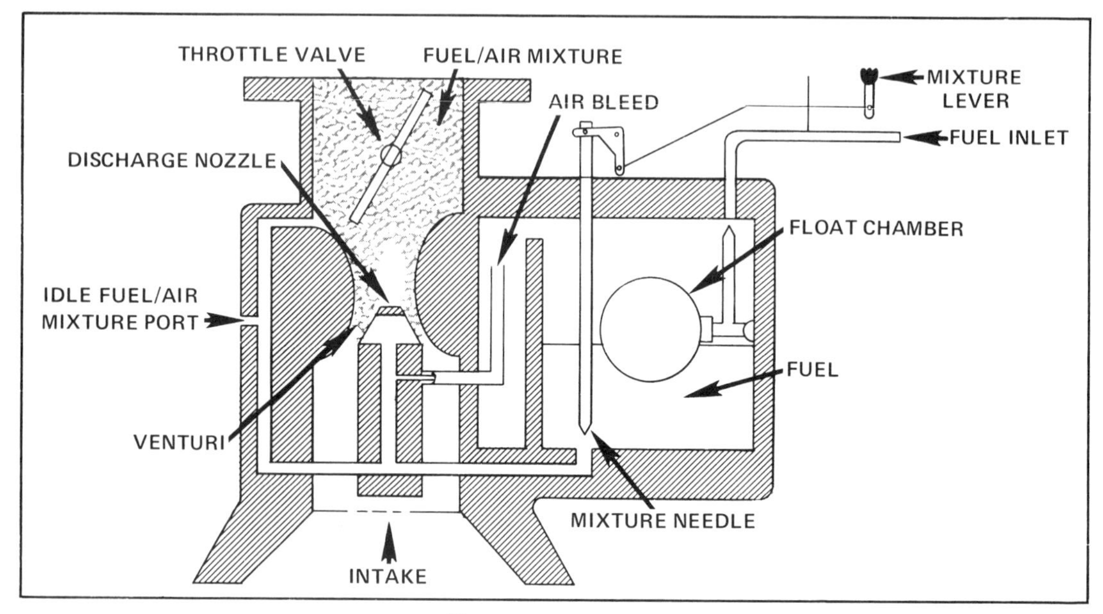

Fig. 1-9. Float Carburetor

CARBURETOR ICE

The cooling effect of fuel vaporization and the air pressure drop across the venturi (necked) portion of the carburetor throat can chill intake air considerably. If the air is humid, moisture may condense and accumulate in the carburetor and intake duct as frost or ice. Even a slight amount of ice will reduce power and a continued accumulation can lead to engine failure. A pilot should become familiar with the particular icing characteristics of each aircraft flown. In general, however, aircraft equipped with carburetors will have the following characteristics.

1. Carburetor ice can occur without visible moisture; however, the danger of carburetor ice is greatest in humid air, at outside air temperatures of 45° to 60° Fahrenheit.
2. The indication of carburetor ice in an aircraft with a fixed pitch propeller is a decrease in r.p.m. and in an aircraft with a constant-speed propeller is a decrease in manifold pressure.
3. Application of carburetor heat allows the warmth of the exhaust system to heat incoming carburetor air, but it *may be ineffective* if carburetor ice is severe enough to cause a large loss of power.
4. Carburetor ice is much more likely to occur at low power settings, so icing symptoms may not be detectable on extended descents.

CARBURETOR HEAT

To combat carburetor ice, a heat system is installed, as depicted in figure 1-10. Outside air is ducted through a heater muff (structure surrounding the exhaust manifold) where the air is heated sufficiently to prevent ice formation or melt ice already formed. After the air has been heated, it is routed to a valve that is controlled by a carburetor heat control located on the instrument panel. Pulling out the carburetor heat knob opens the valve, allowing heated air to pass through the carburetor where it mixes with the fuel, melts any ice, and permits the mixture to continue into the engine cylinders.

Application of carburetor heat causes an immediate, small reduction in power. If ice *is not* present in the carburetor, the power reduction remains constant. When ice *is* present, application of carburetor heat causes a small reduction in power followed (within 10 to 20 seconds) by an increase in power.

MIXTURE CONTROL

The amount of fuel entering the carburetor depends on the volume and *not* the weight of air. Therefore, if the position of the mixture control remains unchanged as the flight altitude increases, the amount of fuel entering the carburetor remains approximately the same for any given throttle setting. Since air is less dense at higher altitude, the fuel/air ratio changes and the mixture becomes richer as altitude increases.

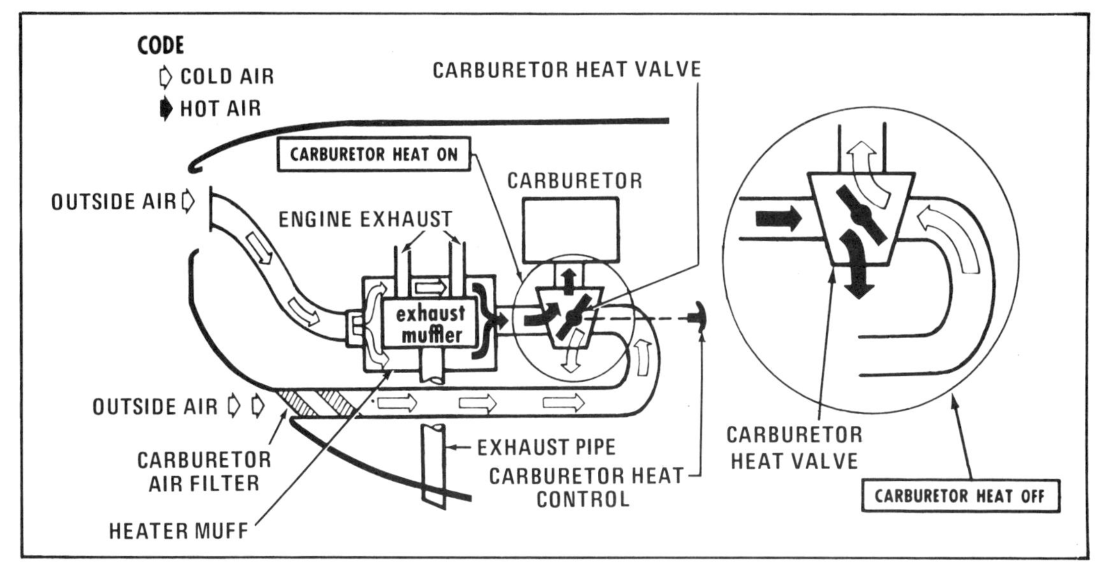

Fig. 1-10. Carburetor Heat System

A mixture control knob is provided to meter the amount of fuel that passes through the main jet in the carburetor and to enable the pilot to adjust the *ratio* of the fuel/air mixture that goes into the cylinders. If the fuel/air mixture is too lean (too little fuel for the amount of air in terms of weight), rough engine operation, sudden cutting out, backfiring, detonation, overheating, and appreciable loss of engine power may occur. If the fuel/air mixture is too rich, rough engine operation and loss of engine power may occur.

To maintain the correct fuel/air ratio, the pilot must adjust the amount of fuel mixed with the incoming air as his altitude increases. He accomplishes this by leaning the mixture. Reference should be made to the aircraft operating manual for proper fuel leaning technique.

FUEL INJECTION SYSTEM

Fuel injection replaces the conventional carburetor with a low-pressure, continuous-flow system which injects fuel without interruption into the intake manifold. The system illustrated in figure 1-11 is typical of those in use in many modern light aircraft.

The main components are a fuel injector pump (item 1) to provide fuel pressure, a control unit (item 2) to meter the fuel, and a manifold (item 3) to distribute the fuel to the cylinders. These components may be combined or arranged differently by different manufacturers, and various fuel metering devices may be used.

In figure 1-11, the fuel flow can be traced from the tank or boost pump to the fuel injection pump (item 1) where vapor is separated and returned to the tank. Pressurized fuel then flows to the fuel-air control unit (item 2). Unused fuel is returned to the injector pump via the overflow line (item 4). The connecting linkage (item 5) between the throttle and the fuel metering unit allows a fuel flow proportionate with the throttle position. The fuel manifold (item 3) distributes the fuel evenly to discharge nozzles (item 6) which draw in a small amount of outside air to aid in fuel vaporization. The mixture control may restrict pressure to the fuel metering unit in order to adjust flow to the fuel manifold for leaning the mixture or stopping the engine. The fuel pressure gauge or a fuel flow gauge can be used as an indirect indication of fuel/air ratio while leaning the mixture.

Operation of the Fuel Injection System

A preflight inspection is not required for most injection systems, since there are few moving parts. The tendency for injection engines to flood on starting is more prevalent than in carburetion systems. However, flooding is usually caused by improper starting technique. The aircraft flight manual should be consulted for proper starting techniques to be used when the engine is either hot or cold. During climbing and cruising, there is little difference in operating technique between a carburetor and injection system. The fuel pressure or fuel flow gauge of the injection system is marked to aid the pilot in leaning the mixture. An exhaust gas temperature

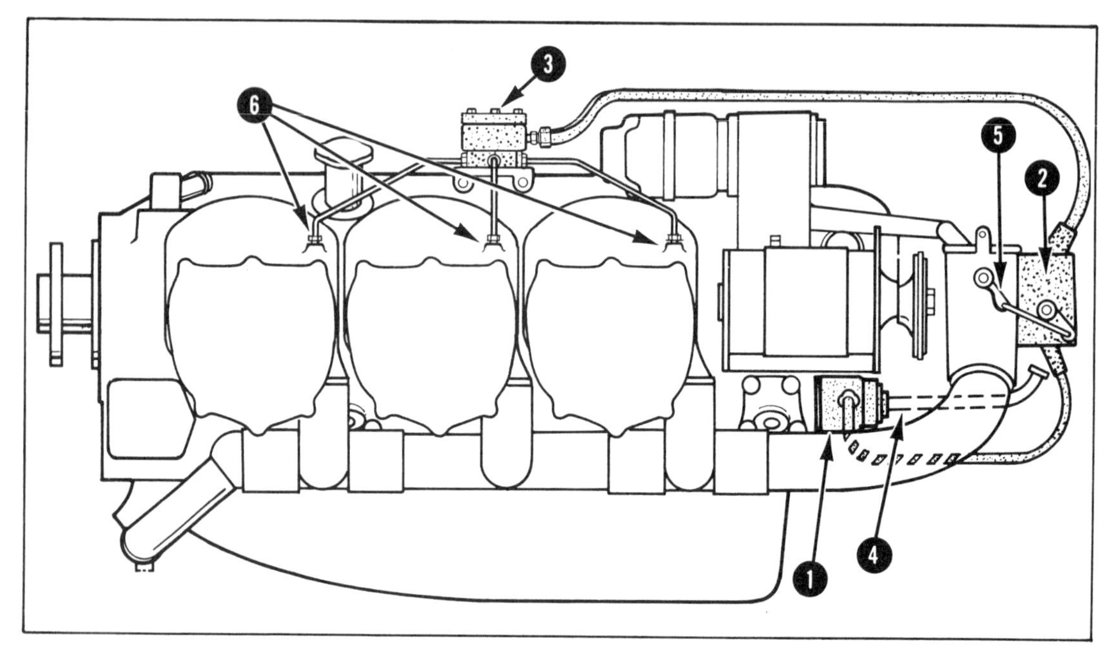

Fig. 1-11. Fuel Injection System

gauge (if installed) allows an even more precise adjustment.

Advantages of Fuel Injection

The fuel injection system provides smoother, more positive control of engine operation. Because of careful design, automatic controls are not required. Thus many components susceptible to maladjustment or dirt are unnecessary. There are several major operating advantages of a fuel injection system.

1. Carburetor icing is eliminated, since only air flows through the throttle body.
2. Even distribution of fuel assures better economy.
3. Elimination of an over-rich tendency with the throttle retarded improves acceleration and reduces sparkplug fouling.
4. Detonation is less likely because of better distribution of fuel additives and the elimination of heated carburetor air.
5. Smooth, continuous operation is assured in any flight condition or attitude.

ELECTRICAL SYSTEM

Electrical energy is supplied by a 14- or 28-volt, direct current system (most light aircraft use a 14-volt system) which may be powered by an engine-driven generator or alternator. (See Fig. 1-12.) The electrical storage battery serves as a stand-by power source, supplying current to the system when the generator or alternator is inoperative. A warning light or ammeter is provided to indicate generator or alternator failure. Control of the charging current and voltage is accomplished by a voltage regulator.

MASTER SWITCH

A master switch controls the entire airplane electrical system, with the exception of the ignition system which gets its electricity from magnetos. Turning the master switch on supplies electrical energy to all electrical equipment circuits in the airplane.

When the battery master switch is in the ON position, electrical current is permitted to flow from the power supply to all components of the electrical system. Newer aircraft use a split, rocker-type master switch.

The right half of this switch, labeled *BAT* (for battery), controls the flow of electrical current to all electrical components of the airplane except to the ignition system. The left half of the switch, labeled *ALT* (for alternator), cuts in or removes the alternator from the electrical system as a source of power.

BUS BAR

The bus bar connects most accessory items to the main electrical system and simplifies the wiring system. Fuses or circuit breakers are used in conjunction with the bus bar. Their purpose is

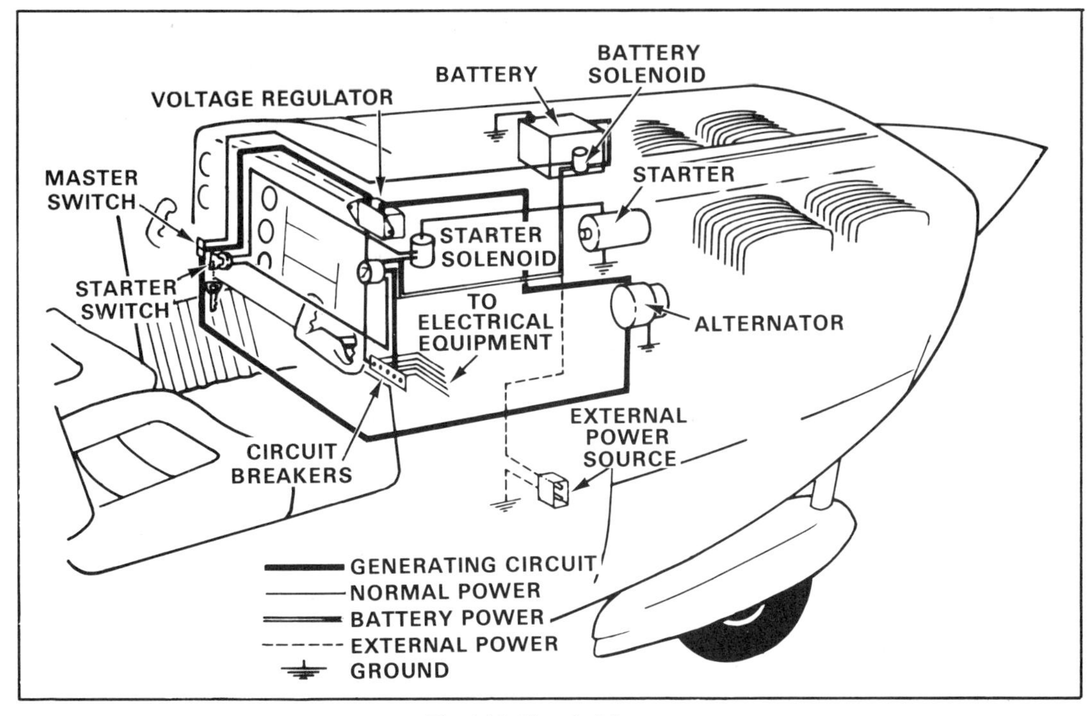

Fig. 1-12. Electrical System

to protect the various electrical circuits from damage caused by alternator or accessory malfunction.

AMMETER

The ammeter is the only instrument which allows the pilot to monitor the electrical system. However, not all aircraft are equipped with an ammeter; some have a light which, when lit, indicates a discharge or alternator/generator malfunction.

FLIGHT INSTRUMENTS

GYROSCOPIC PRINCIPLES

The primary trait of a rotating gyro is *rigidity in space*, or *gyroscopic inerita.* Newton's First Law states in part, "A body in motion tends to move in a constant speed and direction unless disturbed by some external force." The spinning rotor inside a gyro instrument maintains a constant attitude in space as long as no outside forces change its motion. This quality of stability is enhanced if the rotor has great mass and speed. Therefore, gyros in aircraft instruments are constructed of heavy materials and designed to spin quite rapidly (between 10,000 and 15,000 r.p.m.).

ATTITUDE INDICATOR

The attitude indicator is the single most important instrument for determining aircraft attitude since it is the only instrument which provides information regarding the overall attitude of the aircraft. This instrument, depicted in figure 1-13, works on the principle of "rigidity in space."

The miniature aircraft in the attitude indicator depicts the aircraft's attitude relative to the horizon bar in the same manner the pilot views the aircraft in relation to the outside visual reference (natural horizon). The horizon bar always remains parallel to the natural horizon, while the miniature aircraft retains a position identical to that of the actual airplane.

In addition, pitch attitudes are depicted by the miniature aircraft's nose moving up or down in relation to the horizon bar. The movement indicates the relationship of the actual airplane's nose to the natural horizon. Four pitch reference lines are incorporated into the instrument. Two are below the artificial horizon bar and two are above. The pitch limit is approximately 60°.

The bank pointer, located at the top of the instrument, indicates the degree of bank during

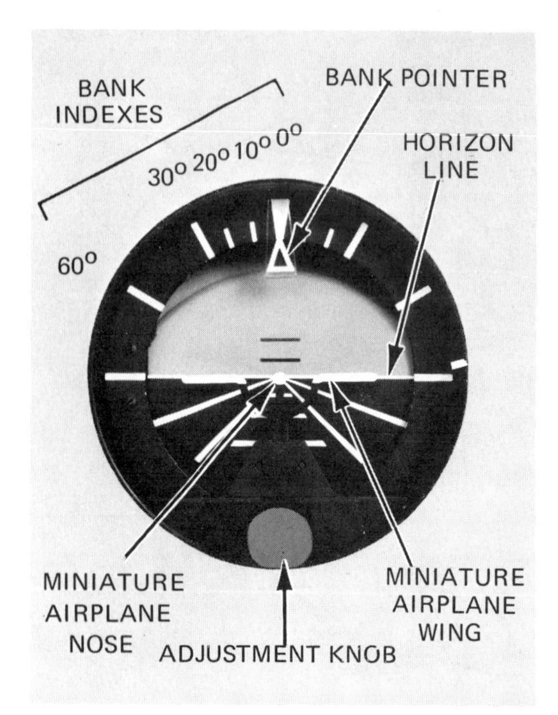

Fig. 1-13. Attitude Indicator

turns through the use of index marks. These marks are spaced at 10° increments from 0° through 30°, with larger marks placed at 30°, 60°, and 90° bank positions. The bank limit is approximately 100°.

PRINCIPLE ATTITUDE INDICATOR ERRORS

Acceleration Error

As the airplane accelerates, precession of the gyro causes the horizon bar to move down, indicating a slight nose-high attitude. This error is normally a very minor consideration.

Deceleration Error

Similar to the acceleration error, deceleration causes the horizon bar to move up, indicating a false nose-low attitude. The amount of error is directly related to the rate of deceleration.

Turn Error

During a normal coordinated turn, or a skid, centrifugal force causes the gyro to precess toward the *inside* of the turn. This gyro precession increases as the bank increases and, therefore, is greatest during the actual turn. Upon rolling out at the end of a 180° turn, using a normal roll-out rate, the instrument will indicate a slight climb and a bank opposite the direction of the turn.

HEADING INDICATOR

The *heading indicator*, formerly called the directional gyro, uses the principle of gyroscopic rigidity to provide a stable heading reference. The instrument is depicted in figure 1-14. Bearing friction and rotation of the earth cause the heading indicator to "drift." An error of no more than three degrees in 15 minutes is acceptable for normal operations.

Due to the precessional error, the heading indicator should be regularly compared to the magnetic compass during flight. However, to make accurate comparisons, the aircraft must be in straight-and-level, unaccelerated flight.

Fig. 1-14. Heading Indicator.

RATE OF TURN INDICATOR

There are two basic types of rate-of-turn indicators in use — the *turn coordinator* and the *turn-and-slip indicator.* Both of these gyroscopic instruments indicate the *rate* at which the airplane is turning. The turn coordinator incorporates a miniature airplane to show when the actual airplane is in a turning condition; on the other hand, the turn-and-slip indicator, utilizes a vertical needle which deflects in the direction the aircraft is turning. Figure 1-15 depicts these two instruments.

In the turn coordinator, the standard rate indexes are located just below the wings-level position of the miniature airplane. As the aircraft is rolled into a turn, the small aircraft banks in the same direction. A standard rate turn (three degrees per second) is indicated

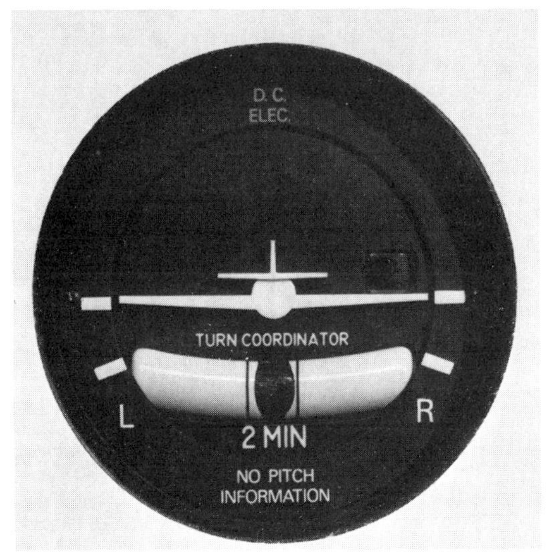

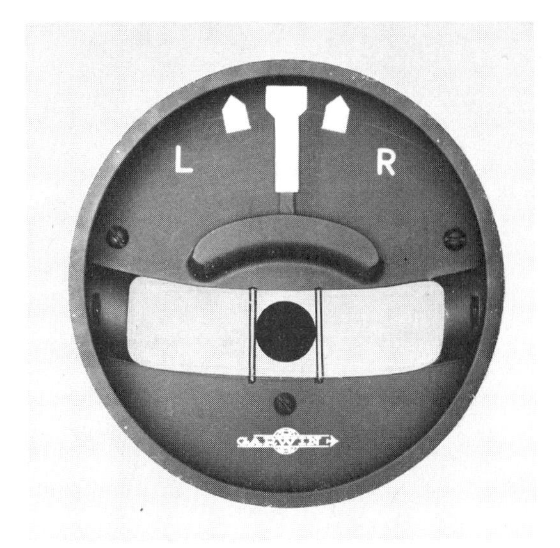

Fig. 1-15. Rate of Turn Indicators

when the miniature aircraft wing is on a standard-rate index marking.

The miniature aircraft is completely independent of the *ball* or *inclinometer*. The ball is used to indicate the *quality* of the turn. When the miniature aircraft indicates a turn and the ball is not centered, an uncoordinated condition exists.

RATE OF TURN

If two aircraft are turning at the same angle of bank, the slower aircraft has the shorter turning radius and a greater rate of turn. A common *misconception* is that the faster airplane completes a 360° turn in the least time. For example, a jet in a 20° bank flying at a true airspeed of 350 knots requires approximately 5.3 minutes to complete a 360° turn. On the other hand, a slower aircraft also has a 20° bank but a true airspeed of only 130 knots. This aircraft requires just two minutes to complete a 360° turn.

The radius of turn also increases with an increase in airspeed. The actual radius of the turn varies as the square of the true airspeed. Since 350 knots (speed of the jet) is almost three times 130 knots (speed of the other aircraft), the radius of turn for the faster aircraft is approximately nine times that of the slower aircraft.

COMPASS CONSTRUCTION

The aircraft's magnetic compass is a very simple self-contained instrument which consists of a sealed outer case containing a pivot assembly and a float containing two or more magnets. (See Fig. 1-16.) A compass card is attached to the float with the cardinal headings (north, east, south, and west) shown by corresponding letters on the face of the card. Between the cardinal headings, each 30° increment is shown as a number with the last zero removed. For example, 30° is shown as the numeral 3. The compass card is free to rotate and it is also free to tilt up to 18°.

The case is filled with an acid-free white kerosene which helps to dampen oscillations of the float and lubricate the pivot assembly. In addition, the pivot assembly is spring-mounted to further dampen aircraft vibrations so the compass heading may be read more easily. A glass face is mounted on one side of the compass case with a lubber or reference line in the center. Compensating magnets are located within the case in order to correct the compass reading for the presence of magnetic lines of force generated by components of the aircraft.

MAGNETIC DIP

The lines of force in the earth's magnetic field pass through the center of the earth. They exit at both magnetic poles of the earth, bending around to reenter at the opposite pole. Near the Equator, these lines become almost parallel to the surface of the earth. However, as they near the poles, they tilt toward the earth until in the immediate area of the magnetic poles they actually point sharply into the earth. Since the poles of a compass tend to align themselves with the magnetic lines of force, the magnet within the compass tends to tilt or dip toward the earth

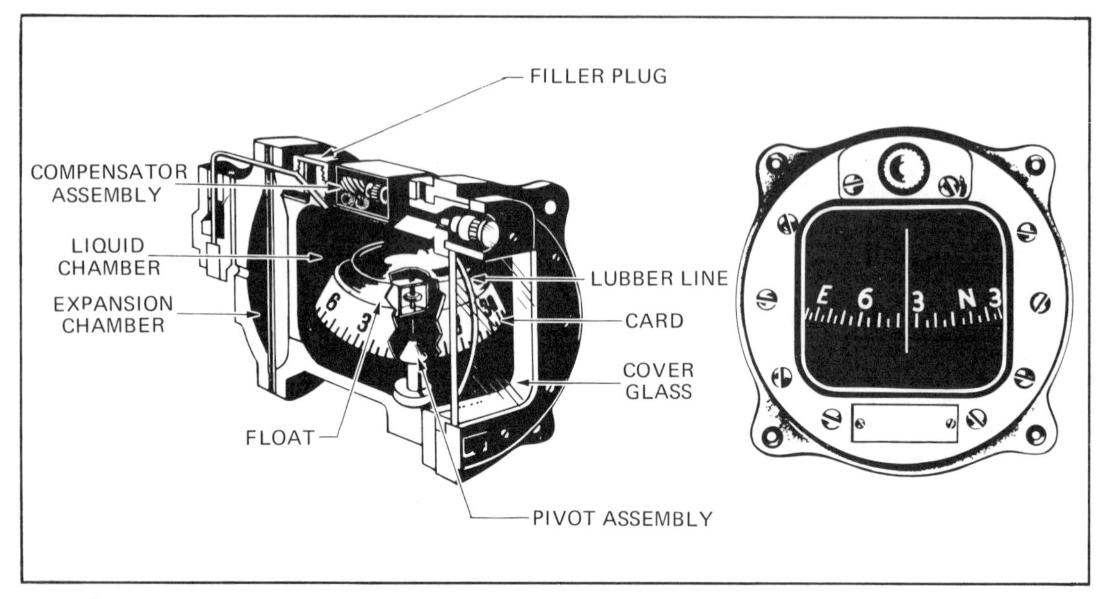

Fig. 1-16. Magnetic Compass Construction

in the same manner as the magnetic lines of force.

NORTHERLY TURNING ERROR

One result of magnetic dip is northerly turning error. This causes the magnetic compass to provide improper indications when turning to or from north or south headings.

When making a turn *from a heading of north* in the Northern Hemisphere, the compass briefly gives an indication of a turn in the *opposite* direction. If this turn is continued east or west, the compass card begins to indicate a turn in the correct direction, but lags behind the actual turn at a diminishing rate until within a few degrees of east or west. If the pilot makes a very gradual and shallow banked turn (less than five degrees) from a compass indication of north, it is possible to change the actual heading of the aircraft by 20° or more while still maintaining an indication of north on the compass.

When making a turn *from a heading of south*, the same set of forces is in effect but the indications appear quite different. The compass gives an indication of a turn in the *correct* direction, but at a much *faster* rate than is actually being experienced. As the turn is continued toward east or west, the compass indications continue to *precede* the actual turn, but at a diminishing rate until within a few degrees of west or east. The effects of northerly turning error are illustrated in figure 1-17.

ACCELERATION ERROR

Acceleration error is another effect caused by a combination of inertia and the vertical component of the earth's magnetic field. The compass card, because of its pendulous-type mounting and magnetic dip error is tilted during changes of speed. The momentary tilting of the card from the horizontal, results in an error which is most apparent on headings of east and west.

When accelerating on either of these headings, the error causes an indication of a turn to the north, and when decelerating, the error causes an indication of a turn to the south. (A memory aid which helps clarify this relationship between airspeed change and the direction of error is the acronym *ANDS*— Accelerate North, Decelerate South.)

OSCILLATION ERROR

Erratic swinging of the compass card, which may be the result of rough air or certain maneuvers is called *oscillation error*. This error can be compensated for by "averaging" the compass indications.

FLYING BY THE COMPASS

The magnetic compass, if its errors and characteristics are thoroughly understood, offers the pilot a reliable means of determining the direction in which his airplane is headed. To accurately read the compass when determining direction, the pilot must be certain the airplane is flown as steady as possible, and is at a constant airspeed.

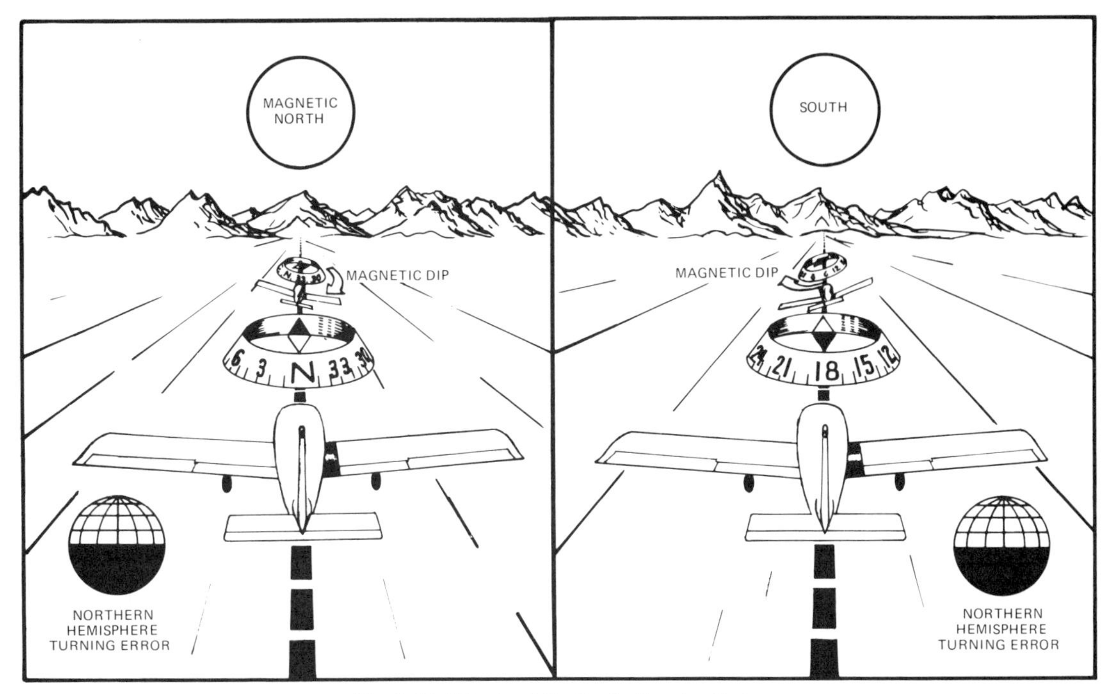

Fig. 1-17. Effects of Northerly Turning Error

TURNS USING THE MAGNETIC COMPASS

When utilizing the magnetic compass for reference during turns, a 15° bank angle should be used. By employing 15° of bank angle in turning north or south, the amount of roll-out lead is proportional to the latitude. For example, in turning to north from south in the 30° north latitude area, a pilot should start a roll-out 30° before reaching north. In addition, a five degree lead should be allowed for roll-out from the banking attitude. For purposes of this discussion, it is assumed that a five degree lead is needed to roll the airplane to wings level flight. The roll-out heading should be initiated at either 035° or 325° depending on direction of turn. To roll-out on a heading of south, the aircraft should be flown past south the number of degrees of latitude, minus the five degree lead. For example, in a right turn from north to south at 30° north latitude with 15° of bank, the roll-out should be initiated at 205° (180° + 30° N latitude — 5° roll-out lead = 205°).

PITOT-STATIC SYSTEM

The pitot-static system has two major parts — the impact pressure chamber and lines, and the static pressure chamber and lines. The impact chamber contains ram air pressure supplied by the pitot tube, which is usually located on the leading edge of the wing. Since the pitot tube faces forward, an increase in aircraft speed increases ram air pressure. The static air vents, which are usually located on the sides of the aircraft, supply atmospheric pressure (air pressure at flight altitude) to the pressure instruments. The pitot-static system is the source for operation of three vital instruments — airspeed indicator, vertical velocity indicator, and the altimeter. As indicated by the diagram in figure 1-18, the pitot tube provides dynamic ram pressure for the airspeed indicator and the static source supplies static pressure to all three instruments. Additionally, some airplanes are equipped with a pitot heater to prevent ice formation from blocking the pitot tube, should icing conditions be encountered in flight.

AIRSPEED INDICATOR

The airspeed indicator, which is vented to both pitot and static lines, reacts to any change between ram (dynamic) air pressure and static (passive) air pressure. The greater the differential between these two pressure readings, the greater the airspeed indication.

Basic Aircraft Speeds

Indicated airspeed (IAS) is read directly from the airspeed indicator. However, this speed will seldom match the actual airspeed of the airplane throughout the speed and altitude ranges. Indicated airspeed reflects actual (true) airspeed only when corrected to standard atmospheric

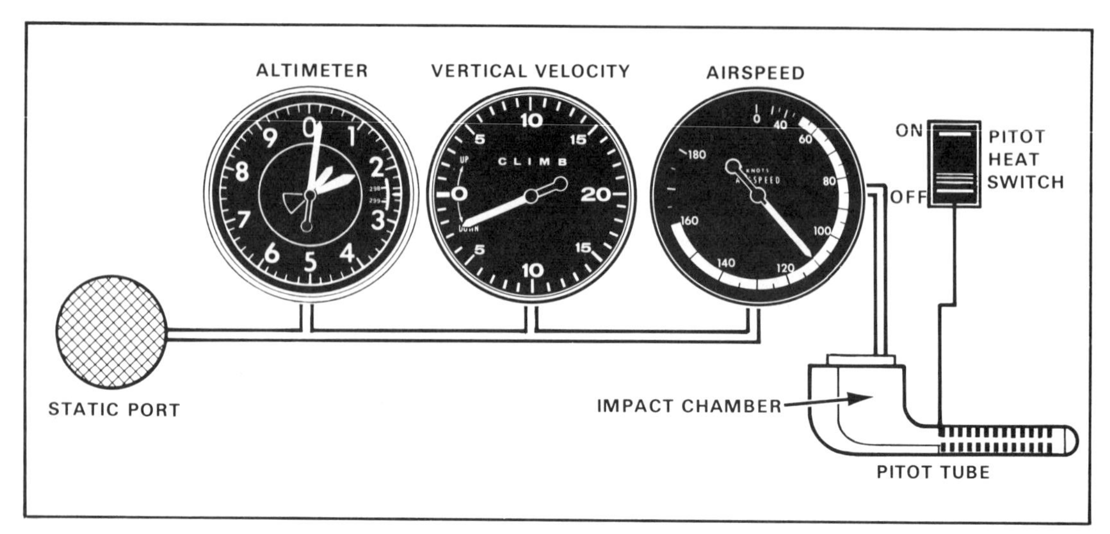

Fig. 1-18. Pitot-Static System

conditions (59° Fahrenheit, 29.92 inches of mercury at sea level).

Calibrated airspeed (CAS) is indicated airspeed corrected for errors resulting primarily from the position of the static source and, to a much lesser degree, from pitot tube locations. This error is due mainly to differences in airflow over the static port at varying angles of attack. The error is usually greatest in the low and high speed ranges and smallest at normal operating speeds. Calibrated airspeed tables can be found in the airplane owner's manual.

True airspeed (TAS) is determined by converting indicated airspeed under actual conditions, to an airspeed at a standard temperature and pressure. This determination is required because the pitot-static system is designed to accurately measure the difference between ram and static air pressure at sea level only when the actual barometric pressure is 29.92 inches of mercury and the temperature is 59° Fahrenheit (standard conditions). The airspeed indicator interprets less dense air (higher altitude) as an airspeed decrease, when in reality the airplane's actual speed increases in the less dense air at higher altitude. This happens because fewer particles of air create less drag. Some airspeed indicators incorporate a TAS computer on the dial face, like the one shown in figure 1-19, enabling the pilot to read TAS directly from the indicator.

AIRSPEED COLOR MARKINGS AND V-SPEEDS

The face of the airspeed indicator is usually marked in both statute and nautical miles per hour, as shown in figure 1-19. It also has colored arcs to show important aircraft speed limits and operating ranges. Figure 1-19 also points out various V-speeds relating to the color markings on the instrument. It must be remembered that all colored markings and airspeed limitations on the airspeed indicator are expressed in calibrated airspeeds.

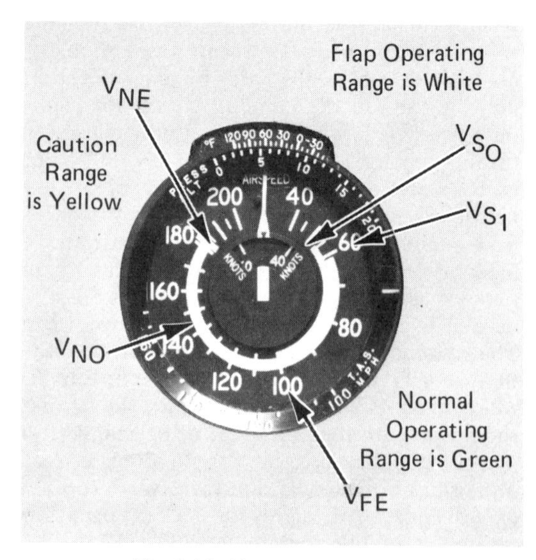

Fig. 1-19. Airspeed Indicator

Each of these V-speeds is shown on the airspeed indicator. The white arc on the airspeed indicator designates the flap operating range. The green arc shows the normal operating range, and the yellow (caution) arc signifies the smooth air cruising range. This speed range should never be used during gusty or turbulent atmospheric conditions.

V-SPEED	AIRSPEED INDICATOR MARKINGS
V_{NE}	*never exceed* (red line)
V_{NO}	*maximum for normal operations* (high speed end of green arc)
V_{FE}	*maximum for flap extension* (high speed end of white arc)
V_{S_1}	*stall, power-off, clean configuration* (low speed end of green arc)
V_{S_O}	*stall, power-off, landing configuration* (low speed end of white arc)

ALTIMETER

The altimeter senses the normal decrease in air pressure that accompanies an increase in altitude. The air-tight instrument case is vented to the static port. With an increase in altitude, the air pressure within the interior of the case decreases and a sealed aneroid barometer (bellows) within the case expands. The barometer movement is transferred to the indicator via mechanical linkage. The indicator is calibrated in feet and may have two or three pointers.

ALTIMETER SETTING WINDOW

Since the aneroid barometer cannot differentiate between actual altitude changes and changes in the barometric pressure of the airmass itself, a calibration unit is provided. An altimeter face with an altimeter setting window is shown in figure 1-20. This window allows adjustment of the current altimeter setting on a small scale calibrated in inches of mercury. As a result, the indicator will respond only to altitude change, provided the altimeter setting is accurate and the pilot "updates" the current setting as reports become available.

ALTITUDE DEFINITIONS

Vertical separation of aircraft is based on local altimeter settings. Since pressure variations en-route require changes in the altimeter setting, and TAS computations are based on temperature and pressure conversions, the following definitions should be understood.

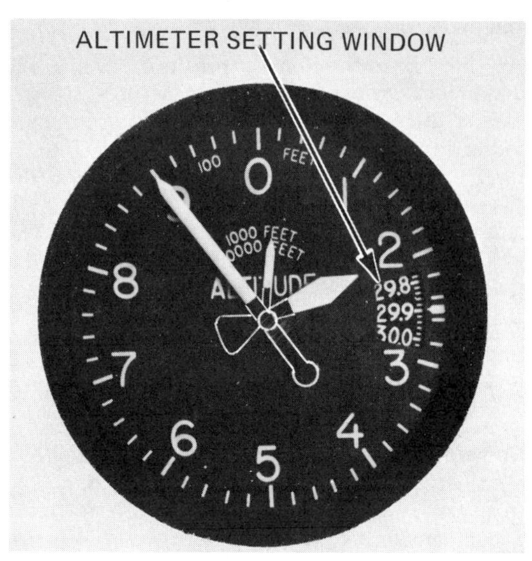

Fig. 1-20. Altimeter

1. *Indicated Altitude* — read directly from the altimeter when set to current barometric pressure.
2. *Pressure Altitude* — read from the altimeter when set to standard barometric pressure of 29.92 in. Hg.
3. *Density Altitude* — pressure altitude corrected for nonstandard temperature.
4. *True Altitude* — exact height above mean sea level.
5. *Absolute Altitude* — actual height above the earth's surface.

VERTICAL VELOCITY INDICATOR

The vertical velocity indicator has a sealed case connected to the static line through a calibrated leak (restricted diffuser). Inside the case, a diaphragm, attached to the pointer by a system of linkages, is vented to the static line without restrictions, as depicted in figure 1-21.

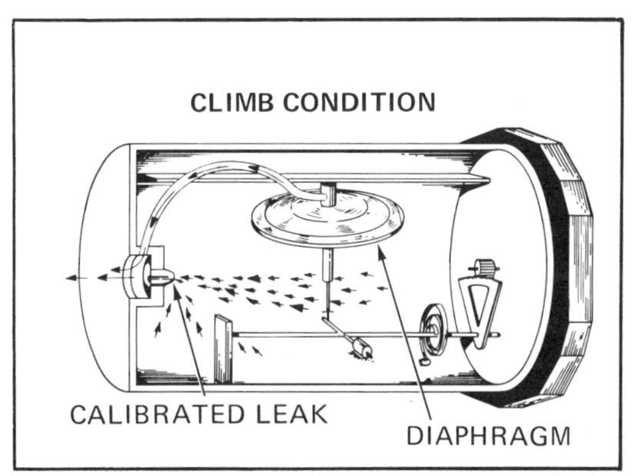

Fig. 1-21. Vertical Velocity Indicator Case

OPERATION

As the aircraft climbs, pressure within the diaphragm drops faster than case pressure can escape through the restrictor, resulting in climb indications; the reverse is true during descent. If level flight is resumed, pressure equalizes in the case and diaphragm within six to nine seconds and the pointer returns to zero rate of climb. The vertical velocity indicator has 100-foot calibrations with numerals to indicate each 500 feet of altitude, as shown in figure 1-22.

Fig. 1-22. Vertical Velocity Indicator

ATTITUDE INSTRUMENT FLYING

Control of an aircraft during VFR flight is considered *attitude flying* since the establishment of a specific pitch and bank attitude, accompanied by a particular power setting, results in predictable aircraft performance. When pitch, bank, and power are established by reference to the flight instruments, and the desired performance is confirmed by instrument indications, the result is *attitude instrument flying.*

Attitude instrument flying has three basic ingredients — *scan, interpretation,* and *aircraft control.* Proper instrument scan is of *extreme* importance to the instrument pilot. As in visual flying, instrument flying requires that certain attitude references be used more often during one maneuver than during another. For instance, during a constant airspeed climb, the altimeter is of less importance than the airspeed indicator. If the airspeed in a visual climb is slower than desired, a check of the aircraft's pitch attitude will quickly confirm that the aircraft nose is too high in relation to the natural horizon. Under instrument conditions, the attitude indicator is used to determine the aircraft's pitch and to reestablish an attitude which will increase the airspeed to the desired value.

Because the attitude indicator replaces the normal outside visual references, it is considered the *principal attitude control instrument used in instrument flight.* The attitude indicator is considered the hub of a wheel, with the supporting instruments located around the rim. The pilot should incorporate this concept into the instrument scan. As shown in figure 1-23, the instruments are not connected to each other; rather, they are connected to the attitude indicator. This method utilizes the supporting instruments to *confirm* the performance indicated by the attitude indicator.

To aid in the development of proper scanning technique, each of the following illustrations in this chapter indicates, the relative amount of attention each instrument should receive during the maneuver. The approximate amount of total scan is represented by the boldness of the scan arrow. A heavy arrow indicates the majority of the scan, the thin arrow a lesser amount, and the open arrow an intermittent amount of scan.

The second important ingredient in instrument flying is proper instrument *interpretation.* The attitude indicator provides the pilot with an *artificial horizon* to replace the natural horizon; hence, proper interpretation of this principal *attitude control instrument* is extremely important. Reference to the other instruments is essentially the same under instrument conditions as under visual conditions.

The last ingredient, *aircraft control,* is actually the result of scan and interpretation. After proper instrument scan and interpretation is accomplished, it is simply a matter of applying the proper control pressures to attain the desired airplane performance. Since the human body is subjected to sensations which in some cases may be unreliable when interpreting the airplane's actual attitude, it is *necessary* that the pilot learn to disregard these sensations. *The aircraft must be controlled through proper scan and interpretation of the flight instruments.*

Instrument flying is essentially a series of small corrections. Even the best instrument pilot cannot maintain an exact heading or altitude.

However, through continuous, small corrections, precision instrument flight is possible

STRAIGHT-AND-LEVEL FLIGHT

As in all maneuvers, the attitude indicator is the key instrument during straight-and-level flight. The instruments that confirm the attitude indicator during straight-and-level flight are the heading indicator, vertical velocity indicator, and the altimeter. The attitude indicator shows general pitch and bank information, while the heading indicator provides *specific* heading information, the vertical velocity indicator provides the *actual* pitch trend, and the altimeter provides *specific* altitude information.

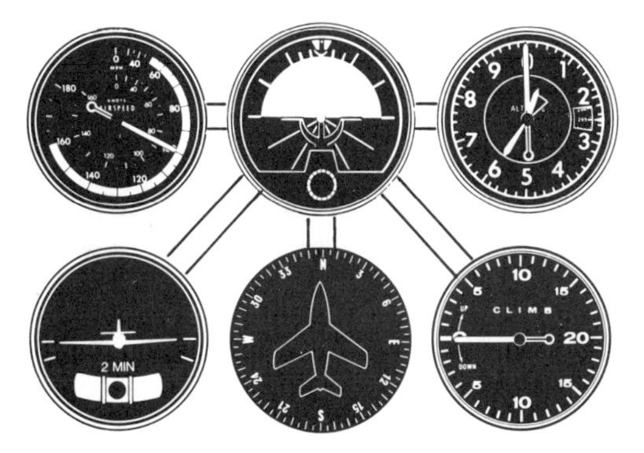

Fig. 1-23. Scan Pattern

To establish straight-and-level flight, the miniature airplane in the attitude indicator is placed over and aligned with the horizon. Figure 1-24 indicates the proper scanning procedure for straight-and-level flight. Before reference can be made to another flight instrument, the attitude indicator must be scanned.

The next two instruments in order of scanning attention, are the heading indicator, which confirms whether or not the airplane is in straight flight, and the vertical velocity indicator, which shows any changing trend in altitude. Although it takes several seconds for the vertical velocity indicator to display an actual rate of altitude change, it displays the trend of the change immediately.

If the miniature airplane in the attitude indicator portrays a slight wing-low situation with no change in aircraft heading, this is most likely the result of rudder pressure being applied unnecessarily. Reference to the ball in the turn indicator will confirm this. To return the ball to center, rudder pressure is applied on the side to which the ball has moved; in other words, *"step on the ball."* If the airplane is equipped with a rudder trim tab, it is adjusted to relieve the pressure from the foot after the ball has been centered.

During straight-and-level flight, a constant power setting results in a constant airspeed. If power is increased, the airplane will accelerate and start climbing. Therefore, it is necessary to maintain constant attitude and power settings in order to hold the desired altitude, heading, and airspeed.

STRAIGHT CLIMBS

Figure 1-25 depicts a straight climb attitude. Once again, the attitude indicator is the key instrument. To enter a climb from straight-and-level flight, the miniature airplane should be raised approximately one bar width above the horizon. This attitude coupled with full power produces a climb at approximately the best rate-of-climb airspeed. As the pitch attitude is changed, power is increased to the climb power setting. It is important that the wings remain level during the transition and climb. Again, the airplane is trimmed in order to relieve all pressures on the controls.

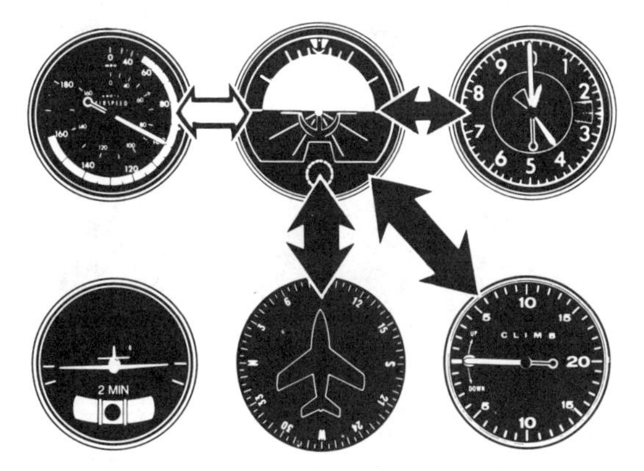

Fig. 1-24. Straight-and-Level Flight

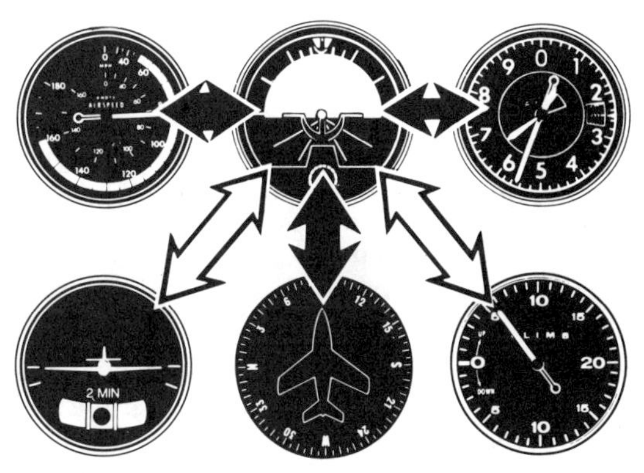

Fig. 1-25. Straight Climb

Whenever the aircraft is in a climb attitude with full power, the aircraft tends to turn left. This tendency is the result of *left turning forces* caused by low airspeed and high angle of attack. Right rudder pressure is necessary to maintain a constant heading. Heading information is derived from the heading indicator and, as in straight-and-level flight, it is used in conjunction with the attitude indicator. If the heading indicator shows a change from the desired heading, a correction is accomplished by using a gentle, coordinated turn back to the desired heading.

Since the best rate-of-climb airspeed is desired, the airspeed indicator should be scanned in conjunction with the attitude indicator. If the airspeed is greater or less than desired, a change of one-half bar width at a time in the pitch attitude is made on the attitude indicator to obtain the proper airspeed.

Throughout this maneuver, a constant power setting is assumed. Hence, when power remains constant, the airspeed is controlled only by pitch.

Leveling off is accomplished by lowering the nose to a level flight attitude on the attitude indicator when approaching the desired altitude. Leading the desired altitude by approximately 10 percent of the vertical speed is sufficient. For example if the vertical speed is 800 f.p.m., the level-off is started when approximately 80 feet below the desired altitude. Climb power is maintained while the miniature airplane is held on the horizon bar until the desired cruise airspeed is obtained. During the transition, all control pressures are relieved by adjusting the trim. *After* cruise airspeed is obtained, the power is reduced to the cruise power setting, and the trim adjusted to relieve control pressures necessary to maintain the selected pitch attitude.

STRAIGHT DESCENT

To establish a descent the nose of the miniature airplane in the altitude indicator is placed just below the horizon bar. As this is being accomplished, power should be reduced.

When the power is reduced, the aircraft has a tendency to turn to the right. Therefore, it may be necessary to apply slight left rudder pressure to maintain straight flight. Once again, heading corrections are accomplished by gently banking in the proper direction, using the attitude indicator until the aircraft returns to the desired heading, as indicated by the heading indicator. Small changes in pitch are used to maintain the desired rate of descent while small power changes are used to maintain the desired airspeed. With a little practice, the pilot should be able to determine the power settings and pitch attitudes required to provide a given rate of descent at the desired airspeed. A 500 f.p.m. rate of descent is used for a cruise descent. *If the power is constant, pitch is the only method of changing airspeed or rate of descent. If power is a variable, it should be used to control the airspeed.* This method works especially well during an instrument landing approach.

When approaching the desired altitude, level off is initiated using the 10 percent of vertical speed rule.

The nose of the miniature airplane is raised until it is over the horizon bar. At the same time, power is increased to the cruise power setting. The wings are held level throughout the maneuver and the airplane is retrimmed for straight-and-level flight when the desired altitude is reached.

LEVEL STANDARD-RATE TURNS

As mentioned previously, a standard-rate turn is at a rate of three degrees per second. Also, the rate of turn is a function of the angle of bank and the true airspeed. Figure 1-26 depicts the amount of bank required to establish a standard rate turn at 150 m.p.h. *As airspeed increases, it is necessary to use a greater angle of bank to maintain the same rate of turn.*

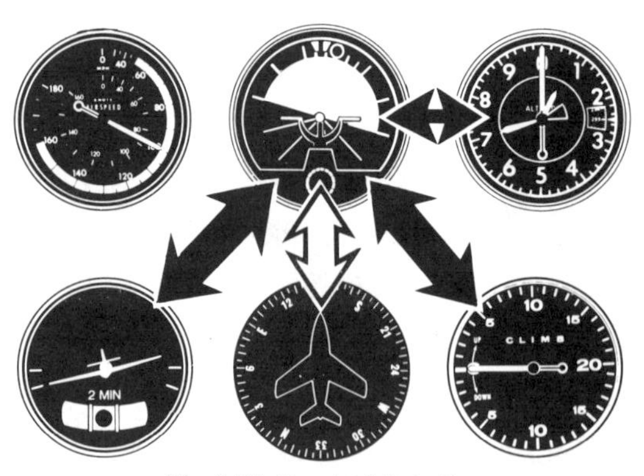

Fig. 1-26. Standard-Rate Turn

To enter a standard-rate turn, first the approximate angle of bank that is required must be determined. A general rule is to drop the last numeral from the airspeed and add five. For example, 150 m.p.h. minus the zero is 15; an addition of 5 equals a 20° angle of bank (15 + 5 = 20°).

At the beginning of the turn, the proper angle of bank is established by using the attitude indicator bank pointer at the top of the instrument. After the initial angle of bank is established, the turn coordinator (or turn-and-slip indicator) is checked to confirm the performance of a standard-rate turn. Any bank corrections are made and then maintained by reference to the attitude indicator.

As in visual flight, back elevator control pressure is required to maintain altitude after the aircraft is banked. Therefore, the nose of the miniature aircraft in the attitude indicator should be raised slightly above the level flight position. Throughout the turn, all *general* pitch and bank information is derived from the attitude indicator. The scan throughout the turn includes the altimeter and vertical velocity indicator to confirm that a constant altitude is being maintained. Reference is made to the turn coordinator to ensure a standard-rate turn. The heading indicator should be cross-checked occasionally in order to roll out on the desired heading.

CLIMBING TURNS

A climbing turn combines both straight climb procedures and turn procedures into one maneuver. As before, the attitude indicator is the main instrument for establishing proper airplane attitude. The maneuver begins with a straight climb. After the climb is initiated, the aircraft is rolled into the proper bank for a standard-rate turn. Since the airspeed is reduced, the angle of bank required for a standard-rate climbing turn is less than the bank required for a level flight turn.

The airspeed indicator is scanned to confirm the pitch attitude, and the turn coordinator is scanned to show if a standard-rate turn is being obtained. A cross check of the heading indicator will show when to begin the roll-out. Again, a lead of approximately one-half the bank angle in advance of the desired heading is sufficient. As the aircraft approaches the desired altitude, the nose of the miniature airplane is lowered to the horizon bar. The aircraft should be allowed to accelerate to cruise airspeed before the power is reduced to the cruise setting. Throughout the maneuver, the aircraft is trimmed so control pressures are relieved.

DESCENDING TURNS

Like the climbing turn, the descending turn also combines two maneuvers — a straight descent and a turn. To begin the maneuver, a straight descent is entered. As this is accomplished, a bank is initiated in the desired direction using the attitude indicator as the reference, as illustrated in figure 1-27. Scanning the turn coordinator indicates whether the proper turning rate of three degrees per second is being obtained. Likewise, the airspeed and vertical velocity indicators reveal if the pitch attitude selected is sufficient for the descent power setting. If the rate of descent is too great, the pitch attitude should be adjusted, as necessary, by referring to the miniature airplane in the attitude indicator. Airspeed adjustments are accomplished by either adding or reducing power. Corrections usually require small pitch and power changes.

Fig. 1-27. Descending Turns

As the desired heading is approached, as shown by the heading indicator, roll out is commenced using a lead equal to one-half the bank angle. Approximately 50 feet above the desired altitude, the nose of the miniature airplane is raised to the horizon bar, and the power is simultaneously increased to cruise setting. Throughout the maneuver, the airplane is trimmed to aid in maintaining the desired altitude.

STEEP TURNS

Aircraft control during steep turns is maintained by reference to the attitude indicator as the main pitch and bank instrument. The maneuver is entered by rolling into a 45° bank, as established by the bank pointer at the top of the attitude indicator. This is illustrated in figure 1-28.

Entry and recovery do not require greater control deflection than that required in a shal-

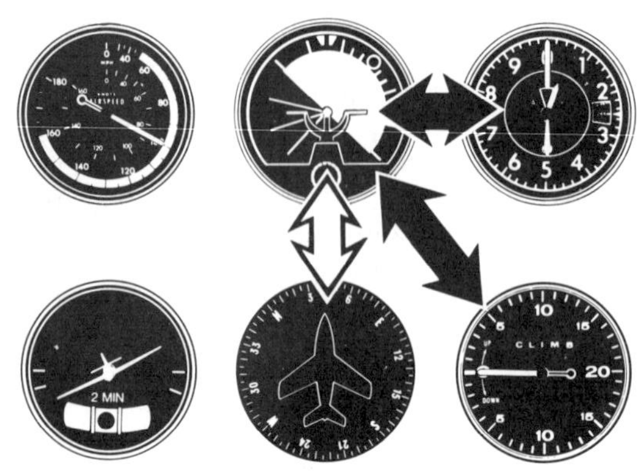

Fig. 1-28. Steep Turns

low turn. The steep bank is the result of holding control pressure for a longer time period.

The altimeter and vertical velocity indicator are checked immediately after entry and recovery. This aids in establishing the pitch attitude required for level flight. During the steep turn, it is necessary to increase the pitch attitude to compensate for centrifugal force.

Therefore, the nose on the miniature airplane is considerably higher in a steep turn than in a standard-rate turn. Throughout the turn, constant reference to the attitude indicator must be maintained in order to confirm that a 45° bank is maintained. Reference to the altimeter and vertical velocity indicator is necessary to confirm the altitude and the correct pitch attitude.

Since a 45° bank is being used, the roll-out lead must be considerably larger. Using one-half the angle of bank for roll-out lead time, a lead of approximately 20° is sufficient. Hence, at a point 20° before the desired heading, coordinated rudder and aileron pressures are used to roll the miniature airplane level on the attitude indicator. Since a considerable amount of back pressure is necessary to maintain altitude during the turn, the nose of the miniature airplane is lowered to the horizon bar during the roll-out. With practice, roll-out on the desired heading and altitude can be accomplished with little difficulty.

STALLS

Prior to practicing stalls, two 90° clearing turns should be executed. At the completion of the second turn, the stall maneuver is entered immediately.

APPROACH STALLS

Approach stalls are practiced in the normal approach configuration for the aircraft being flown. In most light, single-engine aircraft, this is with the gear down and flaps retracted.

Carburetor heat is applied, power reduced to approach setting, and pitch attitude increased by reference to the attitude indicator in order to initiate a descent at approach speed. When the aircraft is established in a constant-rate, straight-ahead descent at approach speed, the pitch attitude is increased approximately three bar widths above the horizon. As the airspeed decreases toward stalling speed, the selected pitch attitude and initial heading are maintained.

Recovery is initiated as aerodynamic buffeting begins. This is accomplished by lowering the miniature aircraft to the horizon on the attitude indicator, adding full power, and removing carburetor heat. To maintain the original heading, right rudder pressure is increased to counteract the left turning tendency caused by adding power. Maintaining the level flight attitude causes an increase in airspeed. As the obstacle clearance airspeed is attained, a climb to the entry altitude is initiated.

TAKEOFF AND DEPARTURE STALLS

Straight Ahead

For performance of takeoff and departure stalls power is reduced and altitude maintained by slowly increasing the pitch attitude, using the attitude indicator, vertical velocity indicator, and altimeter as references. As the airspeed decreases to liftoff speed, a wings-level straight-ahead climb at full power is initiated, using a pitch angle which results in a power-on straight ahead stall. This is accomplished by raising the nose of the miniature aircraft approximately three bar widths above the horizon and simultaneously adding full power.

As airspeed decreases, back elevator pressure is increased to hold the pitch attitude, and right rudder pressure increased to counter torque and P-factor. This configuration is maintained until a stall buffet occurs.

When the buffet occurs, the nose of the miniature aircraft is lowered to the horizon and the aircraft allowed to accelerate. Recovery altitude is maintained and the maneuver completed on the entry heading. As cruise airpseed is attained, power should be reduced to the cruise power setting.

Turning

The turning takeoff and departure stall is initiated in the same manner as the straight-ahead departure stall. Power is reduced and altitude maintained. As the turn is completed and airspeed decreases to liftoff speed, pitch is increased to the second pitch reference line, full power is applied, and a 15° bank is made in either direction. This climbing turn attitude is maintained on the attitude indicator until a stall buffet occurs.

Recovery is affected by lowering the pitch attitude to the horizon bar, leveling the wings, and maintaining the recovery altitude and heading. As the airspeed approaches cruise, the power is reduced to the cruise setting.

CRITICAL ATTITUDE RECOVERY

Critical or unusual attitudes are usually the result of improper instrument flying technique, distraction, or turbulence. In some cases they may be a combination of all three. Although the vast majority of instrument pilots are never confronted with the necessity of a critical attitude recovery, this practice develops confidence and promotes safety. The entry to the critical attitude in an airplane is accomplished by an instructor. The pilot is directed to remove his hands and feet from the controls and close his eyes. The instructor then puts the aircraft into any number of critical attitudes, such as spiraling dives or approaches to stalls. After the critical attitude is established, the pilot is instructed to open his eyes, take over the controls, and make the recovery. The recovery is made to straight-and-level flight using the attitude indicator as the primary reference.

While scanning the attitude indicator, there are two possibilities which must be determined immediately — whether the nose is excessively high or excessively low. If the nose is excessively high, as in figure 1-29, it indicates the aircraft is rapidly approaching a stalled attitude. If the nose is excessively low, critical airspeed will soon be obtained.

Recovery from a "nose-high" critical attitude is performed as follows:

1. lower aircraft nose to the horizon bar,
2. apply full power, and
3. roll wings level.

Fig. 1-29. Nose High Critical Attitude

To recover from a "nose-low" critical attitude the pilot should:

1. reduce power,
2. roll wings level, and
3. gently raise aircraft nose to the horizon bar

The primary reason for different recovery techniques in nose-high and nose-low critical attitudes is the objective of each recovery. In a nose-high attitude, the objective is to avoid stalling the aircraft. *First*, the aircraft nose is lowered (to reduce angle of attack), then full power is applied (to increase airspeed). As experience is gained, nose-high critical attitude recoveries are initiated by simultaneously lowering the aircraft nose and adding full power.

In a nose-low attitude, the objective is to avoid a critically high airspeed and load factor. In this case, power is reduced (to stop the power dive), and the wings are leveled (to reduce load factor).

Recovery from a critical attitude under *emergency* panel conditions requires the same basic techniques. The main difference is that the turn coordinator (or turn-and-slip indicator) is used to level the wings and stop the turn, and the airspeed indicator, altimeter, and vertical velocity indicator are utilized for pitch attitude information. As the movements of the altimeter, airspeed, and vertical velocity needles decrease, the aircraft attitude is approaching level flight. When the needles stop movement and reverse direction, the aircraft is passing through level flight. At this point, control pressure is neutralized to maintain this attitude.

It must be remembered that critical attitude recovery is made through prompt, positive, and smooth application of the controls. As the airplane is returned to approximately straight-and-level flight, all control movements are ma[illegible] progressively smaller.

2 METEOROLOGY

INTRODUCTION

The numerous changes in weather conditions occurring throughout the world from day to day are caused by two principal physical conditions. These two governing factors are unequal heating of the atmosphere and the earth's rotation.

GENERAL WIND SYSTEMS

The idealized circulation pattern due to thermal differences on a nonrotating earth, is shown in figure 2-1. This airflow is the result of air being heated at the Equator, then flowing northward and southward to the poles. After cooling, it settles and flows along the surface to the Equator where it again is heated and continues the process. However, this simple two-cell pattern is greatly altered by the earth's rotation.

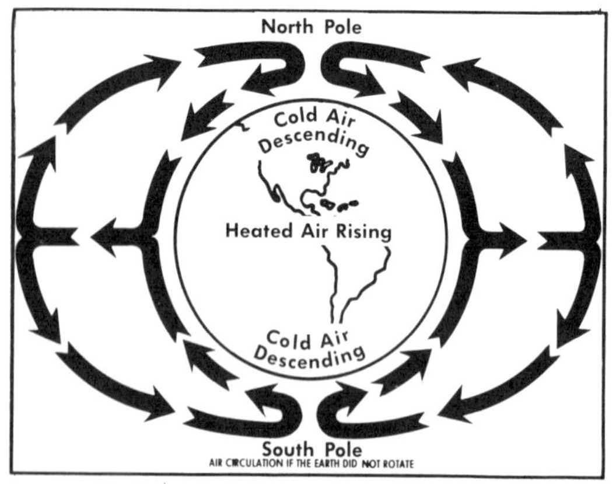

Fig. 2-1. World Air Circulation.

CORIOLIS EFFECT

The earth's rotation, from west to east, introduces an effect called Coriolis force. Coriolis force is the apparent force that causes objects free of the earth to be deflected to the right in the Northern Hemisphere and to the left in the Southern Hemisphere.

Coriolis effect can be explained by using a turntable, placing a flat disc on it, and rotating the disc counterclockwise (opposite the direction of a turntable on a record player). While the disc is rotating, a straight edge is placed just above the rotating disc in a position from the center of the turntable to the edge. While the disc is rotating, a line is drawn along the straight edge from the center of the disc to the outside edge. The path of the line drawn along the straight edge is curved to the right on the rotating disc, as shown in figure 2-2. This is what happens to any free body, including the atmosphere, in the Northern Hemisphere. The prevailing westerly winds at about 30° to 35° north latitude are the result of Coriolis force.

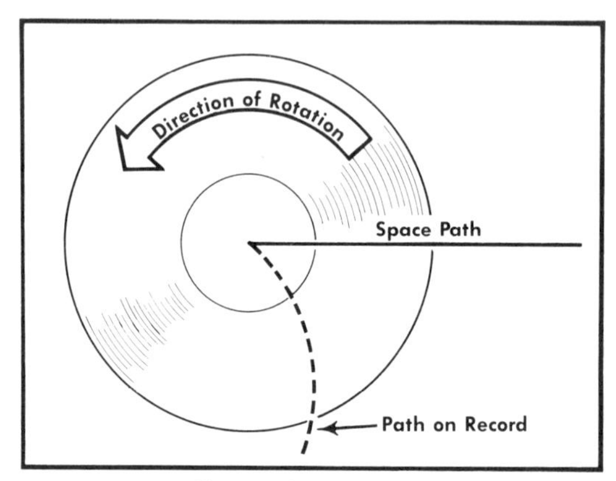

Fig. 2-2. Coriolis Effect

CENTRIFUGAL FORCE

Centrifugal force tends to cause a rotating body to be propelled away from the center of rotation. This, coupled with Coriolis force, balances pressure gradient and causes the wind to flow counterclockwise around a low and clockwise around a high in the Northern Hemisphere, as shown in figure 2-3.

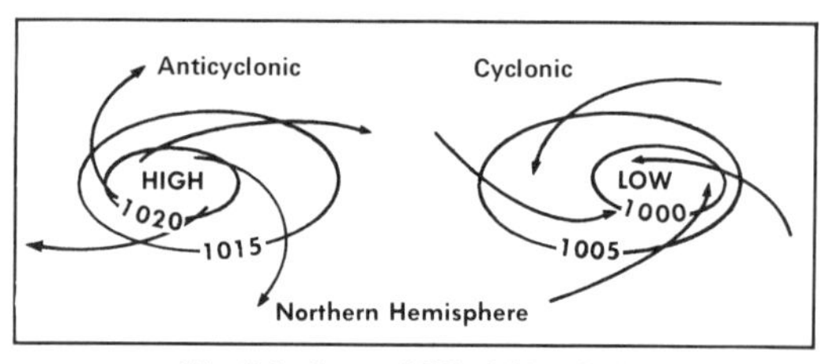

Fig. 2-3. General Wind Circulation

AIRMASSES

An airmass is a thick and extensive body of air whose temperature and humidity characteristics are approximately the same in a horizontal direction. Airmasses are formed over extensive areas of water or land, called source regions, that have almost identical characteristics, as shown in figure 2-4. Airmasses can be thousands of square miles in area and usually vary from three to eight miles in depth.

NORTH AMERICAN SOURCE REGIONS

The two airmasses normally found within the North American continent are called polar and tropical. The polar airmasses of the Northern

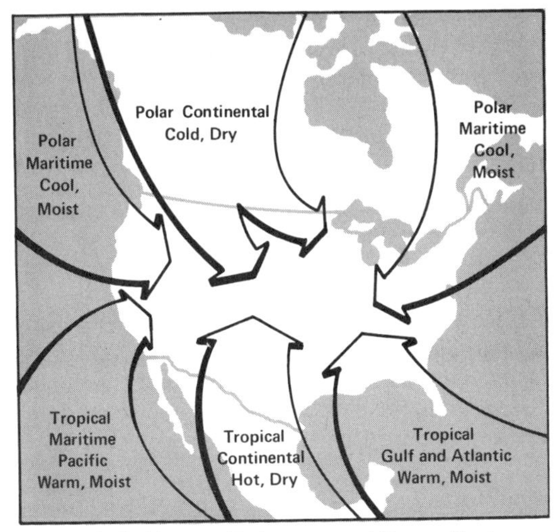

Fig. 2-4. Airmass, Trajectory, and Source Regions

Hemisphere are formed in the area near 60° north latitude. Continental polar airmasses form from the Hudson Bay west to the Rockies. Maritime polar airmasses form over water near the Gulf of Alaska and off the coast of Greenland. Tropical airmasses are formed over land or water in the regions about 20° to 30° north latitude. On rare occasions, parts of the United States may be affected by arctic or equatorial airmasses which form over the arctic or equatorial regions.

Both polar and tropical airmasses are described as continental or maritime. The term *continental* is used to indicate that the airmass has formed over a land area. *Maritime* indicates that it was formed over a large body of water. These descriptive terms are used to indicate the relative moisture content of the airmass. A moist body of air moving into an area with a large temperature differential will form greater amounts of clouds than will dry air, and may become unstable and turbulent.

AIRMASS FORMATION

An airmass must remain relatively stationary over the source region for several days to several weeks to acquire definite characteristics. Pressure differences then begin to move the airmass out of its source region. A polar continental airmass, flowing from Canada between Hudson Bay and the Rockies, normally moves in a southeasterly direction across the United States. This movement brings cool air to the central and eastern parts of the country. Maritime tropical air from the Gulf of Mexico and the Caribbean usually moves in a northerly direction across the United States. These are the two main airmasses affecting the central, southern, and eastern parts of the United States. Maritime polar air from the Pacific Ocean has the greatest effect on the western part of the United States.

SEASONAL EFFECT

To a large extent, the seasons determine the airmasses found over the United States. Except for the Gulf coast and Florida, continental polar air usually covers the area from the Rocky Mountains eastward during the winter. During the summer months maritime tropical or continental tropical air is more likely to be found from the Rocky Mountains eastward. These airmasses advance and recede in connection with circulation patterns set up by pressure centers. The heaviest precipitation and most violent weather usually occurs during the spring and fall when these two contrasting airmasses are most often in conflict.

From the Rocky Mountains westward, maritime Pacific air is the predominant influence on weather. As a result, a maritime climate prevails along the western coast bringing warm, wet winters and, for the most part, cool, dry summers.

AIRMASS MODIFICATION

After an airmass leaves a source region, it is modified by the terrain over which it passes. For example, a cold airmass is warmed from below as it moves south. The warming from below, in the form of a temperature inversion, causes turbulent convective currents and may create clouds if sufficient moisture is present. Conversly, a warm airmass will be cooled as it moves over a colder surface. This temperature distribution inhibits convection and produces stable air and smooth flying conditions.

Modifications also occur when air is forced to rise over hills, mountains, or other airmasses. When sufficient moisture is present this action causes cooling of the rising air and formation of clouds. *Frontal* systems are formed when two contrasting airmasses are brought together by pressure systems and circulation patterns.

THE POLAR FRONT

The polar front, separating polar and tropical airmasses, forms along an irregular line that extends around the earth. This "front," or line of discontinuity in temperature, wind, and moisture, exists throughout the year. Its latitudinal location varies with the seasons.

SHIFTING OF THE POLAR FRONT

The polar front shifts with the seasons, lagging about one month behind the sun's seasonal northward and southward movement. In other words, it follows the sun. In the spring when the sun moves north of the Equator, the polar front moves north. By middle or late summer, the polar front is located at about the Canadian border. When the sun begins to move southward, the polar front follows it so that, in the middle or late winter, it is located near the southern extremity of the United States. The polar front, which does not form a straight or well-defined line, can be considered the southern extremity of polar airmasses.

Cyclonic storms are produced by the interaction of warm and cold airmasses along the front. In the temperate zone, these cyclonic storms carry tropical airmasses far to the north and polar airmasses far to the south of their source regions. The process of cyclonic development is termed *cyclogenesis.* Cyclones will be discussed later, since they produce most of the adverse weather in the United States.

CLOUD TYPES

According to their appearance, clouds generally are described as stratiform or cumuliform. *Stratiform* clouds are formed in strata or layers. *Cumuliform* clouds have vertical development and indicate convective currents, instability, and turbulence.

These two types of clouds are further divided into three categories.

1. High clouds — cirrus, cirrostratus, and cirrocumulus
2. Middle clouds — altostratus and altocumulus
3. Low clouds — stratus, nimbostratus, cumulonimbus, and stratocumulus

The term *nimbo* or *nimbus* indicates a cloud from which precipitation is falling.

CLOUD ALTITUDES

The term *ceiling* is defined as the lowest height above the ground where more than one-half the sky is covered by opaque clouds. High clouds generally are located from 20,000 feet to 40,000 feet MSL and are mostly ice crystal in nature. Middle clouds normally are found from 6,000 feet to 20,000 feet AGL and are composed mainly of water droplets; however, certain forms may consist of ice crystals. Low clouds generally have bases below 6,000 feet AGL and consist mostly of water droplets. Figure 2-5 illustrates cloud types and elevations.

CLOUDS AND ASSOCIATED FLIGHT CONDITIONS

When sufficient moisture is present, stratus clouds form in stable air. These clouds indicate relatively smooth flying and continuous precipitation of uniform intensity. Low stratus clouds are a hazard because they often obscure the tops of hills and mountains. Surface visibility is usually poor beneath these low clouds. The type of structural icing most often encountered in stratiform clouds is rime icing.

Cumuliform clouds produce showers and thunderstorms. In a thunderstorm, severe or extreme turbulence, heavy rain, and hail can produce extremely hazardous flying conditions. The large water droplets of cumuliform clouds are favorable for the formation of clear ice on airplanes at altitudes above the freezing level. Surface visibility is generally good under cumuliform clouds except during precipitation, blowing snow, or dust.

Altocumulus and stratocumulus clouds, possessing characteristics of both stratus and cumulus types, usually produce weather of less intensity than that produced by cumulus clouds. The following table gives some characteristics of clouds.

	Cumuliform	Stratiform
Size of Water Droplets	Large	Small
Stability of Air	Unstable	Stable
Flying Conditions	Rough (Turbulent)	Smooth
Precipitation	Showery	Continuous (of uniform intensity)
Surface Visibility	Good, except in precipitation or blowing dust or snow	Usually Poor
Airplane Icing	Predominantly Clear Ice	Predominantly Rime Ice

THE ELEMENTS OF WEATHER

The elements of weather (those factors which are the direct source of weather) are moisture (water vapor), temperature, pressure and winds. Temperature and moisture content usually are the most obvious elements; however, pressure and winds also are important.

When the term "atmosphere" is used, it generally denotes the troposphere. This is where most weather occurs. The depth of the troposphere

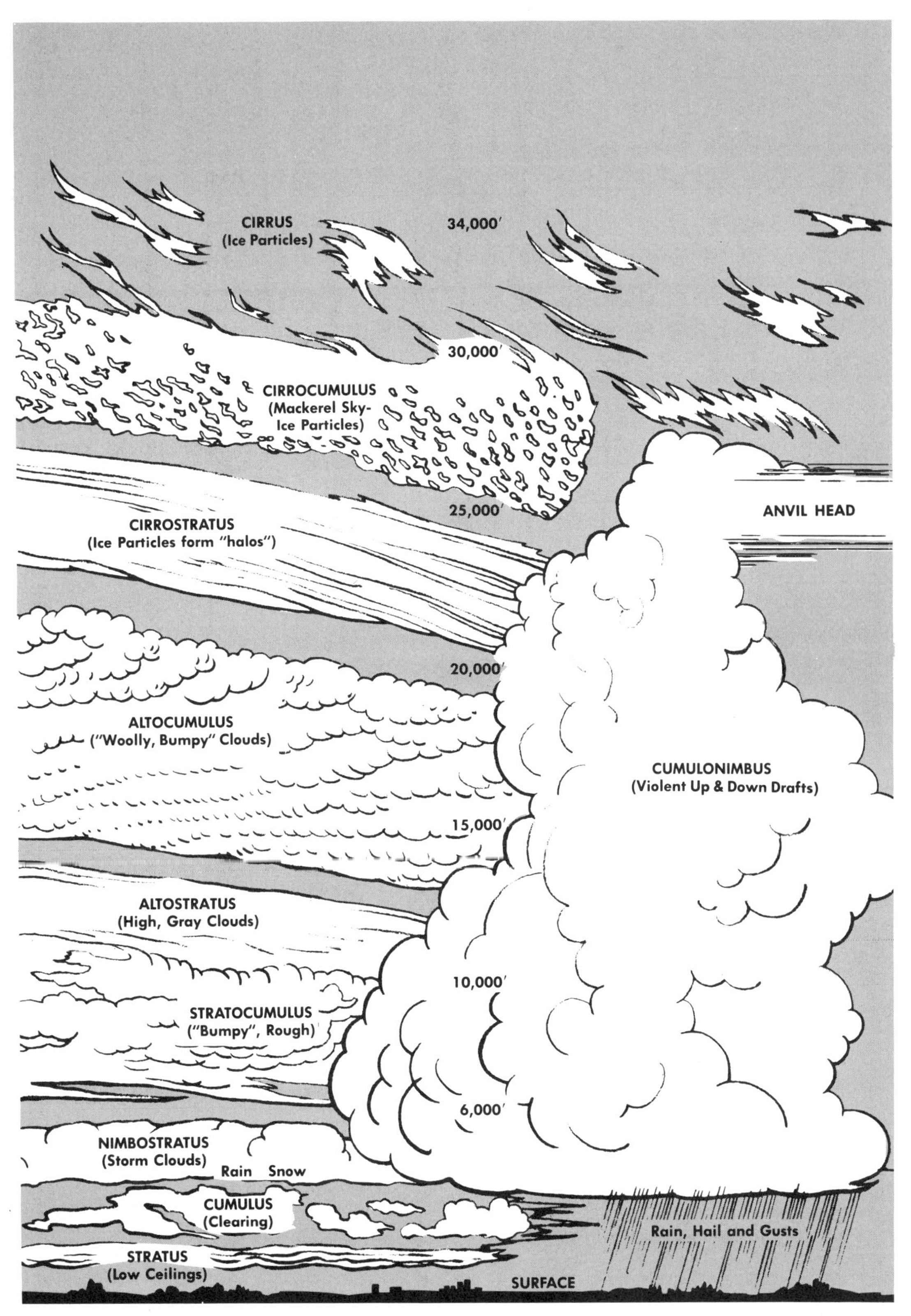

Fig. 2-5. Cloud Types and Elevations

varies from approximately six miles near the poles to about 10 miles at the Equator, and changes with the seasons.

MOISTURE

Most water vapor present in the atmosphere has evaporated from water surfaces such as the oceans. Some water vapor is gained from plants and the earth's surface.

Water vapor present in the atmosphere may vary from a trace to approximately five percent by volume. The source of energy for storms comes from the *latent heat of condensation*, which is the heat energy in the water vapor. The significance of this latent heat is that it is released into the atmosphere wherever condensation occurs, producing heat energy, and directly affecting the temperature of the air. Conversely, evaporation requires heat energy and consequently causes a cooling effect. Water vapor is probably the main factor in weather, since moisture is the source of all precipitation and clouds.

TEMPERATURE

Due to a decrease in pressure with altitude and the resulting cooling by expansion, temperature normally decreases with altitude. The rate of temperature loss is called the *lapse rate.*

STANDARD LAPSE RATE

The average lapse rate throughout the world is approximately 3.5° Fahrenheit for each 1,000-foot increase in altitude, and the average sea level temperature is approximately 59° Fahrenheit. These temperature conditions have been established as "standard."

Since standard conditions are seldom present, the standard lapse rate of 3.5° Fahrenheit per 1,000 feet is representative and can be used only as a rule of thumb for *estimating* temperatures aloft. In figure 2-6, the solid line shows a condition with a lapse rate of 5 degrees Fahrenheit per 1,000 feet. This emphasizes that the standard value of 3.5° Fahrenheit per 1,000 feet in altitude is an *approximate average* and actual conditions may vary greatly from this value.

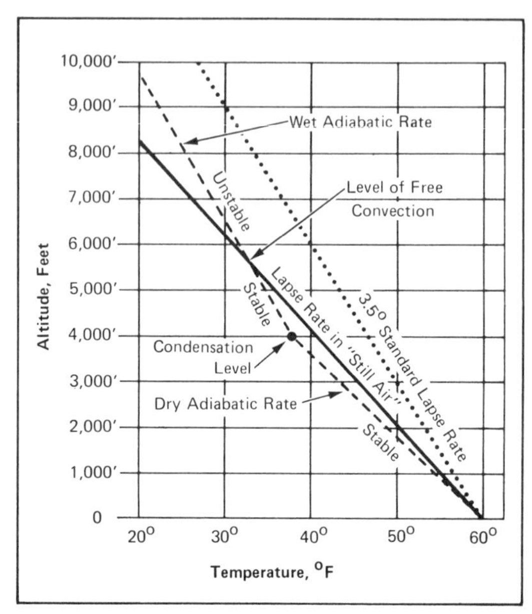

Fig. 2-6. Normal Lapse Rate with Condensation Level

DRY ADIABATIC LAPSE RATE

When dry, unsaturated air is forced to rise, it expands due to the decrease in weight of air above it and cools at approximately 5.5° Fahrenheit per 1,000 feet. Conversely, air contracts and is heated at this same rate when it descends. The dry adiabatic lapse rate is the change in temperature which occurs with altitude without addition or withdrawal of sources of outside heat. The change takes place entirely within the airmass itself, and the rate of change is always constant for dry air, as shown in figure 2-7.

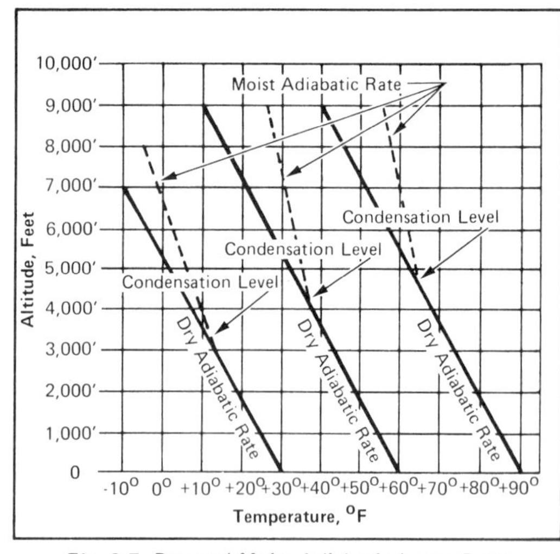

Fig. 2-7. Dry and Moist Adiabatic Lapse Rates

MOIST ADIABATIC LAPSE RATE

The moist adiabatic lapse rate varies with altitude, depending on air saturation and condensation. At lower altitudes, the average figure for the moist adiabatic rate is three degrees Fahrenheit for each 1,000-foot increase in altitude, as shown in figure 2-7. Temperature decreases with altitude more rapidly than normal when condensation is taking place. In the case of saturated air, the moist adiabatic rate is a factor only while the air is ascending, because the water vapor condensed during the upward motion of the air is not absorbed as the air descends.

INVERSIONS

An *inversion* is present when temperature *increases* with an increase in altitude, as shown in figure 2-8. When warm air flows in over a colder area, cooling of the air is rapid at the surface. Since air is a poor conductor of heat, the heavier, cold air tends to remain at the surface causing an inversion to persist for days or weeks before the upper air is cooled. Because of this, an inversion may cause smog or fog to lie in an area for days when there is no horizontal movement of the air. Inversions may occur at or above the surface when warm air overlies a colder layer of air, as shown in the right-hand portion of figure 2-8.

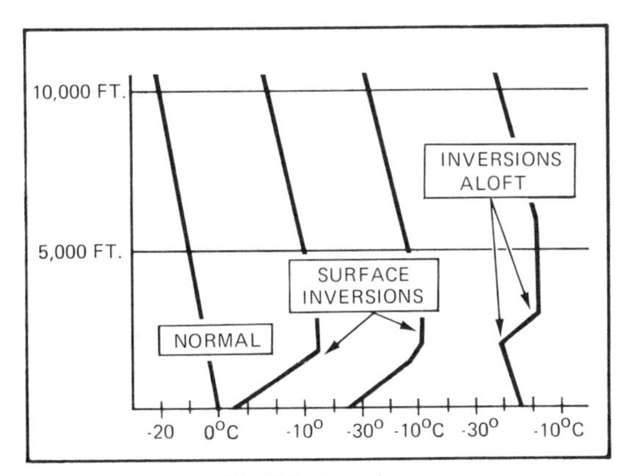

Fig. 2-8. Inversions

WIND

Unequal heating of the earth's surface causes the differing pressure systems that generate air movement. Wind, which is simply air in motion, tends to blow from a high-pressure area toward an area of relatively lower pressure. A semipermanent low-pressure area prevails over the Gulf of Alaska and a semipermanent high-pressure area generally is found near the Azores, over the South Atlantic Ocean.

ATMOSPHERIC PRESSURE

Atmospheric pressure is measured in inches of mercury, millibars, or pounds per square inch. The approximately average air pressure at sea level, over the world, has been established as the standard pressure. This value is 29.92 inches of mercury, 1013.2 millibars, or 14.7 pounds per square inch. Under standard conditions at sea level, the weight (pressure) of air is sufficient to force mercury to a height of 29.92 inches in a vacuum tube, as shown in figure 2-9. In the lower levels, as altitude increases the pressure of air decreases by approximately one inch of mercury for each 1,000 feet.

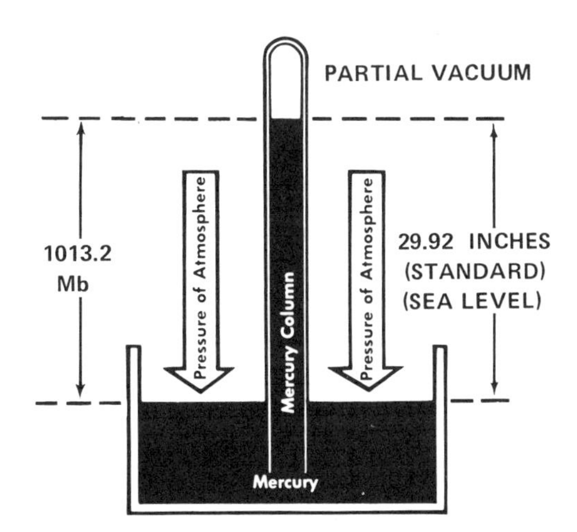

Fig. 2-9. Mercury Barometer

PRESSURE GRADIENT

On surface weather maps, lines marking points of equal pressure are called *isobars*. These lines usually are drawn for every three or four millibars of pressure change, as shown in figure 2-10. The pressure difference and the distance between these isobars determine *pressure gradient*. When the isobars are spaced closely, the pressure gradient is steep and wind velocity is strong; conversely, when the isobars are widely spaced, the pressure gradient is shallow and the wind velocities are light. Pressure gradients result in the horizontal transfer of warm or cold air by *advection*.

Advection and convection (vertical transfer) both are essential to the formation of weather systems. Advection is of greater overall important than convection, since it is the means by which all horizontal transport of heat, cold and moisture is accomplished. Convective transport is associated with local pressure systems and most local weather.

HIGH AND LOWS

The diagram in figure 2-11 represents the semipermanent high-pressure and low-pressure systems and associated wind belts. Air circulates clockwise around a high-pressure system, flowing away from the center. In a low-pressure system, the air circulates counterclockwise, toward the center of the system. This circulation

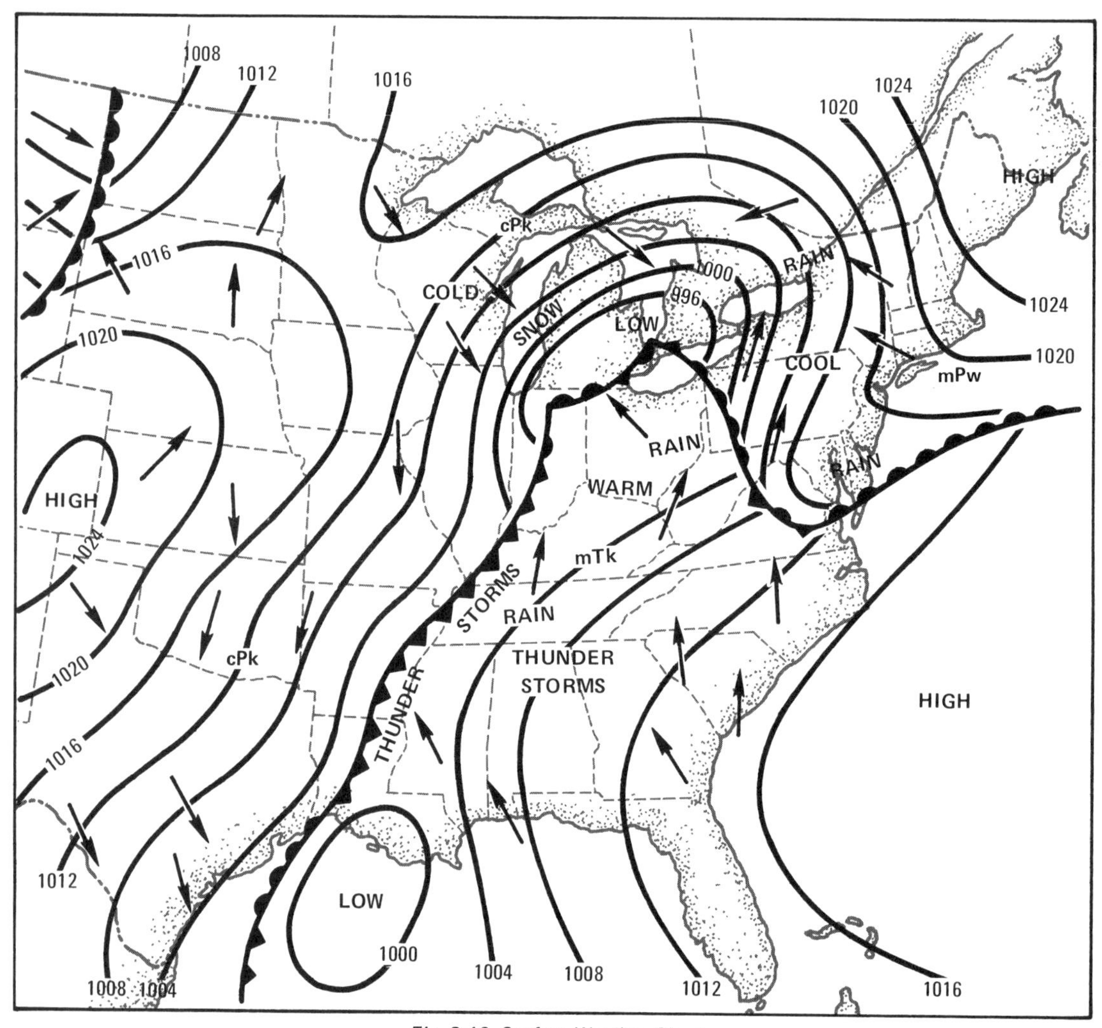

Fig. 2-10. Surface Weather Chart

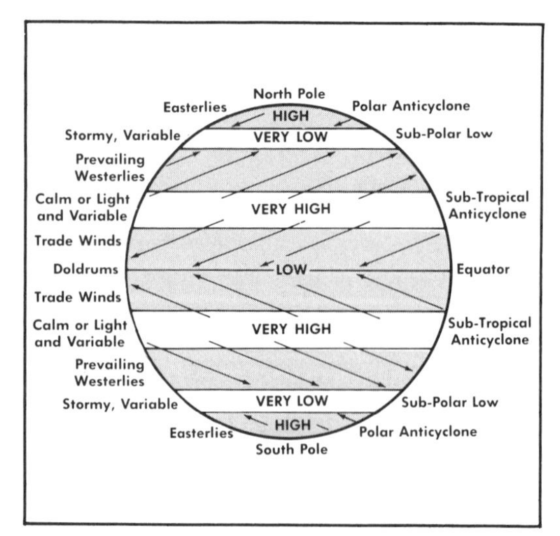

Fig. 2-11. High-Low Pressure Belts

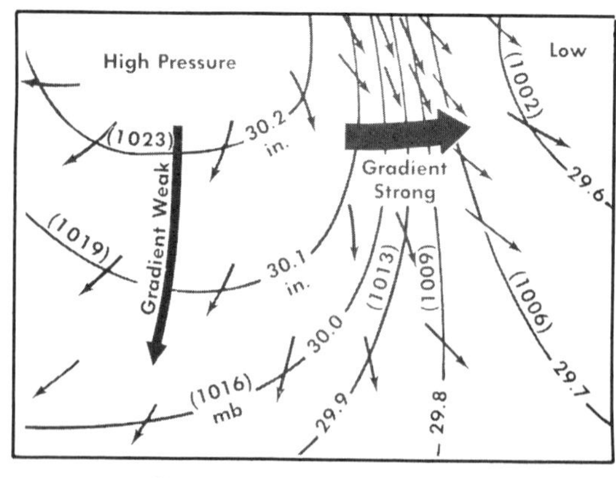

Fig. 2-12. Gradient "Strength"

pattern is true of both semipermanent and local high-pressure and low-pressure systems, as shown in figure 2-12.

TERRAIN AND FRICTIONAL EFFECTS ON WINDS

General wind circulation near the surface of the earth is affected by the friction between the surface and the air as shown in figure 2-13. In areas where there are no hills, mountains, or trees, the wind tends to flow fairly constantly with relatively high velocities. For example, in the Great Plains and western Texas, winds are usually fairly strong (20 to 30 miles per hour) and persistent, usually from a westerly direction.

Friction effect is due to the drag created at the surface. The smoothness of the surface greatly affects this drag. This effect is noticeable from the surface to about 2,000 feet and causes the wind to veer with altitude. Normal veer (clockwise shift) is 10° over water and from 30° to 50° over land, depending on the roughness of terrain and vegetation present.

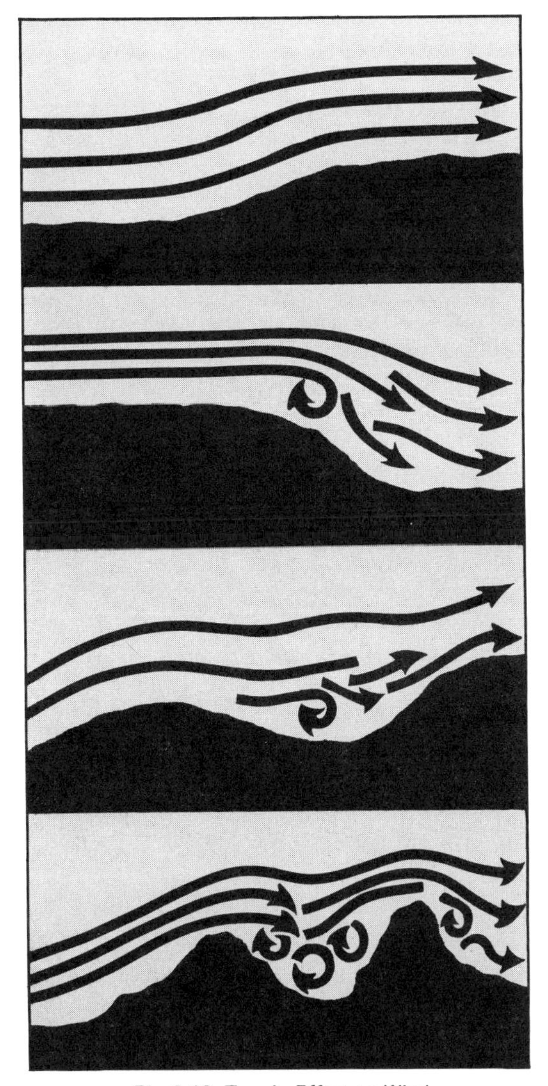

Fig. 2-13. Terrain Effect on Wind

MOUNTAIN AND VALLEY WINDS

In hilly areas or mountains, the slopes of the land produce mountain and valley winds. During the daytime, the heating of the hillsides facing the sun produces convective currents which flow through the valley (valley winds), as shown in figure 2-14. Valley winds normally have a velocity of less than 20 knots. At night, the cooling of air against the barren mountains or hillsides causes the air to become very dense and sink down the mountainside toward the valley floor. This produces mountain winds that flow down to the lower elevations. (See Fig. 2-15.) The effect of these winds should be considered prior to mountain takeoffs and landings. Downslope winds of a warm, dry nature are Foehn winds, or Chinook thawing winds, of the Rockies. Foehn winds are heated by compression at the dry adiabatic lapse rate and are relatively warm.

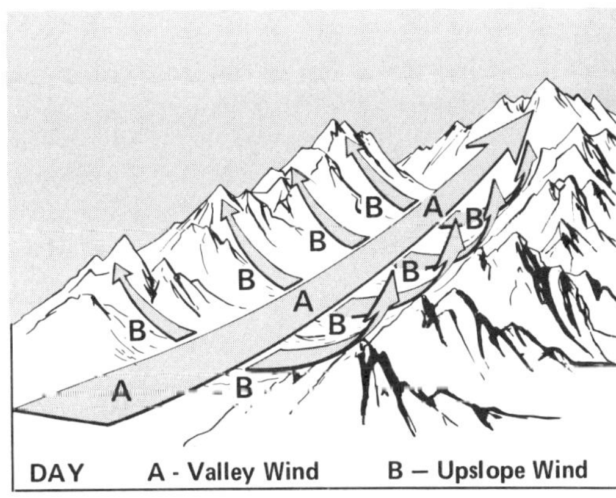

Fig. 2-14. Day Valley Wind

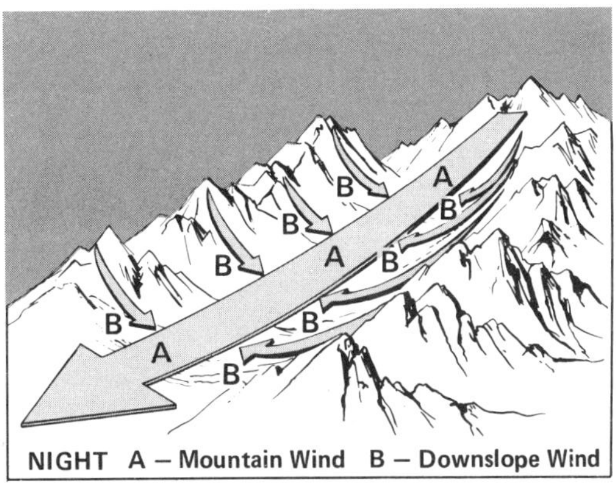

Fig. 2-15. Night Mountain Wind

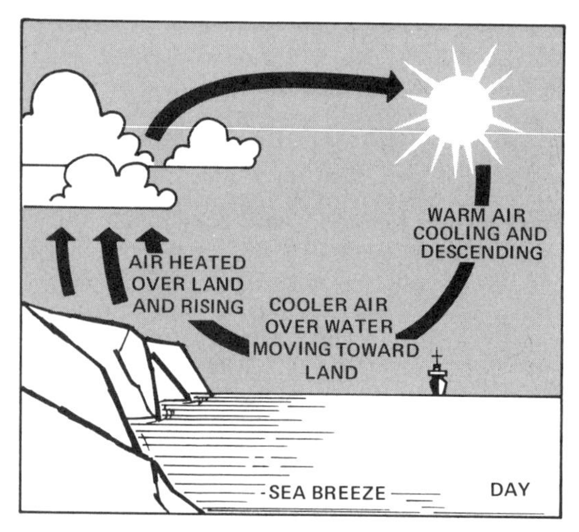

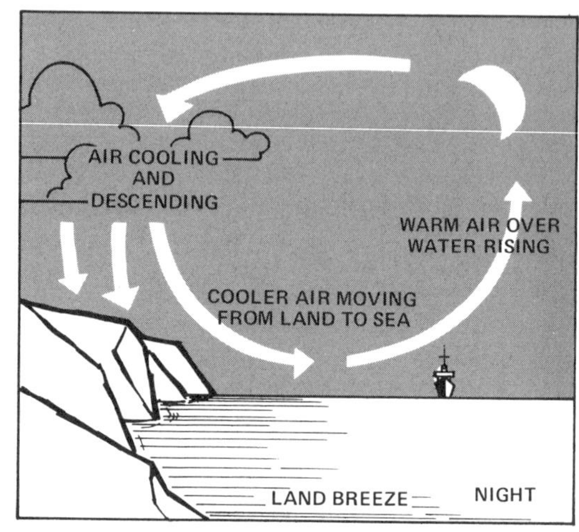

Fig. 2-16. Sea and Land Breezes

LAND AND SEA BREEZES

Land and sea breezes are noticeable along seashores and large bodies of water. On a warm summer day, the sun heats land more rapidly than water. This causes a relatively low-pressure area with rising currents to form over land, with a relatively high-pressure area over water. In this situation, wind flows from high to low pressure, or from the sea to the land, and may be fairly strong for some distance inland. At night, the reverse is true. Land cools faster than water and, as the wind blows out to sea, land breezes are established. (See Fig. 2-16.)

ALTITUDE, TEMPERATURE, AND PRESSURE

The rule commonly used to approximate pressure with increase in altitude is that for every 1,000 feet of altitude, pressure decreases by one inch of mercury. This is primarily true for altitudes below 10,000 feet. At higher altitudes, the pressure decreases more gradually. For example, between 18,000 feet and 35,000 feet, an average climb of 2,150 feet is required to reduce the pressure one inch.

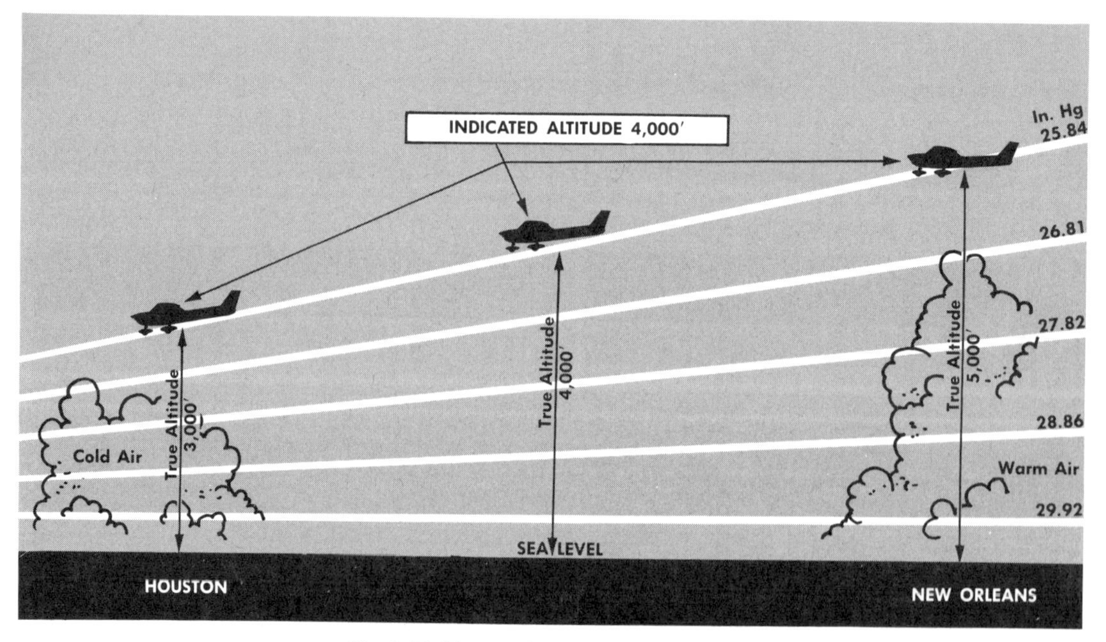

Fig. 2-17. Warm to Cold Air Pressure Change

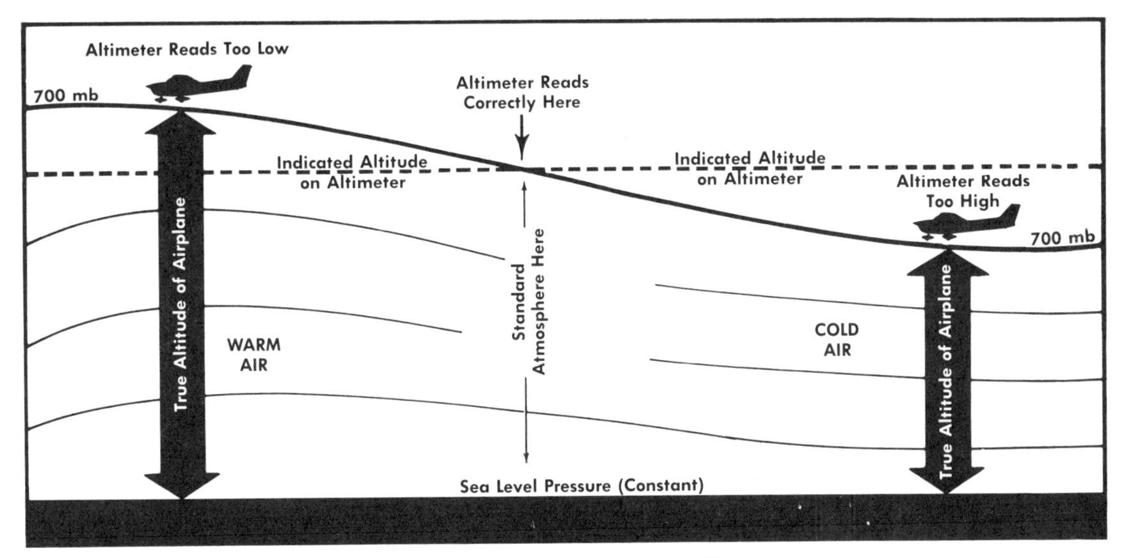

Fig. 2-18. Cold to Warm Air Pressure Change

Since cold air is more dense than warm air, if the altimeter setting is not adjusted, the vertical distance to the surface decreases during a flight toward cold air and increases when flying toward warm air, as shown in figures 2-17 and 2-18. Likewise, in flying from high to low pressure with a constant altimeter setting, the airplane is flying along a constant pressure surface and loses altitude, as seen in figure 2-19. A helpful rule to follow is that when flying "from hot to cold or high to low, lookout below."

Normally, the pressure difference between the centers of lows and highs and their outside edges or margins is several tenths of an inch or 10 to 20 millibars. (See Fig. 2-20.) Winter lows have been known to show as much as 35 millibars, or one inch of pressure difference from the center to the outer edge. High-pressure systems are usually associated with fair or good weather. Lows, on the other hand, are associated with marginal or bad weather. An elongated low is often referred to as a trough, while an elongated high is referred to as a ridge.

FRONTS AND FRONTAL WEATHER

COLD FRONTS

A front occurs when two air masses of contrasting temperature and moisture content come together. The cold front occurs when cold air displaces warm air and causes it to rise over the

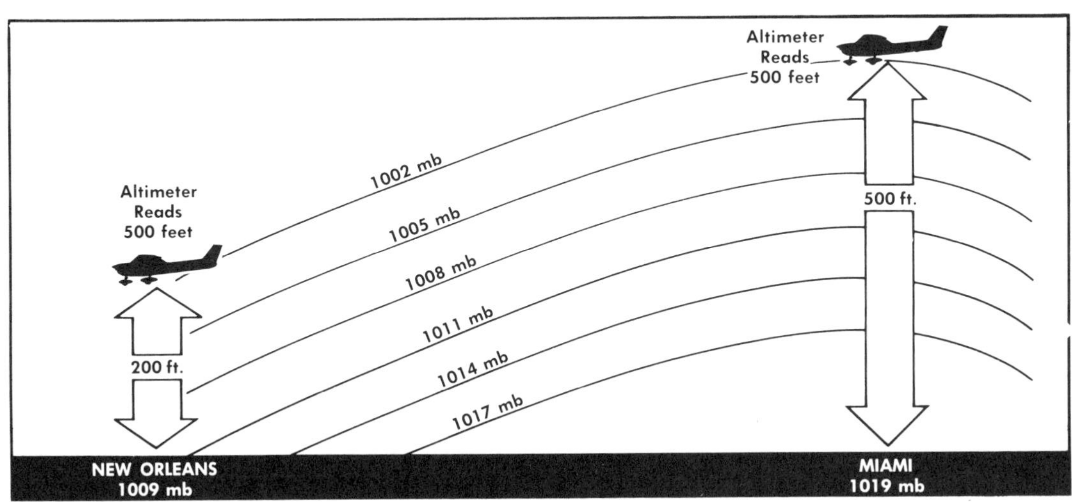

Fig. 2-19. Effect on Altimeter of Changes in Pressure Altitude Levels

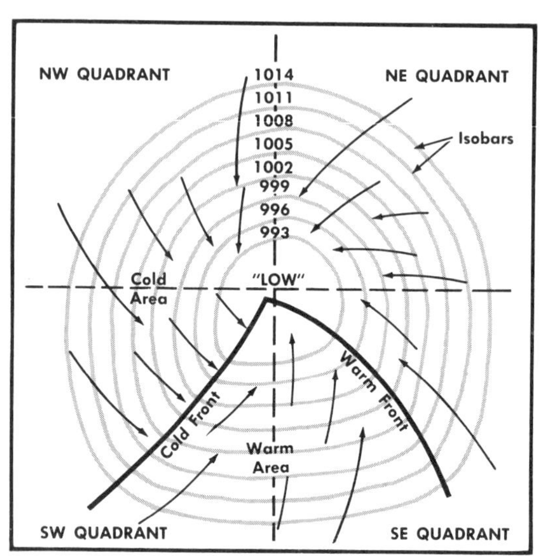

Fig. 2-20. Millibar Change Around a Low

advancing cold air. The actual front may vary in width from a small area defined by a sharp windshift to an area covering several miles. The weather associated with the cold front is of two types—extremely stormy weather, in the case of fast-moving cold fronts with unstable air, and moderate to mild weather, in the case of slow moving cold fronts with stable air.

When a fast-moving cold front displaces moist, unstable air, cumulus and cumulonimbus clouds normally result (See Fig. 2-21). Thunderstorms often are associated with fast-moving cold fronts and can form within a few miles of the front or several hundred miles ahead in a line called a *squall line.* Thunderstorm cells may be so close together that it is impossible to penetrate them safely in a light airplane. Severe or extreme turbulence can cause structural damage or structural failure. Vertical currents up to 6,000 feet per minute, or approximately 60 knots, have been noted in thunderstorms. In addition, hail and icing, as well as heavy rain, may be encountered.

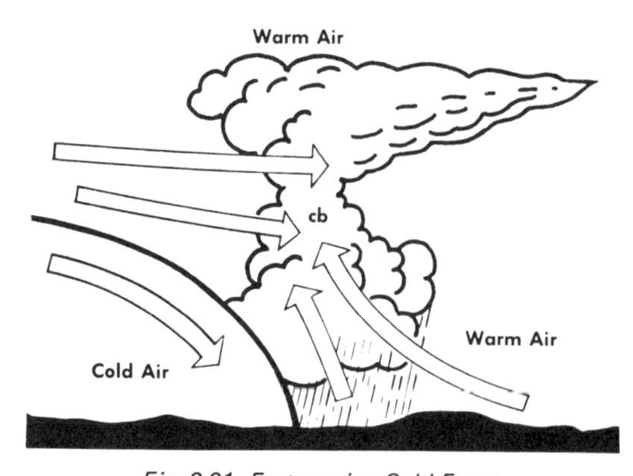

Fig. 2-21. Fast-moving Cold Front

Adverse weather generally precedes a rapidly moving cold front and clearing results shortly after its passage. A clockwise wind shift of 45° to 180° can be expected with the passage of the front. In the United States, winds usually shift from a southeasterly direction to a north or northwest direction with cold front passage and temperature drops markedly behind the front.

In the case of a slow-moving cold front which is replacing stable moist air, as shown in figure 2-22, clouds are normally of the stratus and nimbostratus type. Poor visibility and widespread, gentle rain normally are associated with this type of front. Icing can be expected with a cold front in areas where the temperature approaches freezing (32° Fahrenheit or 0° Celsius).

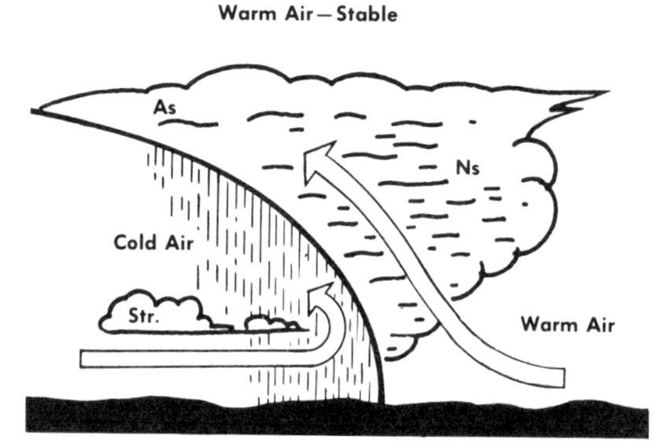

Fig. 2-22. Slow-moving Cold Front

WARM FRONTS

A warm front moves about half the speed of a cold front and is preceded by high cirrus, altostratus or altocumulus, stratocumulus, and nimbostratus clouds in sequence. In a well-developed warm front with stable air, cirrus clouds may appear as far in advance of the surface front as 600 to 1,000 miles and may extend to 30,000 feet or higher above the surface, as shown in figure 2-23.

Turbulence in a warm front is usually light, rains are gentle, but widespread, and visibility is extremely poor, with ceilings lowering as the front is approached. Clear ice can be expected if the temperature is near or below freezing and rain is falling from the warm sector aloft into the cold air below.

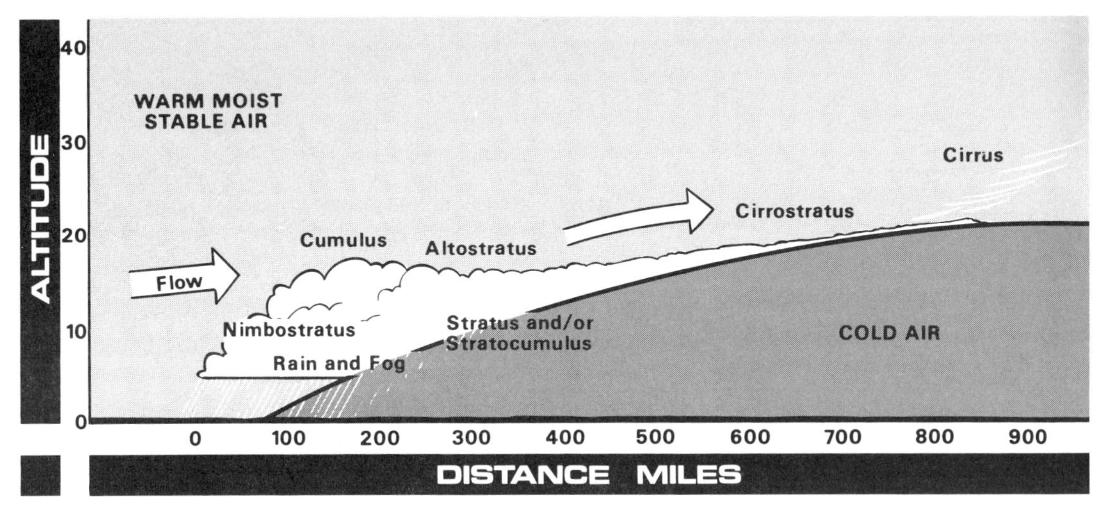

Fig. 2-23. Warm Front – Stable Airmass

A warm front with unstable air may have cumulonimbus clouds imbedded in the nimbostratus clouds, creating extremely dangerous flying conditions. (See Fig. 2-24.) These cumulonimbus clouds usually contain thunderstorms accompanied by extreme turbulence and other adverse weather associated with such storms. Where cumulonimbus clouds are associated with the front, rain and turbulence can be expected ahead of and along the front.

Passage of a warm front is followed by a marked rise in temperature, a light, clockwise wind shift of approximately 45°, clearing weather, and a rising barometer. There are several important factors which need consideration when flying into a warm front.

1. Possible presence of cumulonimbus clouds
2. Condensation level (cloud bases)
3. Freezing level
4. Turbulence
5. Conditions reported by pilots who have flown recently in the area (PIREPs issued by the FSSs)

LIFE CYCLE OF A CYLCONE

The cyclonic storm produces a vast amount of adverse weather in the middle latitudes. Figure 2-25 depicts the development and dissipation of this type of storm. It develops where a cold airmass from the polar areas comes in contact with a warm, tropical air mass. Vertical sections and a horizontal plan view are depicted in figure 2-26 showing wind circulation and the general

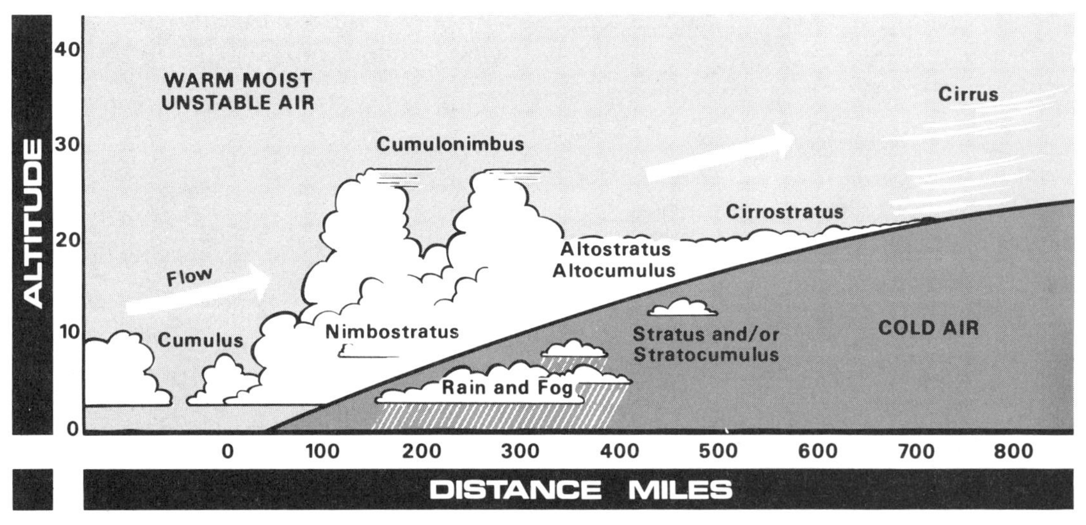

Fig. 2-24. Warm Front – Unstable Airmass

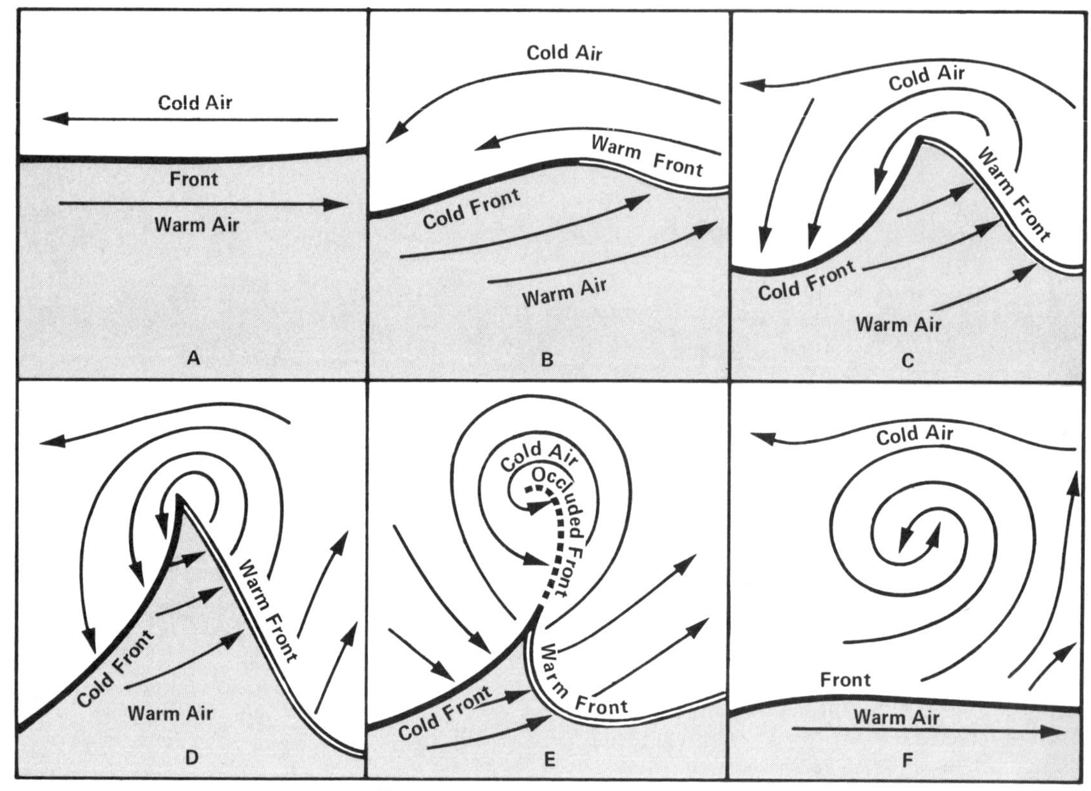

Fig. 2-25. Cyclonic Storm Development

arrangement of isobars about the low. The low-pressure area is the focal point. A cyclonic storm consists briefly of a warm front followed by a cold front, both radiating outward from the center of a low-pressure system. Figure 2-26 shows a diagrammatic sketch of a cyclonic storm system. Since these storms develop in the area of interaction of cold and warm airmasses, or the polar front area, their activity is greatest in the United States in the winter time. In the winter, the polar front is displaced southward, due to the southward movement of the sun. Cyclonic storms follow the general paths indicated in figure 2-27, from west to east. This happens because cyclonic storms form in the westerly

Fig. 2-26. Cyclonic Vertical Section and Fronts

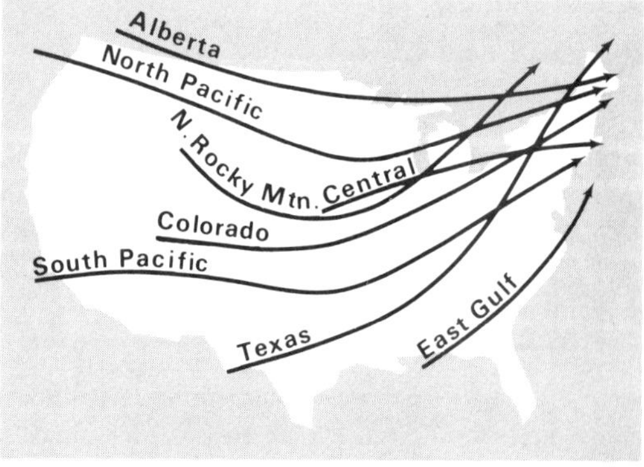

Fig. 2-27. Cyclonic Storm Paths

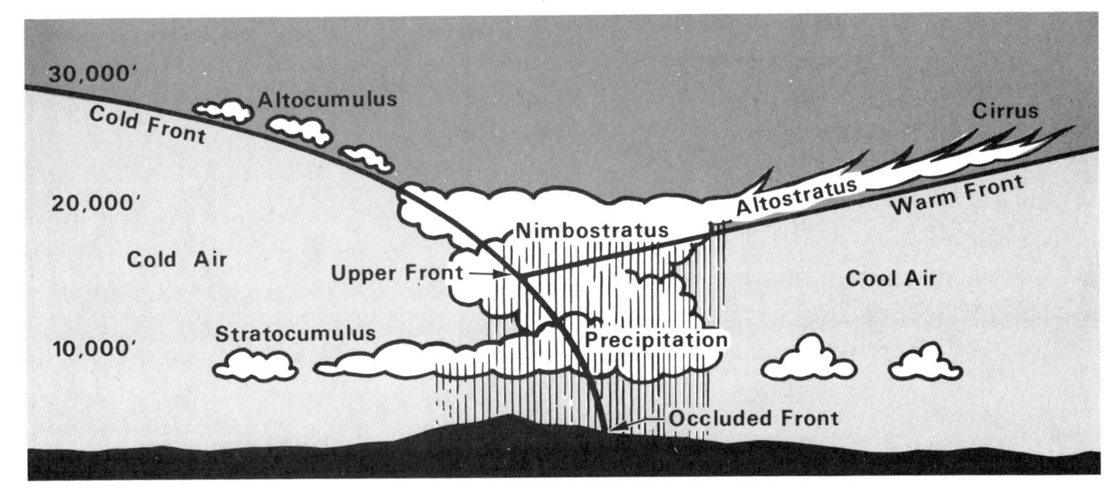

Fig. 2-28. Clouds of Cold Front-Type Occlusion

winds belt. These storms move at a rate of 20 to 30 miles per hour or 480 to 720 miles per day. There is often a concentration of the cylconic storms in the northwest and in the northeastern areas of the United States. Most of these storms follow the more northerly paths, swinging southeastward in an arc that terminates in a northeasterly path near the east coast.

OCCLUDED FRONTS

The occluded front develops out of a mature cyclonic storm. Occluded fronts are of two types, cold and warm, designated by the front that develops at the surface. The occlusion starts at the center of the low and progresses outward as the cold front overtakes the warm front. If the air behind the cold front is colder than the air ahead of the warm front, the air ahead of the warm front is forced aloft as the cold front overtakes the warm front and a cold type occlusion occurs at the surface, as shown in figure 2-28. The weather associated with the cold occlusion is a composite of both warm and cold fronts. The weather ahead of the warm front aloft is similar to that of the true warm front, but is usually more moderate. This weather usually consists of widespread rain and cloud cover with poor visibility and possible icing. (See Fig. 2-28.)

In the case where the air behind the cold front is warmer than the air ahead of the warm front, the advancing cold front is forced aloft and a warm front occlusion results, as shown in figure 2-29. The weather associated with the warm front occlusion is similar to that in a normal warm front, but with some cold front characteristics aloft and ahead of the occluded front, as indicated in figure 2-29.

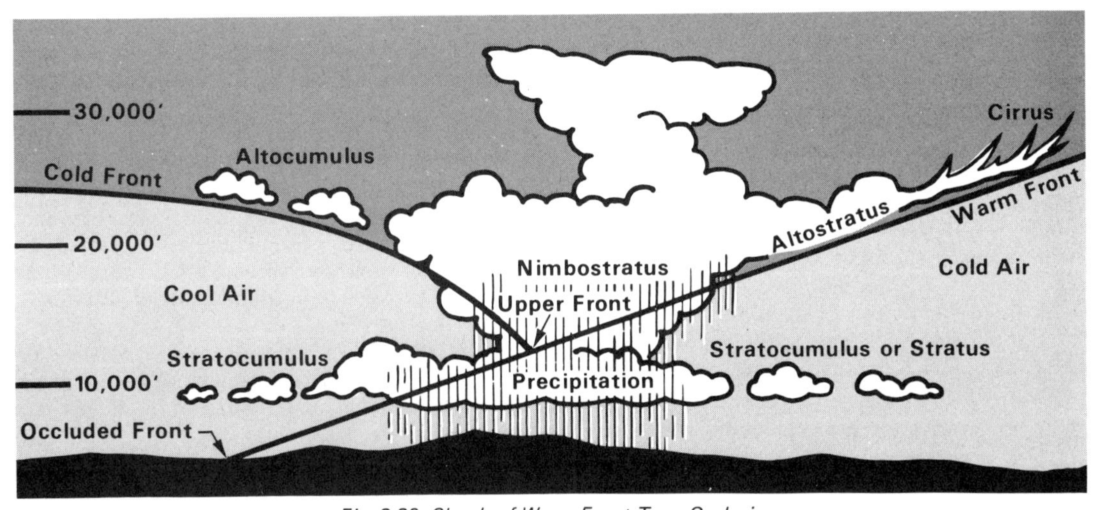

Fig. 2-29. Clouds of Warm Front Type Occlusion

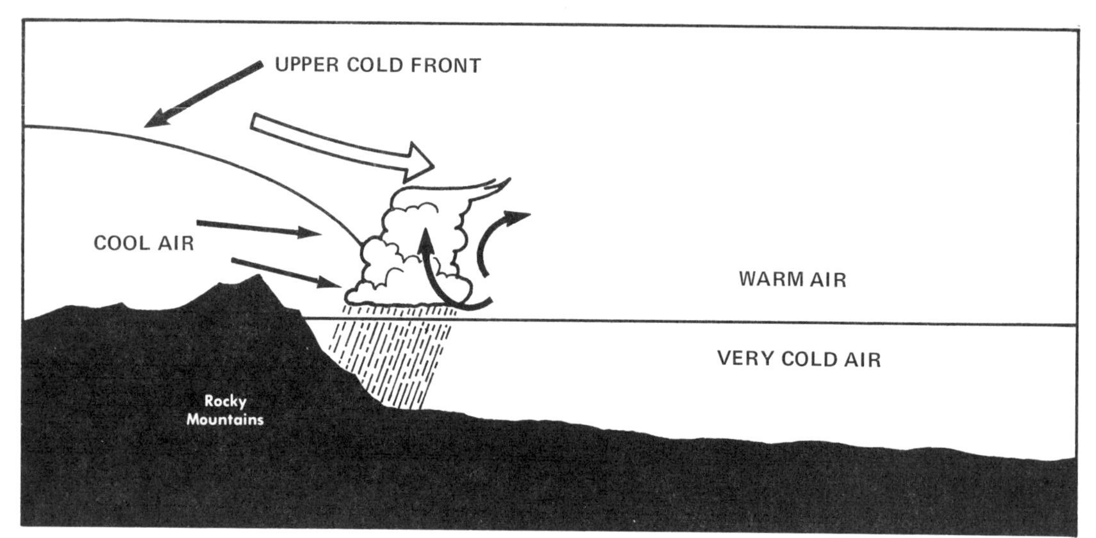

Fig. 2-30. Upper Front Near Mountains

UPPER FRONTS

In general, there are two types of upper fronts—those occuring where the warm or cold front is forced aloft by the coldest air associated with the occlusion and those occurring where an extremely cold layer of air is laying up against the leeward side of a mountain with two opposing dissimilar airmasses meeting above the cold layer, as illustrated in figure 2-30. These fronts follow the same general rules for cold or warm fronts formed at the surface. In an upper front, severe icing may be encountered in the cold air below the front, if the temperature is approaching freezing. Frontal weather is often hazardous and should be approached cautiously. However, fronts may exist with little or no adverse weather, so the pilot must check with the weather station to determine the severity of the weather associated with a particular front.

STATIONARY FRONTS

A stationary front results when neither the cold or warm airmass is in motion. Although areas of the two opposing airmasses have intermingled along the frontal surface or slope, if general horizontal movement is absent, the resultant weather resembles that of a warm front. The stationary front produces widespread low ceilings and moderate precipitation that may persist in a given area for days. This situation can hamper or almost eliminate flying in the area.

THUNDERSTORMS

Thunderstorms may occur in any part of the United States; however, they are most likely in the central and southern portions. Although all thunderstorms are similar, they may be classed as airmass types and frontal types. To produce thunderstorms, the air involved must be lifted by some force. Forces that cause lifting are winds blowing over mountain barriers, or similar slopes, and air rising due to convectional heating. These forces produce airmass-type thunderstorms. In the frontal type thunderstorm, air rises due to being forced up the frontal slope. Thunderstorms develop in three stages-cumulus, mature, and dissipating.

In the cumulus stage, a cumulus cloud develops with definite updrafts throughout the entire cell or cells, as shown in figure 2-31. As this stage develops, turbulence may be present at varying distances from the cumulus clouds, as illustrated in figure 2-32. Maximum updrafts may exceed 50 feet per second (3,000 f.p.m.) as the cumulus cloud approaches maturity. These clouds must be avoided. In the mature stage, thunderstorms develop downdrafts and rain at the surface, as can be seen in figure 2-33. Water droplets have developed to such a large size that they are no longer supported by the rising currents and begin to fall. Strong downdrafts develop near columns of rising currents creating the shearing effect often present in thunderstorms. When the descending air reaches the surface, strong, gusty, shifting winds occur. Within or near the cloud, rain, snow, heavy freezing rain, or hail may be encountered. Tops above 60,000 feet have been observed during the mature stage.

In the dissipating stage, updrafts diminish, then disappear, turbulence decreases rapidly, precipi-

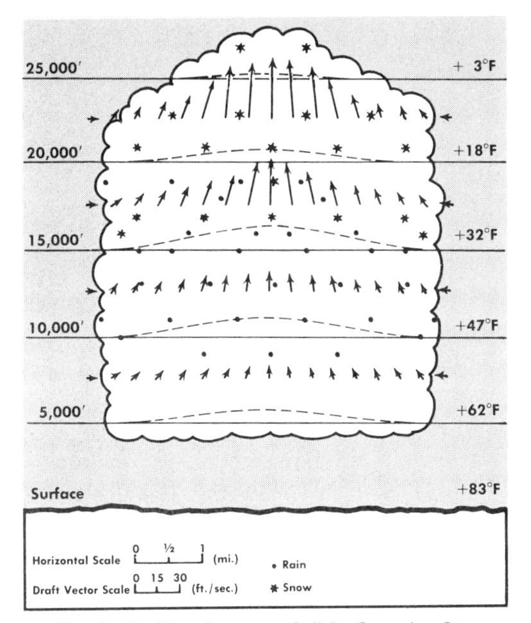

Fig. 2-31. Thunderstorm Cell in Cumulus Stage

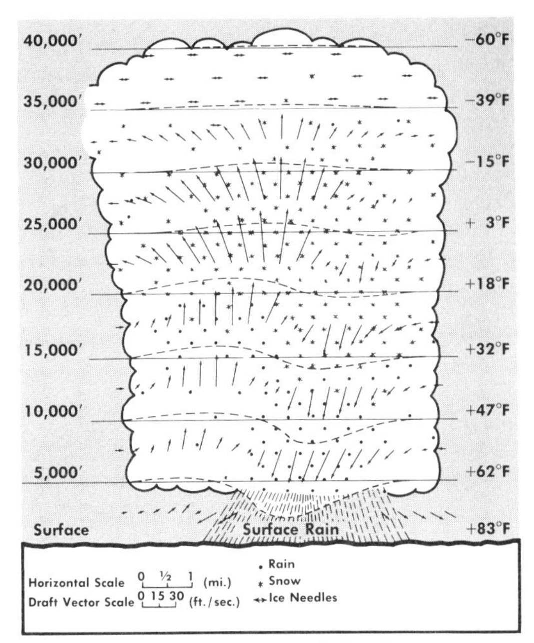

Fig. 2-33. Mature Stage of Thunderstorm

tation becomes spotty and ceases, and clouds may disintegrate into layered stratus. Figure 2-34 illustrates the beginning of this process.

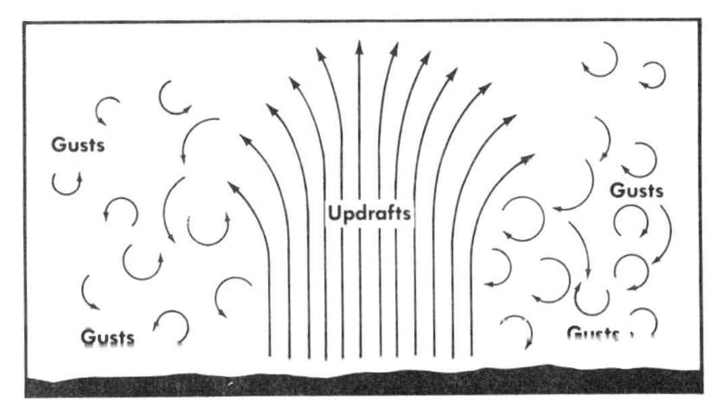

Fig. 2-32. Turbulence Near a Thunderstorm

A thunderstorm may progress from the cumulus stage through the dissipation stage in approximately 1 hour and 30 minutes. However, there may be many cells in a given storm at different stages of development, so a storm may last several hours. Considering the short period of time it takes thunderstorms to develop, it should be remembered that an area of little or no turbulence along a route near developing cumulus clouds may develop severe or extreme turbulence a few minutes later.

SQUALL LINES

Squall lines, or lines of thunderstorms, often develop from 50 to 300 miles ahead of a fast-moving cold front which is overtaking warm, moist (maritime tropical) air. All of the characteristics of thunderstorms are intensified and some of the severest flying conditions known, including tornadoes, are found in and near squall lines. They may range from 50 to several hundred miles in length. Although some squall lines travel as fast as 60 m.p.h., most proceed between 25 to 40 m.p.h. Their rate of movement depends upon the speed of the associated front and the winds aloft. Frequent or continuous lightning usually accompanies a mature squall line.

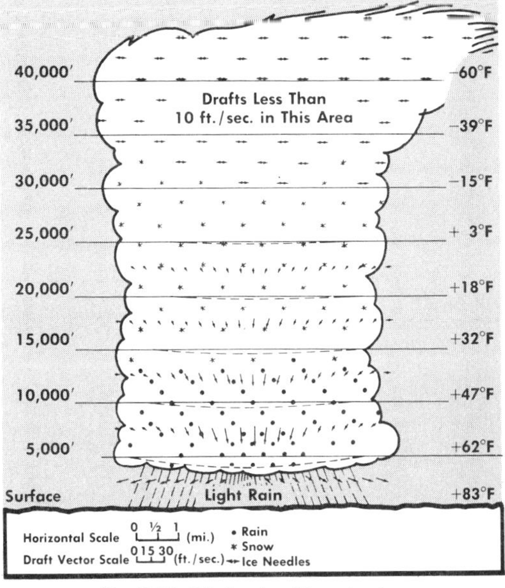

Fig. 2-34. Dissipating Thunderstorm

TORNADOES

Tornadoes are by far the most violent and destructive weather phenomena. Although these storms are usually relatively narrow at the ground, averaging 300 yards, they may vary from about 30 yards up to approximately one mile in width. They develop at the base of thunderclouds (cumulonimbus clouds) and have vertical velocities in excess of 200 miles per hour and horizontal velocities between 300 and 500 miles per hour. The funnel, or rope-like cloud, of a tornado may be visible or entirely obscured by other low clouds in heavy rain. The conditions necessary for the tornado to form are warm, humid air flowing into the cyclone from the south or southwest and a mass of cold air advancing from the west into the warm air, at a level above the ground. Tornadoes are prevalent in Kansas, Oklahoma, Texas, and the Gulf states; however, they may occur in many other places if conditions are favorable.

ST. ELMO'S FIRE

St. Elmo's Fire is a form of static electricity that discharges from the propellers, wingtips and other protruding parts of the airplane. It occurs when there is a great deal of motion within dry airmasses, particularly shearing air currents. The static electricity, generated by friction of the air, discharges from one area to another and produces a bluish-gray light which, although not dangerous, may be disturbing to the pilot.

ICING

There are three types of ice with which a pilot should be familiar—frost, rime, and clear. Rime and clear ice may form whenever visible moisture is present, and the temperature is near or below freezing. Frost usually forms in fair weather.

FROST

Frost is formed by sublimation (water vapor turns directly to ice crystals without going through the liquid stage). Most often, it forms on the surfaces of parked airplanes on clear nights when the temperature is near or below freezing and moisture content of the air is relatively high. However, it may also form on airplanes descending from areas of cold air to areas of warm, moist air. Even a small amount of frost disrupts air flow and should be removed before takeoff, as it reduces lift and increases drag.

RIME ICE

Rime ice is a granular, whitish opaque, rough deposit of ice formed from tiny, supercooled water drops often found in stratiform type clouds. It may be found in the upper areas of thunderstorms where snow and supercooled water droplets are present, or near the surface when precipitation is snow mixed with rain.

Rime ice usually forms on the leading edges of aircraft surfaces in flight, entrapping large amounts of air which produces the opaque appearance. Rime ice is brittle due to the entrapped air, and relatively easy to remove with normal de-icing equipment. It may occur at temperatures from near 0° to —40° Celsius but is most often encountered at air temperatures

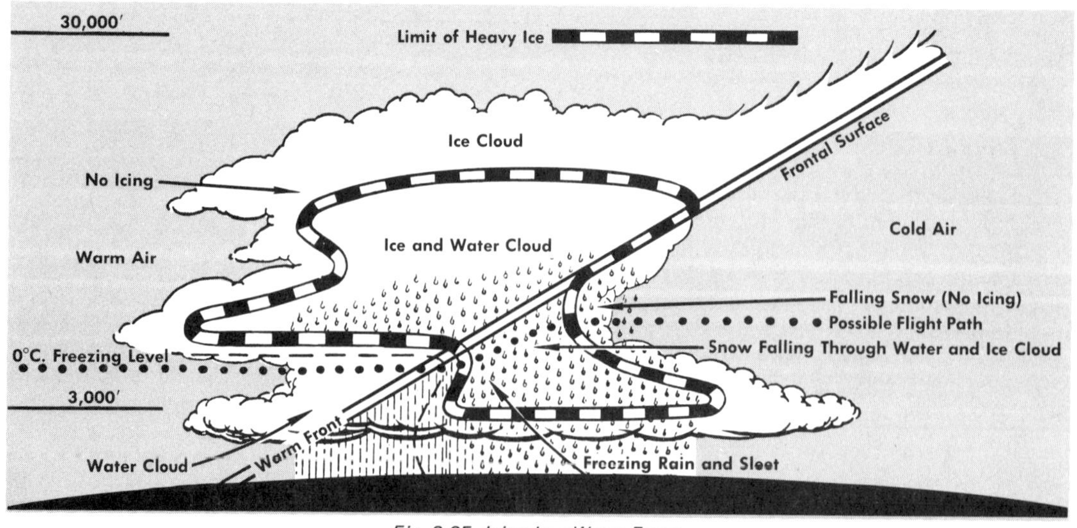

Fig. 2-35. Icing In a Warm Front

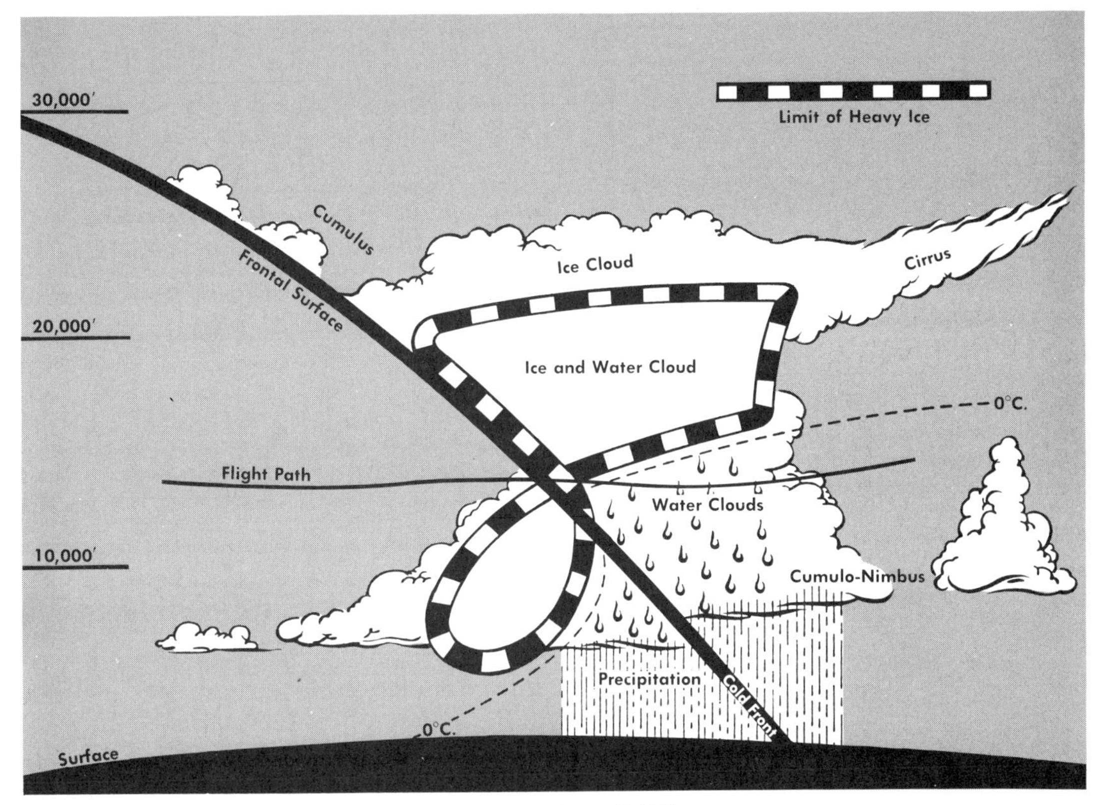

Fig. 2-36. Icing Along a Cold Front

from −10° to −20° Celsius. Figure 2-35 depicts areas where icing occurs in a warm frontal zone.

CLEAR ICE

Conditions favorable for clear ice formation are large water droplets such as those found in cumuliform clouds, and temperatures near or below freezing. Clear ice is formed by the relatively slow freezing of water droplets which strike the leading edge of airfoils and flow rearward. It changes the aerodynamics of the airfoils, including the propeller, and adds weight to the airplane. A heavy buildup will cause a considerable decrease in lift and an increase in drag. Thus, the aircraft will stall at a higher speed. Clear ice adheres more tightly and is more difficult to remove than rime ice. Heavy clear ice should be expected in cumulus clouds when temperatures are near or below freezing or whenever freezing rain is occurring. Figure 2-36 shows areas of icing in a cold front and figure 2-37 shows icing in cumulus clouds over mountains. Slush and water on runways in near freezing weather may also cause problems by freezing the brakes, the control hinges and the landing gear.

TURBULENCE AND ASSOCIATED HAZARDS

Turbulence results from three principal causes. All of these factors may be acting at the same time.

1. Vertical currents from convective activity
2. Disturbed air currents due to obstructions or terrain features
3. Disturbances due to wind shear

The uneven heating of the ground due to the difference in vegetative cover or lack of it may also cause turbulence. Bare ground heats faster than grass or forest covered land; therefore, when flying over bare ground, crops of grain, and plowed fields, as indicated in figure 2-38, the pilot may experience a series of bumps due to rising and descending currents. This turbulence can extend to several thousand feet above the surface and may develop, even when insufficient moisture is present for cloud formation. However, cumulus-type clouds are normally an indication of turbulence.

In the fall or early winter, an outbreak of cold air flowing over the warm surface may cause air currents up to several thousand feet, due to

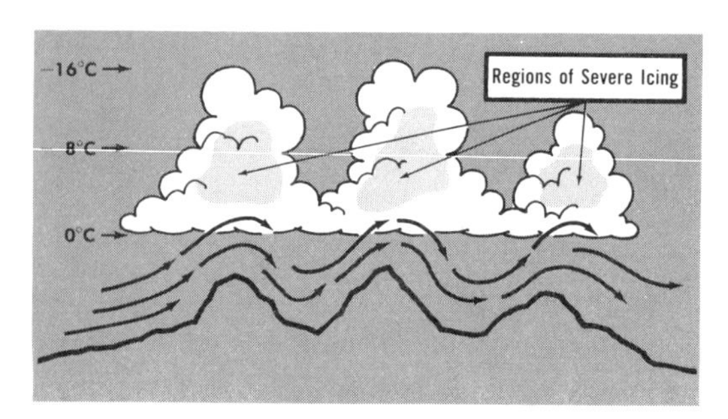

Fig. 2-37. Icing Over Mountains

heating from below and rising unstable air. Orographic lifting over mountains or high hills may cause air to become unstable and create vertical currents extending to several thousand feet above the mountains, creating turbulence. In these areas, altocumulus clouds usually are present.

MOUNTAIN WAVES

Mountain waves are associated with relatively strong winds blowing across mountain ranges. The effects of these waves are extremely hazardous to airplanes, and may extend to thousands of feet above mountain ranges. Additionally, eddies on the lee side may cause an airplane to be forced into the mountainside. Mountain waves are mainly a wintertime phenomenon but may occur during any month of the year.

The mountain wave is often marked by the presence of lenticular (lens shaped) clouds on the leeward side of the mountains with cap clouds on the mountaintops. As shown in figure 2-39, rollclouds, level with or below and to the lee side of mountain ranges, also may be present. However, this is not always the case. If insufficient moisture is present in the atmosphere, the clouds do not develop, but an extremely intense mountain wave may still be present. The turbulence may extend 30 miles or more downwind from the range.

When flying in an area of possible mountain waves, the pilot should maintain an altitude at least 50 percent higher than the vertical distance between the base and the top of the range. Mountain waves should be anticipated whenever the winds aloft exceed 35 knots.

WIND SHEAR

Wind shear is an effect caused by winds blowing in different directions or at different velocities. This condition is sometimes found at the boundary of an inversion. When the temperature contrast is relatively great, the wind direction and velocity change suddenly. This shear effect also is present along sharply defined frontal surfaces where there is a sudden wind shift. Cold air in a valley with warm air on top in motion often produces wind shear due to the sudden change in wind speed.

WAKE TURBULENCE

All airplanes in flight produce some degree of wake turbulence, the intensity of which is dependent on the weight and speed of the aircraft. The greater the weight and slower the speed (resulting in a high angle of attack), the greater the wake turbulence generated. Since extended flaps decrease the turbulence, it can be determined that wake turbulence is greatest behind heavily loaded commercial jet airplanes in a clean configuration immediately after liftoff and just prior to landing. As the airplane moves forward, air tends to flow from the bottom of

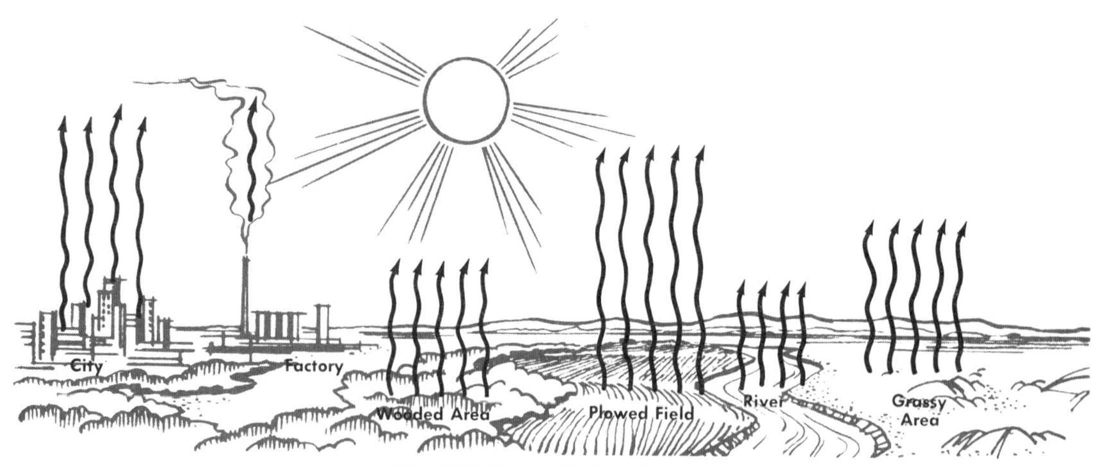

Fig. 2-38. Uneven Surface Heating

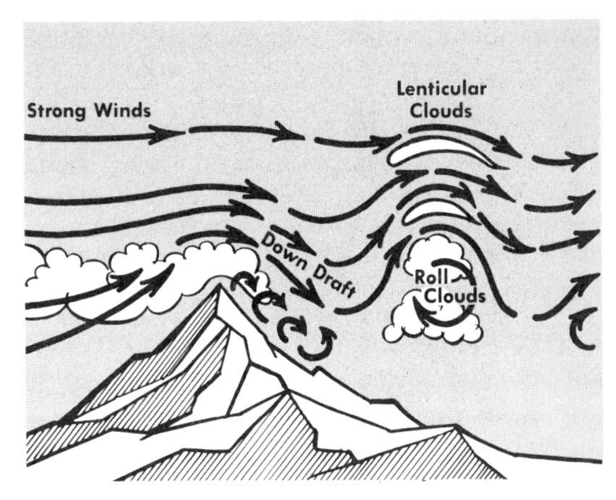

Fig. 2-39. Mountain Waves and Associated Clouds

the wing over the wingtip into the low pressure area on top of the wing. This creates a vortex behind each wingtip as illustrated in figure 2-40. These vortices tend to sink at about 400 to 500 feet per minute. When close to the ground, each vortex moves slowly outward from the departure path at about five knots.

Wake turbulence is not generated unless the wing is producing lift; therefore, to avoid turbulence when landing behind a commercial aircraft, the pilot should make the final approach above that of the preceding airplane and land beyond its touchdown point. A pilot departing directly behind a large aircraft should liftoff prior to the liftoff point of the preceding aircraft and climb out above and on the upwind side of its flight path. Also, when taking off behind a landing aircraft and landing behind a departing aircraft, the pilot should allow three or four minutes for the turbulence to dissipate. Although controllers may offer advice or a warning, it is the pilot's responsibility to avoid wake turbulence.

RESTRICTIONS TO VISIBILITY

Stable air is favorable for the formation of low clouds, fog and light rain that restrict visibility. Haze and smoke are trapped in stable layers of the atmosphere and also restrict visibility. Unstable air with rising currents tends to dissipate fog, smoke and haze but may produce blowing snow, dust, sand and heavy rain showers that also reduce visibility. Visibility is normally used to describe horizontal surface visibility unless otherwise noted.

FOG

Fog is the suspension of minute water particles, or ice crystals within the atmosphere, close to the earth's surface. It is basically a cloud touching the surface. Figure 2-41 shows areas of the United States most affected by fog. Conditions favorable for fog are high relative humidity, an abundance of condensation nuclei and light surface winds. Fog is prevalent, in most areas of the world, in the colder months and is more likely near bodies of water. Inland fog is more persistent near industrial areas due to the availability of smoke for condensation nuceli, and may form when the relative humidity is below 100 percent. Light breezes tend to thicken fog and stronger ones lift and dissipate it.

TYPES OF FOG

There are several types of fog—radiation, advection, upslope, frontal, and steam. The conditions that tend to produce fog also are favorable to inversions.

RADIATION FOG

Radiation fog, sometimes called ground fog, forms on clear nights when the earth's heat is lost very rapidly to the atmosphere. The air next

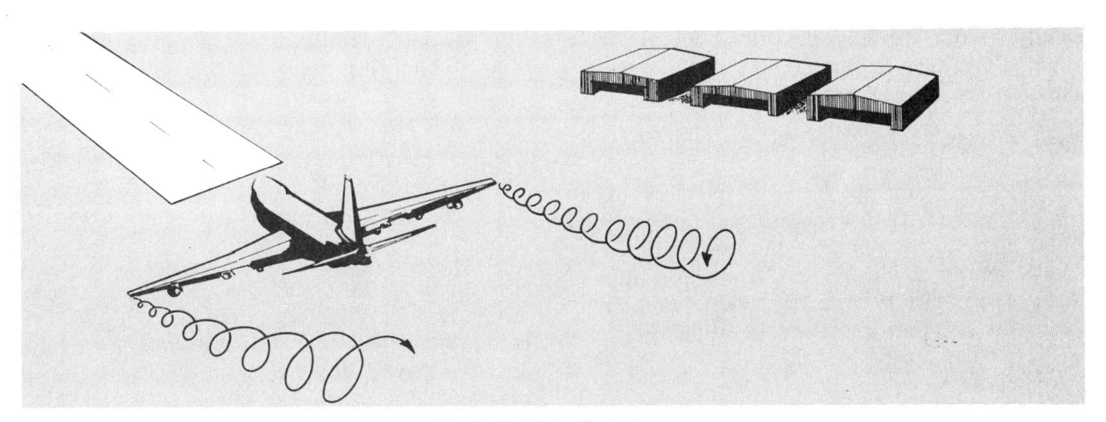

Fig. 2-40. Wake Turbulence

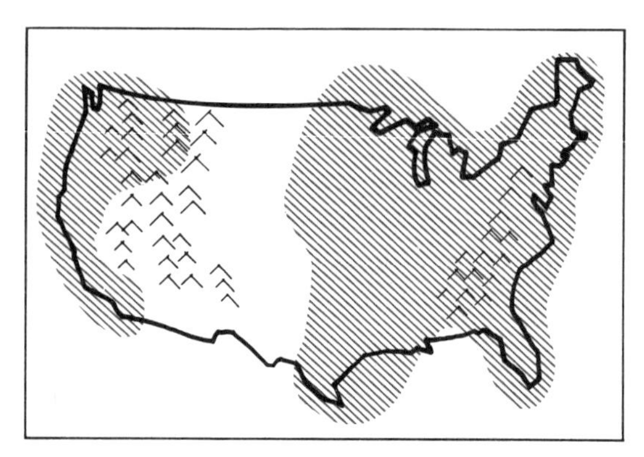

Fig. 2-41. Fog Areas in the United States

to the ground is cooled by conduction and, when relative humidity is high, condensation occurs. Light winds of approximately five knots tend to deepen fog.

ADVECTION FOG

Advection fog is produced by warm, moist air flowing over a colder surface. Winds up to 15 knots may deepen the fog. Above this velocity, the fog generally lifts to form stratus clouds. Advection fog is common along the Gulf Coast in winter and along the East Coast in summer. It also is quite common along the coast of California in summer and throughout the lower Mississippi and Ohio Valleys in winter, when tropical airmasses move northward over the cold land surfaces.

UPSLOPE FOG

Upslope (orographic) fog is formed by the movement of stable air up a sloping land surface. The air rises up the slope, cools by expansion and, if sufficient moisture is present, fog develops. An upslope wind is necessary for the formation and maintenance of this fog. This type of fog is frequently observed along the eastern slope of the Rockies and forms when an easterly wind prevails, provided the relative humidity is high enough for condensation to take place with the cooling of the rising air.

FRONTAL FOG

Frontal fog forms in the cold airmass of a frontal system. It occurs most often in the winter near slow-moving or stationary fronts. It forms rapidly under warm front conditions where precipitation from the warm air above evaporates as it falls into the cold air below.

STEAM FOG

Steam fog forms when water vapor from a relatively warm water surface flows into air which is cold enough to produce condensation. Steam fog is formed near rivers and lakes in the temperate zone during the fall of the year. It is most common in the Arctic area, particularly in the Aleutian Chain, where the moist, warm, unstable air rises from the surface. In this area, steam fog is called sea smoke because it resembles smoke.

HAZE, SMOKE, BLOWING DUST AND BLOWING SNOW

Haze is fine dust and/or salt particles trapped and concentrated in a stable layer of the atmosphere, usually an inversion. The greatest restriction to visibility in this condition occurs when the pilot is looking toward the sun.

Smoke restricts visibility when trapped in a stable layer of air (inversion). It is usually most dense on the downwind side of industrial areas. Visibility in smoke is similar to that in haze. Since smoke particles are nuclei upon which water vapor condenses, smoke and fog together often produce *smog* in industrial areas.

Blowing dust and sand (generally found in barren or arid areas) also cause restrictions to visibility. Strong winds and vertical air currents may lift fine dust to 8,000 feet or higher, and cause it to spread over thousands of square miles. Blowing snow occurs over snow-covered regions where the winds are strong and the air is unstable. "Whiteouts," often associated with snow-covered ground and blowing snow, may be hazardous during landings because they obscure the horizon.

Clouds and any form of precipitation also restrict visibility. Wet snow, especially at night, may make runway alignment difficult.

WEATHER RECOGNITION

The weatherwise pilot is constantly alert for any clues to changing weather. He is also well versed on the general weather conditions associated with each type of front and with each geographical area in which he flies. Seasonal weather trends are also vital factors which must be considered.

Frontal weather may range from perfectly clear conditions to thunderstorms, severe turbulence, icing, and low ceilings and visibilities. The moisture content and degree of stability of the air being forced aloft by a cold front provides an excellent indication of what weather conditions should be expected. The more unstable the air, the greater the likelihood of cumuliform clouds and their associated weather. On the other hand, very stable air will lead to the formation of stratiform clouds with large areas of continuous precipitation.

Shallow frontal surfaces often produce thunderstorms which are embedded in the stratiform cloud mass. A thorough preflight briefing aids the pilot in foreseeing such hidden obstacles.

Cloud formations also provide good indications of weather conditions. Altocumulus clouds indicate light turbulence and small amounts of icing, while standing lenticular altocumulus clouds provide outstanding indications of very strong winds aloft with moderate to severe turbulence. The possibility of strong surface winds is always good when these clouds are present. Towering cumulus and cumulonimbus clouds should always be avoided, since they are indicative of severe or extreme turbulence, hail, strong winds, and icing.

Nimbostratus and stratus clouds are generally free of turbulence. However, they may cause severe icing if temperatures are at or near freezing. The pilot should plan to fly above small cumulus clouds associated with fair weather to avoid turbulence that often exists below these clouds.

The temperature-dewpoint spread is a general indication of the possibility of fog. If the spread is less than five degrees Fahrenheit, the surface winds are approximately 5 to 15 knots, and there is sufficient moisture, fog is almost certain to form. Fog may also occur when relatively warm precipitation falls through cold temperatures near the surface or when low stratus clouds cover a very large area.

Radiation, or ground, fog usually "burns off" rather rapidly after sunrise, although it may persist until midday. Other types of fog may be much more persistent and deepen as wind speed increases up to about 15 knots. A stronger wind will lift the fog into a layer of low stratus or stratocumulus clouds which may cover vast areas and be very persistent.

The pilot should be aware that vertical visibility downward through fog may be very good but as the fog is entered horizontal visibility may become zero. Also, while the weather at one airport may be below minimums due to fog, another field only a few miles distant may be VFR.

The pilot should develop a thorough knowledge of meteorology and its practical application to flight operations. Such knowledge will help insure safe and enjoyable flights.

GLOSSARY OF METEOROLOGICAL TERMS

Adiabatic Process— A process in which cooling occurs due to expansion of air and heating occurs due to the compression of air without the addition or withdrawal of heat by an outside source.

Advection— The process of transfer of heat by horizontal motion of air.

Airmass— A term applied to an extensive body of air within which the conditions of temperature and moisture, in a horizontal plane, are essentially the same.

Anticyclone— An area of high barometric pressure as related to surrounding areas.

Atmosphere— The whole mass of air surrounding the earth. It includes the troposphere, tropopause, and the stratosphere.

Backing Wind— The shifting of wind in a counterclockwise direction.

Barometer— An instrument for measuring the pressure of the atmosphere. The two principal types are the *mercurial* and the *aneroid* (latter used in airplanes).

Barometric Tendency— The change in barometric pressure within a specified time (usually three hours) before the observation.

Buys-Ballot's Law— In the Northern Hemisphere, if a person faces the wind, the low-pressure area is to his right.

Calorie— The amount of heat required to raise the temperature of a gram of water one degree Celsius at sea level.

Ceiling— The height above the earth's surface of the lowest layer of clouds or obscuring phenomena that are reported as "broken", "overcast", or "obscuration" and not classified as "thin" or "partial."

Cap cloud— A cap-like cloud crowning a mountain summit, or another cloud, especially a mass of cumulonimbus.

Cold Front— A cold front occurs when a cold airmass is displacing warmer air.

Condensation— The process by which a vapor becomes a liquid.

Convection— The upward or downward movement of a limited portion of the atmosphere caused by difference in air density.

Convergence— The condition that exists when winds converge into a narrower path causing an upward movement of air.

Cyclogenesis— The term applied to the process which creates or develops a new, or deepens a pre-existing, cyclone.

Cyclone— An area of low barometric pressure with its attendant counterclockwise inflowing winds system.

Deepening— The occurrence of decreasing pressure in the center of a pressure system.

Depression— A cyclonic area, or region, where the pressure is lower than that of the surrounding area—a weak, unorganized low.

Dew— Atmospheric moisture condensed in liquid form upon objects cooler than the air, occurs frequently at night.

Dewpoint— The temperature at which, under ordinary conditions, condensation begins in a cooling mass of air.

Discontinuity— The term applied in a special sense by meteorologists to a zone within which there is a rapid change of the meteorological elements.

Divergence— The condition that exists when winds diverge, or spread out laterally; usually accompanied by clearing or fair weather.

Drizzle— Precipitation consisting of numerous tiny droplets of water falling at a relatively slow rate.

Dry Adiabatic Lapse Rate— The rate at which a body of unsaturated air will cool off due to expansion when rising or warm up due to compression when descending. It is 5.5° Fahrenheit per 1,000 feet.

Fog— A cloud at the earth's surface.

Frontogenesis— The term used to describe the formation of a front or to the process which increases the intensity of a pre-existing front.

Frontolysis— The term used to describe the process which tends to destroy a pre-existing front.

Frost— Atmospheric moisture deposited by sublimation upon objects in the form of ice crystals. Also called hoarfrost.

Gradient— Change of value of a meteorological element, usually pressure or temperature, per unit of distance.

Gradient Wind— A wind that flows parallel to the isobars and has a velocity such that the pressure gradient, Coriolis and centrifugal forces acting on the air, are in balance.

Gusts— A sudden brief increase in the force of the wind, most winds near the surface display alternate gusts and lulls.

Humidity— The degree to which the air is charged with water vapor. This may be expressed in several ways. Absolute humidity expresses the mass of water vapor per unit volume of air; relative humidity is the percentage of saturation; specific humidity expresses the mass of water vapor contained in a unit mass of moist air.

Isolation— Solar radiation (heat) as received by the earth or other planets from the sun; also, the rate of delivery of the same, per unit of horizontal surface.

Instability— The condition that exists when a particle of air is warmer or colder than the surrounding air. If placed in motion upward or downward, depending on temperature, it will continue to move away from its original position at an increasing rate. Often produced by thermal activity (heating).

Inversion— An abbreviation for "inversion of the vertical gradient of temperature." The temperature of the air is ordinarily observed to decrease with increasing altitude, but occasionally the reverse is true, and when the temperature increases with altitude there is said to be an "inversion."

Isobar— A line on a chart, or diagram, connecting points of equal pressure.

Isotherm— A line on a chart, or diagram, connecting points having equal temperature.

Jet Stream— A concentrated core of high velocity winds imbedded in the general flow, usually found at altitudes of 20,000 feet or above.

Land and Sea Breezes— The breezes that, on certain coasts and under certain conditions, blow from land by night (land breeze) and from water by day (sea breeze).

Lapse Rate— The rate of decrease of temperature in the atmosphere with altitude (not to be confused with adiabatic).

Low— An area where the barometric pressure is lower than that of the surrounding area and where the wind circulation is counterclockwise and inward.

Millibar— A unit of pressure equal to a force of 1,000 dynes per square centimeter. 1,013.2 millibars is standard atmospheric pressure at mean sea level.

Nuclei— A particle upon which condensation of water vapor occurs in the free atmosphere in the form of a water drop or ice crystal.

Occluded Front— Occurs when a cold front overtakes a warm front forcing the warm air aloft. The result is an occluded front caused by a combining of the two.

Occlusion— The term used to denote the process whereby the air in the warm sector of a cyclone is forced upward from the surface to higher levels.

Polar Front— The surface of discontinuity separating an airmass of polar origin from one of tropical origin.

Pressure— Used in meteorological work to denote the weight of the atmosphere at a given time and place. The units of millibars and inches of mercury are used to measure pressure—29.92 inches of mercury and 1,013.2 millibars denote standard pressure at sea level.

Pressure Gradient— The decrease in barometric pressure per unit of horizontal distance in the direction in which the pressure decreases most rapidly.

Ridge— An elongated anticyclone or high-pressure area.

Squall Line— A more or less continuous line of thunderstorms often 10 to 20 miles wide and as much as 200-300 miles ahead of a cold front.

St. Elmo's Fire— A luminous bluish discharge of electricity from elevated or protruding objects, such as propellers or wing and tail surfaces of airplanes.

Saturated Adiabatic Lapse Rate— The rate of temperature decrease with altitude of saturated air.

Saturation— The point at which air holds all the water vapor it can hold at a given temperature.

Stable Air— A state in which the vertical distribution of temperature is such that an air particle will resist displacement from its level and, if displaced, will tend to return to its original level.

Standard Lapse Rate— 3.5° Fahrenheit or approximately 2° Celsius per 1,000 feet of altitude.

Standard Sea Level Pressure— 29.92 inches of mercury.

Standard Sea Level Temperature— 59° Fahrenheit or 15° Celsius.

Stratosphere— An upper region or layer of the atmosphere in which the temperature is practically constant in a vertical direction.

Tropopause— The surface in the atmosphere at which the decrease in temperature with increasing altitude ceases. This surface marks the base of the stratosphere.

Troposphere— The lower region of the atmosphere from the ground to the tropopause, in which the average condition is typified by a more or less regular decrease of temperature with increasing altitude, and where most weather occurs.

Trough— An elongated area of low barometric pressure.

Visibility— The distance an object can be seen with the naked eye, usually given in miles.

Warm Front— Borderline between the forward edge of relatively warm air that is displacing a retreating colder airmass.

3

WEATHER FORECASTS AND REPORTS

TELETYPE REPORTS

The National Weather Service and the Federal Aviation Administration collect and disseminate weather information for pilots. Observation stations, meteorological centers, forecast offices, and weather service outlets combine efforts to provide essential weather forecasts and reports. The forecasts and reports discussed in this chapter include area forecasts, SIGMETs and AIRMETs, terminal forecasts, surface aviation weather reports, winds aloft forecasts, and pilot reports (PIREPs). These reports, plus the various weather charts available, provide a total weather picture for flight.

Area and terminal forecasts provide fairly detailed weather information for the first 18 hours. A simple categorical outlook is provided for the remainder of the forecast period. The categories used are as follows:

1. LIFR (Low IFR) — Ceiling less than 500 feet and/or visibility less than 1 mile.
2. IFR — Ceiling 500 to less than 1,000 feet and/or visibility 1 to less than 3 miles.
3. MVFR (Marginal VFR) — Ceiling 1,000 to 3,000 feet and/or visibility 3 to 5 miles inclusive.
4. VFR — Ceiling greater than 3,000 feet and visibility greater than 5 miles; includes sky clear.

In addition, the cause of the weather restrictions are given as shown in the following examples:

1. LIFR CIG — Low IFR due to low ceiling.
2. IFR F — IFR due to visibility restricted by fog.
3. MVFR CIG H K — Marginal VFR due both to ceiling and to visibility restricted by haze and smoke.
4. IFR CIG R WIND — IFR due both to low ceiling and to visibility restricted by rain; wind expected to be 25 knots or greater.

AREA FORECASTS

The United States is divided into several major areas for weather forecasting purposes. These major forecast offices issue *area forecasts* specially designed to indicate flying conditions anticipated for the following 18 hours. Since the wording of an area forecast is abbreviated, it may be necessary to use the decoding book or have a forecaster provide a plain language translation.

Figure 3-1 shows an area forecast for Montana, Idaho, Nevada, Utah, and Arizona. All heights are ASL (above sea level) values unless otherwise noted. Ceilings are AGL (above ground level) values. The forecast is decoded as follows.

SYNOPSIS: Surface high pressure over the forecast area weakening with cold front extending from western Washington through western Oregon and moving to the central part of Montana, southeastern Idaho, northwestern Utah, and

```
SLC FA 2Ø124Ø.
13Z TUE-Ø7Z WED.
OTLK Ø7Z WED-19Z WED.

MT ID NV UT AZ...

HGTS ASL UNLESS NOTED...

SYNS...
SFC HI PRES OVR AREA WKNG WITH CDFNT WRN WA WRN OR MVG TO CNTRL
MT SE ID NW UT AND N CNTRL NV BY Ø7Z. AMS BCMG MOIST AND UNSTBL
WITH AND BHD FNT WITH AREAS OF MID AND HI LVL MSTR AHD FNT.

SIGCLDS AND WX...
NRN AND SW ID MT W DVD...
9Ø-12Ø BKN SPRDG OVR AREA WITH SCT TO BKN LYRS ABV TO 18Ø. FEW
AREA OF RAIN OR RSHWRS WITH MTNS AND PASSES OBSCD AT TIMES. CLDS
LWRG WITH AND BHD FNT TO 7Ø-1ØØ OVC LYRD CLDS TO 25Ø AND AREAS OF
RAIN AND SNW DVLPG WITH MTNS AND PASSES MSTLY OBSCD. OCNLY GSTY
SFC WNDS. SCT SHWRS CONTG BHD FNT WITH TPS TO 18Ø. OTLK... VFR TO
MVFR CIG WIND.

ICG... MDT ICGIC BLO 15Ø SPCLY IN FNTL ZN. FRZ LVL 7Ø-1ØØ AHD FNT
AND 5Ø-7Ø BHD FNT XCPT RMNG 1ØØ-12Ø SRN NV SRN UT AZ.
```

Fig. 3-1. Area Forecast

north central Nevada by 0700Z. Airmass becoming moist and unstable with and behind the front with areas of mid and high level moisture ahead of front.

SIGNIFICANT CLOUDS AND WEATHER: Northern and southwestern Idaho mountains west of the continental divide, 9,000 to 12,000 broken ceiling spreading over the area with scattered to broken layers above, to 18,000. Few areas of rain or rain showers with mountains and passes obscured at times. Clouds lowering with and behind the front to ceilings of 7,000 to 10,000 overcast with layered clouds to 25,000 and areas of rain and snow developing with mountains and passes mostly obscured. Occasionally gusty surface winds. Scattered showers continuing behind the front with tops to 18,000. Outlook, VFR to marginal VFR due to ceilings. In addition, winds in excess of 25 knots are expected.

ICING: Moderate icing in the clouds below 15,000, especially in the frontal zone. Freezing level 7,000 to 10,000 ahead of the front and 5,000 to 7,000 behind the front. Freezing level over southern Nevada, southern Utah, and Arizona is at 10,000 to 12,000 feet.

SIGMETs AND AIRMETs

These advisories are issued when significant and potentially hazardous changes occur in existing or forecast weather. The intent of the advisory is to provide enroute pilots with information that may not have been available prior to flight. It is then the pilot's responsibility to evaluate the information on the basis of individual experience and aircraft operational limits. Generally, SIGMETs are more severe and apply to *all* aircraft while AIRMETs are less hazardous and may apply only to *light* aircraft.

CONVECTIVE SIGMETs are issued for the more severe phenomena as follows.

1. Tornadoes
2. Lines of thunderstorms
3. Embedded thunderstorms
4. Hail three-fourths of an inch or more in diameter
5. Isolated intense thunderstorms
6. Areas of thunderstorms

NONCONVECTIVE SIGMETs are issued by the National Weather Service when *current* or *forecast* conditions include severe and extreme turbulence, severe icing, and dust or sandstorms which lower visibility below three miles.

AIRMETs are issued when one or more of the following conditions are expected.

1. Moderate icing
2. Moderate turbulence
3. Extensive areas of visibilities less than three miles, or ceilings less than 1,000 feet, including mountain ridges and passes
4. Sustained winds of 30 knots or more at the surface

The first advisory issued after midnight local standard time is designated ALFA 1. The next advisory of the ALFA series is designated ALFA 2. In cases where a potentially hazardous situation develops in a second sector, a second series of advisories is issued and designated BRAVO 1, BRAVO 2, etc. Figure 3-2 depicts an AIRMET issued by the Kansas City forecast office, valid from 2230Z on the twenty-fourth to 0500Z on the twenty-fifth, as indicated in the heading.

The advisory is decoded "...Flight precaution. Mountains of Wyoming and north and central Colorado occasionally obscured above 8,000 MSL to 11,000 MSL in clouds and showers with scattered thunderstorms. Conditions continuing most sections beyond 0500Z. Continue AIRMET beyond 0500Z."

TERMINAL FORECASTS

A terminal forecast (FT) is issued for a specific airport and encompasses an area within a five-mile radius of the runway complex. In the remarks section, a 10-mile radius applies. Figure 3-3 shows an excerpt from a group of terminal forecasts.

```
MKC WA 24223Ø
24223Ø-25Ø5ØØZ

AIRMET CHARLIE 1. FLT PRCTN. MTNS WY AND N AND CNTRL CO OCNL
OBSCD ABV 8Ø-11Ø IN CLDS AND SHWRS WITH SCT TSTMS. CONDS CONTG
MOST SECS BYD Ø5Z. CONT AIRMET BYD Ø5Z.
```

Fig. 3-2. AIRMET

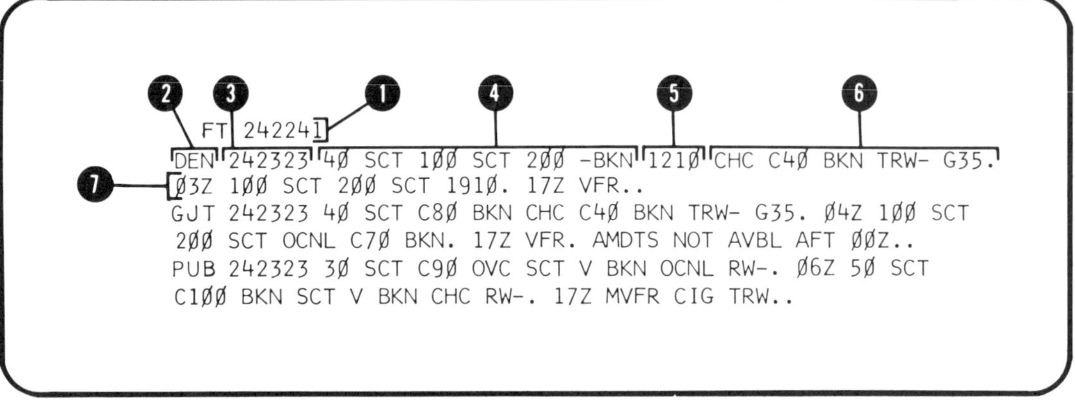
FT 242241
DEN 242323 4Ø SCT 1ØØ SCT 2ØØ -BKN 121Ø CHC C4Ø BKN TRW- G35.
Ø3Z 1ØØ SCT 2ØØ SCT 191Ø. 17Z VFR..
GJT 242323 4Ø SCT C8Ø BKN CHC C4Ø BKN TRW- G35. Ø4Z 1ØØ SCT
2ØØ SCT OCNL C7Ø BKN. 17Z VFR. AMDTS NOT AVBL AFT ØØZ..
PUB 242323 3Ø SCT C9Ø OVC SCT V BKN OCNL RW-. Ø6Z 5Ø SCT
C1ØØ BKN SCT V BKN CHC RW-. 17Z MVFR CIG TRW..

Fig. 3-3. Terminal Forecast Excerpt

The heading (item 1) indicates this is an FT transmitted on the twenty-fourth day of the month at 2241Z. The first forecast is for Denver, Colorado (item 2), while item 3 indicates that the valid period is from 23Z on the twenty-fifth. The last six hours of the forecast are simply a categorical outlook.

The Denver forecast for the period beginning at 2300Z on the twenty-fourth (item 4) indicates scattered clouds at 4,000 and 10,000 feet and a thin broken layer of clouds at 20,000 feet. Visibility is forecast to be more than six miles with winds 120° at 10 knots (item 5). There is a chance of broken ceilings at 4,000 with thunderstorms, light rain showers, and wind gusts to 35 knots (item 6). Item 7 indicates that different conditions are forecast by 0300Z.

Three items to remember about terminal forecasts are listed below.

1. Ceilings are always indicated by a "C" before the height.
2. Visibility is included only when it is expected to be six miles or less.
3. Winds are included only if they are forecast to be 10 knots or more.

SURFACE AVIATION WEATHER REPORTS

Weather observation stations throughout the United States and in many foreign countries report existing weather on the hour. These surface aviation weather reports frequently are referred to as hourly weather reports. Each station transmits the existing weather over the teletype to all other stations on the circuit. These reports normally arrive in the weather station above five minutes past the hour and are broadcast at 15 minutes past each hour by all flight service stations having voice facilities. If some special weather condition at a station needs to be reported, that station transmits a "special report." The reports are standard and always follow the same format. To help pilots read and understand these reports, figure 3-4 shows a typical surface aviation weather report with an explanation of all symbols used.

The section showing sky cover and ceiling often shows several layers of clouds. A ceiling is the distance from the ground to the bottom of the lowest opaque layer of clouds which covers more than one-half the sky. A ceiling cannot be a scattered or thin layer of clouds. It must be either broken or overcast opaque clouds, or the sky must be totally obscured by a surface condition such as fog or precipitation.

WINDS ALOFT FORECASTS

Winds aloft forecasts (FD) show the forecast wind direction and speed at various altitudes and the forecast temperature in degrees Celsius. FDs are issued for the following levels: 3,000; 6,000; 9,000; 12,000; 18,000; 24,000; 30,000; 34,000; and 39,000 feet MSL.

The first level reported for a particular station is 1,500 feet or more *above the station elevation.* The temperatures are forecast for all wind levels that are 2,500 feet or more above the station with the exception of the 3,000-foot level. Minus signs are deleted preceding the temperatures for 30,000, 34,000, and 39,000 feet.

The first line after the heading in figure 3-5 begins with "FT" (feet), and denotes the levels for which the forecast is made. Denver (DEN) has no forecast for 3,000 feet or 6,000 feet because both altitudes are less than 500 feet above the station elevation.

The group of numbers (0706) adjacent to HOU (Houston) for the 3,000-foot level is decoded as

EXPLANATION OF TELETYPEWRITER WEATHER REPORTS

LOCATION IDENTIFIER AND TYPE OF REPORT*	SKY AND CEILING	VISIBILITY WEATHER AND OBSTRUCTION TO VISION	SEA-LEVEL PRESSURE	TEMPERATURE AND DEW POINT	WIND	ALTIMETER SETTING	RUNWAY VISUAL RANGE	CODED PIREPS
MKC	15SCT M25 OVC	1R-K	132	/58/56	/18Ø7	/993/	RØ4LVR2ØV4Ø	/UA OVC 55

NOTAM INFORMATION

(continuation of weather report)

→ MKC 4/4 NDB OTS ↙ C 4/1 ↘ 4/2 4/3

SKY
Sky cover contractions are in ascending order. Figures preceding contractions are heights in hundreds of feet above station.

Sky cover contractions are:

CLR Clear: Less than 0.1 sky cover.
SCT Scattered: 0.1 to 0.5 sky cover.
BKN Broken: 0.6 to 0.9 sky cover.
OVC Overcast: More than 0.9 sky cover.
\- Thin (When prefixed to the above contractions.)
-X Partial obscuration: 0.1 to less than 1.0 sky hidden by precipitation or obstruction to vision (bases at surface).
X Obscuration: 1.0 sky hidden by precipitation or obstruction to vision (bases at surface).

CEILING
Letter preceding height of layer identifies ceiling layer and indicates how ceiling height was obtained. Thus:

E Estimated height
M Measured
W Indefinite
V Immediately following numerical value, indicates a variable ceiling.

VISIBILITY
Reported in statute miles and fractions. (V=Variable)

WEATHER AND OBSTRUCTION TO VISION SYMBOLS

A	Hail	IC	Ice Crystals	S	Snow
BD	Blowing Dust	IF	Ice Fog	SG	Snow Grains
BN	Blowing Sand	IP	Ice Pellets	SP	Snow Pellets
BS	Blowing Snow	IPW	Ice Pellet Showers	SW	Snow Showers
D	Dust	K	Smoke	T	Thunderstorms
F	Fog	L	Drizzle	T+	Severe Thunderstorm
GF	Ground Fog	R	Rain	ZL	Freezing Drizzle
H	Haze	RW	Rain Showers	ZR	Freezing Rain

Precipitation intensities are indicated thus:

— Light; (no sign) Moderate; + Heavy

EXPLANATION OF TELETYPEWRITER WEATHER REPORTS

WIND

Direction in tens of degrees from true north, speed in knots. ØØØØ indicates calm. G indicates gusty. Peak speed of gusts follows G or Q when gusts or squalls are reported. The contraction WSHFT followed by GMT time group in remarks indicates windshift and its time of occurrence. (Knots x 1.15 = statute mi/hr.)
EXAMPLES: 3627 = 360 Degrees, 27 Knots;
3627G40= 360 Degrees, 27 Knots Peak speed in gusts 40 Knots.

ALTIMETER SETTING

The first figure of the actual altimeter setting is always omitted from the report.

RUNWAY VISUAL RANGE (RVR)

RVR is reported from some stations. Extreme values during 10 minutes prior to observation are given in hundreds of feet. Runway identification precedes RVR report.

CODED PIREPS

Pilot reports of clouds not visible from ground are coded with ASL height data preceding and/or following sky cover contraction to indicate cloud bases and/or tops, respectively. UA precedes all PIREPS.

NOTAM INFORMATION

→ ADP Automatic Data Processing code indicates NOTAM content.

↙ NTF NOTAM To Follow precedes each subsequent new or cancelled NOTAM.

C Cancelled NOTAM

↘ CNI Current NOTAM Indicator, the CNI code is followed by the serial number of previously transmitted current NOTAM(s). Consult the NOSUM for current NOTAM data.

4/3 The 4 indicates month of April and the 3 indicates third NOTAM issued in the month of April.

DECODED REPORT

Kansas City: Record observation, 1500 feet scattered clouds, measured ceiling 2500 feet overcast, visibility 1 mile, light rain, smoke, sea level pressure 1013.2 millibars, temperature 58°F, dewpoint 56°F, wind 180°, 7 knots, altimeter setting 29.93 inches. Runway 04 left, visual range 2000 ft. variable to 4000 feet. Pilot reports top of overcast 5500 feet. Kansas City NOTAM 4/4 NDB out of service ↙ NOTAM 4/1 cancelled ↘ NOTAM 4/2 and 4/3 current.

*TYPE OF REPORT

The omission of type-of-report data identifies a scheduled record observation for the hour specified in the sequence heading. An out-of-sequence, special observation is identified by the letters "SP" following station identification and a 24-hour clock time group e.g., "PIT SP Ø715 - X M1 OVC." A special report indicates a signicant change in one or more elements.

Fig. 3-4. Explanation of Teletype Writer Weather Reports

```
FDUS3 KWBC 211945
DATA BASED ON 2112ØØZ
VALID 2212ØØZ FOR USE Ø9ØØ-15ØØZ. TEMPS NEG ABV 24ØØØ

FT   3ØØØ     6ØØØ      9ØØØ      12ØØØ     18ØØØ     24ØØØ     3ØØØØ    34ØØØ    39ØØØ

DEN                    2315+Ø4   2516+ØØ   2519-16   2523-27   252741   25345Ø   25446Ø
HOU  Ø7Ø6   99ØØ+12    99ØØ+Ø6   31Ø9+Ø1   3Ø22-11   3Ø25-24   3Ø354Ø   3Ø3948   3Ø3755
JOT  1714   19Ø8+14    24Ø8+Ø8   2612+Ø2   2825-13   2935-24   2Ø4638   3Ø5147   3Ø4856
```

Fig. 3-5. Winds Aloft Forecast Excerpt

follows. The first two numbers (07) represent a true wind direction of 070°. The last two numbers (06) indicate a wind speed of six knots. The Joliet (JOT) winds at 24,000 feet are forecast to be from 290° at 35 knots. The temperature is forecast to be -24°C.

It is often necessary for a pilot to interpolate between given levels of a winds aloft forecast for flight planning. When interpolating, it is usually most convenient to determine the change per 1,000 feet. To determine the winds at 20,000 feet over Joliet (JOT), the pilot first checks the winds aloft forecast and finds that the winds at 18,000 feet are from 280° at 25 knots and at 24,000 feet they are from 290° at 35 knots. Within an altitude span of 6,000 feet, the wind direction changes 10°, from 280° to 290°. This 10° change of direction in 6,000 feet averages approximately 1.6° per thousand feet. Thus, the wind direction at 20,000 feet is approximately 283°.

The wind speed also is calculated to increase at a rate of approximately 1.6 knots per 1,000-foot rise in altitude. Therefore, the wind speed at 20,000 feet is determined to be about 28 knots. Wind speeds between 100 and 198 knots are shown by adding 50 to the wind direction code and subtracting 100 from the speed. For example, 7520 denotes a wind from 250° at 120 knots. When using this coding method, a *speed* of 99 indicates 199 knots or more. It should not be confused with 9900, which indicates light and variable winds.

PILOT WEATHER REPORTS

Whenever ceilings are at or below 5,000 feet, visibilities are at or below five miles, or thunderstorms are reported or forecast, FSS stations are required to solicit and collect *pilot weather reports* (PIREPs) which describe conditions aloft. Pilots are urged to cooperate and volunteer reports of cloud tops, upper cloud layers, thunderstorms, ice, turbulence, strong winds, and other significant flight condition information. Such conditions observed between weather reporting stations are vitally needed. The PIREPs should be given to the FAA ground facility with which communication is established, i.e., FSS or air route traffic control center. In addition to complete PIREPs, pilots can materially help round out the in-flight weather picture by adding to routine position reports, both VFR and IFR, the following phrases, as appropriate.

ON TOP
BELOW OVERCAST
WEATHER CLEAR
MODERATE (or SEVERE) ICING
MODERATE (or EXTREME)
  TURBULENCE
FREEZING RAIN (or DRIZZLE)
THUNDERSTORM (location)
BETWEEN LAYERS

If pilots are not able to make airborne PIREPs, they should report the in-flight conditions encountered to the nearest flight service station or National Weather Service Office immediately upon landing. Some of the uses made of the reports are indicated below.

Airport traffic control towers use the reports to expedite the flow of air traffic in the vicinity of the field and also forward reports to other interested offices.

FAA flight service stations broadcast the reports so that other airmen may benefit from them. They also deliver them to other enroute airplanes.

FAA air route traffic control centers use the reports to expedite the flow of enroute traffic and determine most favorable altitudes.

National Weather Service Forecast Offices find pilot reports very helpful in issuing advisories of hazardous weather conditions. These offices also use the reports to brief other pilots, and in forecasting.

Local Weather Service Offices use the reports in briefing other pilots and in forecasting.

WEATHER CHARTS

National Weather Service Offices and flight service stations are equipped with facsimile machines. These machines reproduce charts that are transmitted by wire from central locations, where they are drawn periodically by specialists or plotted by computers. The pictorial weather charts enable pilots and briefers to see graphic illustrations or "pictures" of weather conditions.

Weather charts of any kind are most effective when used to learn large-scale weather trends and general patterns. A briefer who is familiar with local aviation weather should be consulted for details and local variations. The most important pictorial charts for low-altitude flight planning are the *surface analysis charts, weather depiction charts, radar summary charts,* and *low-level prognostic charts.* The symbols used on these charts are easy to read and somewhat similar to each other. Of necessity, there are some variations, however.

SURFACE ANALYSIS CHART

The *basic* weather chart is the surface analysis chart. Weather observations are taken simultaneously throughout the United States, Canada, and over the Northern Hemisphere. This information is brought together at the National Meteorological Center and plotted on a single surface analysis chart. This chart provides the meteorologist, weather briefer, and pilot a pictorial view of weather throughout the continent, as it existed at one point in time.

The collection, plotting, analyzation, transmission, and distribution of the data require about two hours; thus, by the time the pilot sees a surface analysis chart, the information will be a *minimum* of two hours old. The time indicated on the chart is the time of the observation, and should *always* be noted by the pilot.

The information displayed on a surface analysis chart includes the surface wind, direction, and velocity. Additionally, temperature, dewpoint, and various other weather data, including the surface atmospheric pressure patterns, the positions of any fronts, and areas of high or low pressure, are shown. The symbols used on surface analysis charts are shown in figure 3-6.

In figure 3-7, there is a low center over Cleveland, Ohio, with a cold front southward then southwestward to southwestern Texas, then northwestward to western New Mexico. Three high centers on the chart are located over Georgia, Missouri, and eastern Idaho. The individual stations have weather information plotted according to the station model and symbols shown in figure 3-6.

WEATHER DEPICTION CHART

This chart shows a simplified picture of aviation weather over the conterminous 48 states. Much information is presented by use of a very few symbols. All areas with ceilings of less than 1,000 feet and/or visibilities of less than three miles are enclosed by a smooth, solid line. Normally, only instrument flights can be conducted within those areas. Areas with marginal VFR weather are enclosed by a scalloped line, and within those areas, ceilings are greater than 1,000 feet, but less than 3,000 feet; visibilities are three to five miles. The lines may enclose only one station or several states. No lines are drawn around areas where the ceiling is greater than 3,000 feet and the visibility is in excess of five miles.

At each station on the chart, a station model is used to depict the total sky cover, ceiling or cloud heights, visibility, weather, and obstructions to vision. As depicted in figure 3-8, the circle around each station is marked to symbolize the sky coverage, such as: clear conditions (○), scattered (◔), broken (◕), overcast (●), overcast with breaks (⦶), or obscured (⊗) . Immediately to the left of each circle, weather symbols show the significant weather, if any. (See figure 3-9.) Farther to the left is the visibility in miles. No visibility value is given if the visibility is greater than six miles.

The weather depiction chart is released every three hours, one hour after the surface weather chart. These two charts, and the latest aviation weather reports, should be consulted to get an up-to-date weather picture, since the latest chart may be as much as four hours old.

The legend in figure 3-9 should be studied so the weather depiction charts can be read and interpreted easily. Figure 3-8 is a portion of a weather depiction chart.

LOW-LEVEL PROGNOSTIC CHART

These charts show pictorial *forecasts* for the 48 conterminous states. Each chart contains four panels. A prognosis, or forecast, for a time 12 hours after the data collection time is pictured in the left-hand panels. The right-hand panels depict the weather outlook for a time 24 hours after the data collection time. These prognostic charts show conditions as they are forecast to

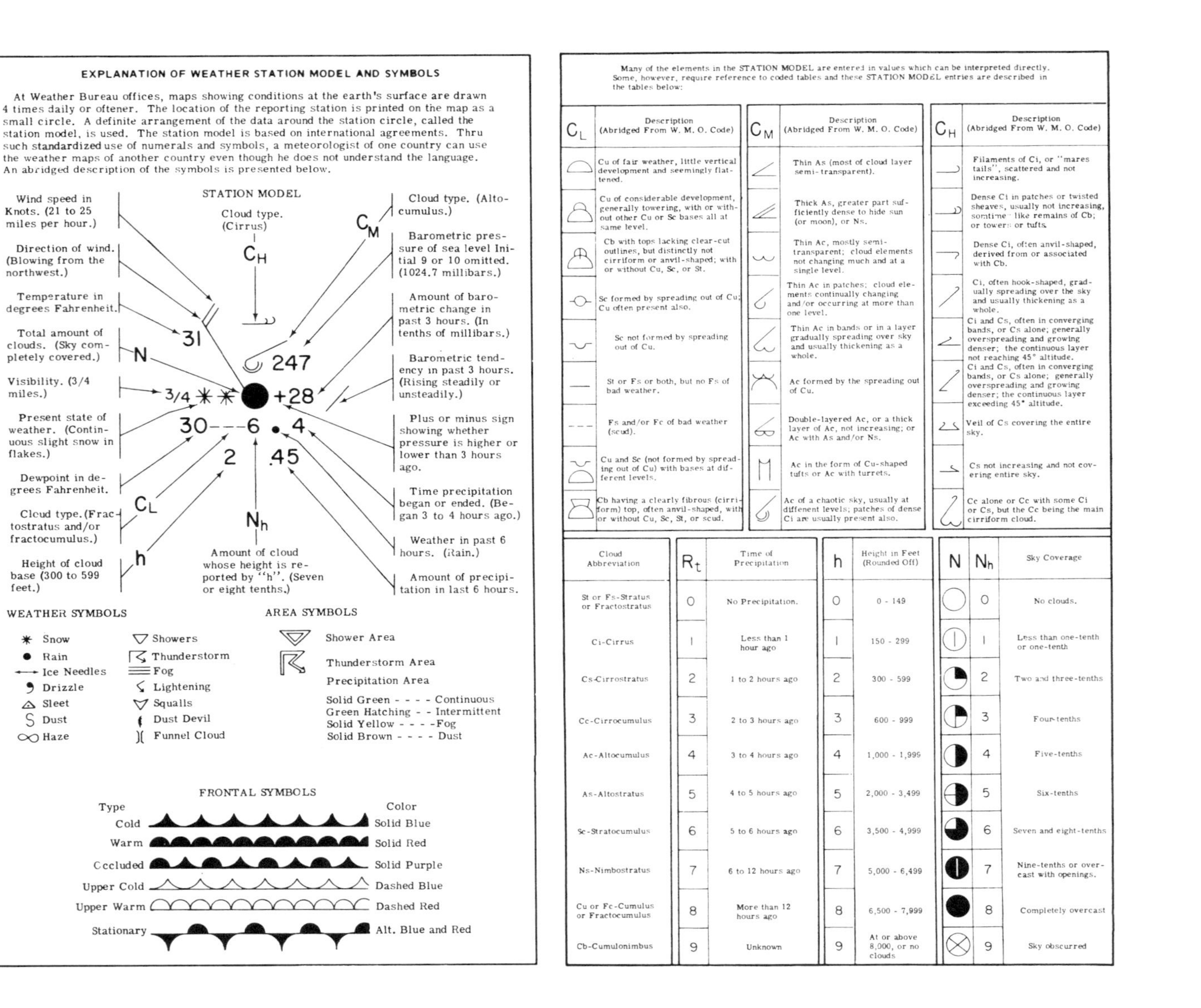

Many of the elements in the STATION MODEL are entered in values which can be interpreted directly. Some, however, require reference to coded tables and these STATION MODEL entries are described in the tables below:

C_L	Description (Abridged From W. M. O. Code)
	Cu of fair weather, little vertical development and seemingly flattened.
	Cu of considerable development, generally towering, with or without other Cu or Sc bases all at same level.
	Cb with tops lacking clear-cut outlines, but distinctly not cirriform or anvil-shaped; with or without Cu, Sc, or St.
	Sc formed by spreading out of Cu; Cu often present also.
	Sc not formed by spreading out of Cu.
	St or Fs or both, but no Fs of bad weather.
	Fs and/or Fc of bad weather (scud).
	Cu and Sc (not formed by spreading out of Cu) with bases at different levels.
	Cb having a clearly fibrous (cirriform) top, often anvil-shaped, with or without Cu, Sc, St, or scud.

C_M	Description (Abridged From W. M. O. Code)
	Thin As (most of cloud layer semi-transparent).
	Thick As, greater part sufficiently dense to hide sun (or moon), or Ns.
	Thin Ac, mostly semi-transparent; cloud elements not changing much and at a single level.
	Thin Ac in patches; cloud elements continually changing and/or occurring at more than one level.
	Thin Ac in bands or in a layer gradually spreading over sky and usually thickening as a whole.
	Ac formed by the spreading out of Cu.
	Double-layered Ac, or a thick layer of Ac, not increasing; or Ac with As and/or Ns.
	Ac in the form of Cu-shaped tufts or Ac with turrets.
	Ac of a chaotic sky, usually at different levels; patches of dense Ci are usually present also.

C_H	Description (Abridged From W. M. O. Code)
	Filaments of Ci, or "mares tails", scattered and not increasing.
	Dense Ci in patches or twisted sheaves, usually not increasing, sometime like remains of Cb; or towers or tufts.
	Dense Ci, often anvil-shaped, derived from or associated with Cb.
	Ci, often hook-shaped, gradually spreading over the sky and usually thickening as a whole.
	Ci and Cs, often in converging bands, or Cs alone; generally overspreading and growing denser; the continuous layer not reaching 45° altitude.
	Ci and Cs, often in converging bands, or Cs alone; generally overspreading and growing denser; the continuous layer exceeding 45° altitude.
	Veil of Cs covering the entire sky.
	Cs not increasing and not covering entire sky.
	Cc alone or Cc with some Ci or Cs, but the Cc being the main cirriform cloud.

Cloud Abbreviation	R_t	Time of Precipitation	h	Height in Feet (Rounded Off)	N	N_h	Sky Coverage
St or Fs-Stratus or Fractostratus	0	No Precipitation.	0	0 - 149		0	No clouds.
Ci-Cirrus	1	Less than 1 hour ago	1	150 - 299		1	Less than one-tenth or one-tenth
Cs-Cirrostratus	2	1 to 2 hours ago	2	300 - 599		2	Two and three-tenths
Cc-Cirrocumulus	3	2 to 3 hours ago	3	600 - 999		3	Four-tenths
Ac-Altocumulus	4	3 to 4 hours ago	4	1,000 - 1,999		4	Five-tenths
As-Altostratus	5	4 to 5 hours ago	5	2,000 - 3,499		5	Six-tenths
Sc-Stratocumulus	6	5 to 6 hours ago	6	3,500 - 4,999		6	Seven and eight-tenths
Ns-Nimbostratus	7	6 to 12 hours ago	7	5,000 - 6,499		7	Nine-tenths or overcast with openings.
Cu or Fc-Cumulus or Fractocumulus	8	More than 12 hours ago	8	6,500 - 7,999		8	Completely overcast
Cb-Cumulonimbus	9	Unknown	9	At or above 8,000, or no clouds		9	Sky obscurred

Fig. 3-6. Station Model and Weather Symbols

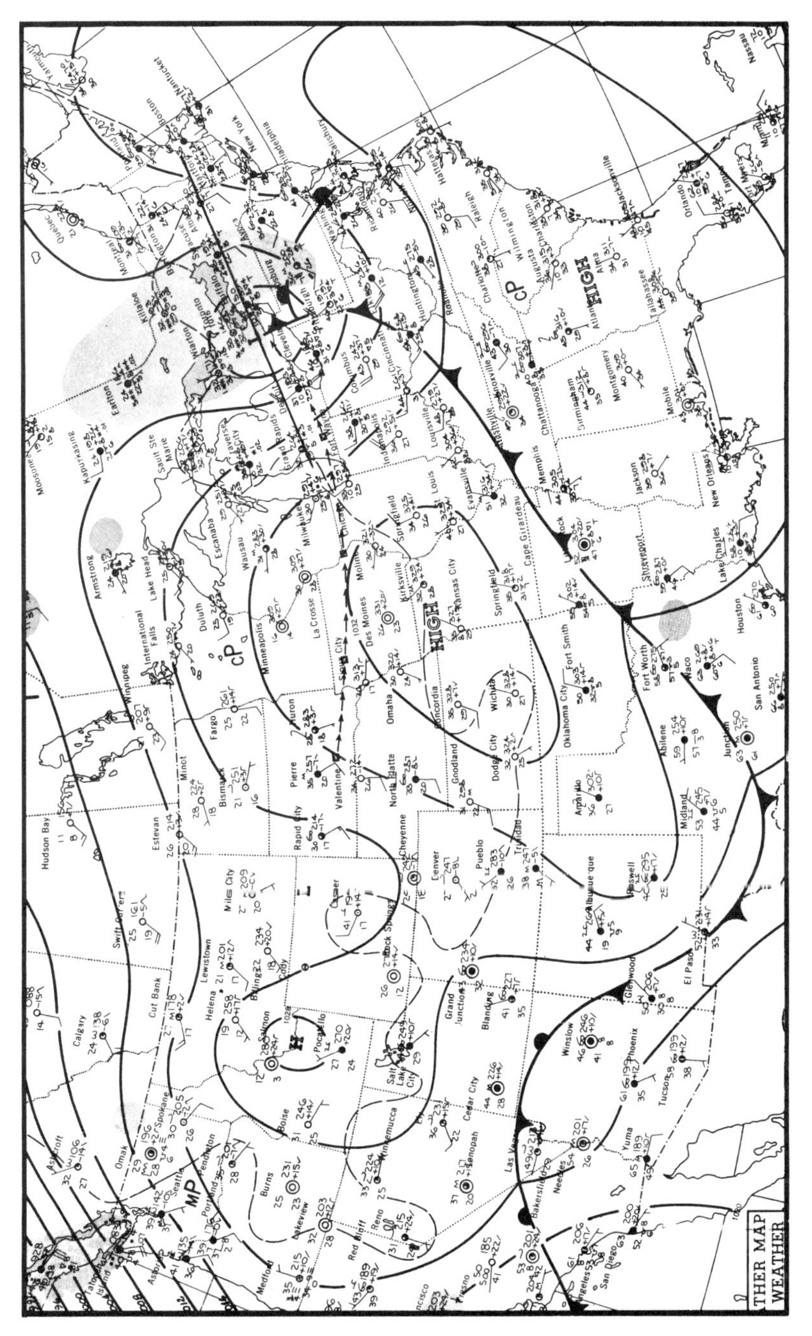

Fig. 3-7. Surface Analysis Chart

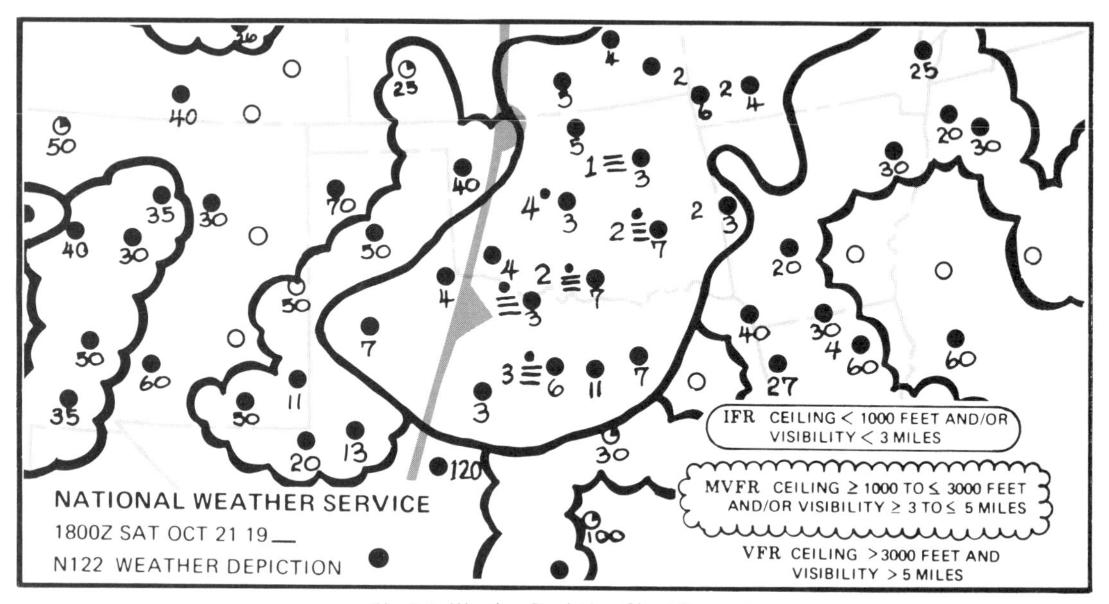

Fig. 3-8. Weather Depiction Chart Excerpt

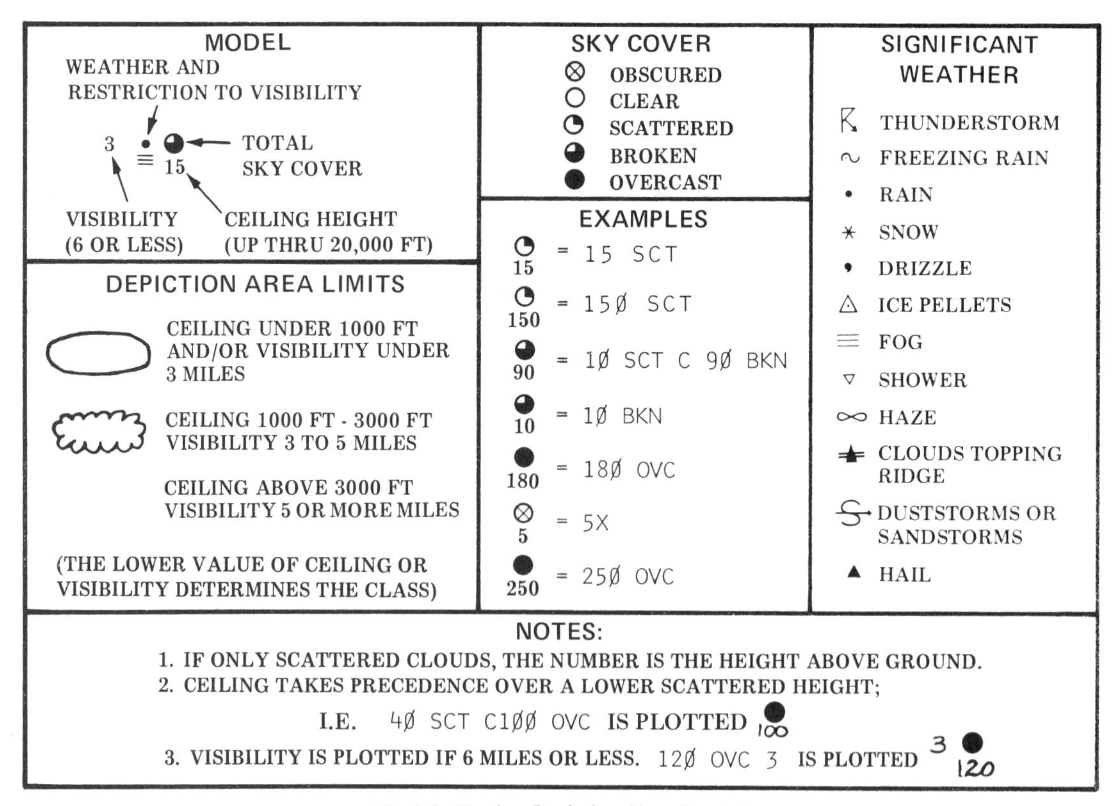

Fig. 3-9. Weather Depiction Chart Symbols

be at the valid times of the chart. Figure 3-10 portrays the four-panel, 12-hour and 24-hour low-level prognostic charts, together with their symbols.

The two upper panels portray ceiling, visibility, turbulence, and freezing level (the MSL altitude where the temperature is exactly 0°C). The two lower panels picture the predicted positions of

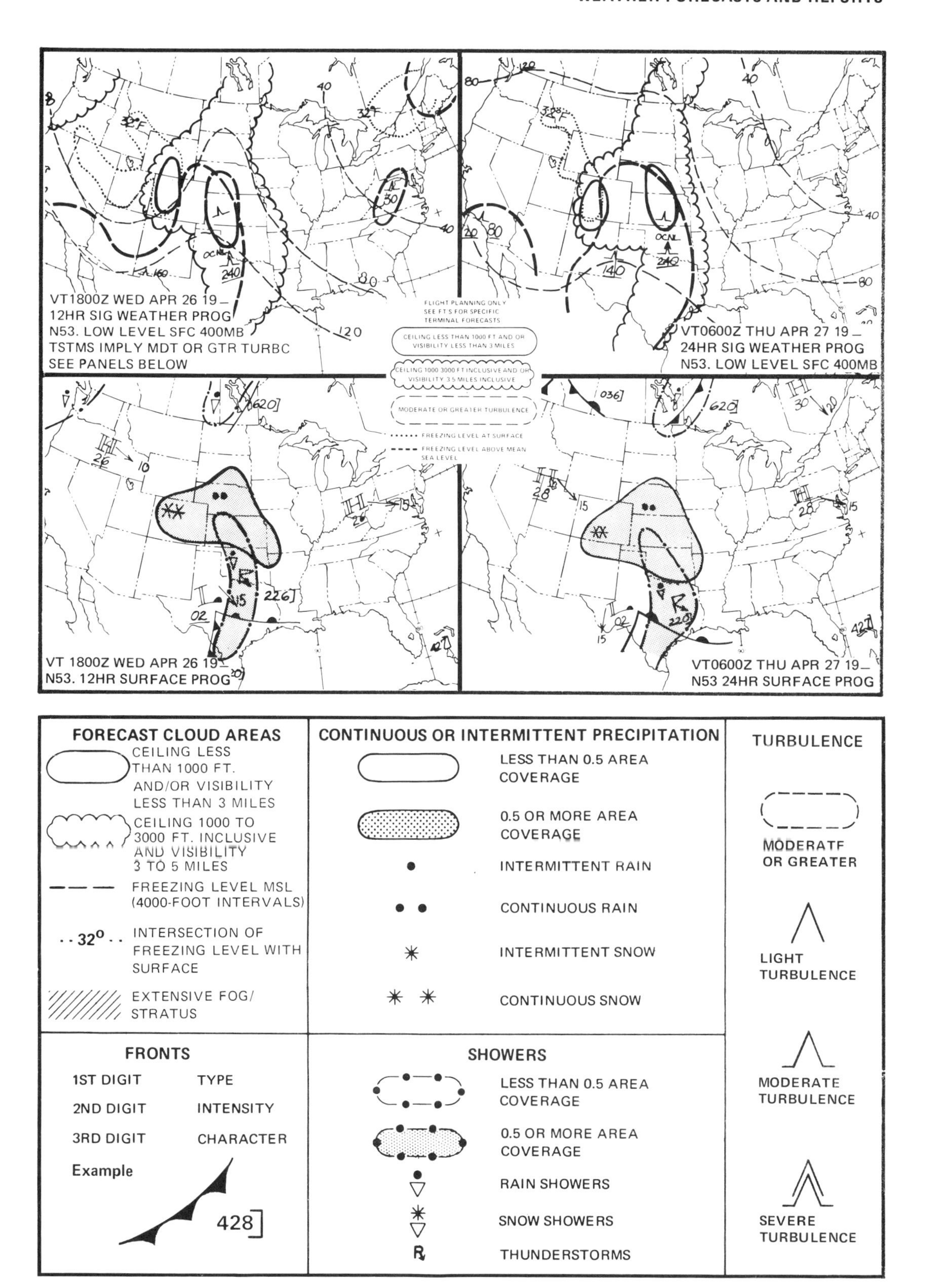

Fig. 3-10. Low-Level Prognostic Charts and Symbols

major weather systems such as fronts, highs, and lows, as well as areas of precipitation.

The "valid time" of each panel appears in the lower left-hand corner of that panel. For example, the weather picture in a left-hand panel containing "VT 1800Z APR 26" represents a prediction of the weather situation that is expected to exist at 6:00 p.m. Zulu Time on April 26.

On the lower chart panels, an area that is expected to have showers or thunderstorms will be enclosed by a line of dots and dashes. If precipitation is expected to be more persistent (continuous or intermittent), a solid line is used. When more than half the area will be affected, the enclosed area will be shaded. Arrows may be used to show the predicted direction of movement of a pressure center. The speed of the center (in knots) is shown near the arrows. Symbols are used to indicate where turbulence and thunderstorms are expected. Thunderstorm markings on surface prognostic charts always imply at least moderate turbulence in the storms, even though a general area of turbulence may not be outlined. Cloud bases and tops are sometimes shown by a horizontal line with the forecast heights of the cloud tops written above the line and the cloud base heights below the line. Heights are given in hundreds of feet, as in most other reports and forecasts. For example, 130-140 stands for forecast tops of 13,000 feet to 14,000 feet above MSL. In this case the cloud *bases* are not forecast.

On the two upper panels, scalloped lines enclose areas where a *ceiling* is expected to be from 1,000 feet to 3,000 feet above ground level (AGL), and/or the visibility is expected to be three to five miles. The points of the scalloped lines always point *toward* the area with the forecast ceiling. A ceiling is defined as the lowest cloud layer covering six-tenths or more of the sky, except for thin layers or partial obscurations. A smooth solid line is drawn around an area with a forecast ceiling below 1,000 feet and/or visibility less than three miles.

A dotted line broken only by the figure "32°" indicates the predicted location of the freezing isotherm at the surface. An isotherm is a line of equal temperature (just as an isobar is a line of equal pressure). This freezing isotherm is a line either enclosing or north of which all temperatures are below freezing at the earth's surface. The broken lines represent heights, in hundreds of feet, of the forecast freezing level. That is, a line broken by a "40" shows where the freezing level will be at 4,000 feet MSL. All air temperatures will be below freezing at altitudes above 4,000 feet at all locations along the broken line. A pilot planning to fly above 4,000 feet, north of that line, should expect to encounter icing if flying in visible moisture.

Symbols are used to portray forecast fog and low stratus clouds and to depict areas of turbulence. A spike (_/_) with 160 printed near the spike means that turbulence is predicted below 16,000 feet MSL in the area of the spike symbol.

The legend in figure 3-10 is used with both the upper and the lower panels. Many of the symbols are explained on the sample chart panels.

RADAR SUMMARY CHART

Another excellent weather briefing aid is the radar summary chart, shown in figure 3-11. It is a graphic illustration of weather observed by radar. Special Weather Service radar is used to detect and track thunderstorms, tornadoes, and hurricanes. These storms contain very heavy concentrations of liquid moisture that reflect radar signals back to the antenna. These "echoes" are then analyzed and plotted on the chart. Weather symbols noted on this chart are important to pilots because the conditions that cause strong

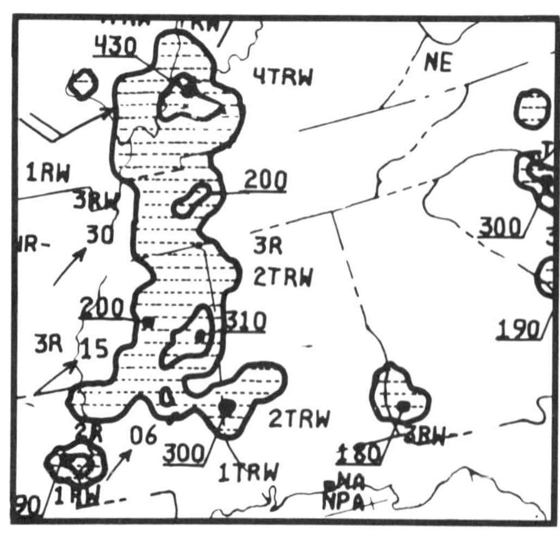

Fig. 3-11 Radar Summary Chart

echoes often are associated with heavy rain, hail, severe icing, and severe turbulence.

Hourly radar summary facsimile charts and plain language radar summaries are prepared each day. This information is transmitted on Service A teletypewriters and to subscribers on the facsimile circuit. The charts have symbols and abbreviations which are easily identified. The most common ones are summarized in the accompanying legend.

RADAR SUMMARY CHART	
Symbol	Meaning
[symbol]	Intensity 1 Light and moderate
[symbol]	Intensity 3 (Second Contour) Heavy and very heavy
[symbol]	Intensity 5 (Dark Area) Intense and extreme
[dashed box]	Dashed lines define areas of severe weather
——→ 20	Echoes are moving at 20 knots
[barbed arrow]	Line or area is moving at 20 knots (10 knot barbs)
240/80	Echo top 24,000' MSL echo base 8,000' MSL
INTENSITY AND TREND OF PRECIPITATION	
+	Increasing precipitation
–	Decreasing precipitation
ECHO COVERAGE	
Symbol	Meaning
SLD	Solid, over 9/10 coverage
BKN	Broken, 6/10 to 9/10 coverage
SCT	Scattered, 1/10 to 5/10 coverage
NE	No echo – Equipment operating but no echoes
NA	Observation not available
OM	Equipment out for maintenance
●	Strong cell identified by one station

Areas of echoes are surrounded by irregular lines. Speed and direction of movement of the echo areas or individual cells are given. One of the important features of the radar summary chart is that it shows the height of cloud tops. However, some weather phenomena, such as thunderstorms, can change so rapidly that the radar summary cannot be depended on for specific details.

Any important surface precipitation is given in the same terminology as teletype surface aviation reports. For example, "RW" means "rain showers," "S" means "snow," and "A" means "hail." Numbers may precede these symbols to indicate how much of the area is affected. A notation of "3RW," for instance, indicates that 3/10 of the area is covered by rain showers.

WEATHER BRIEFINGS

Various sources of weather information exist for use by the aviation community. If at all possible, the pilot should obtain an in-person weather briefing. Sometimes, however, this may not be possible. Should this be the case, the National Weather Service has provided several other means of obtaining preflight weather, such as person-to-person telephone briefings, the pilot's automatic telephone weather answering service (PATWAS), the automatic aviation weather service (AAWS), and transcribed weather broadcasts (TWEBs).

IN-PERSON WEATHER BRIEFING

The in-person weather briefing is the *preferred* way of obtaining weather information. This personal visit will assist the pilot in receiving a *total* picture of the weather, since he will be provided with numerous sources of weather information. He can inspect the charts and read area, terminal, and winds aloft forecasts, plus the surface aviation weather reports.

During flight planning, the pilot should make use of the expert help available at weather service offices and flight service stations. At these locations, forecasters or attendants are available to provide necessary weather information.

TELEPHONE WEATHER BRIEFING

Person-to-person telephone briefing and PATWAS also are commonly used methods of obtaining a weather briefing. These briefings may be obtained from an FAA flight service station or a National Weather Service Office. The telephone numbers for the flight service stations and weather service offices are listed in telephone directories, or Jeppesen J-AID.

The *pilot's automatic telephone weather answering service* (PATWAS) is a transcribed service which includes flight precaution information, a synopsis of current weather, a 12-hour forecast of air route weather for 200 to 350 miles around the station, and a winds aloft forecast. The *automatic aviation weather service* (AAWS) is similar to PATWAS, but is more general in terms and includes an outlook for up to 24 hours.

WEATHER BRIEFING CONSIDERATIONS

Standard procedures should be used whether the weather briefing is obtained in person or by telephone. The weather briefer needs the following information in the order listed.

1. Aircraft identification
2. Type of airplane
3. Estimated time of departure
4. Proposed route, flight altitude, and destination
5. Estimated time of arrival at destination or time enroute
6. Intended flight—IFR or VFR

In addition, a weather checklist will save time and assure that the pilot has all necessary information when the briefing is concluded. Before calling or entering the station for a briefing, the pilot should be prepared to write the following items on a weather log.

1. *Weather synopsis*, including positions of fronts and pressure centers that are influencing the weather along the route
2. *Current departure and destination weather*, with a description of current enroute weather
3. *Forecast* of enroute and destination weather, including cloud tops, freezing level and precipitation
4. *Alternate routes and alternate destination weather*, if necessary
5. *Forecast* or *reported hazardous weather*
6. *Winds aloft* forecast
7. Pilot reports and NOTAMs

The pilot *should request* any necessary or desired information omitted by the briefer. The use of the weather checklist is comparable in importance to the pretakeoff airplane checklist.

Of special interest to the instrument pilot is the weather in the areas to each side of the route. In case of deteriorating weather conditions at the destination and alternate, or an in-flight emergency, this knowledge is extremely valuable. An instrument pilot always must know where favorable weather can be found and how to get there.

For example, a communications and navigation radio failure necessitates an immediate diversion to an area of VFR weather. With only a navigation radio failure, an area with weather minimums that permit a radar or DF approach must be chosen. If an engine of a multi-engine airplane must be feathered while enroute, the pilot may need to divert to an area where he is sure the weather will not cause a missed approach.

PILOT'S WEATHER LOG

By using the surface analysis charts, the hourly aviation weather reports, and area, terminal, and winds aloft forecasts, the pilot can develop an overall view of weather that may be encountered during a flight. When properly completed, the weather log, shown in figure 3-12, will serve as a checklist.

By using the surface weather maps, hourly weather reports, area forecasts and winds aloft reports, the pilot can get a general estimation of weather that will be encountered during a flight. Even though the pilot can estimate the weather, he should *always use the services of a professional forecaster, when available.*

BASIC WEATHER ANALYSIS USING FORECASTS AND REPORTS

An example of weather analysis is shown in figures 3-13 and 3-14. The surface weather map shows that a cold front passed St. Louis at 0700, moving eastward at approximately 40 knots. Therefore, the weather associated with the front can be determined by using the 0700 hourly weather for St. Louis. In addition, by considering the speed of the cold front, a pilot can estimate the weather that will exist at the reporting stations three to four hours later.

WEATHER LOG

	DEPARTURE	ENROUTE	DESTINATION	ALTERNATE
REPORTED WEATHER	DEN E22 BKN 15 45/30 2215 / 997	DEN-AKO-LBF C20-30 BKN LWRG TO 15 OVC AKO, 10 OVC 4H LBF	GRI M12 OVC 6 1612/002	OMA E15 OVC 8 48/38 1210/006
FORECAST WEATHER	NO CHANGE	GRDLY LWRG NEBR. PTN TO C6-10 OVC 2-4L-F	8 OVC 2L-F 1612	8 SCT 12 OVC 6H 1412 SCT V BKN OCNL 3L-F
WINDS ALOFT	DEN 90-2615 +02 120-2720 -08 GKC 60-2216 +12 90-2420 +04 ONL 60-2418 +08 90-2520 +00			
ICING & TURBULENCE	FRZG LVL 100 COLO TO 110 ERN NEB. ICGIC ABV FRZG LVL. NO TURB.			
CLOUD TOPS	TOPS 120 DEN SLPG TO 80 TO 90 ERN NEB.			
SIGMETS & AIRMETS	CIGS OCNLLY BLO 10 ERN HALF NEB.			
PILOT & RADAR REPORTS	AKO 1400Z TOP OVC 100 TEMP +10 NO ICE AT 90.			

Fig. 3-12. Weather Log

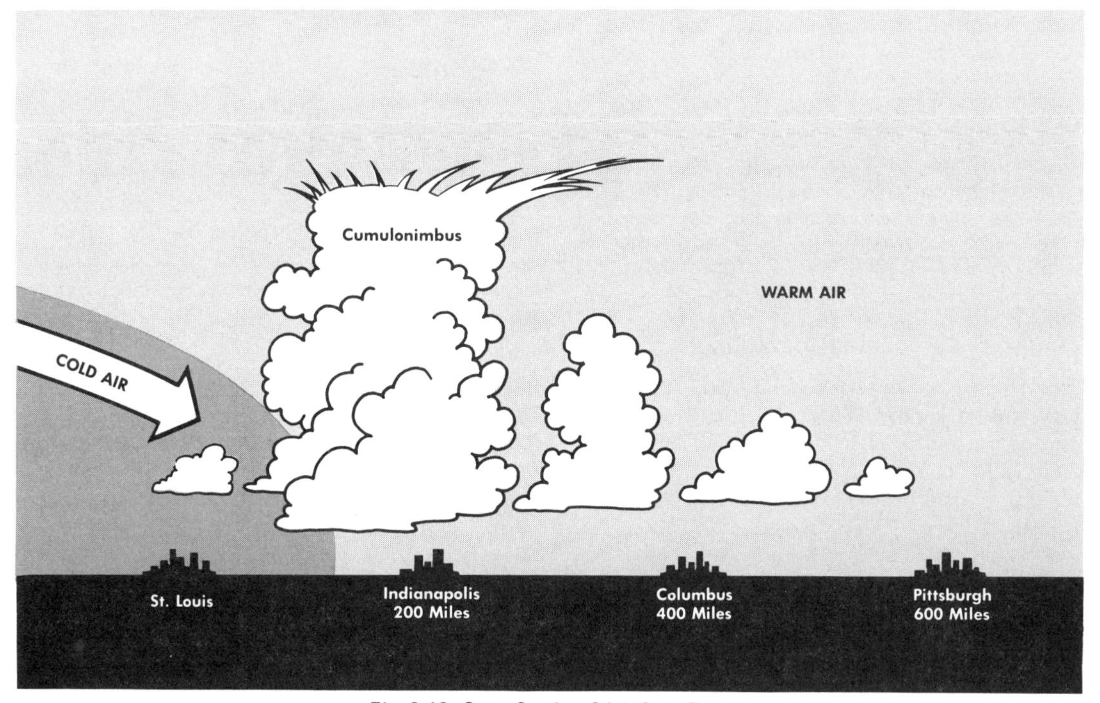

Fig. 3-13. Cross-Section Of A Cold Front

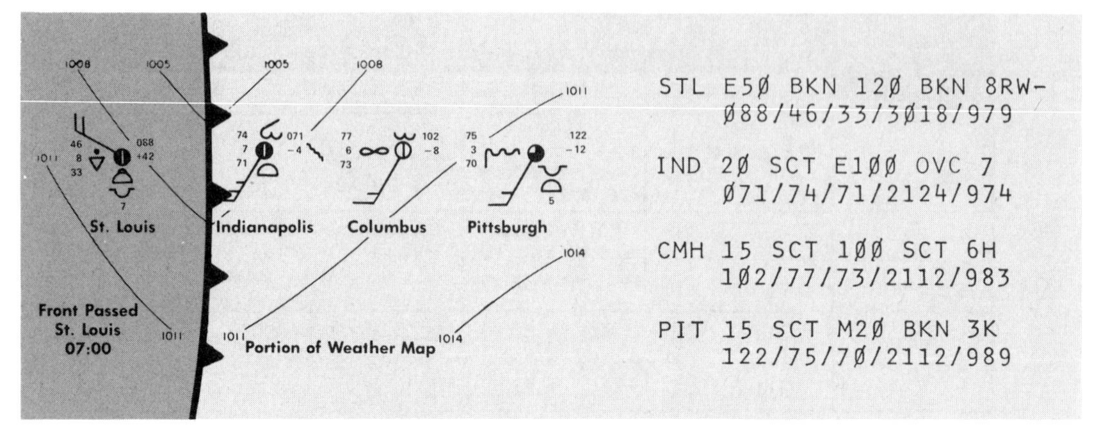

Fig. 3-14. Weather Map and Weather Reports Associated With Cold Front

Figures 3-15 and 3-16 show the cross-section of a warm front for the same stations. Analyzing the sequence reports, the pilot can get a good picture of the frontal movements and the weather associated with it.

A pilot should always remember that weather is a changeable factor and that even professional forecasts are only a prediction of what may exist. Most of the predictions are reasonably accurate, but pilots must be alert for changes during a flight, since unobserved and unreported weather conditions may exist between weather stations. The Weather Service will provide all available weather information so the pilot can decide whether or not to make the flight.

IN-FLIGHT BRIEFING

Several sources provide the pilot with updated weather information while in flight. The FAA provides transcribed weather broadcasts, scheduled weather broadcasts, request-and-reply service, and enroute flight advisory service.

TRANSCRIBED WEATHER BROADCASTS

The transcribed weather broadcast (TWEB) is very similar to PATWAS. The TWEB is broadcast over nondirectional beacons and some VOR facilities. Transcribed weather broadcasts over nondirectional beacons have an average reception distance of 75 statute miles. However, the reception distance for broadcasts over VOR facilities will vary according to the altitude of the airplane.

The current weather portion of the TWEB contains the surface aviation weather reports at selected locations within a 400-mile radius of the station. The reports of the stations surrounding the transcribing station are given in sequence, usually clockwise beginning at the north or northeast quadrant.

SCHEDULED WEATHER BROADCASTS

All flight service stations having voice facilities on continuously operated VORs or radio beacons, broadcast weather reports and other airway information at 15 minutes past each hour. The broadcast consists of weather reports from the stations within approximately 150 miles of the broadcast station.

SIGMETs and AIRMETs are broadcast on an unscheduled basis at the time of issuance. For the first hour thereafter, SIGMETs are broadcast at 15-minute intervals beginning on the hour, and AIRMETs are broadcast at 30-minute intervals beginning at 15 minutes past the hour. For the remainder of the valid time, the advisories are identified at 15 and 45 minutes past the hour, and can be obtained on the FAA request-and-reply service.

REQUEST-AND-REPLY SERVICE

Weather information may be received in flight through radio contact with FAA flight service stations by utilizing the request-and-reply service. The flight service specialist interprets the briefing material supplied by the National Weather Service when answering an in-flight request.

ENROUTE FLIGHT ADVISORY SERVICE

The FAA has developed an *enroute flight advisory service* (EFAS) to provide specific

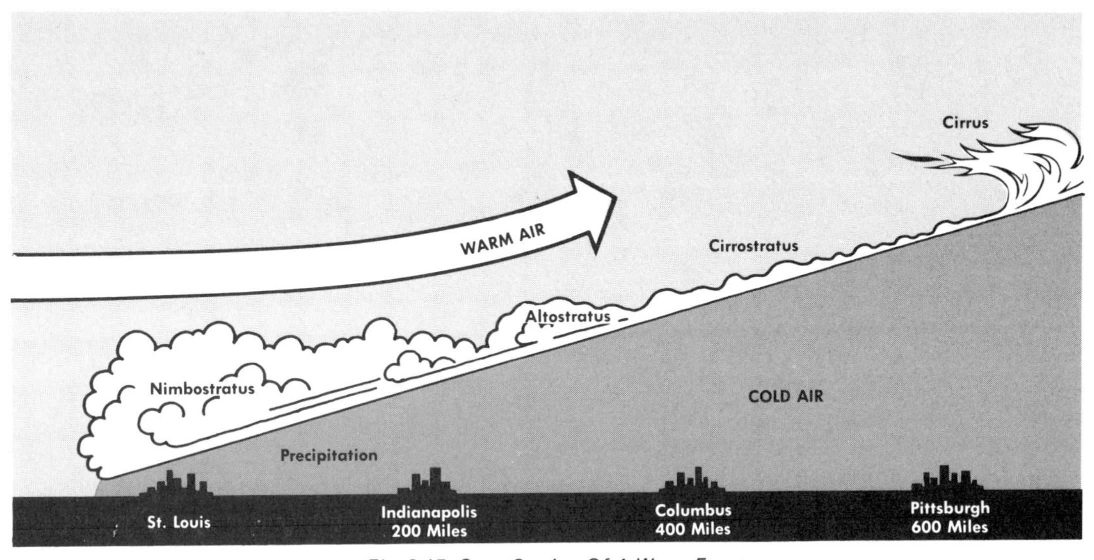

Fig. 3-15. Cross-Section Of A Warm Front

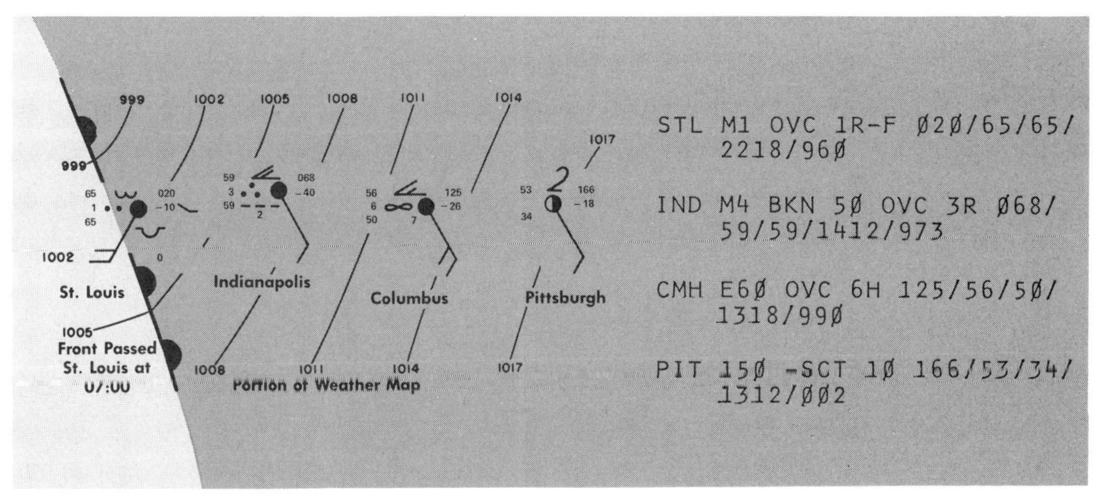

Fig. 3-16. Weather Map & Weather Reports Associated With Warm Front

enroute weather information. This service is available to all pilots from selected flight service stations throughout the United States.

Each of the selected flight service stations provides EFAS in its own geographical area and the area served by its remote communication outlets. The communication frequency used for this service is 122.00 MHz. To request this service the pilot should call the flight service station by using the FSS name followed by flight watch; for example, "*Oakland Flight Watch.*"

EFAS enables a pilot to obtain routine weather information plus current reports on the location of thunderstorms and other hazardous weather, as observed and reported by pilots or noted on radar. To increase the efficiency of this service, all pilots are encouraged to report weather encountered in flight to the nearest flight watch, or flight service station.

4 NATIONAL AIRSPACE SYSTEM

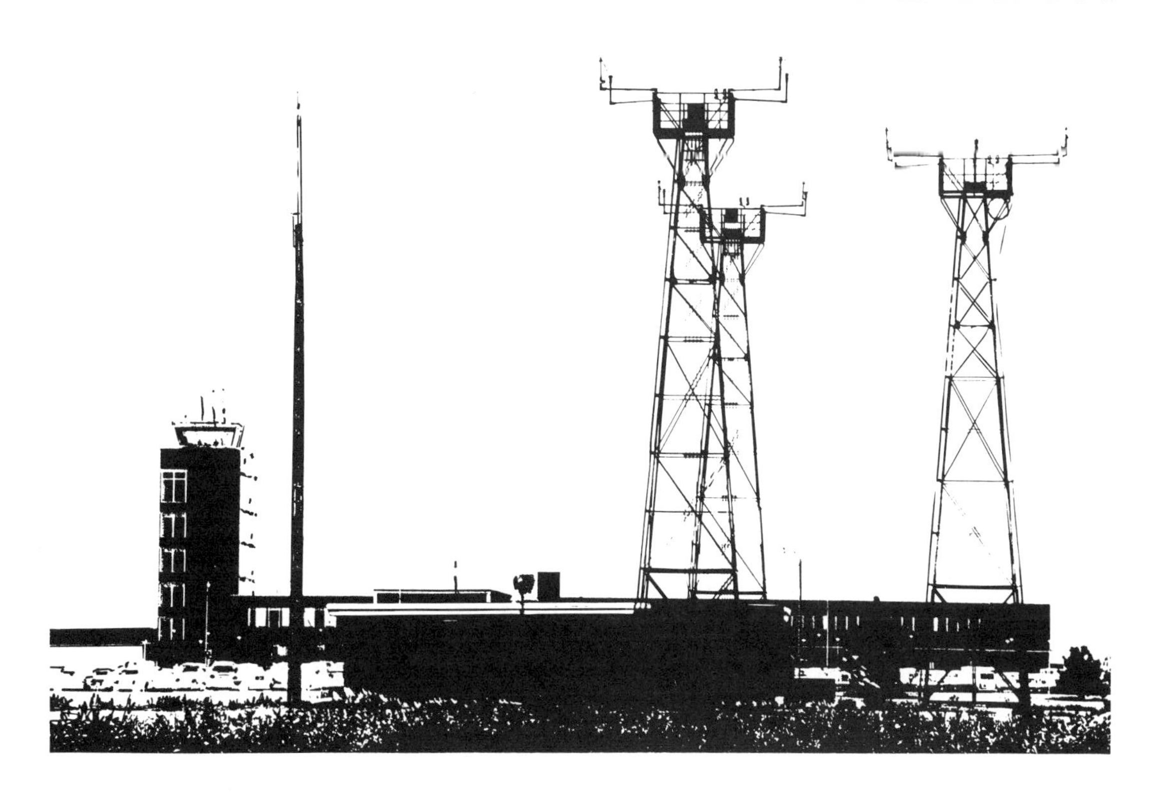

CONTROLLED AIRSPACE

Controlled airspace refers to airspace regulated to provide an orderly and safe flow of traffic. Initially, controlled airspace came into existence to provide safe separation for instrument flights by excluding noninstrument flights from the area. However, since that time, air traffic has increased to the point where the control of airspace is necessary for safe execution of all flight functions. Controlled airspace includes airspace designated as positive control areas, continental control areas, control areas, transition areas, control zones, and terminal control areas. An ATC clearance is required for flight within these areas when IFR conditions exist.

On low altitude enroute charts, controlled airspace is depicted in white. Uncontrolled airspace is shown by blue shading on Jeppesen charts and brown shading on NOS charts.

POSITIVE CONTROL AREA

A *positive control area* encompasses airspace extending from 18,000 feet MSL to 60,000 feet MSL, as shown in figure 4-1, item 1. In order to operate within this airspace, the pilot must be instrument rated, file an IFR flight plan, and meet the requirements for instrument flight, as stated in FAR Part 91.

CONTINENTAL CONTROL AREA

The *continental control area* encompasses airspace at and above 14,500 feet MSL, as shown in figure 4-1, item 2. This area extends throughout the United States, District of Columbia, and a portion of Alaska. Except as listed in FAR Part 71, prohibited or restricted areas and airspace less than 1,500 feet above the surface are excluded from this control area. Rules governing VFR flight at altitudes above 10,000 feet MSL, up to, but not including, 18,000 feet, require a pilot to have at least five miles visibility and maintain one mile horizontal and 1,000 feet vertical separation from all clouds.

CONTROL ZONES

Control zones are usually circular areas five miles in radius with the necessary extensions for instrument departures and arrivals. A control zone extends from the surface up to the base of the continental control area, as shown in figure 4-1, item 3. Control zones are shown on Jeppesen low altitude charts by green dashed lines, and on NOS charts by blue dashed lines. To conduct VFR flight in a control zone, the visibility must be at least three miles and the ceiling at least 1,000 feet. When the ceiling and visibility are below minimums, a special VFR clearance can be obtained at many control zones. Special VFR minimums are one mile visibility and clear of the clouds. At night, a special VFR clearance is issued only to instrument rated pilots flying aircraft that meet instrument flight requirements.

AIRPORT TRAFFIC AREA

An *airport traffic area* exists at each airport with an *operating* control tower. In this case, the area within five statute miles of the airport and up to, but not including, 3,000 feet above the airport elevation will constitute the airport traffic area, as shown in figure 4-2. To check the existence of an airport traffic area, the pilot should determine the availability of a control tower by reference to the *Airport Facility Directory*, Jeppesen J-AID, or the appropriate aeronautical chart.

TERMINAL CONTROL AREAS

Terminal control areas (TCAs) have been established at busy air terminals to promote safety by regulating the flow of all air traffic. Generally, TCAs are tiered layers of airspace shaped something like an inverted wedding cake. They may have extensions, gaps, or corridors where necessary to expedite the flow of air traffic.

Since an IFR clearance through a TCA also constitutes a TCA clearance, if needed, the instrument pilot is primarily concerned with equipment requirements. Therefore, the geographic area of a TCA is depicted by blue shading on the enroute low altitude and area charts. The more detailed information needed by VFR pilots is shown on sectional charts and special VFR terminal control area charts. Pilot, equipment, and procedural requirements for fixed-wing airplanes are summarized in figure 4-3. The procedural and equipment requirements vary according to aircraft type and operation. Pilot familiarity with the rules governing specific operations is essential.

AIRWAYS

The Victor airways extending between navigation aids are another form of controlled airspace. When operating within this area, the pilot has the same visibility and cloud clearance requirements found in other controlled airspace. Therefore, if VFR minimums are not prevalent, the

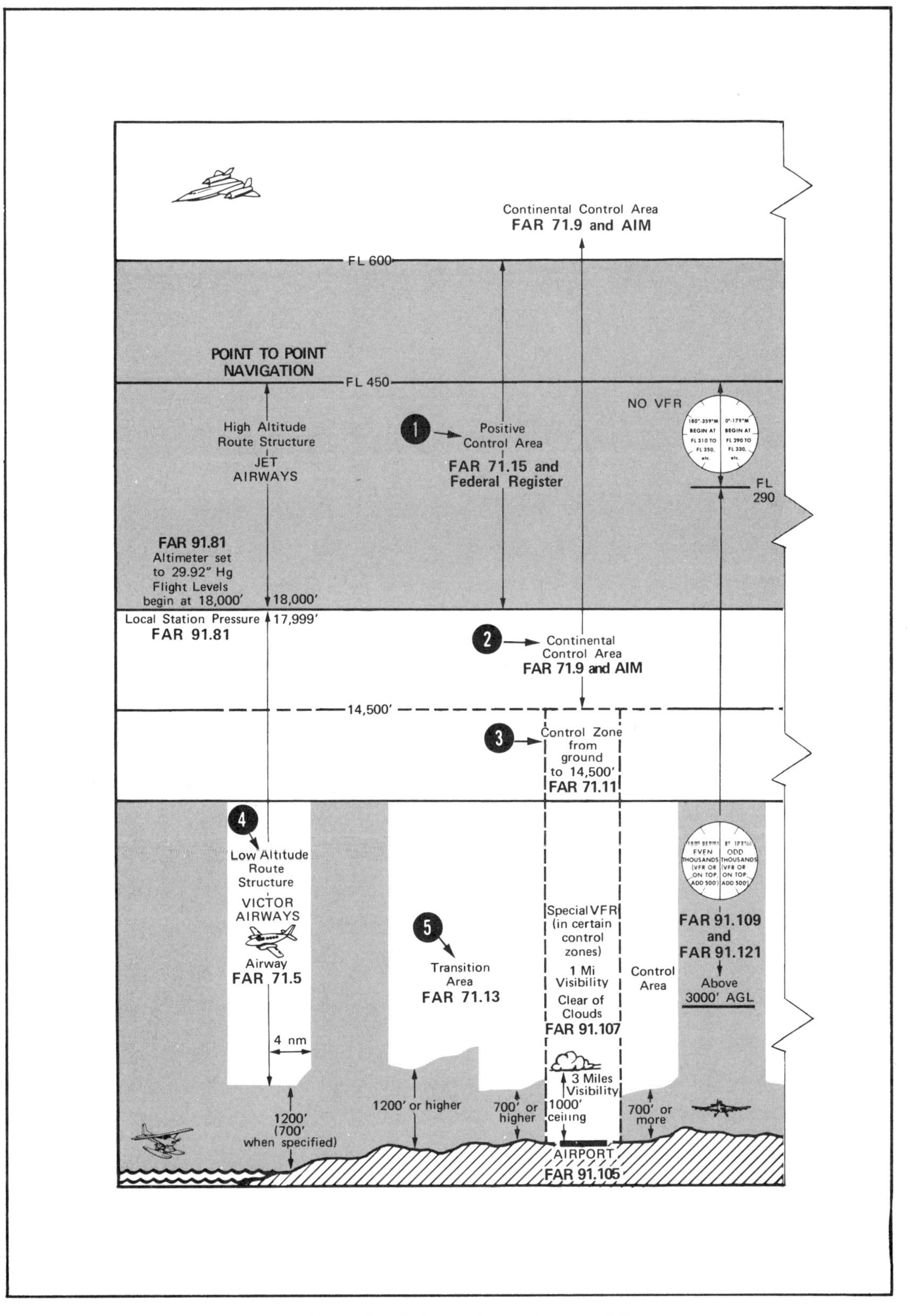

Fig. 4-1. National Airspace System (page 1 of 2)

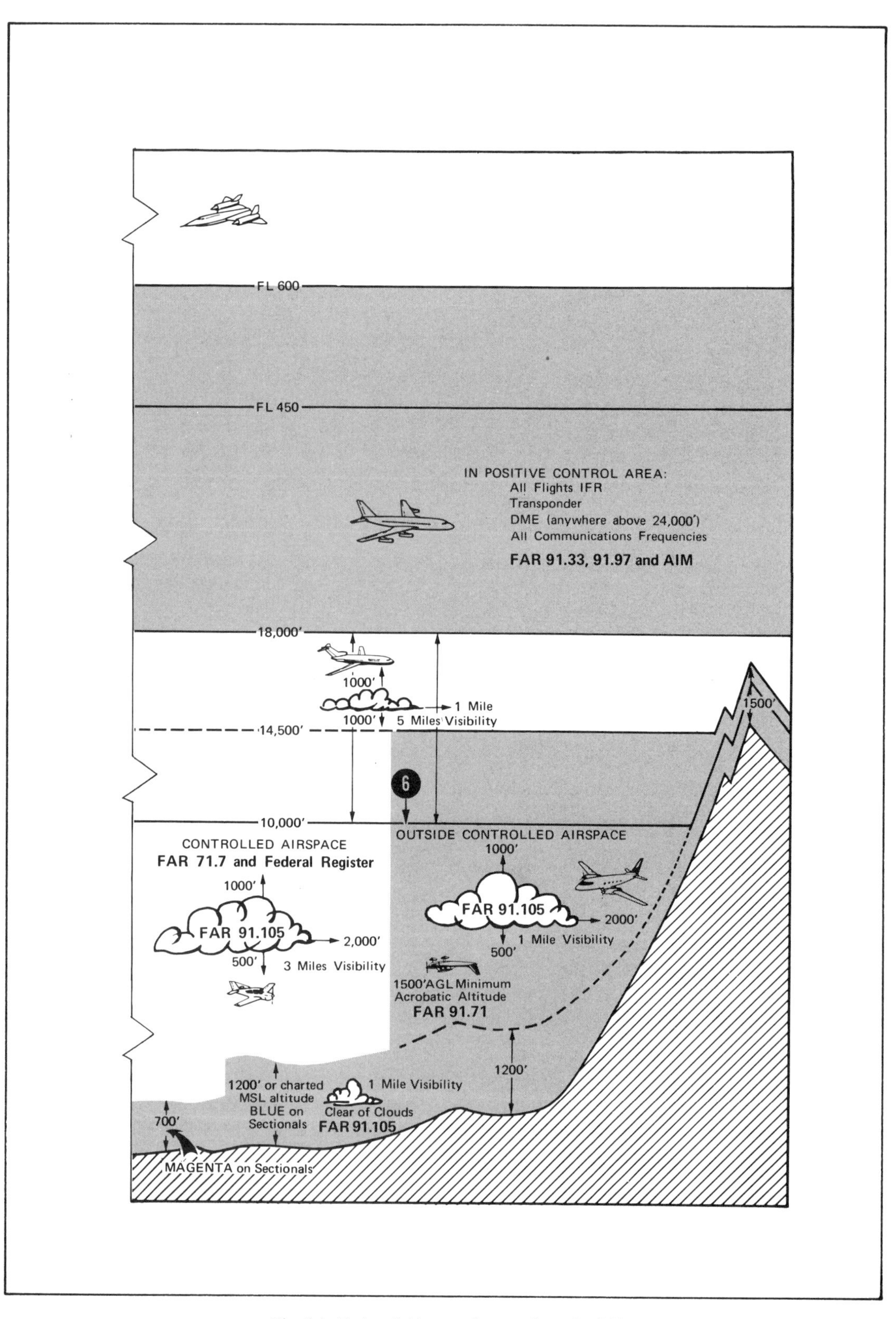

Fig. 4-1. National Airspace System (page 2 of 2)

Fig. 4-2. Airport Traffic Area

pilot must file for and receive an IFR clearance *prior* to operation within the Victor airway.

The airways are eight nautical miles in width for the first 50.8 nautical miles out from the forming navaids. Beyond that distance, they expand outward from each edge of the airway at an angle of 4.5°. In addition, they extend from 1,200 feet AGL upward to, but not including, 18,000 feet MSL.

TRANSITION AREAS

Transition areas have been formed to enable an instrument pilot to remain in controlled airspace when flying from the terminal to enroute operations, or vice versa. Transition areas extend upward from 700 feet or more above the surface when designated in conjunction with airports having instrument approach procedures. However, when the transition area is used in conjunction with an airway, as shown by item 5 of figure 4-1, the base altitude begins at 1,200 feet or more above the surface. The upper limit terminates at the base of the overlying controlled airspace (unless otherwise specified).

TERMINAL CONTROL AREAS (TCA)

FARs 91.24, 91.70, 91.90

Requirements Listed Below are Minimums for Civilian, Fixed-Wing Airplanes

<table>
<tr><th colspan="2"></th><th>Group I</th><th>Group II</th><th>Group III</th></tr>
<tr><td colspan="2">Pilot Requirements</td><td>Private pilot certificate</td><td colspan="2">None specified</td></tr>
<tr><td colspan="2" rowspan="3">Operating Rules</td><td colspan="3">Maximum speed below floor of TCA or through VFR corridor is 200 knots I.A.S. unless a minimum safe speed is higher. Maximum speed in TCA is 250 knots I.A.S. below 10,000'.</td></tr>
<tr><td colspan="2">Unless otherwise authorized by ATC, large turbine aircraft operating to or from the primary airport shall operate at or above the designated floors while within the lateral limits of TCA</td><td>Not specified</td></tr>
<tr><td colspan="2">Prior ATC authorization required</td><td>Prior ATC authorization not required if aircraft is equipped with 4096-code transponder and encoding altimeter</td></tr>
<tr><td rowspan="5">AVIONICS REQUIREMENTS</td><td>Navigation</td><td colspan="2">VOR receiver</td><td>None specified</td></tr>
<tr><td>Two-Way Communi-cations</td><td colspan="2">Required on appropriate ATC frequencies.</td><td>Not required if aircraft is equipped with 4096-code transponder and encoding altimeter.</td></tr>
<tr><td rowspan="2">4096-Code Transponder</td><td>Required for all flights</td><td>Required for all flights to airports within the TCA</td><td rowspan="3">Not required if two-way communications are maintained.</td></tr>
<tr><td colspan="2">Not required for IFR approach transitions to near-by airport if transition penetrates TCA.</td></tr>
<tr><td>Encoding Altimeter</td><td>Required for all flights</td><td>Not required for any flights</td></tr>
</table>

ATC deviations may be authorized:
1. Immediately in the case of an inoperative transponder.
2. Immediately with an operating transponder but without an operating encoding altimeter.
3. With four hours notice for flight without a transponder.

For other authorized deviations refer to FAR's 91.24, 91.70, and 91.90.

NOTE: A 4096-code transponder and encoding altimeter is required for flight that is within controlled airspace above 12,500 MSL excluding the airspace at and below 2500 AGL.

Fig. 4-3. Terminal Control Area

UNCONTROLLED AIRSPACE

Although most of the airspace in the United States is controlled, there is still some airspace designated as *uncontrolled.* (See Fig. 4-1, item 6.) For IFR operations within this airspace, both pilot and equipment must meet the same requirements set forth for IFR operations within controlled airspace.

Since no traffic control or separation is provided for flight in uncontrolled airspace, it is unnecessary for the pilot to obtain an ATC clearance or file an IFR flight plan. However, the pilot should exercise extreme caution, since other pilots may be operating under IFR conditions within the same airspace.

COMMUNICATIONS

Communication with ground facilities is a vital part of the IFR flight. There are three basic types of communication ranges.

1. The ultrahigh frequency (UHF) range is used primarily for military communications, DME, and transponder operations.
2. The very high frequency (VHF) range is used in navigation and communications facilities in the United States.
3. The low and medium frequency (L/MF) range applies to certain navigation aids and some navigation system components.

VHF and UHF radio waves are transmitted along line of sight, as shown in figure 4-4; that is, they do not follow the curvature of the earth. When using these aids, the higher the aircraft is above the ground, the greater the reception range becomes.

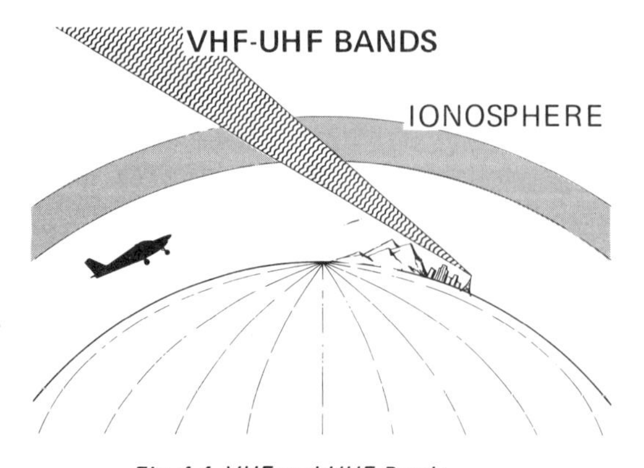

Fig. 4-4. VHF and UHF Bands

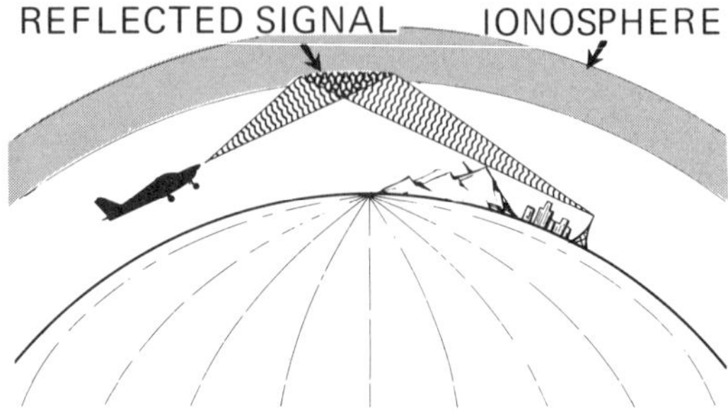

Fig. 4-5. L/MF Bands

Low/medium frequency waves are not transmitted along line of sight; that is, they are reflected back and forth between the ionosphere and the earth, as shown in figure 4-5. Low/medium frequency ranges, however, are limited by atmospheric conditions such as weather, daylight, or darkness.

Most aircraft incorporate at least a 90-channel VHF radio. These radios have a range from 118.00 through 126.90 MHz, with 100 kHz spacing between channels; however, most well-equipped instrument aircraft have 360- or 720-channel transceivers, with frequencies up to 135.975 MHz. The spacing between the channels is 50 and 25 kHz, respectively.

FLIGHT SERVICE STATIONS

Flight service stations are important to both VFR and IFR pilots due to the varied services they provide. These include weather briefings, processing flight plans, scheduled weather broadcasts, relayed air route traffic control center communications, airport advisory service, and initiation of search and rescue procedures.

WEATHER BRIEFING

Flight service stations throughout the country provide weather information and briefings for pilots. This information can be obtained in person, by telephone, or by radio while enroute.

Enroute flight advisory service (EFAS) can be obtained from selected flight service stations on the frequency of 122.00 MHz. This frequency is a discrete frequency assigned only for weather advisory service.

FLIGHT PLANS

Another primary function of a flight service station is processing of flight plans. VFR flight plans are helpful in the event search and rescue operations become necessary. IFR flight plans are necessary for ATC to provide traffic separation and control. Although a pilot may file a flight plan enroute, the preferred method is either in person or by telephone.

WEATHER BROADCASTS

Besides providing weather briefings, the flight service station broadcasts current area weather over the VOR system at 15 minutes past the hour. If a significant weather change occurs during the hour, a special weather report is issued. Besides the regular and special weather reports, SIGMETs, AIRMETs, and NOTAMs also are conveyed to pilots by the FSS.

ARTCC COMMUNICATIONS RELAY

When pilots are operating from uncontrolled airports, the FSS relays their IFR clearances. The FSS also transmits messages from the controller to the pilot, and vice versa, in some geographical areas where communication facilities are not adequate. It can be seen that the FSS acts as a go-between when IFR communications between ARTCC and pilots is impossible because of limited communications.

AIRPORT ADVISORY SERVICE

At airports offering airport advisory service, information normally is provided on wind, active runway, altimeter setting, and any known traffic. Normally, ATC will instruct the pilot conducting an instrument approach when to contact the flight service station for an airport advisory.

SEARCH AND RESCUE

If a pilot files a VFR flight plan with an air traffic control facility and fails to report by 30 minutes after his estimated time of arrival (15 minutes for a jet), search and rescue operations will be initiated. IFR flight plans are closed automatically when the flight terminates at a controlled field, but the pilot is responsible for closing the flight plan at *all other times.*

FREQUENCIES

Flight service stations have been allotted several VHF communication frequencies, as well as the emergency frequency of 121.50 MHz. Normally, a pilot may contact a flight service station on one or more of the frequencies listed in figure 4-6.

FSS FREQUENCIES

PILOTS TRANSMIT ON	PILOTS RECEIVE ON
*121.50	121.50
+122.00	122.00
122.05	VOR FREQUENCY
122.10	VOR FREQUENCY
†122.20	122.20
122.30	122.30
122.60	122.60
123.60	123.60
123.65	123.65

**Emergency Only*
+Weather Frequency (selected FSS's)
†Standard FSS Frequencies

Fig. 4-6. FSS Frequencies

AIR TRAFFIC CONTROL

ATIS

Automatic terminal information service (ATIS) is provided at busy terminal airports. The information transmitted by this service is broadcast continuously, and includes such information as runways and instrument approaches in use, tower frequencies, surface winds, etc. These broadcasts are updated whenever conditions change, and each new broadcast is labeled with a phonetic letter, such as alpha, bravo, etc. On initial callup to the tower, approach control, or ground control, the pilot should indicate which ATIS information he has received by repeating the phonetic designation. ATIS may be assigned a certain VHF tower frequency or may be broadcast over the voice feature of a navaid near the airport.

CLEARANCE DELIVERY

In order to relieve congestion on the ground control frequencies at large airports, a discrete clearance delivery frequency is provided. Some clearance delivery facilities allow a pilot to receive an IFR clearance 10 minutes prior to taxi.

GROUND CONTROL

Ground control is exercised at tower airports on either the specified tower or ground control frequency. At many IFR airports served by a tower, ground control has the function of delivering clearances to pilots. Ground control frequencies extend from 121.60 through 121.90 MHz; however, 121.90 MHz is the most commonly used frequency.

AIRPORT CONTROL TOWER

The control tower has the responsibility of maintaining a safe, smooth, orderly flow of air traffic within the airport traffic area. Pilots operating on instrument flight plans normally contact the tower when ready for departure, receive their takeoff clearance, and are "handed off" to departure control.

Similarly, an IFR flight usually is handed off from approach control to the tower when crossing the final approach fix inbound for landing. If no approach or departure control facility exists at the airport, the handoffs are made directly between the air route traffic control center and the tower.

APPROACH AND DEPARTURE CONTROL

Air traffic control for arrivals and departures is exercised by approach and departure control, as shown in figure 4-7. Since these facilities coordinate closely with the control tower and the air route traffic control center (ARTCC) in handling air traffic, they act as go-betweens for the pilot when he transitions from the terminal to enroute operations, or vice versa.

Approach and departure control have the primary responsibility of sequencing and separating IFR aircraft. In addition, separation is provided between IFR traffic and known VFR traffic.

Fig. 4-7. Approach and Departure Control

The individual frequencies utilized for approach and departure control are between 118.00 MHz and 135.95 MHz. These frequencies are found in the Airport/Facility Directories and on the various IFR charts. The availability of radar service can also be determined from the same sources.

AIR ROUTE TRAFFIC CONTROL CENTERS

Air route traffic control centers (ARTCCs) are the central authority for issuing IFR clearances and assuring the safe and efficient movement of all enroute IFR traffic. There are about 20 centers in the United States with associated radar and communication facilities. Each of these centers are further divided into sectors to provide adequate communication and radar surveillance for enroute IFR aircraft.

The IFR flight plan filed by the pilot is processed by the ARTCC in which the flight originates. Due to the time involved in checking the route of flight for other air traffic, weather hazards, and issuing the appropriate clearance, the pilot should file his flight plan at least 30 minutes before the intended departure time.

Center frequencies are shown on both Jeppesen and NOS enroute charts in boxes along specific IFR routes. The frequencies for these centers range from 118.00 through 135.975 MHz.

AIRPORT MARKINGS

Airport markings vary from airport to airport. These markings are determined by the type of runway, its direction, and the various services which may be associated with it.

BASIC RUNWAY

The basic runway is used for operations under visual flight rules. Its markings include the runway number and centerline. Runways are numbered to the nearest 10° in relation to magnetic north, with the last digit omitted. For example, a runway with a magnetic heading of 084° is marked with the numeral 8. The opposite end of the runway is marked with the reciprocal of that heading (26 in this case). Three parallel runways are designated by the same number with the addition of an "L," "C," and "R" to indicate the left, center, and right runways, respectively. The centerline marking consists of a white, dashed line, as shown in figure 4-8.

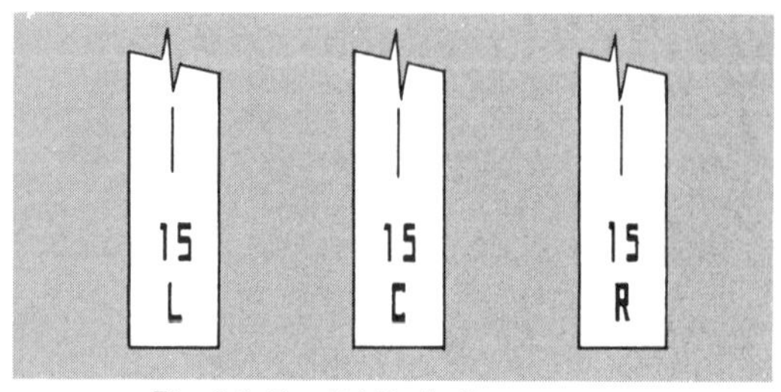

Fig. 4-8. Parallel Basic Runways

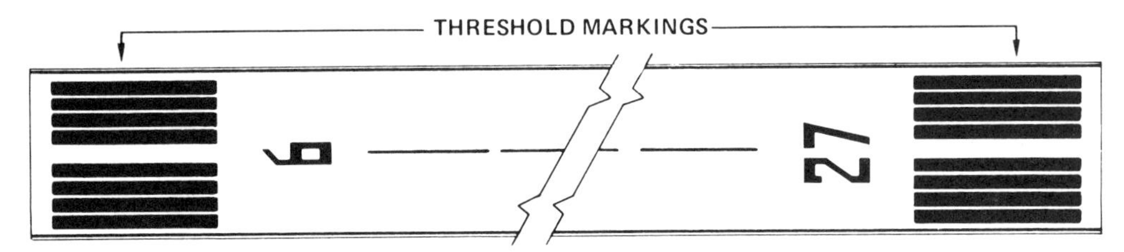

Fig. 4-9. Nonprecision Instrument Runway

NONPRECISION INSTRUMENT RUNWAY

A nonprecision instrument runway incorporates basic runway markings and threshold markings, as shown in figure 4-9. A nonvisual, nonprecision navigation aid (VOR, NDB) serves this type of runway.

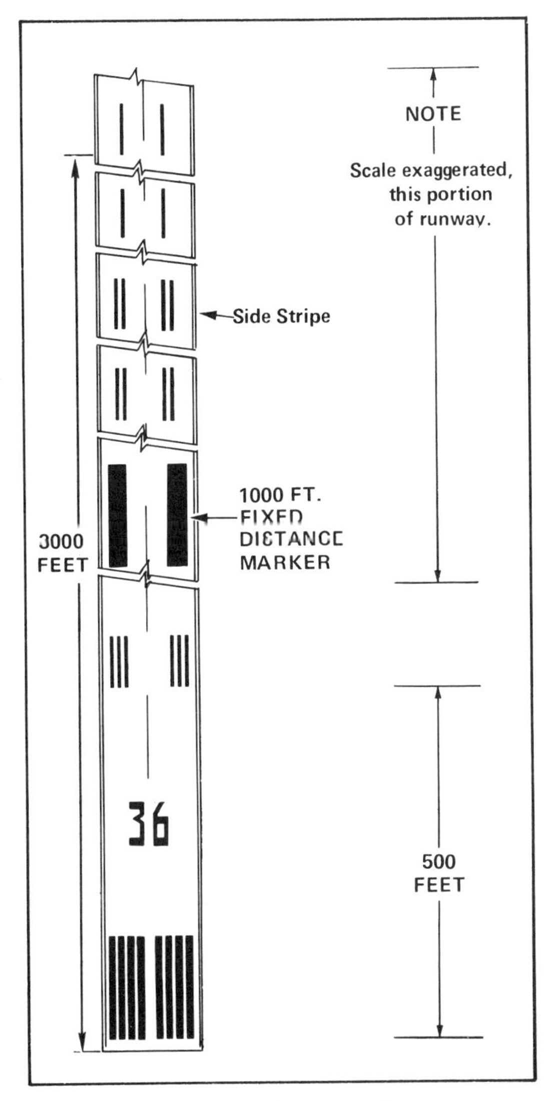

Fig. 4-10. Precision Instrument Runway.

PRECISION INSTRUMENT RUNWAY

Precision instrument runways are served by nonvisual precision approach aids, such as the instrument landing system (ILS) or precision approach radar (PAR). Both provide glide slope information. In addition to the nonprecision instrument runway markings, precision runways incorporate touchdown zone markings, fixed distance markings, and side stripes, as shown in figure 4-10.

SPECIAL PURPOSE AREAS

Closed or overrun/stopway areas are designated as special purpose areas. These terms refer to any surface or area which appears usable, but which, due to the nature of its structure, is unusable as shown in figure 4-11.

The stopway is an area which may be used to decelerate an aircraft in case of an aborted takeoff. This area will support the aircraft, but, due to its structural nature, it is not designed for continuous operational use.

In addition to normal runway markings, short takeoff and landing (STOL) runways have the letters "STOL" painted on the approach end of the runway and a touchdown aim point designated, as shown in figure 4-12. These runways, as the name infers, are to be used for aircraft designed for short takeoffs and landings.

Taxiway centerlines are marked with a continuous yellow line. Taxiway edges are marked with two continuous lines six inches apart. Holding lines consist of two continuous lines and two dashed lines perpendicular to the taxiway centerline. Standard holding lines will appear 50 to 150 feet from the runway edge. Holding lines for runways authorized for Category II operations will be at least 200 feet from the runway edge, as shown in figure 4-13.

NOTE: *A Category II approach is a precision approach that requires specific equipment and authorization and allows descents below normal ILS minimums.*

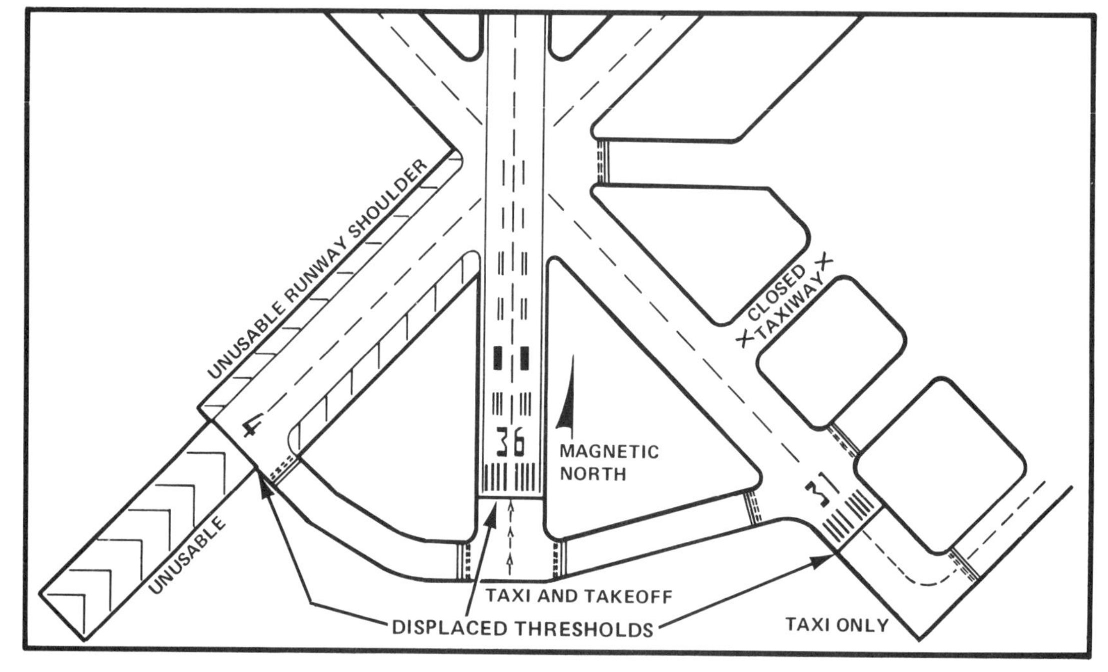

Fig. 4-11. Special Airport Markings

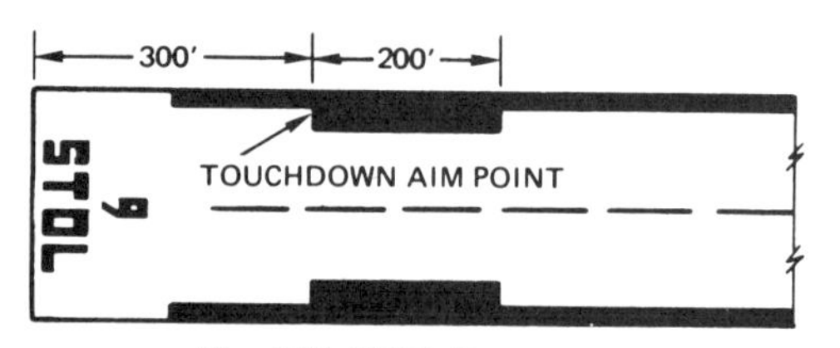

Fig. 4-12. STOL Runways

The Category II holding line is made up of two lines drawn across the taxiway with several sets of double lines drawn parallel to the taxiway, contacting the two lines perpendicular to the taxiway. The Category II holding line need not be observed when Category II approaches are not being made. A standard holding line will always accompany the Category II holding line unless the two are collocated.

DISPLACED THRESHOLDS

A *threshold marking* is a line perpendicular to the runway centerline designating that portion of a runway usable for landing. Displaced threshold markings are used at the beginning of the full-strength runway pavement to show restrictions to the use of the first portion of the runway.

The different types of displaced threshold markings are shown in figure 4-11. The pilot should land beyond the displaced threshold marking

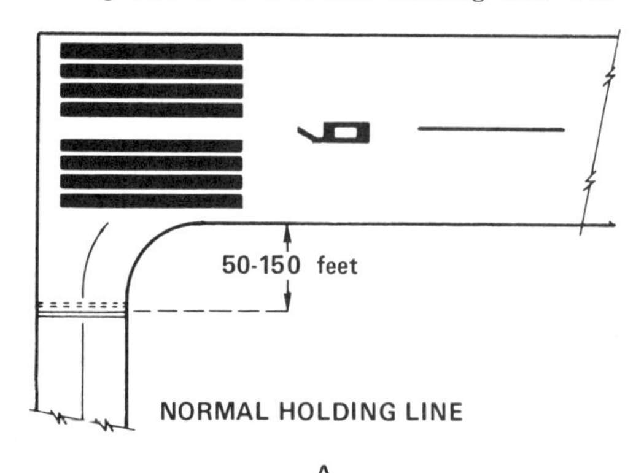

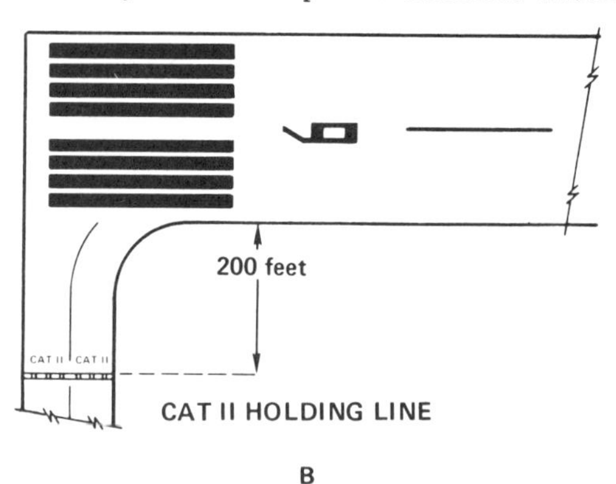

Fig. 4-13. Holding Lines

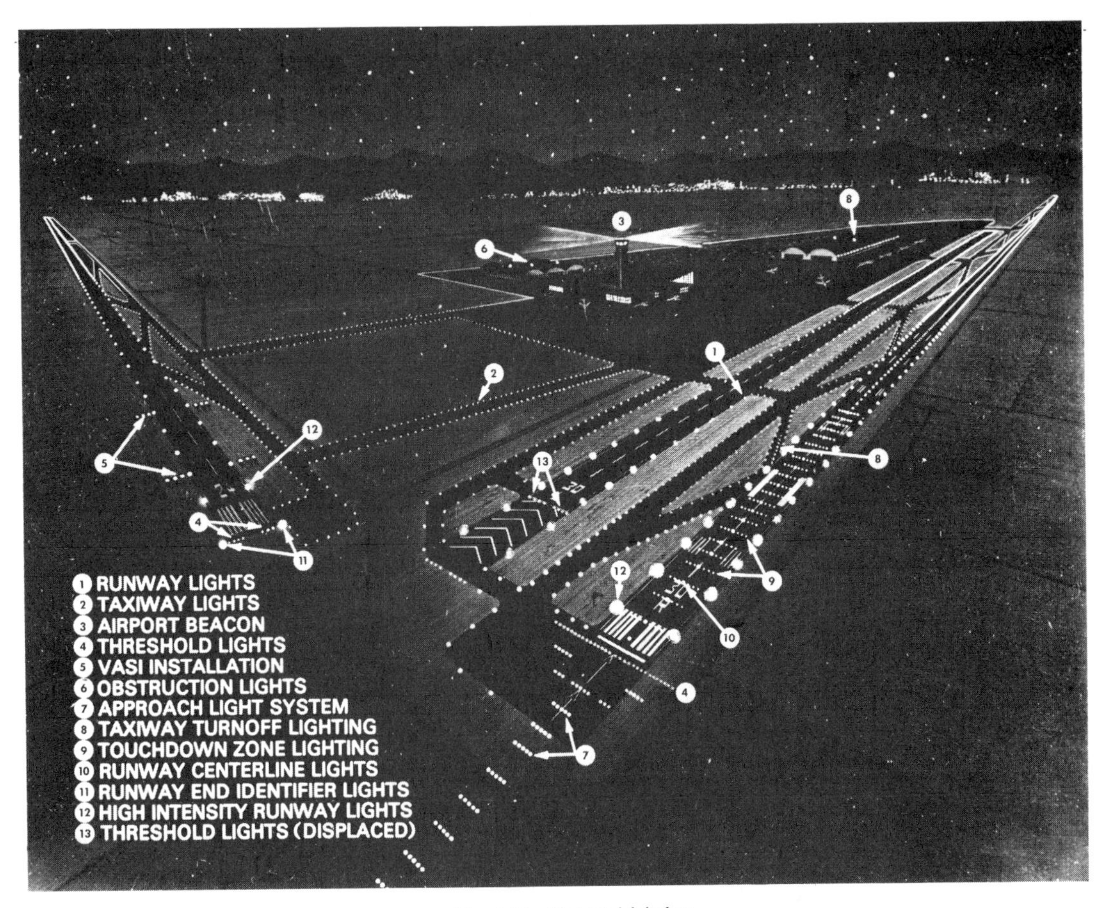

Fig. 4-14. Airport Lighting

because it either denotes the presence of *obstructions* in the aircraft approach path, or that the designated area *may not support* the aircraft's weight. If the displaced threshold area is marked by arrows leading to the full-strength runway, as on runway 36 in figure 4-11, the pilot may use the area for taxiing or takeoff only. Should the area be marked with chevrons, as on runway 4, the area may only be used for blastpad, overrun, or stopway purposes, and may *not* be used for taxiing, takeoff, or landing.

AIRPORT LIGHTING

Lighting and marking systems are employed to help the pilot locate and define the runway and general airport environment. The following discussion refers to the numbered callouts in figure 4-14. Each numbered callout follows the subheading in the remainder of this chapter.

RUNWAY EDGE LIGHT SYSTEMS

The airport runway edge lights are white, and are used to outline the runway during periods of darkness or restricted visibility. These light systems are classified according to the brightness they are capable of producing. The different classifications are high intensity runway lights (HIRL), medium intensity runway lights (MIRL), and low intensity runway lights (LIRL). The HIRL and MIRL systems have variable intensity controls and the LIRL system normally has only one intensity setting.

When the HIRL system is installed on an instrument runway, the last 2,000 feet of runway (viewed from the takeoff direction) is equipped with bidirectional lights. This provides the pilot with *yellow edge lights*, rather than white, to identify the last 2,000 feet of available runway.

TAXIWAY LIGHTS

Taxiway lights are blue to distinguish them from the runway lights. These lights outline the taxi areas and, at some airports, their intensity or brightness can be adjusted.

AIRPORT BEACONS

Airport beacons incorporate color combinations which indicate various types of airports. For example, white and green alternating flashes denote a lighted land airport; while white and yellow flashes inform the pilot that the beacon serves a lighted water airport. Military airport beacons flash *alternating* white and green, but are differentiated from civilian airport beacons by two quick white flashes between the green flashes.

Operation of the airport rotating beacon during daylight hours indicates that the airport is below VFR landing minimums; therefore, the ground visibility in the airport control zone is less than three miles and/or the ceiling is less than 1,000 feet. This means an ATC clearance is required for landings and takeoffs at airports that are surrounded by controlled airspace. During the hours of darkness, instrument or IFR conditions are indicated by flashing lights outlining the tetrahedron or wind tee.

THRESHOLD LIGHTS

Threshold lights mark the beginning of the landing surface at both ends of the runway. For each end of the runway, they consist of four or more green lights, two located on each side of the runway centerline. When used in conjunction with an approach lighting system, additional threshold lights may be added to the basic configurations.

VASI INSTALLATIONS

The most common visual approach slope indicator (VASI) is the two-bar type. When using this system, the pilot should attempt to fly an approach which will result in the far bars indicating red, and the near bars showing white. When the approach is too low, both bars will appear red; when the approach is too high, both bars will appear white.

Some airports serving long-bodied aircraft are equipped with three-bar VASI systems, providing two visual glide paths to the same runway. The first glide path (when approaching the runway) is the same as that provided by a standard two-bar VASI installation. However, the second glide path which must be used by pilots of long-bodied aircraft, is about one-half of a degree higher than the first.

The far bar in a three-bar VASI installation is located approximately 700 feet upwind of the middle bar. The *upwind glide path* also can be used by pilots of small aircraft to avoid the possibility of *wake turbulence* generated on the lower glide path.

At some airports, tri-colored VASI, consisting of a single light unit, projects a three-color visual approach path. A below glide path indication is red, the above glide path indication is amber, and the on glide path indication is green.

OBSTRUCTION LIGHTS

In the event an obstruction is present which may be hazardous to flight operations in the terminal area, a steady red light is found on the obstruction. A flashing red light indicates the marking of an enroute obstruction.

APPROACH LIGHT SYSTEMS

Instrument approach light systems (ALSs) aid the pilot in transitioning from instrument reference to visual flight during the approach to landing. The ALS makes the runway environment more apparent in low visibility conditions. Although there are many lighting systems in operation, only the more common ones will be illustrated in this section.

The most common approach light system in the United States is the U.S. Standard "ALSF-1" lighting system and is shown in figure 4-14. This system is comprised of 30 light bars spaced 100 feet apart and arranged about the extended runway centerline. A distance marker light bar is located at 1,000 feet from the threshold. At this point, sequence flashing lights (if installed) begin and extend outward into the approach zone 2,000 feet, along with the rest of the standard lighting system.

SEQUENCE FLASHING LIGHTS

Condenser-discharge sequence flashing light systems (SFL) are installed in conjunction with the U.S. Standard ALSF-1 and ALSF-2 systems at some large airports. The SFL system consists of a series of brilliant blue-white bursts of light flashing in sequence along the approach lights. From the pilot's viewpoint, the system gives the impression of a ball of light traveling at high speed toward the approach end of the runway.

MEDIUM INTENSITY APPROACH LIGHTS

The *medium intensity approach lighting system* (MALS) consists of seven light bars spaced 200 feet apart along the extended runway centerline. The entire system extends into the approach zone approximately 1,400 feet. When used in conjunction with runway alignment indicator lights (RAIL), the system is abbreviated MALSR. A description of RAIL will follow in this discussion.

SHORT APPROACH LIGHTS

The *U.S. short approach light system* (SALS) consists of the inner 1,500 feet of the U.S. Standard ALSF-1 approach light system. The SALS also may incorporate the sequence flashing lights. When used in conjunction with RAIL, this system is called SALSR.

RUNWAY ALIGNMENT INDICATOR LIGHTS

The *runway alignment indicator light* system (RAIL) consists of a number of sequence flashing lights installed on the extended runway centerline. The lights are installed at approximately 200-foot intervals and provide directional guidance to the runway.

RUNWAY AND TAXIWAY LIGHTING

TAXIWAY TURNOFF LIGHTS

Taxiway turnoff lights, similar to runway centerline lights, generally are flush-mounted and spaced at 50-foot intervals. They define the curved path of an aircraft from a point near the runway centerline to the center of the intersecting taxiway. Taxiway centerline lights further define the taxiway.

TOUCHDOWN ZONE LIGHTING

Touchdown zone (TDZ) lighting is provided to aid the pilot in easily locating the touchdown zone during reduced visibility situations. TDZ lighting consists of a series of white lights flush-mounted in the runway, beginning 75 to 125 feet from the landing threshold and extending 3,000 feet down the runway.

RUNWAY CENTERLINE LIGHTS

Runway centerline lights (CL) are white lights flush-mounted in the runway. These lights are used primarily during takeoff and landing to aid the pilot in remaining in the center of the runway. The centerline lights are spaced at intervals of 50 feet, beginning 75 feet from the landing threshold and extending to within 75 feet of the opposite end of the runway.

The portion of the runway with between 3,000 feet and 1,000 feet remaining is equipped with alternate bidirectional red and white lights. However, all of the lights in the final 1,000 feet of runway are bidirectional and are seen as *red* from the takeoff direction. Therefore, when landing or departing, the centerline lights appear as all white for the first part of the runway, then change to alternate red and white lights when 3,000 feet of runway remains, and finally to all red in the last 1,000 feet of remaining runway.

RUNWAY END IDENTIFIER LIGHTS

Runway end identifier lights (REIL) are installed at larger terminals to provide positive identification of the approach end of the runway. This system is comprised of two synchronized flashing white lights, one on each side of the approach end of the runway. REIL is especially beneficial when used on runways that have many lights in the general airport vicinity.

HIGH INTENSITY RUNWAY LIGHTS

The high intensity runway lights (HIRL) comprise another runway edge lighting system. The HIRL has variable intensity capability and may be adjusted from the control tower.

DISPLACED THRESHOLD LIGHTING

Due to the reduced visibility encountered in night or IFR operations, displaced threshold markings are extremely difficult, if not impossible, to see. It is for this reason that displaced threshold lighting systems are used.

Standard displaced threshold lighting employs a combination of white, green, red, and blue lights to denote the permitted and restricted operations. No landings are permitted short of the *green* displaced threshold lights. Additionally, the *absence* of runway edge lights in this area indicates that no operations are authorized short of the displaced threshold. When *blue* runway edge lights are used in the area short of the displaced threshold lights, the area may be used only for taxi purposes.

The area short of the displaced threshold lights can be used for takeoff purposes when the runway edge lights appear in one of the following combinations.

1. Red runway edge lights located with visible displaced threshold lights, as viewed from the displaced threshold end of the runway
2. White runway edge lights located with no visible threshold lights, as viewed from the runway end opposite the displaced threshold.

ADVISORY CIRCULARS

Advisory circulars are issued by the FAA to provide aviation information of a nonregulatory nature. Although the contents of most advisory circulars are not binding on the public, they contain information and accepted procedures necessary for good operating practices.

Each of the subjects covered is identified by a general subject number, followed by the number for the specific subject matter, as shown in the following list.

00	General
10	Procedural
20	Aircraft
60	Airmen
70	Airspace
90	Air Traffic Control and General Operations
140	Schools and Other Certified Agencies
150	Airports
170	Air Navigational Facilities

When the volume of circulars in a general series warrants a further breakdown, the general number is followed by a slash and a specific subject number. For example, some of the Airports series (150) is issued under the following numbers.

150/4000	Resource Management
150/5000	Airport Planning
150/5200	Airport Safety — General
150/5210	Airport Safety Operations (Recommended Training, Standards, Manning)
150/5220	Airport Safety Equipment and Facilities
150/5230	Airport Ground Safety System

Each circular has a subject number followed by a dash and a sequential number identifying the individual circular (150/4000-1). Changes to circulars have the notation "CHG 1," "CHG 2," etc. after the identification number on pages that have been changed. The date on a revised page reflects the effective date of that change.

The subjects covered by advisory circulars and the availability of each are contained in the Advisory Circular Checklist, which is printed as part of the Federal Register. Generally, this checklist is issued three times each year and contains a listing of all current circulars, those circulars which have been cancelled, and any additions since the last checklist was printed. This checklist can be obtained free of charge by writing to the following address.

Department of Transportation
Publications Section, M443.1
Washington, D.C. 20590

Figure 4-15 shows an excerpt from the Advisory Circular Checklist. This excerpt contains one circular which may be obtained free of charge and one which must be purchased. It should be noted that after the narrative description of AC 61-64, the price and source are listed. This advisory circular costs 40 cents and can be purchased from the Government Printing Office (GPO) or any of the GPO bookstores. The appropriate GPO addresses are listed in the front of the Advisory Circular Checklist.

61-64A Flight Test Guide: Instrument Pilot Helicopter (5-25-77)

Assists the applicant and the instructor in preparing for the flight test for the Instrument Pilot Helicopter Rating. Contains procedures, and maneuvers relevant to the flight test. ($1.30 Supt. Docs.)
SN 050-007-00404-4.

61-65 Part 61 (Revised) Certification: Pilot and Flight Instructors (9-5-73)

Informs pilots and flight instructors of the changes in Part 61, revised January 23, 1973, their effects, and the standards and procedures which will be used in implementing them.

Fig. 4-15. Advisory Circular Checklist Excerpt

The description of AC 61-65 does not list a price; therefore, this advisory circular is free. To request a free advisory circular or to be placed on the FAA's mailing list for future circulars, a request should be sent to the previously listed address for the Department of Transportation. When requesting future circulars, the subject matter desired must be identified by the subject numbers and titles shown in paragraph 3(b) of the Advisory Circular Checklist; for example, 00 General, 10 Procedural, or 20 Aircraft.

5 RADIO NAVIGATION SYSTEMS

VOR NAVIGATION

The Federal Airway System is based primarily upon the very high frequency omnidirectional range (VOR). This extensive system consists of several hundred ground stations that transmit navigation course guidance signals utilized by aircraft in flight.

ADVANTAGES OF VOR NAVIGATION

The VOR navigational system has many advantages for the IFR and VFR pilot.

> *Freedom from interference* — The VOR transmits in the very high frequency range of 108.00 through 117.95 MegaHertz (MHz) and is relatively free from precipitation static and annoying interferences, which are caused by storms or other weather phenomena.
>
> *Extreme accuracy* — A course accuracy of plus or minus one degree is possible when flying the VOR.
>
> *Automatic wind correction* — Wind drift is compensated for automatically by flying with reference with the course deviation indicator.

VOR signals are transmitted on a line-of-sight basis. Any obstacles (mountains, buildings, or terrain features, including the curvature of the earth) block VOR signals and restrict the distance they can be received at a given altitude.

Certain terrain features may produce areas where VOR navigation signals are unusable. These abnormalities are published in the *Airport/Facility Directory* under "Radio Aids to Navigation."

CLASSES OF VOR FACILITIES

TERMINAL VOR

The terminal VOR (TVOR) normally should not be used more than 25 miles from the station or over 12,000 feet mean sea level. They are used primarily as approach navigation aids for airports and are not part of the VOR enroute structure.

LOW ALTITUDE VOR

Next is the low altitude VOR (LVOR), which may be used reliably at 40 nautical miles and up to 18,000 feet. If, because of terrain features, two LVORs of the same frequency are close enough to give an unreliable signal below 18,000 feet, the enroute charts will show a maximum authorized altitude. This altitude is the highest altitude at which stations with identical frequencies will not interfere with each other.

HIGH ALTITUDE VOR

High altitude VORs (HVORs) are frequency protected to 40 nautical miles below 18,000 feet. Above 18,000 feet, they are protected to 130 miles.

VOR COMPONENTS

Figure 5-1 displays the components that make up the aircraft VOR equipment. These components function together to present the pilot with the display information necessary for accurate navigation.

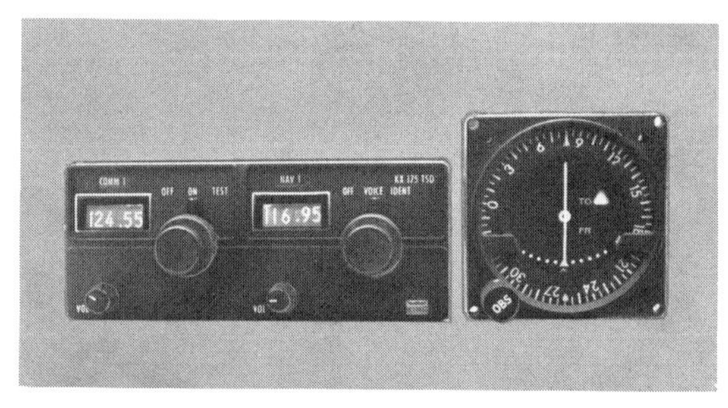

Fig. 5-1. VOR Components

VOR RECEIVER

Frequently, the VOR *receiver* is built in the same case with the VHF communications transceiver. When collocated, the radio is called a nav/com or a one and one-half radio. The radio in figure 5-1 is a nav/com type.

The VOR signals are received by an antenna that normally is located on a vertical stabilizer or on top of the fuselage. This antenna appears to be a "V" that is lying in a horizontal plane. The signals from the antenna are converted by the VOR receiver into the indications that are displayed on the navigation indicator.

NAVIGATION INDICATOR

The VOR *navigation indicator* presents aircraft position information to the pilot. The indicator has three functional components to perform this job — the course selector, the TO-FROM/OFF flag, and the course deviation indicator.

The *course selector*, sometimes called *omnibearing selector* or OBS, rotates the azimuth ring which displays the VOR course selected. This ring may also show the reciprocal of the selected course.

The TO-FROM/OFF flag indicates whether the course selected will take the pilot to or from the station. If the aircraft is out of station range and cannot receive a reliable, usable signal, the TO-FROM/OFF indicator will display OFF. Also, the OFF flag will be displayed when the aircraft is directly over the station or abeam of the station; that is, when crossing a radial which is plus or minus 90° of the course selector setting.

When the aircraft heading is in general agreement with the course selector, the *course deviation indicator* (CDI) shows the pilot his position relative to the course selected and indicates whether the radial is to his right or left. The CDI needle has a 10° spread from center to either side when receiving a VOR signal. Figure 5-2 shows that an aircraft five degrees off course will have the CDI one-half of the way from center to the outside edge. If the aircraft is 10° off course, the needle will be completely to one side. Each dot on the navigation indicator represents two degrees when flying VOR.

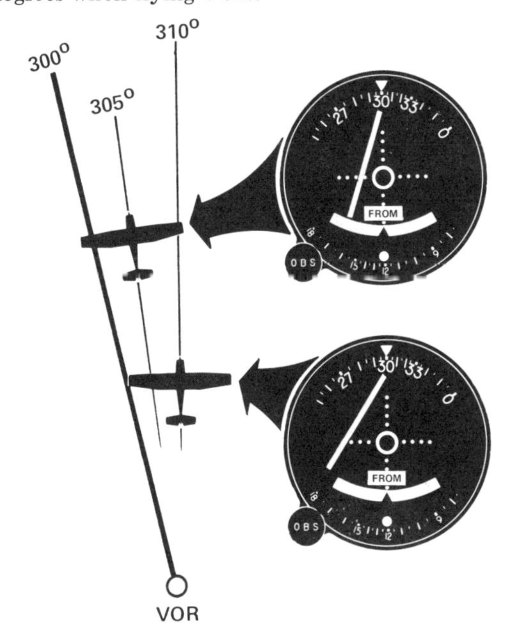

Fig. 5-2. CDI Deflections

VOR ORIENTATION AND NAVIGATION

To effectively use the vast network of federal airways, the pilot must be thoroughly familiar with VOR navigation. Additionally, the instrument pilot should have an understanding of VOR orientation.

COURSE ARROW

Each time a course is selected with the course selector, the area around the VOR station is divided into two halves. It is helpful to think of the dividing line between the two halves or envelopes as a course arrow. The course arrow runs through the station and points in the direction of the selected course. Figure 5-3 illustrates the divisions or segments around a VOR station.

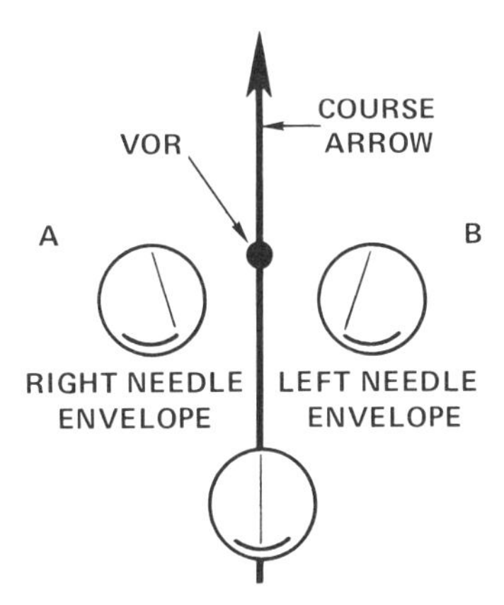

Fig. 5-3. Course Arrow

Once a course is selected, the CDI shows the pilot in which of these two "envelopes" he is located. If the aircraft is located along the course line, the indicator will have a centered CDI needle, as shown by the indicator on the course arrow in figure 5-3. If the aircraft is located at position A, the VOR indicator will display a right CDI indication. Any aircraft on the same side of the course arrow as position A will have a right CDI indication.

The CDI will not move to the other side of the indicator until the aircraft moves into the other area. If an aircraft moves from position A to position B, the indications on the CDI will be the same as B or to the left.

Whenever the pilot changes the course selector, he should visualize an imaginary course arrow placed over the station. In this way, he can look at his CDI and tell in which envelope the aircraft is located.

CROSSBAR

Selecting a course also establishes the position of another line. This line is a crossbar, which runs perpendicular to the course arrow at the station.

The crossbar divides the VOR reception area into two additional envelopes. The area forward of the crossbar is the FROM envelope, and the area aft of the crossbar is the TO envelope. The TO-FROM indicator shows in which envelope the aircraft is located. As shown in figure 5-4, a cross is formed and both aircraft on the same side of the crossbar as the arrowhead of the course arrow will display a FROM indication on the TO-FROM indicator.

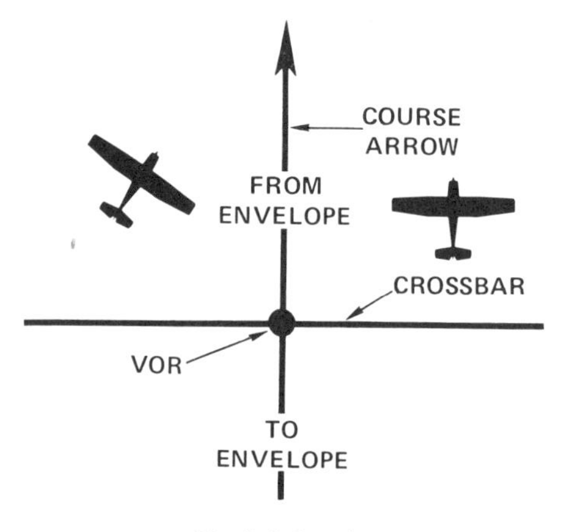

Fig. 5-4. Crossbar

Figure 5-5 shows the indications that aircraft will receive in eight different locations around the VOR station. In position A, the aircraft will show a centered CDI, indicating that it is on course, and the TO-FROM flag will show FROM.

Position B shows a left CDI and a FROM indication. It should be noted that all aircraft located on the arrow side of the crossbar have a FROM indication, and that all aircraft located behind the crossbar have a TO indication.

Aircraft located at positions C and G are over the crossbar and display an OFF indication. This is the area between the TO and FROM sectors in which the VOR indicator will show an OFF flag. In this area, the opposing signals that actuate the TO-FROM indicator cancel each other and produce an OFF indication. A movement across the transition area is indicated by the TO-FROM flag changing from a positive TO or FROM indication to OFF and then to the opposite indication. The farther the aircraft is from the VOR, the longer the OFF flag will be displayed during the TO-FROM transition.

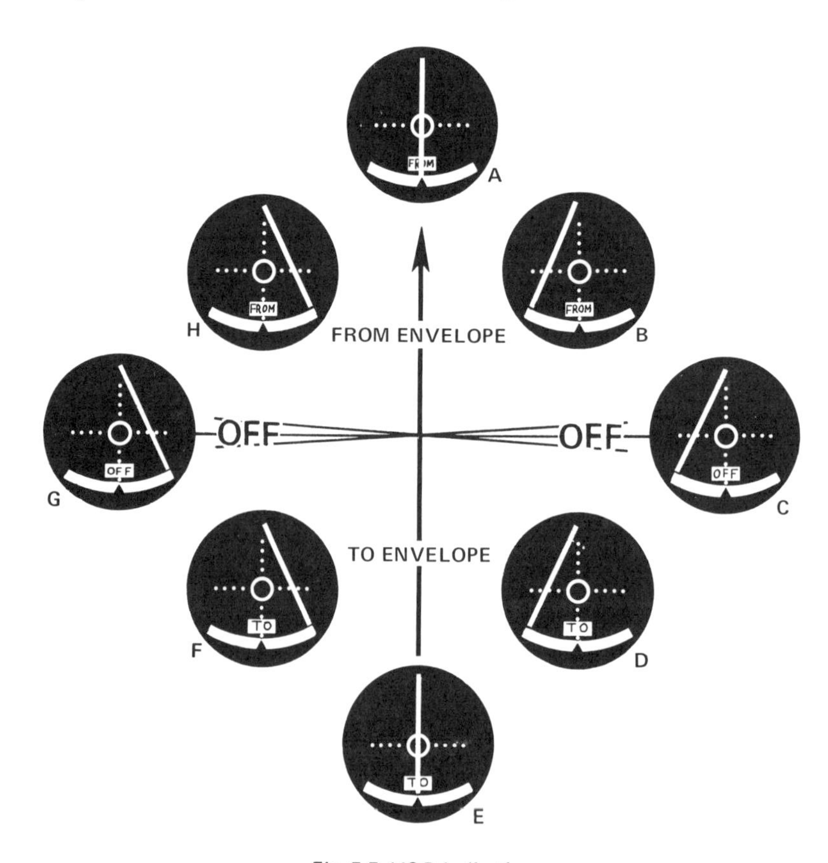

Fig. 5-5. VOR Indications

AIRCRAFT HEADING

Figure 5-6 shows that aircraft heading has absolutely no effect on the VOR indications. Regardless of which direction the aircraft is heading, the pilot will receive the same indication, provided he remains in the same course envelope. This is because the course arrow and crossbar divide the area surrounding the VOR into envelopes or sectors. These sectors are based upon the magnetic course that is indicated on the course selector and not the aircraft heading.

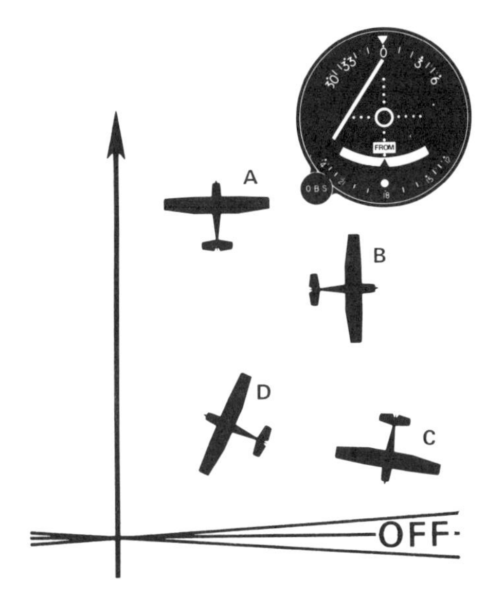

Fig. 5-6. Effect of Heading on VOR Indications

TRACK TO THE STATION

The most common use of VOR navigation is flying on a radial from station to station. The pilot selects the course on the course selector and tracks that course by keeping the CDI needle centered.

By keeping the course selector in general agreement with the heading indicator, the CDI will always point towards the selected course. For example, if the course is to the right, the indicator will point to the right and the pilot must turn in this direction to intercept the course. At this time, it is well to mention that the heading indicator should be checked against the magnetic compass at the beginning of tracking. The VOR indicator only tells the pilot the position of his aircraft along a certain radial; therefore, he must rely upon his heading indicator for aircraft heading information.

As the aircraft passes the VOR station the TO-FROM indicator changes to the opposite reading, such as a TO flag changing to a FROM. When crossing directly over the station the change occurs fairly rapidly. If passing to the side of the station, the change will occur more slowly.

DETERMINING POSITION

Since the VOR gives only direction and not distance from the VOR station, the pilot desiring a position fix must use two stations. By the use of a cross-check with two VOR facilities, the aircraft position can be pinpointed.

The pilot desiring to make a position fix should tune his number one VOR to one of the desired stations and make positive identification. Unless positive identification is made the station should not be used. If a VOR station is shut down for maintenance or the signal is unreliable because of a malfunction, the identification is turned off.

After identifying the station, the pilot should center his CDI needle with a FROM indication on the TO-FROM/OFF flag. At this point, he must determine that he has a positive FROM indication and that there is no indication of an OFF flag.

With the second VOR, the pilot should repeat this procedure using the other VOR station. Then, using his chart, the pilot draws a line outbound from the VORs using the radials indicated by the course selectors. The intersection of these radials will be the aircraft's position.

AUTOMATIC DIRECTION FINDER

One of the major systems of terminal navigation with which pilots should be familiar is the automatic direction finder, or ADF. Airborne ADF equipment consists of a small receiver that receives signals in the low and medium frequency bands (200-1750 kiloHertz), and senses the direction from which the signals are emitted. A needle or pointer on an azimuth dial in the cabin always points *to* the station to which the receiver is tuned. The ADF can be used for limited enroute navigation, position fixes, holding, and instrument approaches.

ADF CHARACTERISTICS

BENEFITS

The ADF system offers two distinct benefits — economical cost of installation and relatively low maintenance. Non-directional beacons (NDBs) are used to provide homing and navigational facilities in terminal areas. Also, the installation of NDBs has enabled smaller airports to acquire an instrument approach which otherwise would not be economically feasible.

The NDB transmits a low frequency signal, which is in the range of 200 to 415 kHz. This signal is not transmitted in a line of sight as with VHF or UHF; rather, it follows the curvature of the earth, providing reception at low altitudes over great distances. In Alaska, Canada, and certain other countries, the ADF is used for primary navigation over long distances.

LIMITATIONS

To obtain maximum utilization from the ADF, the pilot should familiarize himself with the following L/M frequency characteristics and limitations.

1. *Twilight effect* — Radio waves reflected by the ionosphere return to the earth 30 to 60 miles from the station and may cause the ADF pointer to fluctuate. Twilight effect is *most pronounced* during the period just before and just after sunrise and sunset. Generally, the greater the distance from the station, the greater the effect. This effect can be *minimized* by averaging the fluctuation, flying at a higher altitude, or selecting a station with a lower frequency. NDB transmissions on frequencies lower than 350 kHz have very little twilight effect.

2. *Terrain effect* — Mountains or other sharply rising portions of the earth's surface have the ability to reflect radio waves and may have magnetic deposits which can cause indefinite indications. Only strong stations that give definite directional indications should be used.

3. *Shoreline effect* — Shorelines have the ability to refract (bend) low frequency radio waves. The direction of these radio waves may change as they pass from land to water. However, radio waves passing from land to water at an angle greater than 30° will have little or no shoreline effect. Therefore, if the pilot is using a land-based NDB, he should select a bearing (while over water) which crosses the shoreline at an angle greater than 30.

4. *Thunderstorm effect* — In close proximity to electrical storms, the ADF needle will point to the source of lightning flashes rather than to the station being used. Since this electrical activity sends out radio waves which are received by the ADF, the pilot should note the flashes of lightning and not use their indications.

ADF COMPONENTS

The major ADF components are illustrated in figure 5-7. The only component not shown is the sense antenna which, on most light aircraft, is a piece of wire running from an insulator on top of the cabin to the vertical stabilizer.

RECEIVER

Controls on the ADF receiver permit the pilot to tune the desired station and select the *mode* of operation. When the pilot tunes the receiver, he must positively identify the station. This is accomplished by continuous Morse code identifiers, *except* during voice communications. All radio beacons transmit a three-letter identification in Morse code, except the compass locator which transmits a continuous two-letter identifier.

ANTENNAS

The ADF receives signals on both the loop and sense antennas. The loop antenna which is used most often is a small flat antenna without moving parts. Within the antenna are several coils spaced at various angles. The loop antenna senses the direction of the station from the antenna by the strength of the signal on each coil. The sense antenna corrects for the inability of the loop to determine whether the bearing *is to or from the station.* The sense antenna is also used for voice reception.

BEARING INDICATOR

The direction of the station relative to the nose of the aircraft is displayed by the bearing indicator. If the aircraft is flown directly to the station, the bearing indicator will point to zero degrees. It should be emphasized that an ADF with a fixed-card bearing indicator will always represent the nose of the aircraft as zero degrees, and the tail of the aircraft as 180°. In the following discussion, it is assumed the bearing indicator has a fixed card.

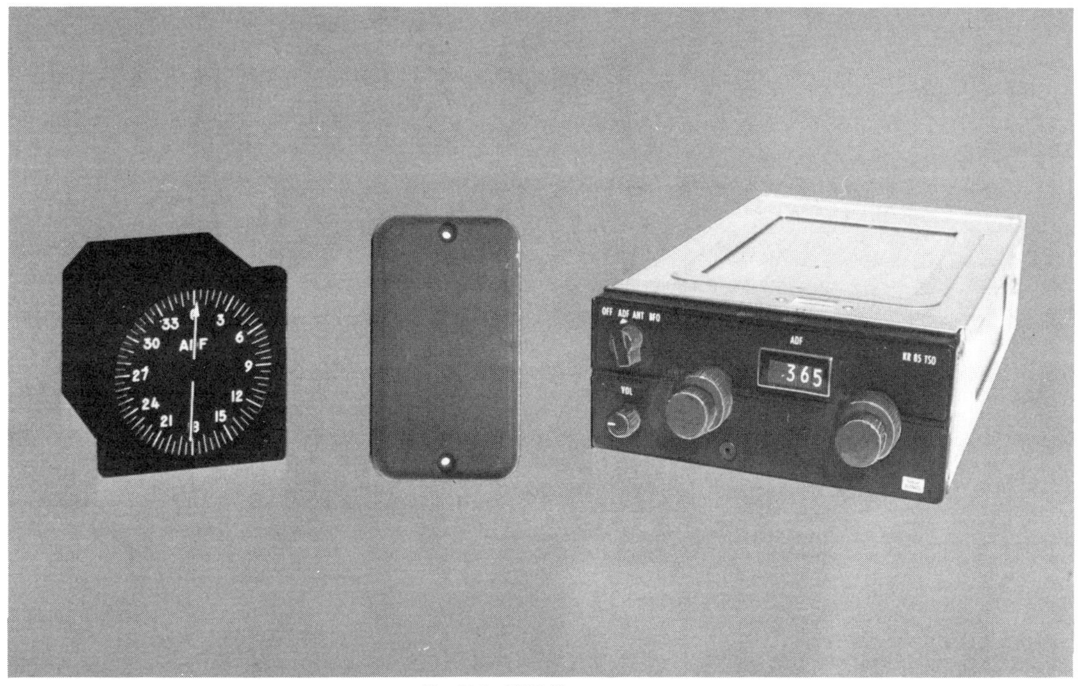

Fig. 5-7. ADF Components

TUNING THE ADF

To tune the ADF, the following steps should be used:

1. Place the function knob in the RECEIVE mode. This turns the set to the mode that provides the best reception. The RECEIVE mode is used for tuning the ADF and for continuous listening when the ADF function is not required.
2. Select the desired frequency band and adjust the volume until background noise is heard.
3. With the tuning controls, tune the desired frequency and then readjust volume for the best listening level and identify the station.
4. To operate the radio as an automatic direction finder, the function knob is switched to ADF.
5. The pointer on the bearing indicator will indicate the bearing to the station in relation to the nose of the aircraft. To check the indicator for proper operation a test switch is employed. By operating the test switch the pointer will move away from the bearing of the selected station. When the switch is released, the pointer will return promptly to the bearing of the selected station. No return, or a return which is sluggish, indicates malfunctioning of the equipment or a signal too weak to utilize.

DEFINITIONS

Relative bearing, as shown in figure 5-8, item 1, is the angle formed by the intersection of a line drawn through the centerline of the aircraft and a line drawn from the aircraft to the radio station. This angle is always measured clockwise from the nose of the aircraft and is indicated directly by the pointer on the bearing indicator.

The *magnetic bearing* illustrated in figure 5-8, item 2, is the angle formed by the intersection

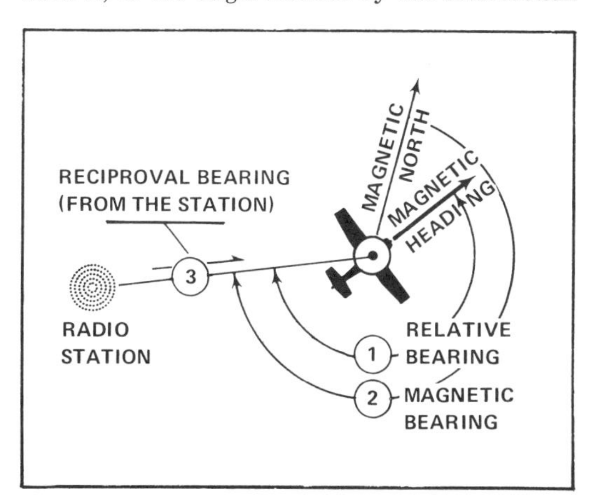

Fig. 5-8. ADF Bearing

of a line drawn from the aircraft to the radio station and a line drawn from the aircraft to magnetic north. A magnetic bearing to the station is obtained by *adding the relative bearing shown on the indicator to the magnetic heading of the aircraft.* For example, if the magnetic heading of the aircraft is 40° and the relative bearing 210°, the magnetic bearing to the station will be 250°.

Reciprocal bearing is the opposite of magnetic bearing and is used when tracking outbound and when plotting fixes. Figure 5-8, item 3, illustrates that a reciprocal bearing is obtained by adding or subtracting 180° from the magnetic bearing. If the bearing is less than 180°, a pilot will find the easiest way to do this is to add 200° and then subtract 20°. Likewise, if the bearing is more than 180° the pilot will add 20° and then subtract 200°. Therefore, in the example shown, the reciprocal bearing would be 070°. By using the "200-20" method the pilot will be able to work these problems quickly and with greater ease.

HOMING

One of the most common uses of the ADF is called "homing to a station." In this procedure, the pilot flies to a station by keeping the bearing indicator needle on zero. The procedures used to home to a station, as shown in figure 5-9, are as follows:

1. Tune the desired frequency and identify the station. Set the *function selector knob* to ADF and note the relative bearing.

2. Turn the aircraft toward the relative bearing until the bearing indicator pointer is at zero degrees.

3. Continue flight to the station by maintaining a relative bearing of zero degrees.

Figure 5-9 illustrates that, if the magnetic heading must be changed to hold the bearing indicator on zero degrees relative bearing, the aircraft is drifting due to a crosswind. If a turn is made only to return to zero degrees relative bearing, crosswind correction is not applied, and the aicraft will fly a curved path to the station. The aircraft in position 2 in figure 5-9 will need to keep changing aircraft heading to maintain the zero degree relative bearing while flying to the station.

TRACKING TO A STATION

However, when flying in a crosswind the pilot must apply a correction. To determine the

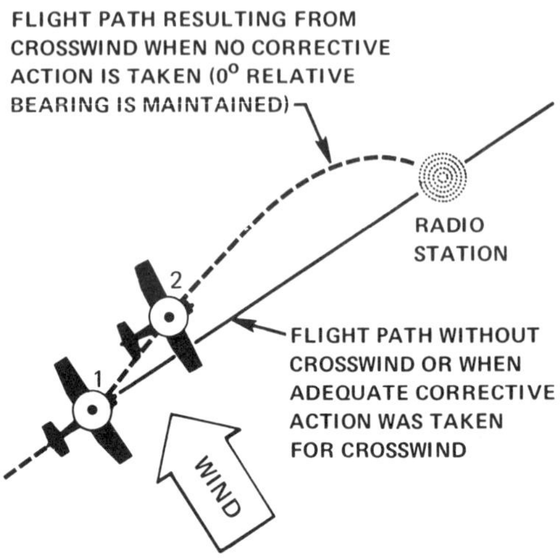

Fig. 5-9. Homing to a Station

amount of correction, the desired magnetic bearing is intercepted and the aircraft is turned until a relative bearing of zero degrees is indicated. While maintaining the *aircraft heading*, drift will become apparent by the change in relative bearing.

As soon as the drift trend is established, a good method of getting back on course is to:

1. Note how many degrees the pointer has moved away from the nose (0° index).

2. Turn into the wind until the pointer is the same number of degrees to the other side of the nose.

3. Maintain that heading until the pointer moves twice as far away from the nose. This shows that the original course has been intercepted.

4. Turn part way toward the station, but leave the nose into the wind by a small amount for drift correction. A good rule is to establish a wind correction angle equal to one half the intercept angle.

It may be necessary to repeat this procedure, on a smaller scale. The pilot must remember that, if the ADF pointer moves toward 0°, the drift correction is too small; if it moves away from 0°, the correction is too great. Once the drift has been stopped, the pointer will stop and the RB will remain constant. In figure 5-10, a 15° wind correction angle has been applied and the airplane is flying inbound on a course of 360°, a relative bearing of 345° and a MH of 015°. The bearing formula can be applied to check whether the aircraft is inbound on the correct bearing.

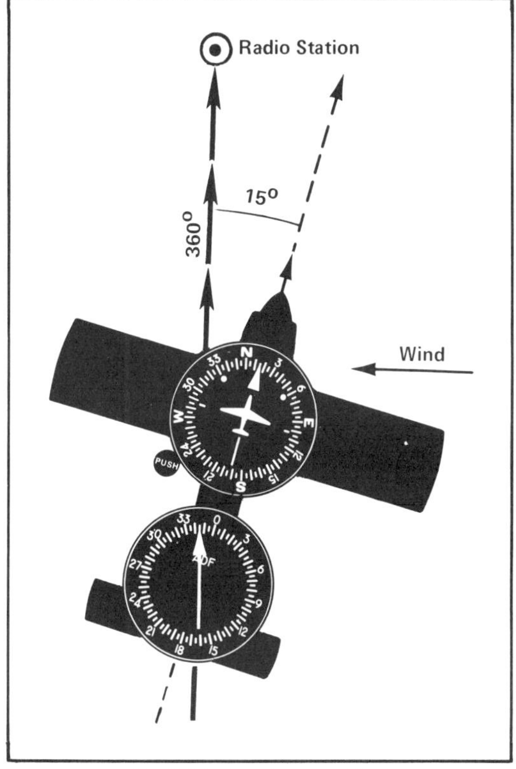

Fig. 5-10. Tracking with Wind

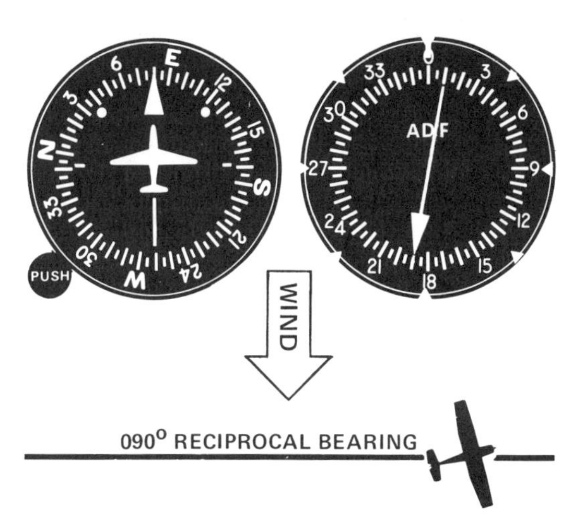

Fig. 5-11. Tracking Outbound

TRACKING FROM A STATION

The principles of tracking a magnetic bearing to a station are also utilized when tracking outbound. Figure 5-11 illustrates an aircraft tracking outbound from a station with a crosswind from the north. The reciprocal bearing is found by subtracting 180° from the magnetic bearing of 270°, equaling 090°. If a 10° wind correction angle is required, the pilot knows he is tracking the desired reciprocal bearing because the heading indicator (080°) and relative bearing (190°) equal the magnetic bearing (270°).

POSITION FIX BY ADF

The ADF receiver can aid the pilot in making a definite position fix by using two or more stations and the process of triangulation. To determine the exact location of the aircraft, proceed as follows:

1. Locate two stations in the vicinity of the aircraft. Note the frequency of each station.

2. With the function selector knob set to RECEIVE, tune in and identify each station.

3. Set the function selector knob to ADF, then note the magnetic heading of the aircraft on the heading indicator. Continue to fly this heading and tune in the stations previously identified, recording the relative bearing for each station.

4. Plot the *reciprocal* of each magnetic bearing on the chart. The aircraft is located at the interception of the bearing lines, as shown in figure 5-12.

RADIO MAGNETIC INDICATOR

The radio magnetic indicator (RMI) is a bearing indicator and a heading indicator incorporated into one instrument. The heading indicator utilizes a "slaved gyro." This term means that that the heading indicator portion of the RMI is connected to a remotely located magnetic compass and is automatically "fed" directional

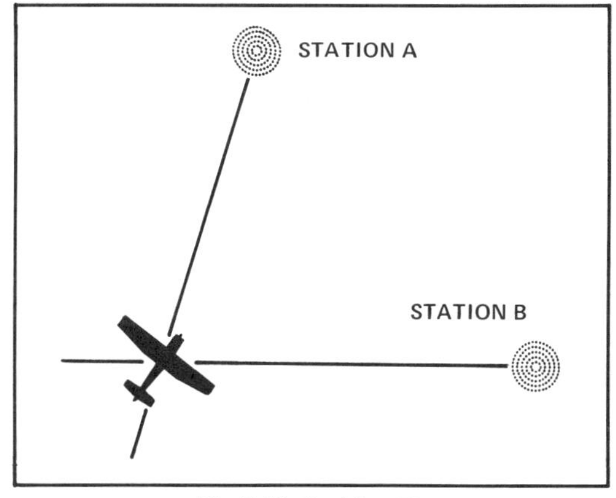

Fig. 5-12. Position Fix

signals. A typical RMI display is shown in figure 5-13. As a result, the heading indicator of the RMI will always read the direction of the aircraft in relation to magnetic north. Therefore, the pointer of the bearing indicator will always display the actual magnetic bearing to the nondirectional beacon. Also, the tail of the pointer indicates the reciprocal bearing. This system lessens the pilot's task and further minimizes the possibility of errors.

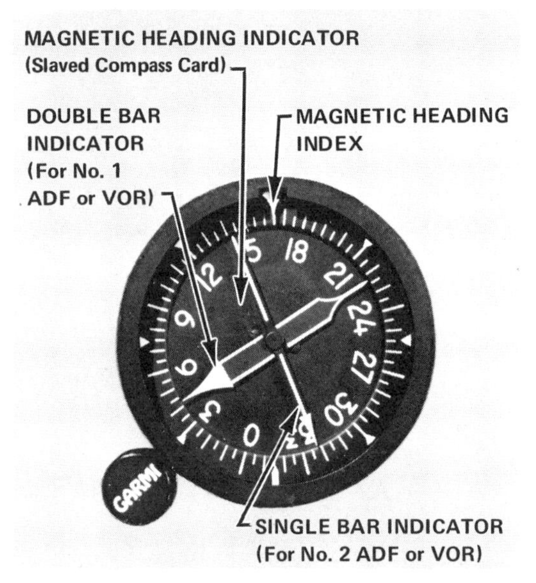

Fig. 5-13. RMI

Tracking to a station is further simplified by the use of the RMI. Instead of referring to two separate instruments, the pilot now *needs to observe only one.* The pilot determines his magnetic heading by looking at the heading index on the azimuth card, and the magnetic bearing to the station by the pointer. The aircraft heading that is used to compensate for wind drift will have no effect on the indication of the magnetic bearing if the airplane remains on the bearing.

DISTANCE MEASURING EQUIPMENT

Distance measuring equipment (DME) consists of ground equipment, usually installed at a VORTAC station, and the airborne equipment installed in the aircraft. The DME provides distance and groundspeed information only from a VORTAC or TACAN ground facility.

The DME in use by the general aviation pilot utilizes the UHF distance signal that is provided by TACAN. For convenience of operation, the FAA has placed most of the TACAN stations on VOR sites. Since there are paired frequencies in use between VORs and TACANs, the pilot using a DME need only tune his radio to the frequency listed for the VORTAC station.

ACCURACY

The DME operates in the ultrahigh frequency (UHF) band and therefore is restricted to line-of-sight transmission. With adequate altitude, the DME signals may be reliably received at distances up to 199 nautical miles with an error of less than one-quarter mile or two percent of the distance, whichever is greater.

DME THEORY

The DME operates by transmitting to and receiving paired pulses from the VORTAC station. The DME transmitter in the aircraft sends out very narrow pulses at a frequency of about 1,000 MHz. These signals are picked up by a receiver at the VORTAC station and trigger a second transmission on a different frequency.

The reply pulses from the VORTAC station are then received by timing circuits in the aircraft's receiver, which measure the elapsed time between transmission and reception. Electronic circuits within the radio convert this measurement to electrical signals which operate the distance and groundspeed indicators.

DME COMPONENTS

The transceiver that sends out the interrogating signal to the VORTAC station incorporates an internal computer to measure the time interval that has passed. The antenna that the DME uses for both transmission and reception is a very small sharks fin antenna that is normally mounted on the underside of the aircraft. The transceiver is shown in figure 5-14.

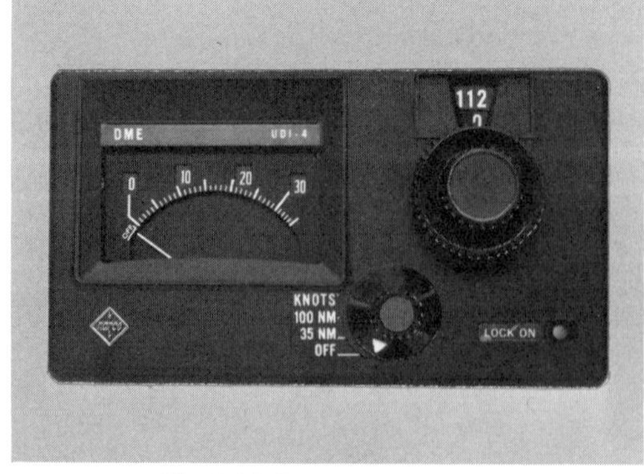

Fig. 5-14. Distance Measuring

The DME displays information in the form of distance to the station and the aircraft's groundspeed. Most DME radios have this display information on the face of the radio. It must be noted that the distance to the station is read as a slant range in nautical miles. For example, if an aircraft is directly over the DME station at 6,076 feet AGL, the distance indicator will read one mile. The DME measures distance to the station as a direct line without regard to aircraft altitude.

DME will give groundspeed readout in knots for the airplane. The groundspeed displayed is accurate only if the aircraft is flying directly to or from the station. Since the DME measures groundspeed by comparing the time lapse between a series of pulses, flight in any direction other than directly to or away from the station will give an unreliable reading. The groundspeed information displayed by the DME provides the pilot with the information necessary to make accurate estimates of time of arrival and accurate checks of his progress along the route of flight.

When the pilot turns the function control knob to groundspeed, he will notice that he does not have an immediate groundspeed readout. This is because of the time needed to make a comparison of the time lapse between a series of pulse signals. The DME must be on the groundspeed function long enough for a comparison of several of these pulse signals to take place.

DME is used to provide position information along the route of flight. A pilot using DME may pinpoint his position using the radial of a VORTAC and the distance information from the same VORTAC, whereas a pilot without DME must use radials from two stations to get a position fix. DME is also used to establish intersections and holding patterns. During holding patterns, the pilot with a DME radio may establish the length of his holding pattern legs by the DME and not by the elapsed time.

Many airports have instrument approach procedures that are based on VOR and DME equipment usage. Normally, an aircraft making this type of instrument approach will have lower minimums when DME is utilized than when only the VOR is used.

AREA NAVIGATION

Area navigation (RNAV) has been developed and designed to provide more lateral freedom and thus more complete use of available airspace. This method of navigation does not require a track directly to or from any radio navigation aid.

The area navigation capability has three principal applications.

1. It permits a direct route between any given departure and arrival point to reduce flight distances and traffic congestion.

2. It permits aircraft to be flown in terminal areas on varied preprogrammed arrival and departure flight paths to assist and expedite traffic flow.

3. It will permit instrument approaches (within certain limitations) at airports not equipped with local instrument landing aids.

There are three RNAV systems presently in use. These are the course-line computer, Doppler radar, and inertial navigation systems. The Doppler radar and inertial navigation systems are not dependent upon ground based navigation aids, but are essentially self-contained units. Since Doppler radar and inertial navigation systems are used primarily aboard airlines and military aircraft, only the course-line computers will be discussed in this text. Course-line computers provide a method of operating on an RNAV basis by utilizing the signals of VORTAC stations.

COURSE-LINE COMPUTERS

The course-line computer (CLC) permits the pilot to set up electronic waypoints (phantom VORTACs) at any location he desires within the reception range of the station. Because the pilot can electronically displace any desired VORTAC station to a different location, he can make his own airways, checkpoints, and airport locators. Also, there are IFR navigation courses and approaches published for use with CLC.

CLC COMPONENTS

In addition to the VHF navigation receiver and the DME in the aircraft, the course-line computer system has the following components. (See Fig. 5-15.)

1. Control panel

2. CLC display

3. Waypoint computer (located behind control panel)

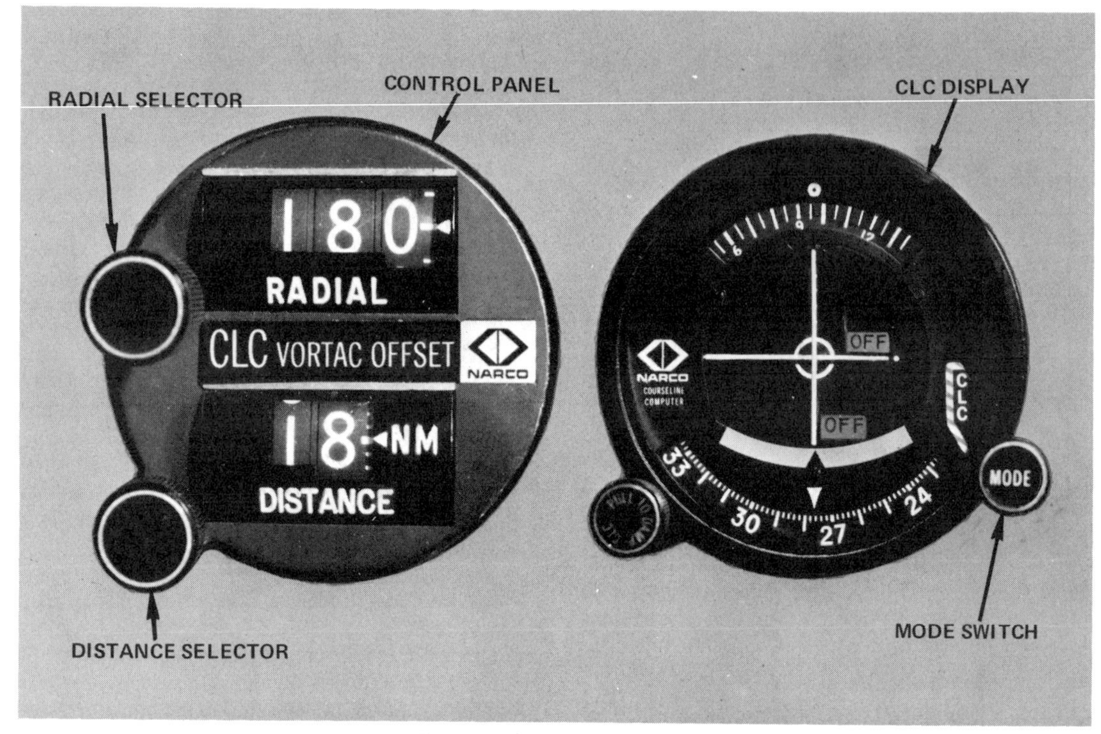

Fig. 5-15. Course Line Computer

PILOT PROCEDURES

The heart of the CLC area navigation system is an analog computer. With two simple settings on the control panel, the analog computer displaces the VORTAC to the desired radial and distance from the original location. It then relays information to the CLC display, providing the pilot the position of the aircraft in relation to this new phantom station.

The CLC display looks much like a standard VOR and ILS indicator; the difference being that it has a mode switch to select CLC or VOR flight. When the CLC display is being used in the VOR mode, the CDI shows displacement from the selected course in degrees. With the CLC mode selected, the CDI displays the nautical miles off course rather than degrees. Each dot displacement of the CDI equals approximately two degrees in the VOR mode and one-half nautical mile in the CLC mode. While using the CLC, the DME indication is in relationship to the new waypoint (phantom VORTAC station) and not the original VORTAC.

To navigate from Belgrade to Haines via the direct route illustrated in figure 5-16, waypoints 1, 2 and 3 must be established.

APPROVED ROUTES

Using course-line computers for VFR navigation affords the pilot great flexibility in that he can determine the routes that he wants and then set up waypoints wherever he desires. But the greater control necessitated in IFR flight caused the development of specific RNAV routes and approach procedures.

RNAV ROUTES

Both NOS and Jeppesen print RNAV enroute high altitude instrument navigation charts for use at and above 18,000 feet MSL. RNAV instrument approach procedure charts also have been prepared for selected terminals and are available from these two sources.

RADAR

Radar is a method of gathering information about distant objects by means of reflected radio waves. When electronically processed, the radio waves can be used to determine the distance and bearing of an object from the radar transmission site. Radar units operated by the Federal Aviation Administration are located

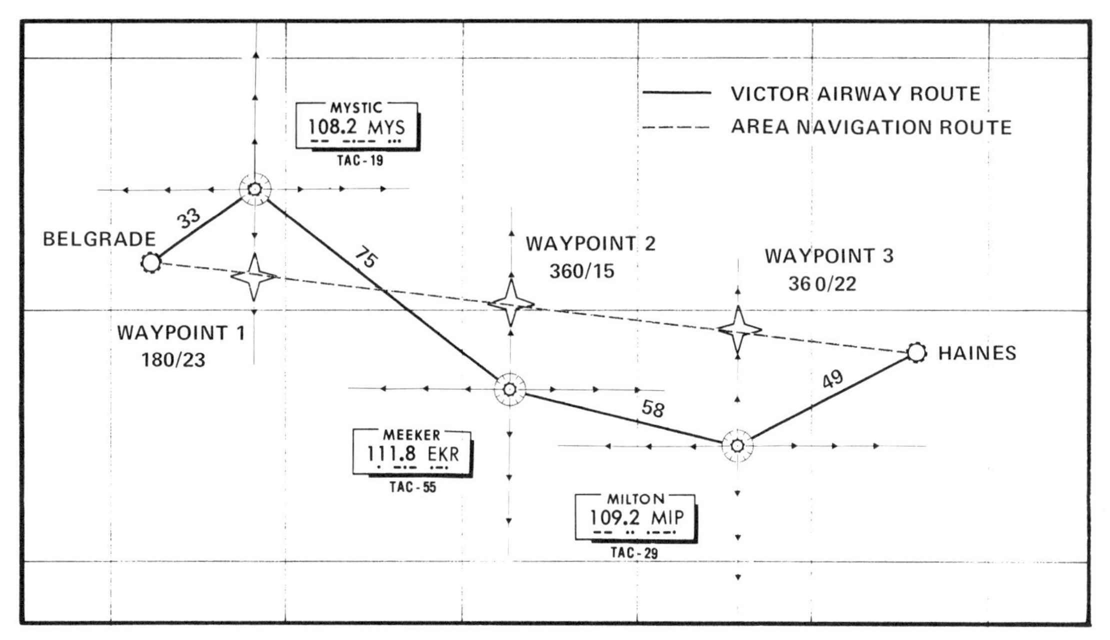

Fig. 5-16. Area Nav Route

throughout the United States and are available to both military and civilian pilots.

Radar utilizes a highly directional rotating antenna to transmit high frequency radio energy. The radio energy emitted by the radar transmitter is either a sequence of short pulses or a continuous wave. The pulses are used for obtaining range (distance) information while the continuous wave transmission is used to measure speed, as in Doppler radar. The radar units to be discussed in this section are those that will be used for obtaining range information.

When a pulse of radio waves is transmitted by a radar unit, it travels from the antenna *at the speed of light*. If an object is within the range of these transmissions, the energy pulse is reflected back from the object to its point of transmission. Generally, the antenna that transmits the pulse waves also receives the reflected signals. The radar receiver then automatically computes the time it took the pulse of energy to travel out and return. Once the time is known, the distance of the object may be electronically computed since a radio impulse travels at a constant rate; thus, the following relationship may be used:

Distance = Rate X Time

When a target is located, the *angle* of the antenna from true north at the time the reflected radio wave is received indicates the azimuth (horizontal direction) of the target.

PRIMARY RADAR

The radar units in operation by the FAA are divided into two categories: primary and secondary radar. Primary radar includes both surveillance and precision radar.

AIRPORT SURVEILLANCE RADAR

Airport surveillance radar (ASR) is designed to provide relatively short range coverage in the general vicinity of an airport and is used to expedite the handling of traffic. ASR is normally used by approach control, departure control, and tower personnel to provide vectors and aircraft separation within the immediate vicinity of the airport. ASR can also be utilized to perform an instrument approach to an airport runway during periods of reduced ceiling and visibility.

AIR ROUTE SURVEILLANCE RADAR

Air route surveillance radar (ARSR) is a long range radar system designed primarily to provide a display of aircraft located in controlled airspace. ARSR is used for air traffic control of cross-country flights. Air traffic control personnel monitor the ARSR scopes to provide guidance, horizontal separation, and other control needed on instrument flights.

PRECISION RADAR

Precision radar is designed for use as a landing approach aid rather than for vectoring and spacing aircraft. Precision approach radar (PAR) equipment may be used as a *primary landing aid*, or it may be used to *monitor* other types of approaches. PAR is designed to display range, azimuth, and elevation information. Two antennas are used in the system — one scanning a vertical plane and the other scanning horizontally.

The radar display scope is divided into two parts. The upper half presents altitude and distance information, and the lower half represents azimuth and distance information. The range of PAR systems is limited to a distance of approximately 10 miles, the azimuth to 20° on each side of centerline, and the elevation to seven degrees above a horizontal plane; therefore, radar coverage is provided only for the final approach area.

SECONDARY SURVEILLANCE RADAR

Secondary surveillance radar is the international term for the air traffic control beacon system. This term refers to the form of radar used only in air traffic control. Secondary radar is a separate system and is capable of independent operation. However, in normal traffic control use, it is interconnected with either the long range air route surveillance radar (ARSR) or the short range airport surveillance radar (ASR). A display of both the primary and secondary radar "targets" is presented on the same radar screen.

Secondary surveillance radar is used to provide greater effectiveness in the airspace system. First, it counteracts some of the shortcomings inherent in primary radar. With secondary radar, aircraft targets will not vary in size as they do with primary radar. Second, the radar display is not degraded by weather conditions, such as precipitation and ground clutter (caused by reflections on the earth's surface), which frequently impair the primary radar display.

Third, the radar beacon transponders transmit definite codes which permit the controller to more easily make positive identification. When only using primary radar, the controller must have the pilot execute a series of turns for positive identification. Fourth, it allows the controller to select aircraft within a certain altitude range or certain area of coverage.

GROUND EQUIPMENT

The ground equipment used with the radar beacon system is divided into three major components: the *interrogator*, *antenna*, and *decoder*. The major function of the interrogator is to trigger the airborne transponder and cause it to reply on the selected code. The interrogations are transmitted on 1030.0 MHz and replies from the airborne equipment are received on 1090.0 MHz.

One section of the interrogator receives replies from the airborne transponder and transmits them to the decoder. The decoder accepts beacon signals from the interrogation unit and displays targets on the radar screen. Interrogations are transmitted and received by a highly directional rotating antenna, such as shown in figure 5-17.

Fig. 5-17. Typical Radar Antenna

AUTOMATIC RADAR TERMINAL SYSTEMS

At certain busy terminal areas around the United States, the new automated radar terminal systems (ARTS-III) are being installed. Computer controlled ARTS-III equipment will automatically provide a continuous display of the aircraft's identification number, actual altitude, groundspeed, and other pertinent information on the radar screen. Computer data processing equipment follows and updates each aircraft position from the time the aircraft enters the terminal approach area until final touchdown.

This system is best used if each aircraft in the area is equipped with a transponder and altitude encoding capability. Sixty-two of these ARTS-III systems are presently being phased into operation at the country's busiest airports.

TRANSPONDER

The transponder is the airborne portion of the air traffic control beacon (secondary radar) system. The coded interrogation signal transmitted by the ground equipment will cause the aircraft transponder to reply automatically and send back a coded signal. This causes the aircraft to appear as a distinctive pattern on the radar controller's scope. Figure 5-18 illustrates a radar scope with the various types of displays which are explained in the following paragraphs. The controller will see transponder-enhanced targets on the scope as blips with a coded signal. All other replies are the normal blips, commonly referred to as "skin paint."

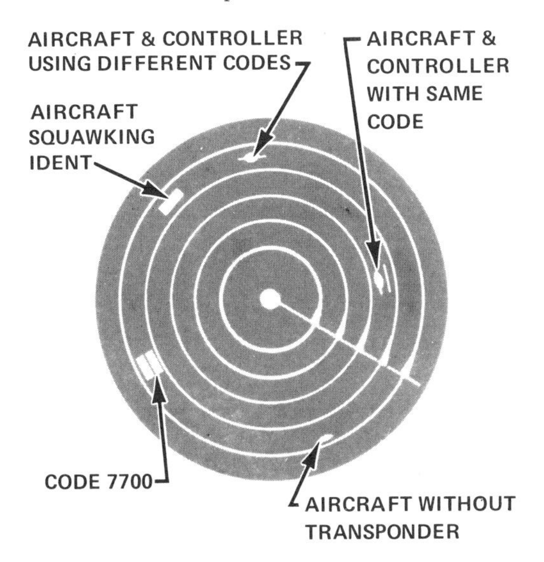

Fig. 5-18. Radar Scope Displays

TRANSPONDER MODES

Most general aviation transponders are equipped for Modes A and A/C operations. Mode A is used by the pilot for most general transponder replies. Mode A/C is for altitude reporting. The altitude reporting program is *being implemented* by the Federal Aviation Administration as the major air terminals are being equipped with ARTS-III capabilities. Altitude reporting equipment is also being installed with the air route traffic control radar equipment.

Most transponders in use today are equipped with the capability of altitude encoding. A transponder equipped with an altitude sensor and encoder will send a signal to the ground radar station that will provide the controller with a visual display of the aircraft's altitude. The display scope of the ground controller will present the aircraft target and altitude.

TRANSPONDER CODES

On the face of the transponder are controls used to select a numerical code requested by the air traffic controller. A typical transponder is shown in figure 5-19. Pilots operating on IFR flight plans will be informed as to the code they must use. For example, the controller might tell the pilot to "Squawk 1100." The term "Squawk 1100" means the pilot should "operate" the transponder on code 1100.

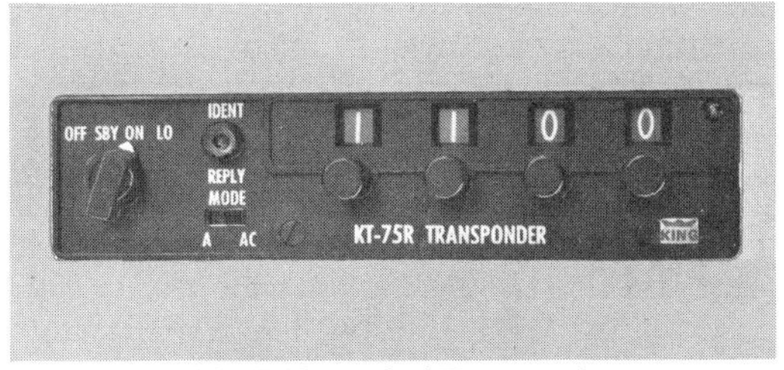

Fig. 5-19. Typical Transponder

When changing transponder codes, pilots should avoid passing through codes 7700 and 0000, which cause a general alarm in ATC facilities. For example, when changing from code 2700 to 7200, first change to 2200 and then to 7200. This will preclude the possibility of passing through the emergency code.

TRANSPONDER CONTROLS

The STANDBY position is used during taxiing to the runway and when the controller instructs the pilot to "Squawk STANDBY." In the STANDBY position, the transponder is warmed up and ready for operation, but it *will not* reply to an interrogating signal. The controller will request STANDBY to eliminate clutter on the scope caused by too many aircraft on the ground or in the immediate vicinity appearing on the screen as distorted blips.

The LO position may be requested by the controller when the aircraft transponder causes the target on the radar scope to extend into a broad arc, making it difficult for the controller to determine the aircraft's actual position. The FAA is now in the process of updating ground facilities with the addition of *side lobe suppression* to eliminate this condition. Because of the side lobe suppression features on ground facilities, some new transponders are being built *without* a LO control position.

NORMAL or ON is the transponder setting normally utilized unless the pilot is instructed otherwise by the controller. On most transponders, the tern ON is used in place of NORMAL. The pilot should switch to ON just prior to takeoff. After completing the landing roll, he should turn the transponder to OFF or STAND-

BY unless instructed to perform this function earlier by the controller.

When asked to "Squawk IDENT," the pilot should press and immediately release the IDENT button. The transponder will send a signal that makes the target presentation "blossom" for a few seconds for positive identification. IDENT should not be "Squawked" unless the controller specifically calls for it.

Most transponders also have a small reply or monitor light. This light blinks each time the transponder replies to a ground interrogation signal, thus informing the pilot the transponder is operating. Many transponders incorporate a *test switch* to test their circuitry; when the pilot pushes the test switch, a signal is sent through the transponder circuitry to evaluate the unit for correct operation. If the transponder is operating properly, the reply monitor light will flash indicating correct operation.

Many major terminal areas have the ARTS III radar system and all air route traffic control centers have the NAS Stage A systems. In both of these, altitude, along with other information is continuously displayed to the controller in a format as shown in figure 5-20. The display of altitude information greatly reduces controller workload and, most importantly, frequency congestion.

When the transponder is coupled to an encoding altimeter, selection of the ALT position (Mode C) will provide ATC with the aircraft altitude. The controller may ask the pilot to verify the accuracy of this information by saying *"verify (altitude his scope shows)"*. In this case, the pilot reports his present altitude.

If the selected altitude is within 300 feet of that displayed on the radar scope, the controller may use the information to provide vertical separation. When a discrepancy of more than 300 feet is noted, he will instruct the pilot to *"stop altitude squawk, altitude differs (number of) feet."*

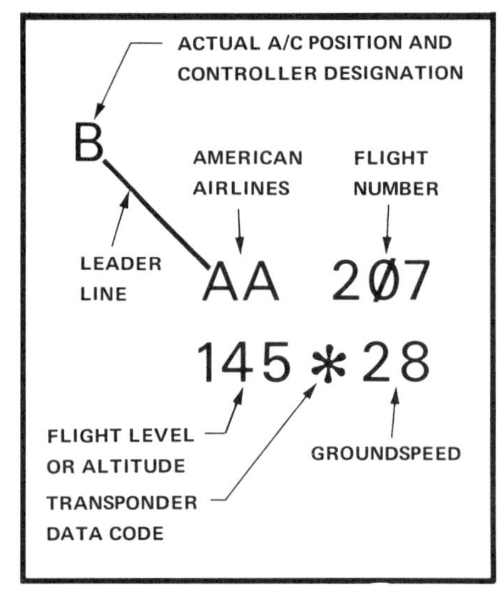

Fig. 5-20 Computer Data Radar Display

INSTRUMENT NAVIGATION CHARTS

6

There are a number of charts used for instrument navigation. In general, there are two sources for these charts — Jeppesen & Co. and the National Ocean Survey. The basic charts used for navigation are the low and high altitude enroute charts, airport diagram charts, area charts, standard instrument departure charts (SIDs), and standard terminal arrival routes (STARs). This chapter is designed to explain each type of chart (both Jeppesen and NOS), its use, and each symbol of the format. Instrument approach charts have been omitted from this chapter and will be reviewed fully in Chapter 7.

LOW ALTITUDE ENROUTE CHARTS

The discussion and comparison of chart symbology will begin with the low altitude enroute charts. Most of the symbols found on the enroute charts are identical to those found on terminal charts; therefore, if the pilot understands the symbology, terminal chart analysis is less complicated.

VOR FACILITY SYMBOLS

The VOR symbol and the location of the station, as depicted on the Jeppesen charts, is shown by a dot in the center of the compass rose which is oriented to magnetic north. On NOS charts, the compass rose is larger and the center is a hexagon which shows where the station is located. These two symbols are shown in figure 6-1.

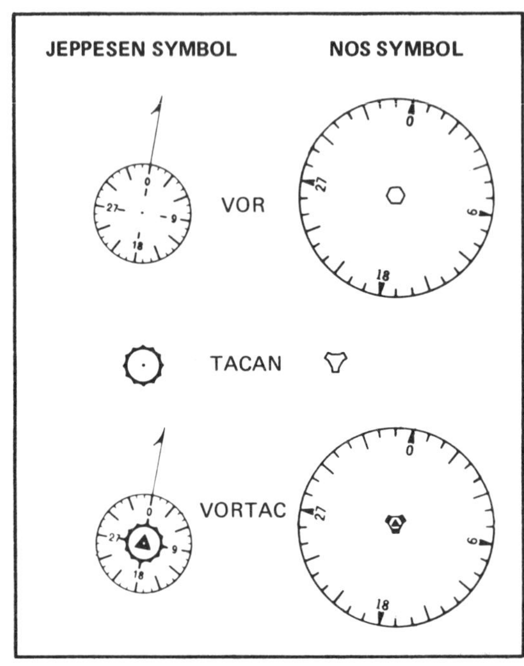

Fig. 6-1. VOR Symbols

If the facility is a TACAN facility, the Jeppesen charts show a circle with a serrated edge. On NOS charts, the symbol resembles a triangle with flat points. TACAN stations are normally, but not always, collocated with VOR facilities. The TACAN symbol is also shown in figure 6-1. When VOR and TACAN are combined, the facility is known as a VORTAC.

NAVIGATION FACILITY BOX

With reference to figure 6-2, information inside the facility box includes the name of the facility, the frequency on which it transmits, and the identification letters in type and in Morse code. On Jeppesen charts, the "D" located in the left side of the box denotes DME available; therefore, the station must be a VORTAC (VOR and TACAN) or VOR-DME facility. On NOS charts, the TACAN capability is shown by indicating the channel number.

The air-to-ground communication frequencies are found above the facility box. Jeppesen charts omit the "1" and "2" before the frequency number, but NOS charts give the complete frequency. Also, on Jeppesen charts, the frequency 122.2 MHz is shown, but is omitted by NOS because it is a standard frequency. The "G" indicates the facility "guards" (listens) on the frequency of 122.1. The NOS symbol is "R" (receives) and indicates the same as "G."

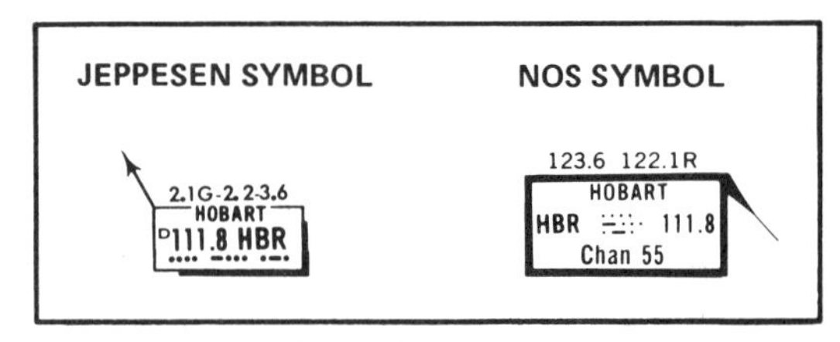

Fig. 6-2. Navigation Facility Box

From the information given in the facility box, it is determined that the navigation frequency is 111.8, it has DME facilities, and the communication frequencies are 122.1G (or "R"), 122.2, 122.6, and 123.6.

VICTOR AIRWAY INFORMATION

The designated routes between two VOR or VORTAC stations are called Victor airways. These airways are formed by the VOR facilities and are designated as a radial outbound from the VOR station. This radial is the angle measured

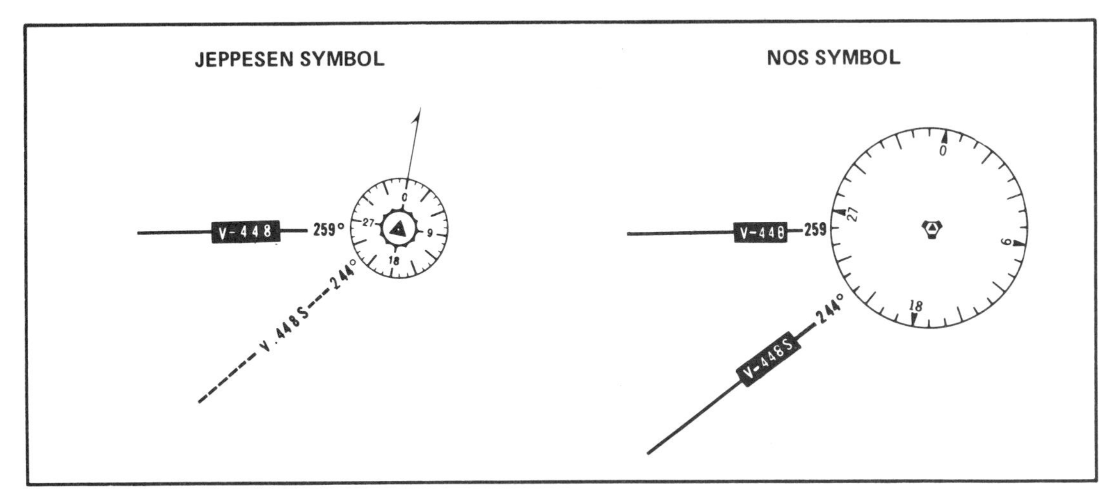

Fig. 6-3. Victor Airways

from *magnetic north*, clockwise to the airway. As shown in figure 6-3 the outbound radial in this example is 259°.

Victor airways are marked with a number that indicates their general direction. Generally speaking, even-numbered airways move traffic in an east-west direction while odd-numbered airways move traffic in a north-south direction. As shown in figure 6-3, V-448 is an east-west airway.

There are two types of airways found on enroute charts — one is the primary airway and the other is an alternate. The alternate airways are marked by the addition of a letter suffix that denotes the general position of the alternate in relation to the primary airway. In the example, V-448S is south of V-448. Additionally, on Jeppesen charts, alternate airways are shown with a dashed centerline.

Alternate airways are used to provide a means of routing aircraft around areas of heavy traffic. These alternates normally are offset 15° from the primary airway.

Victor airways are normally four nautical miles wide from each side of the centerline, thus providing a total airway width of eight nautical miles. This concept is valid until a position 51 nautical miles from the navigational aid. When the airway extends beyond 51 miles from the nearest navaid, it includes the airspace between lines diverging at angles of 4.5° from the centerline of each airway and extending until they intersect the diverging lines of the other navigational aid, as shown in figure 6-4.

The mileages on the enroute chart are expressed in nautical miles. (See Fig. 6-5.) The mileage number *enclosed* within an outlined box indicates mileage between VOR facilities on Jeppesen charts; while on NOS charts, it is a mileage

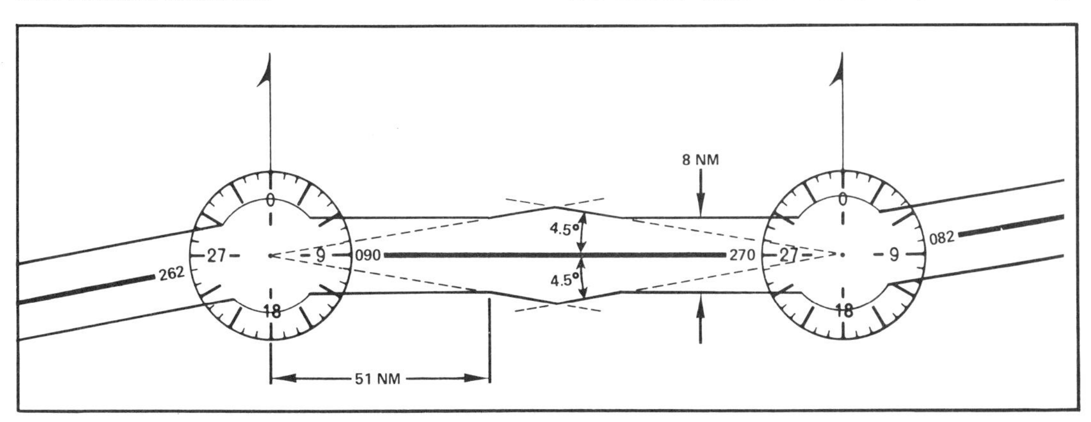

Fig. 6-4. Width of Victor Airways

between VORs or, in some cases, between a VOR and a compulsory reporting point. A mileage number that is *not enclosed* in an outlined box indicates the mileage between intersections, VORs, and mileage breakdown points.

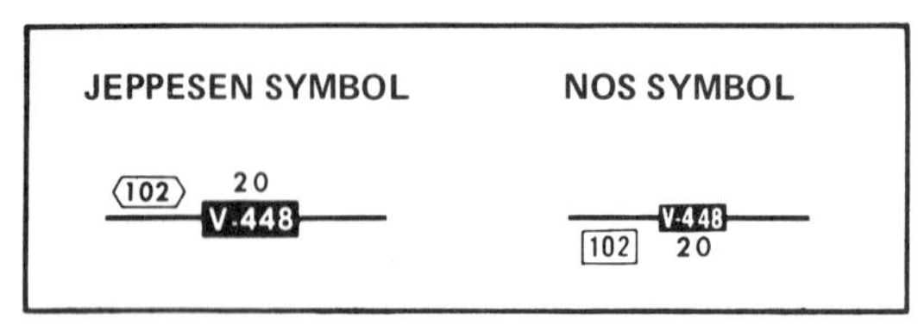

Fig. 6-5. Airway Mileages

A mileage breakdown point is shown on the enroute chart by a small "x" on the airway. This symbol is depicted in figure 6-6, and is found when an airway turns at a place where no intersection name is provided. The small "x" must not be confused with the "X" enclosed in a flag on NOS charts. The flag-enclosed "X" on NOS charts signifies a minimum crossing altitude which will be explained later in the text.

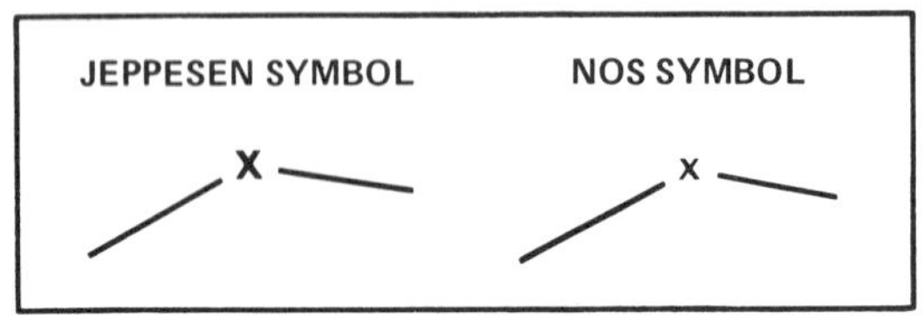

Fig. 6-6. Mileage Breakdown Point

Intersections are checkpoints on airways which provide ATC with definite positions to check the progress of aircraft. The symbol is shown in figure 6-7. They are also located at points where the airway turns and the pilot needs a means to establish his turning point. For ease of identification, intersections are generally named after cities and towns that are found in close proximity.

The actual location of the intersection is generally based on the intersection of two VOR radials; however, other navaids may be used, such as a localizer or nondirectional beacon. If there is any doubt concerning which navaid forms the intersection, a closed arrow is placed between the intersection with the stem toward the navaid forming the intersection. In figure 6-7, the Olney Intersection is established by the 299° radial of the Millsap VOR and the 191° radial of the Wichita Falls VORTAC.

An intersection that is defined by DME is depicted on the Jeppesen charts by a closed

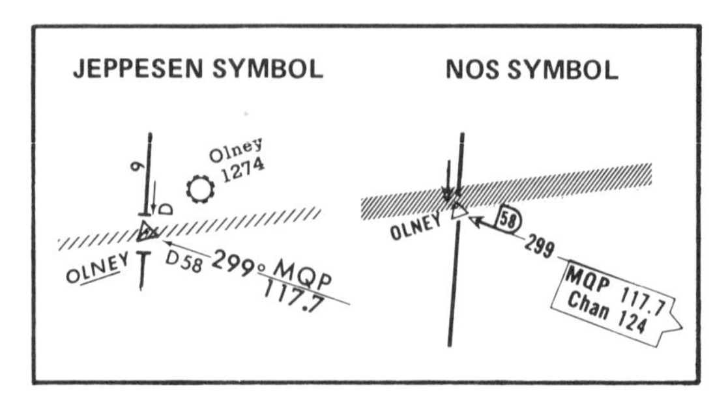

Fig. 6-7. Victor Airway Intersection

arrow and the letter "D." On NOS charts, it is shown by an open arrow. These arrows are located near, and point to, the intersection. The mileage from the navaid to the first intersection is found along the airway and is the standard mileage number. (See Fig. 6-7.)

The mileage along a given airway to the intersection is shown by an arrow and the letter "D," followed by the mileage on Jeppesen charts. On NOS charts, the mileage is enclosed in the letter "D" with an arrow attached. Mileages to these intersections are indicated in this fashion so the pilot will not have to search out and add up mileages between intersections.

All intersections, like airway navaids, may be used as reporting points. These reporting points are divided into two types — first, the *compulsory* reporting point that requires a position report and, second, the *noncompulsory* reporting point where position reports are made only upon request from air traffic control (ATC).

Compulsory reporting points are identified by solid triangles. As a pilot passes over a compulsory reporting point, he must furnish ATC with a position report; however, the position report need not be made over a compulsory reporting point *if ATC advises the pilot that he is in radar contact.* The pilot must resume normal position reporting if ATC advises *"radar contact lost"* or *"radar service terminated."*

Noncompulsory reporting points are identified by an open triangle. These reporting points can become compulsory if the ATC controller so specifies in the clearance. Additionally, as the pilot proceeds on this flight, ATC may request that a position report be made at any specified point.

CONTROL ZONES

Control zones that are depicted on a chart actually may be in effect less than 24 hours a day. Those with specific hours of operation are noted on the chart in the vicinity of the control zone. Hours of operation are depicted on both Jeppesen and NOS charts, as shown in figure 6-8.

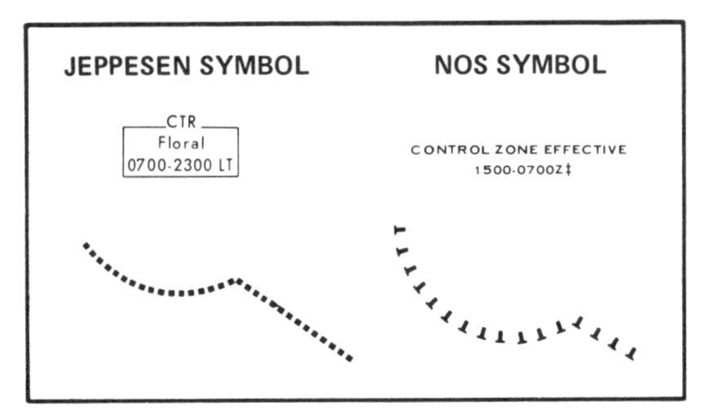

Fig. 6-8. Control Zone Symbols

Certain control zones do not authorize special VFR operations. These are denoted on Jeppesen charts as heavy, dashed lines forming the control boundaries. On NOS charts, this situation is shown with a series of "T's" forming the control zone. Both are shown in figure 6-8.

NONDIRECTIONAL RADIO BEACONS

Figure 6-9 shows the Jeppesen and NOS enroute chart symbols for nondirectional beacons. The nondirectional radio beacon (NDB) is a low-frequency navaid which may be used for navigation and instrument approaches. NDBs operate within a frequency band of 200 to 415 kHz, and transmit a continuous signal to provide identification, except during voice communications. When the NDB is used as a navaid collocated with the outer marker in the instrument landing system, it is called a compass locator (LOM).

The name, frequency, and identification code of an NDB are provided, as indicated by figure 6-9. On NOS charts, an underlined frequency indicates that voice communications are not available. On the Jeppesen charts, when the NDB is used in conjunction with an instrument landing system, the ILS information is included with the NDB information.

FAN MARKERS

Fan markers are beacons that broadcast a continuous signal on the frequency of 75 MHz. The symbols for fan markers on the Jeppesen and NOS charts are shown in figure 6-10.

Originally, fan markers were used as distance markers with the low frequency airway system

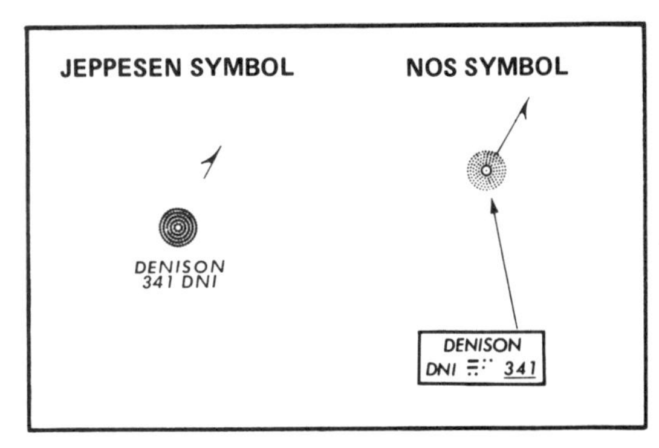

Fig. 6-9. Nondirectional Radio Beacons

which has now been phased out. However, there are still a few fan markers in use around the country on Victor airways.

The outer marker (OM) and the middle marker (MM) beacons use the same symbols except that they are smaller. They also broadcast on 75 MHz, and are used as position fixes during instrument approaches.

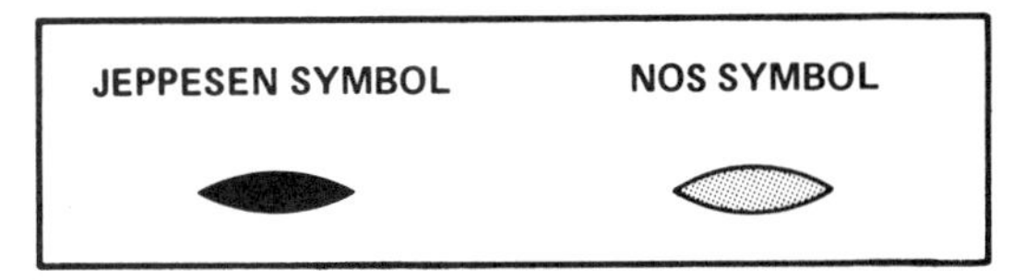

Fig. 6-10. Fan Markers

INSTRUMENT LANDING SYSTEMS

If an instrument landing system (ILS) is available at an airport, it is depicted on the enroute chart by a localizer symbol, as shown in figure 6-11. On NOS charts, a *miniature* symbol is utilized if the localizer is not used to form an enroute fix or airway. Those localizers that are used as fixes for an enroute portion of the airway are depicted on NOS charts by a *large* localizer symbol, feathered on the left side.

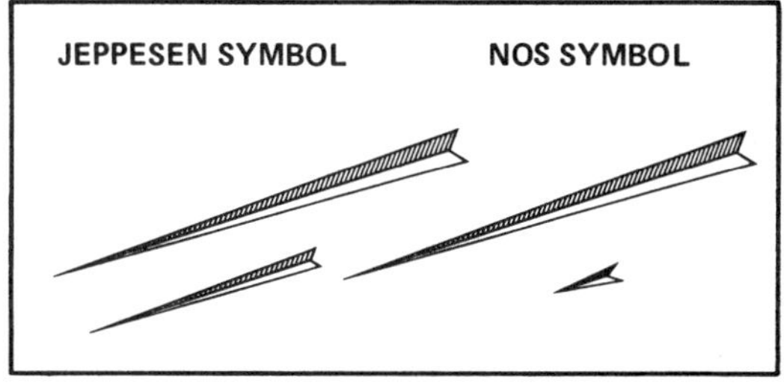

Fig. 6-11. ILS Symbols

On Jeppesen charts, the frequencies of all ILS systems are printed in the associated facility boxes with rounded corners. On NOS charts, the frequencies are provided only for those ILS systems used to form an enroute fix or airway.

COMMERCIAL BROADCAST STATIONS

Certain commercial broadcast stations and associated transmitter site positions are depicted on both Jeppesen and NOS charts. These are stations that can be used as a navigational aid by an aircraft equipped with an ADF receiver. Since commercial broadcast stations do not transmit a continuous identification, they are more difficult to identify. The symbols on each chart are depicted in figure 6-12.

JEPPESEN SYMBOL	NOS SYMBOL
KYU 1190	KYU 1190

Fig. 6-12. Commercial Broadcast Stations

HOLDING PATTERNS

A holding pattern symbol, printed on a navigational aid, depicts the preferred holding pattern to be used at that point. When the pilot receives the holding clearance, the controller may or may not specify the radial on which he is to hold. If the clearance is simply to "*hold*," the pilot should refer to the enroute chart and, if the holding pattern is shown, enter that pattern. If no holding pattern is depicted, the pilot should hold on the inbound radial with standard right-hand turns. Holding pattern symbols are shown in figure 6-13, and a detailed explanation of the procedure is found later in this text.

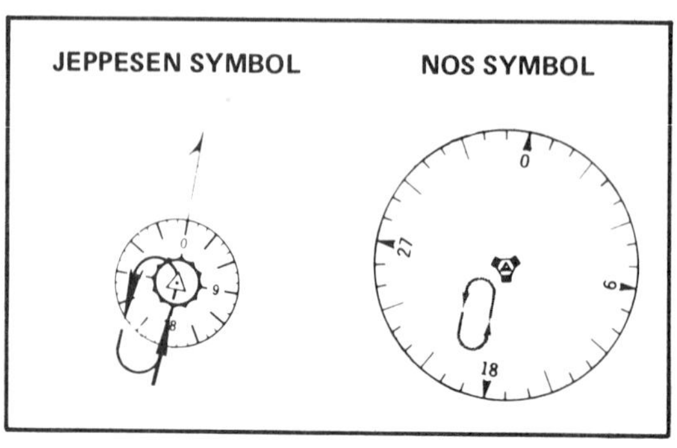

Fig. 6-13. Holding Pattern

MINIMUM ENROUTE ALTITUDES

The minimum enroute altitude (MEA) is usually the lowest altitude that a pilot can fly during a cross-country flight. While planning an IFR cross-country trip he normally *should choose* an altitude above the MEA to correspond to the hemispherical cruising rule. On an easterly course, he should fly at an odd-thousand foot altitude; on a westerly course, he should choose an even-thousand foot altitude.

The MEA guarantees 1,000 feet of clearance above the highest obstacle within five statute miles of the centerline of the airway in non-mountainous terrain. In mountainous terrain, the MEA provides an obstacle clearance criteria of 2,000 feet above the highest obstacle within five statute miles of the centerline. MEA obstruction clearance is illustrated in figure 6-14.

In addition to obstruction clearance, the minimum enroute altitude also guarantees reception

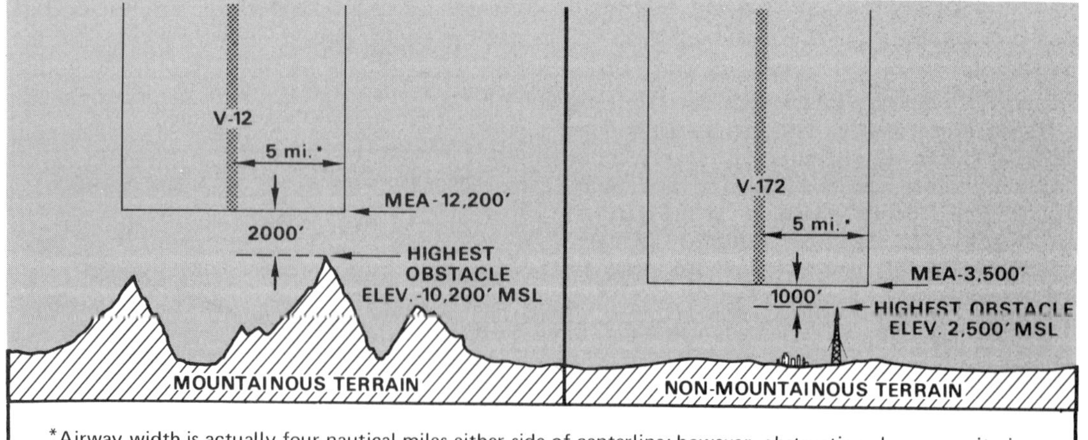

*Airway width is actually four nautical miles either side of centerline; however, obstruction clearance criteria is provided five statute miles on either side of airway centerline.

Fig. 6-14.

of a navigation signal at any point along the airway. The pilot flying at or above this altitude is being assured that the proper terrain clearance and usable navigation signals are available.

As depicted in figure 6-15, on Jeppesen charts the MEA is found below the airway centerline. It is the first of the two altitudes listed if two are given. If only one altitude is noted, it will be the MEA. On NOS charts, the MEA is shown above the airway centerline. Again, it is the first of the two altitudes listed or, if only one is shown, it is the MEA.

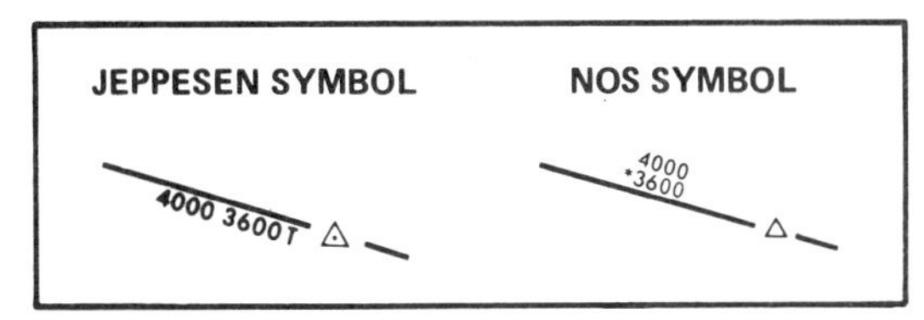

Fig. 6-15. Enroute Altitude Symbols

MINIMUM OBSTRUCTION CLEARANCE ALTITUDE

The minimum obstruction clearance altitude (MOCA) guarantees the same terrain clearance as does the MEA. The major difference between these two altitudes is that the MOCA assures a reliable navigation signal only within 22 nautical miles of the VOR facility, while the MEA guarantees reliable navigation signals throughout the course, as shown in figure 6-16.

When an aircraft is flying more than 22 nautical miles from the VOR station at the MOCA, adequate terrain clearance is assured, but the altitude is not sufficient to receive the line-of-sight transmission from the VOR facility and, therefore, navigate on the airway. The MOCA may be used during an emergency or in a situation when the pilot has either approach clearance or cruise clearance and is within 22 miles of the VOR. The difference between the Jeppesen and NOS symbology is shown in figure 6-17.

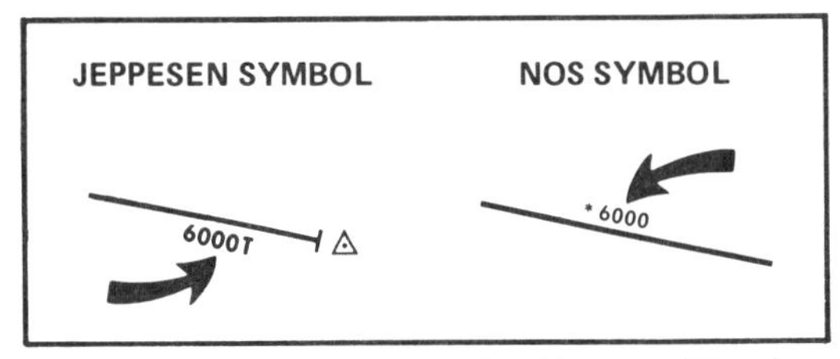

Fig. 6-17. Minimum Obstruction Clearance Altitude

CHANGEOVER POINT

Normally, on an IFR cross-country flight, the pilot changes navigation frequencies midway between the navigational aids. If he is flying outbound from VOR number one, as shown in figure 6-18, both navigational signals cannot be received at midpoint so a changeover point (COP) is indicated on the enroute chart.

These changeover points, as shown in figure 6-19, indicate at which point the pilot should change frequencies to assure a usable navigation signal throughout the entire course. If the pilot does not change from one VOR station to another at the proper changeover point, he may

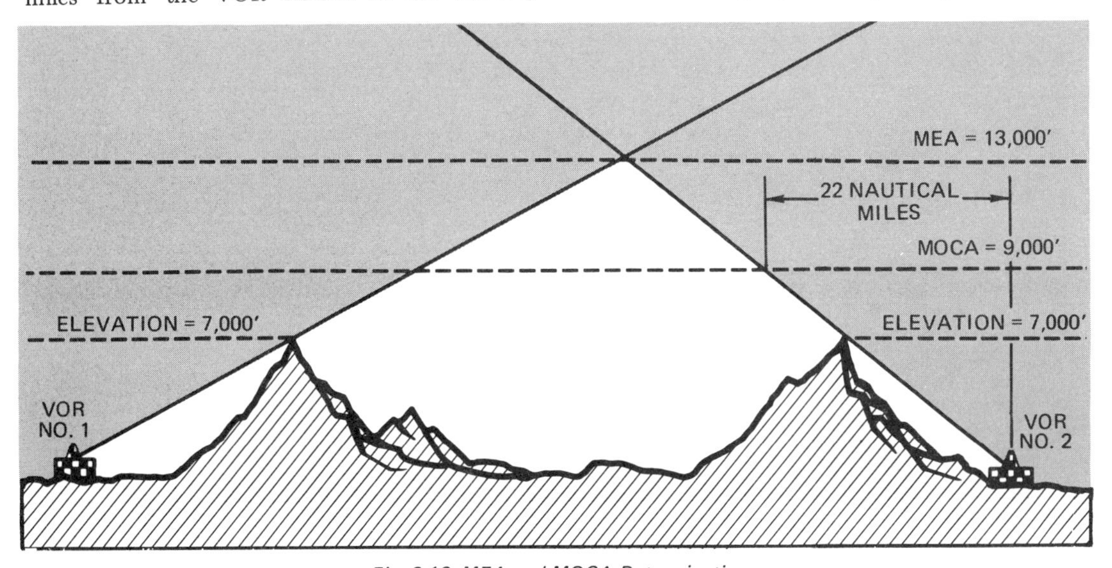

Fig. 6-16. MEA and MOCA Determination

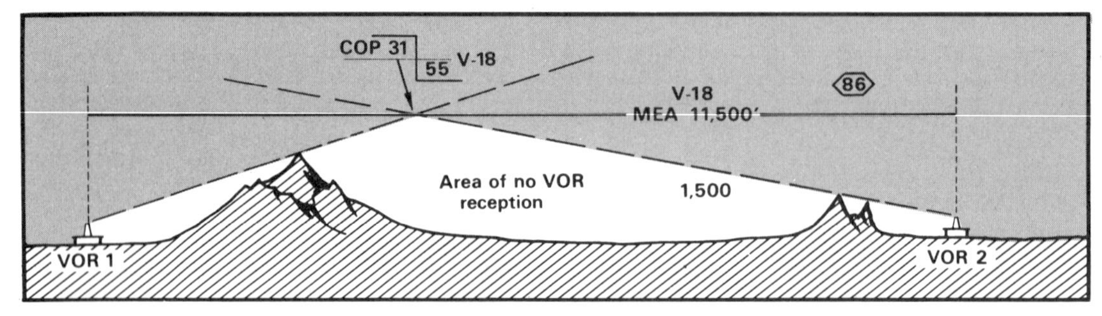

Fig. 6-18.

receive a weak signal or erratic indications on his VOR.

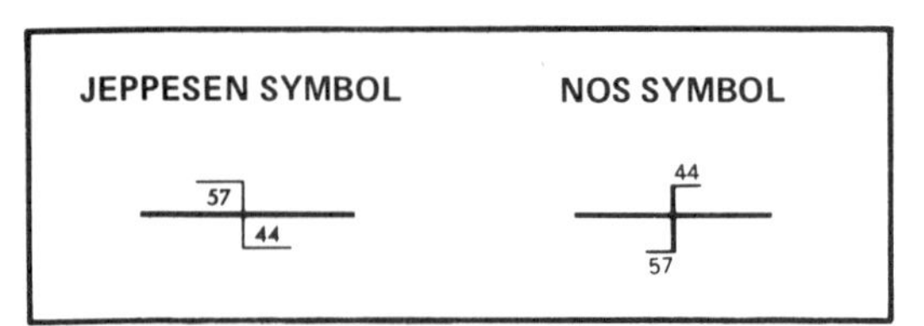

Fig. 6-19. Changeover Point Symbols

CHANGE IN MEA

As illustrated in figure 6-20, a bar crossing an airway at an intersection indicates a change in MEA. Additionally, on NOS charts, it may indicate a change in MOCA or maximum authorized altitude (MAA). A pilot using NOS charts should compare MEAs and MOCAs along the route, or look for a maximum authorized altitude to determine what predicated the change.

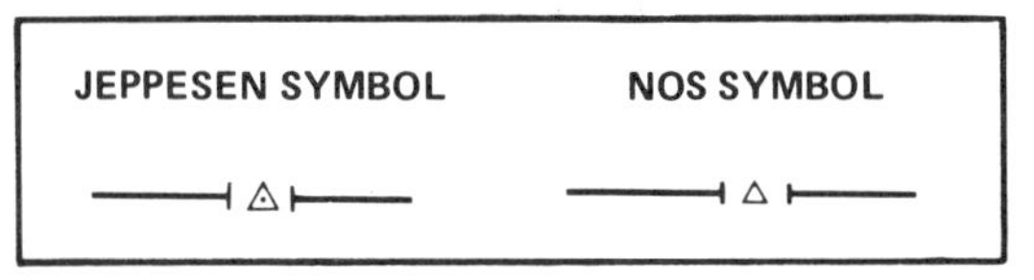

Fig. 6-20. Enroute Altitude Change

MINIMUM CROSSING ALTITUDE

The minimum crossing altitude (MCA) is an obstruction clearance requirement for crossing certain fixes. The MCA is determined by the MOCA of the route segment being flown and the location of obstacles along the route, as illustrated in figure 6-21. Since the mountain in figure 6-21 is situated very close to the Westover Intersection, it can be seen that the pilot must climb to cross the Westover Intersection at or above the MCA of 7,800 feet.

As depicted in figure 6-22, the MCA is shown on Jeppesen and NOS charts by an information

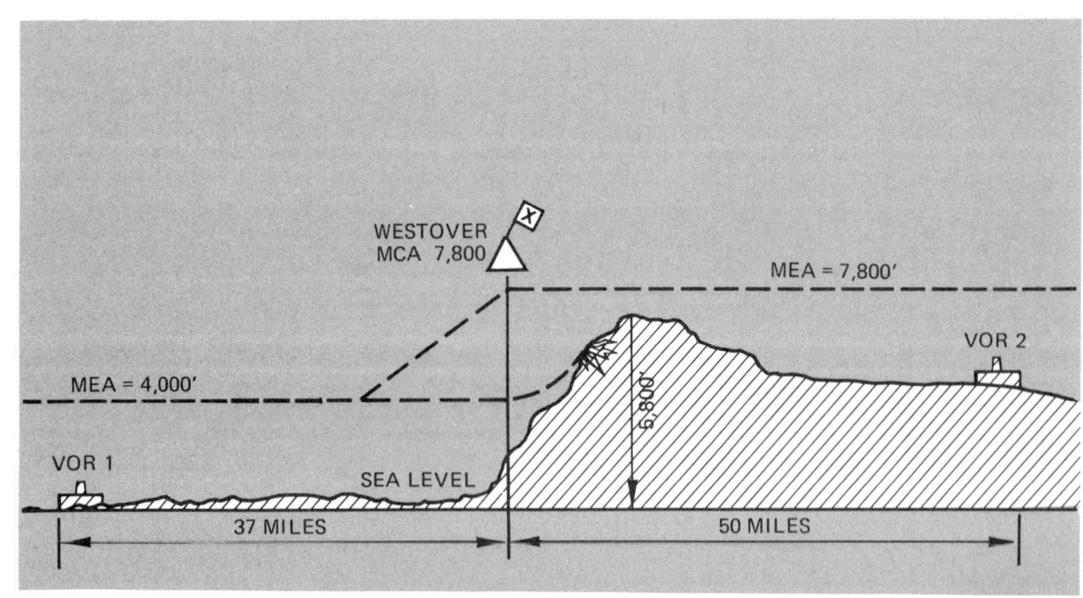

Fig. 6-21. Minimum Crossing Altitude

grouping. This grouping provides the intersection or VOR name, the Victor airway number, the change in altitude required, and the direction of flight for which the change must be made. In the example shown, the pilot proceeding eastbound on V-500 must climb to the MCA of 7,800 feet before reaching Westover Intersection. On Jeppesen charts, which have additional information and symbols around the intersection or VOR that has an MCA, there may be a small number in a circle adjacent to the navaid or fix that is indexed to a minimum crossing box located on the same section of the chart. On NOS charts, a flag containing an "X," originating from the intersection or VOR, indicates a minimum crossing altitude. In this case, the MCA is given adjacent to the flag.

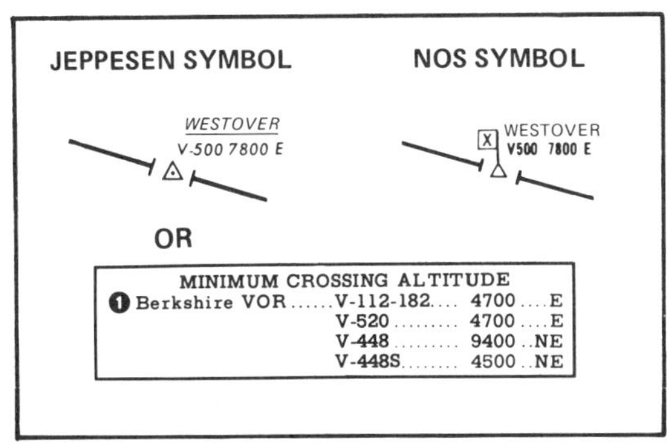

Fig. 6-22. Minimum Crossing Altitude Symbols

MINIMUM RECEPTION ALTITUDE

The minimum reception altitude (MRA) is the lowest altitude which will insure adequate reception of navigation signals forming an intersection or enroute fix. In figure 6-23, the aircraft flying between VOR number one and VOR number two can receive a good signal from these VORs at the MEA of 3,000 feet. However, since VOR number three is located behind the mountain, the aircraft must be at a higher altitude in order to receive station three and, therefore, make a positive identification of the fix. This situation also occurs when the station used to identify the intersection is a considerable distance away, which necessitates more altitude for a strong, accurate signal. The MRA insures reception of navigation signals and not terrain clearance. It is only used if the MRA is above the MEA.

As illustrated in figure 6-24, the MRA is found in the information grouping with the intersection name and will be marked with the letters "MRA," followed by the altitude. NOS charts show the MRA by the letter "R" enclosed in a flag and include the altitude in the associated information grouping. The MRA is of importance to the pilot who is to make a position report over an intersection with a published MRA. If this reporting point is noncompulsory, the pilot must be concerned only with the MRA if he is using it for a navaid.

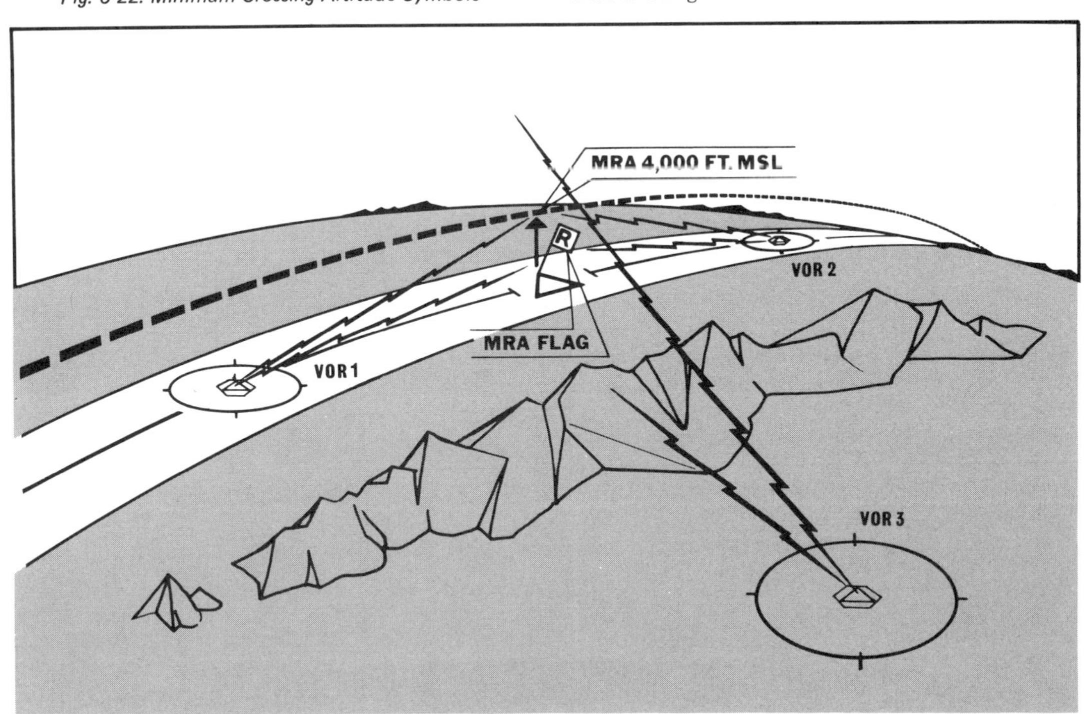

Fig. 6-23. Minimum Reception Altitude

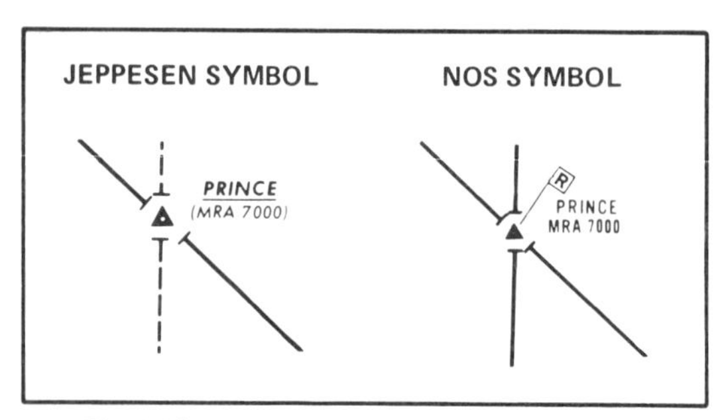

Fig. 6-24. Minimum Reception Altitude Symbols

MAXIMUM AUTHORIZED ALTITUDE

The maximum authorized altitude (MAA) is determined by the transmitting distance of VOR stations which are on the same frequency. Since there are only a limited number of frequencies for VOR transmitters, there is a necessity for duplication. The MAA is established when the distance between two VOR stations on the same frequency is such that both signals can be received at the same time and, therefore, give the pilot unreliable navigation signals. This is illustrated in figure 6-25.

The MAA is shown along the airway on the charts by the letters "MAA," followed by the altitude. Unless the maximum authorized altitude is shown on the chart, VOR stations transmitting on the same frequency are located far enough apart to insure there are no reception problems.

AIR ROUTE TRAFFIC CONTROL CENTER REMOTE SITES

Each air route traffic control center has certain frequencies for communications between air traffic control personnel and pilots on IFR flights in the general vicinity of the center. Additionally, certain remote station sites have been set up throughout the various sectors of each center. These remote sites provide discrete frequencies for communications throughout the area served by the center.

An ARTCC remote communications site is depicted on the chart by a box at the appropriate site location. Included is the ARTCC name, site name, and the discrete frequencies used to establish contact with the center. The air route traffic control center (ARTCC) remote radio site information box, as depicted in figure 6-26, informs the pilot that he can contact the Seattle Center through a remote communications site located near the area of Zinc.

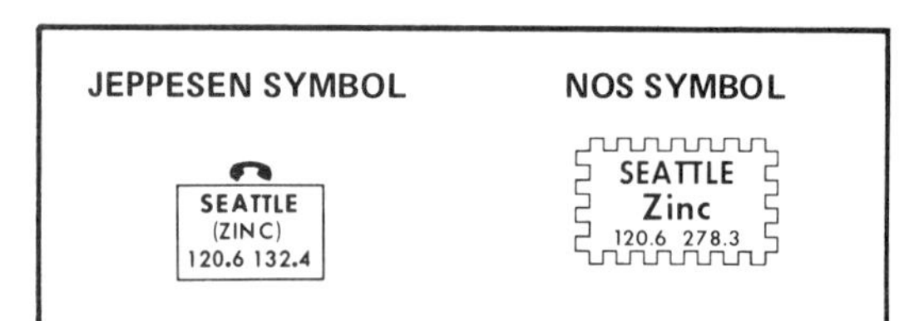

Fig. 6-26. ARTCC Remote Site Symbols

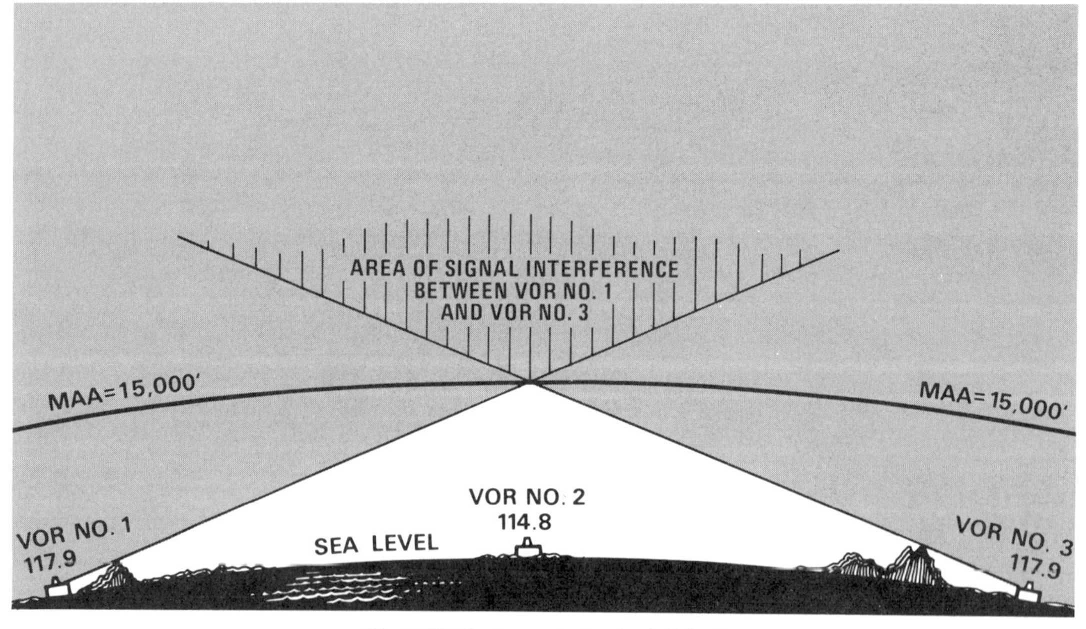

Fig. 6-25. Maximum Authorized Altitude

The symbol shown in figure 6-27, indicates the division line between air route traffic control centers. This is designated on Jeppesen charts by a dotted line and on NOS charts by an irregular line. The name of the controlling center is printed on its respective side, and aircraft within the boundaries are within the center's area of responsibility.

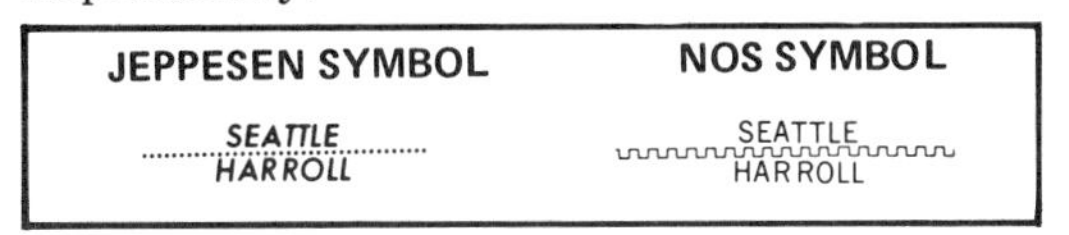

Fig. 6-27. ARTCC Boundaries

SPECIAL USE AIRSPACE

Due to the potential hazards or special activities within particular portions of the airspace, certain areas have been designated as either *restricted, prohibited, warning,* or *alert areas.* Additionally, certain areas have been designated as *military operations areas.* The charts show the following information in, or in close proximity to, the special use airspace.

1. Area boundary
2. Altitude affected
3. Operating time, when the time is other than continuous
4. Weather conditions; i.e., used only during IFR or VFR
5. Name of controlling agency if air-to-ground communications are available

The symbols for these areas are shown in figure 6-28.

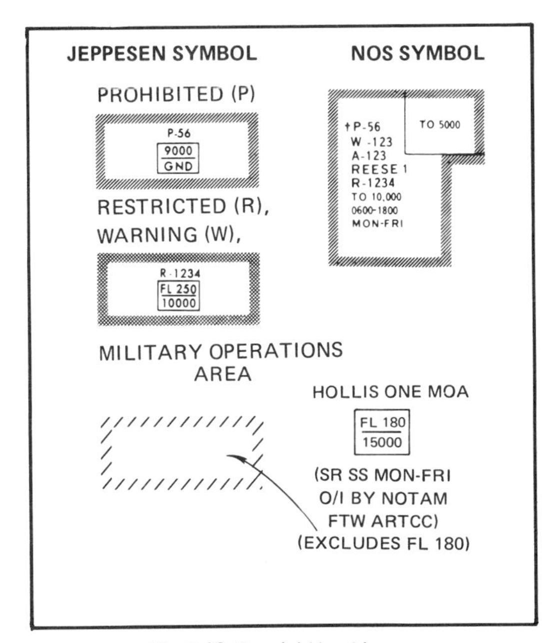

Fig. 6-28. Special Use Airspace

Prohibited airspace means that, for national security reasons, no flights through the area are permitted without prior permission of the controlling or using agency. Because of the nature of the activities within these areas, permission is seldom granted.

Restricted airspace refers to areas of unusual activity, usually of a military nature. Pilots must obtain permission from the using or controlling agency prior to flying through restricted areas. Permission to fly through these areas is usually granted.

A *warning area* is an airspace segment located beyond the three-mile limit along the coast of the United States. Since this airspace is over international waters, the government cannot prohibit or restrict flight within it, but they do issue warnings that flight within the designated area may be extremely dangerous.

Military operations areas (MOAs) consist of special use airspace that has been designated for military flight activities, such as familiarization training, intercept practice, and air combat maneuvers. They include areas previously referred to as *intensive student jet training areas* and *alert areas.* Each is identified by a nickname. For example, names such as Tarheel and Moody 1 are used.

Flight within an MOA is not restricted for VFR aircraft, but pilots are urged to exercise extreme caution. Both the pilots of aircraft participating in activities within the area and pilots transiting the MOA are fully responsible for collision avoidance. Nonparticipating IFR traffic (both civil and military) will be cleared through an MOA if ATC can provide IFR separation service. Otherwise, IFR traffic will be routed around or over the area.

ISOGONIC LINES

Figure 6-29 depicts the symbology for isogonic lines on Jeppesen and NOS enroute charts. The magnetic variation affecting an area is shown on the edge of the charts. There is a small line originating at the periphery of the chart with the variation in degrees printed on the border. These lines become important when solving computa-

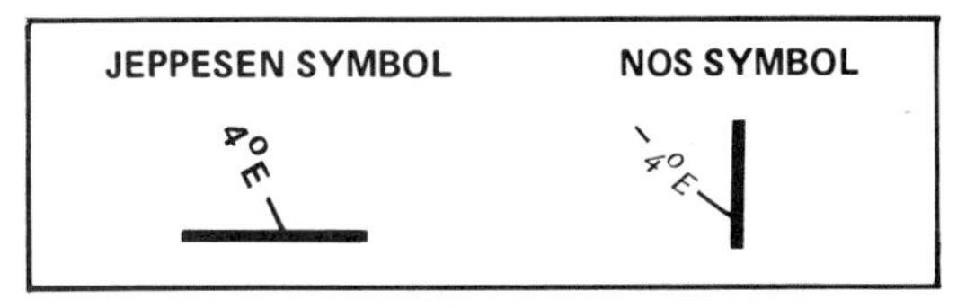

Fig. 6-29. Isogonic Lines

tions such as wind correction and groundspeed. To determine true course on low altitude enroute charts, the pilot must either add or subtract the magnetic variation from his magnetic course.

TIME ZONE BOUNDARIES

Boundaries between time zones and information for conversions to Zulu time are shown on NOS charts as a dotted line with the time zone and conversion data next to the dotted line. The time zones on Jeppesen charts are shown in the enroute chart index located on the front panel. The symbols for the time zones are illustrated in figure 6-30. A time zone change is indicated on Jeppesen charts by a series of the letter "T," with the time zone conversion to Zulu time provided at the top of the enroute chart index.

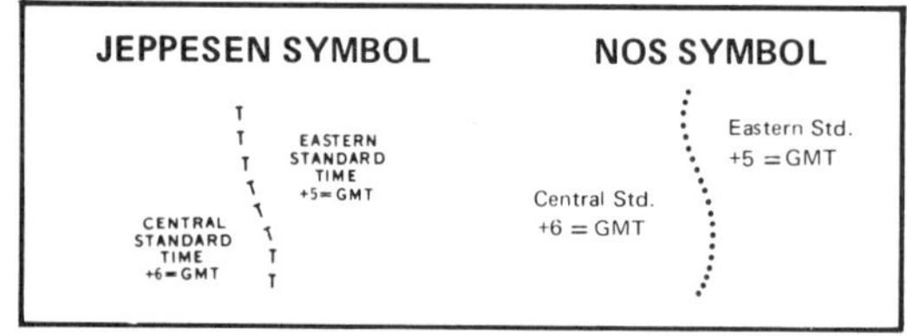

Fig. 6-30. Time Zone Boundaries

AERODROME SYMBOLS

Aerodrome symbols used on both charts are divided into two basic categories — those aerodromes with a published instrument approach procedure and those without such a procedure. On Jeppesen charts, those airports with the name printed in capital letters have an instrument approach procedure available and published. An airport name printed in lower case letters indicates there is not an approved instrument approach chart for the airport. On NOS charts, airports with an instrument approach procedure are printed in blue, while those printed in brown are airports with no instrument approach procedures available.

COMMUNICATIONS FREQUENCIES

Frequencies available for air-to-ground communications are listed on the front panel of Jeppesen and NOS charts. NOS charts list only terminal communication frequencies while Jeppesen charts list terminal communication frequencies plus flight service station frequencies.

As an ATC controller "hands off" the instrument flight to the next controller, the pilot is advised of the frequency to use, however, it will aid the pilot to know what frequency to expect and what additional frequencies he could use. Since the enroute charts are updated frequently, current charts normally provide the correct frequencies.

TERMINAL CHARTS

The four terminal charts that are used for terminal operations are the airport diagram charts, area charts, SIDs, and STARs. Each has a definite application to either arrival or departure operations, and an understanding of the symbology on both the Jeppesen and NOS charts is highly recommended.

AIRPORT DIAGRAM

The airport diagram provides general airport information and is important to arrival, taxi, and departure operations. The Jeppesen airport diagram is located on the reverse side of the first approach chart for each airport. On NOS charts, the airport diagram is located on the face of each chart in the lower right-hand corner. The following items refer to figure 6-31.

1. Airport elevation
2. Geographical name
3. Airport name
4. Latitude and longitude (located on the bottom center of NOS charts)
5. Jeppesen index number
6. Revision date
7. Communications and VOR test facility frequencies
8. Obstructions and physical features
9. Runway number to nearest 10° magnetic heading
10. Approach light configuration
11. Runway length
12. Latitude and longitude scale

AREA CHARTS

Area charts depict certain portions of the low altitude enroute chart where a complicated or congested airway structure exists. For this reason, the area charts are projected in a larger scale so the detailed information may be interpreted easily.

Area charts are identified by the name of the major city in the area covered. The area chart index is found on the front panel of each enroute chart. Jeppesen area charts are published individually; however, all the NOS charts are printed on a single sheet the approximate size of an enroute low altitude chart.

The symbology employed on the area charts is basically the same as on the enroute charts. Therefore, if the pilot is familiar with enroute charts he can easily interpret the area charts.

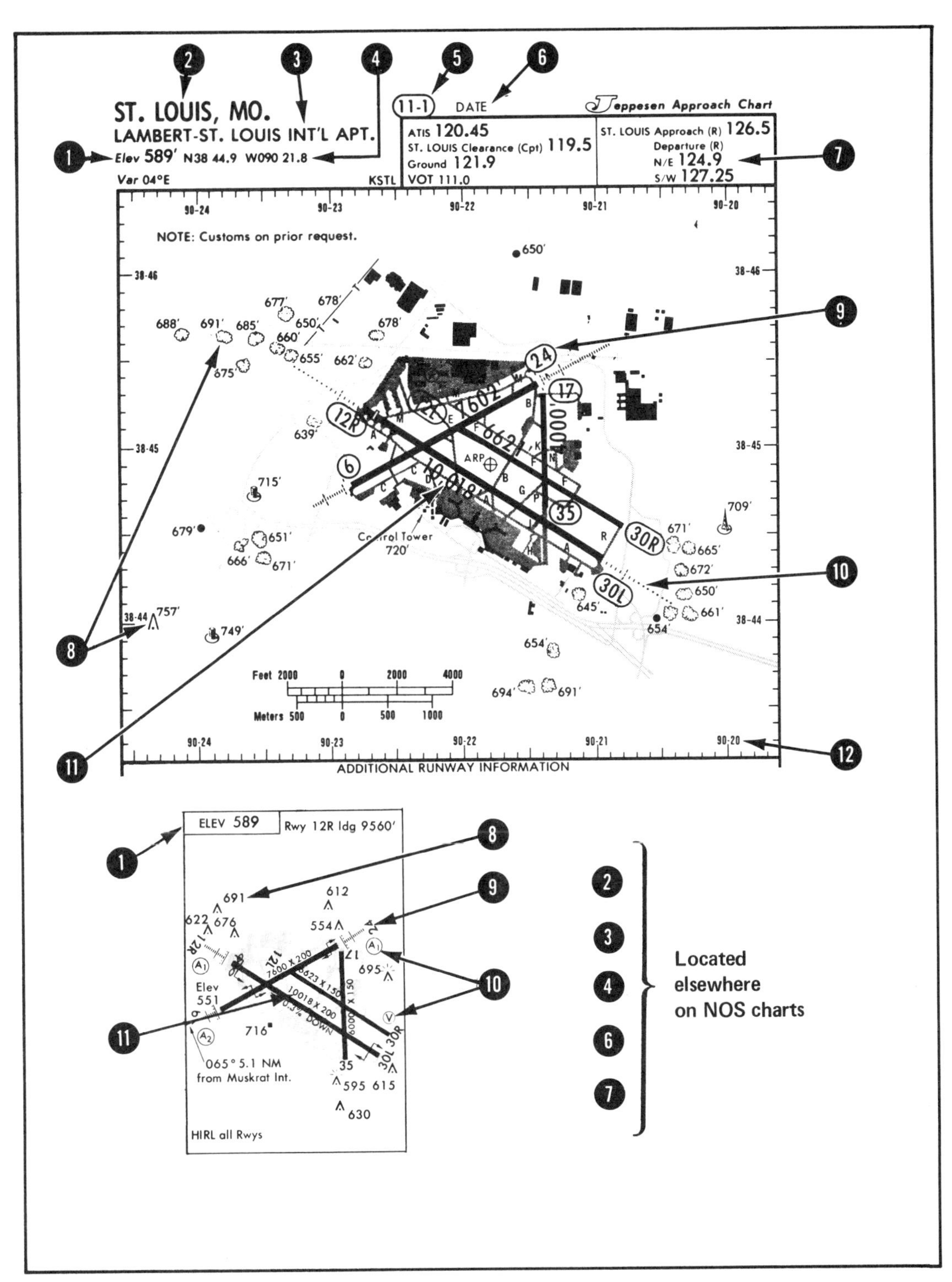

Fig. 6-31. Airport Diagrams

STANDARD INSTRUMENT DEPARTURE CHARTS

Standard instrument departure (SID) charts are published by both Jeppesen and the National Ocean Survey. An example of the NOS SID format is shown in figure 6-32. Both chart types present a narrative summary of the departure, together with a plan view of the procedures and routing. The symbols used on both charts are self-explanatory and conform to the symbols used on the enroute charts. Once a pilot has been cleared to use a departure procedure, *the printed procedure is mandatory* until such a time as departure control or another air traffic control agency clears the pilot to deviate from that procedure.

Some SID procedures specify minimum crossing altitudes and minimum enroute altitudes for the various departure legs. Before accepting a departure clearance, the pilot must ascertain if the specified procedures are within the aircraft's

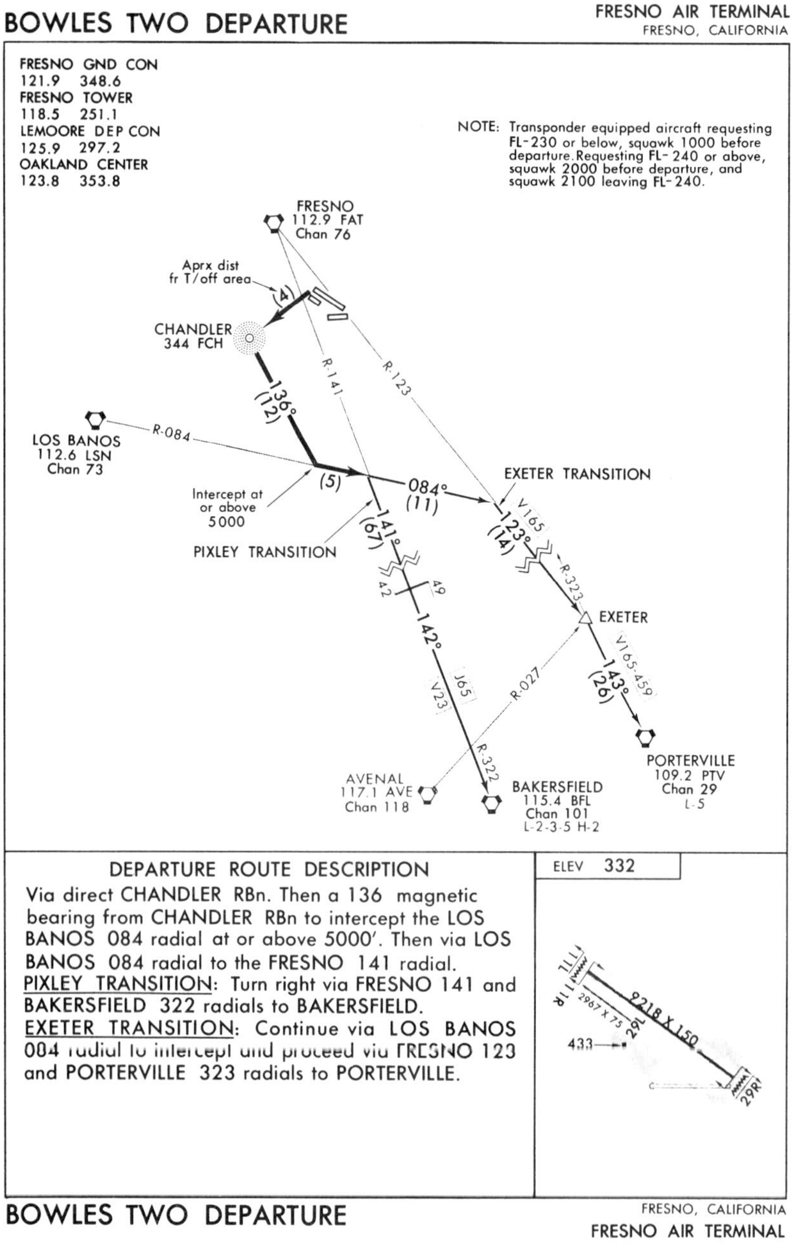

BOWLES TWO DEPARTURE

FRESNO AIR TERMINAL
FRESNO, CALIFORNIA

FRESNO GND CON
121.9 348.6
FRESNO TOWER
118.5 251.1
LEMOORE DEP CON
125.9 297.2
OAKLAND CENTER
123.8 353.8

NOTE: Transponder equipped aircraft requesting FL-230 or below, squawk 1000 before departure. Requesting FL-240 or above, squawk 2000 before departure, and squawk 2100 leaving FL-240.

FRESNO
112.9 FAT
Chan 76
Aprx dist fr T/off area
(4)
CHANDLER
344 FCH
R-141
R-123
136°
(12)
LOS BANOS
112.6 LSN
Chan 73
R-084
Intercept at or above 5000
(5)
084°
(11)
PIXLEY TRANSITION
141°
(67)
42
49
142°
J65
V23
EXETER TRANSITION
V165
123°
(14)
R-323
EXETER
V165-459
143°
(26)
R-027
R-322
AVENAL
117.1 AVE
Chan 118
BAKERSFIELD
115.4 BFL
Chan 101
L-2-3-5 H-2
PORTERVILLE
109.2 PTV
Chan 29
L-5

DEPARTURE ROUTE DESCRIPTION

Via direct CHANDLER RBn. Then a 136 magnetic bearing from CHANDLER RBn to intercept the LOS BANOS 084 radial at or above 5000'. Then via LOS BANOS 084 radial to the FRESNO 141 radial.
PIXLEY TRANSITION: Turn right via FRESNO 141 and BAKERSFIELD 322 radials to BAKERSFIELD.
EXETER TRANSITION: Continue via LOS BANOS 084 radial to intercept and proceed via FRESNO 123 and PORTERVILLE 323 radials to PORTERVILLE.

ELEV 332

11L
11R
2967 X 75
29L
9218 X 150
29R
433

BOWLES TWO DEPARTURE

FRESNO, CALIFORNIA
FRESNO AIR TERMINAL

Fig. 6-32. Standard Instrument Departure (NOS)

capability. If not, the pilot should request an amended clearance immediately. Since SID charts are very simple, it may be desirable to use them in conjunction with area charts. These charts present additional detail in the terminal area and are used to transition to the enroute phase of the flight.

STANDARD TERMINAL ARRIVAL ROUTES

Standard Terminal Arrival Routes (STARs), as well as the SIDs, are utilized as a simple means

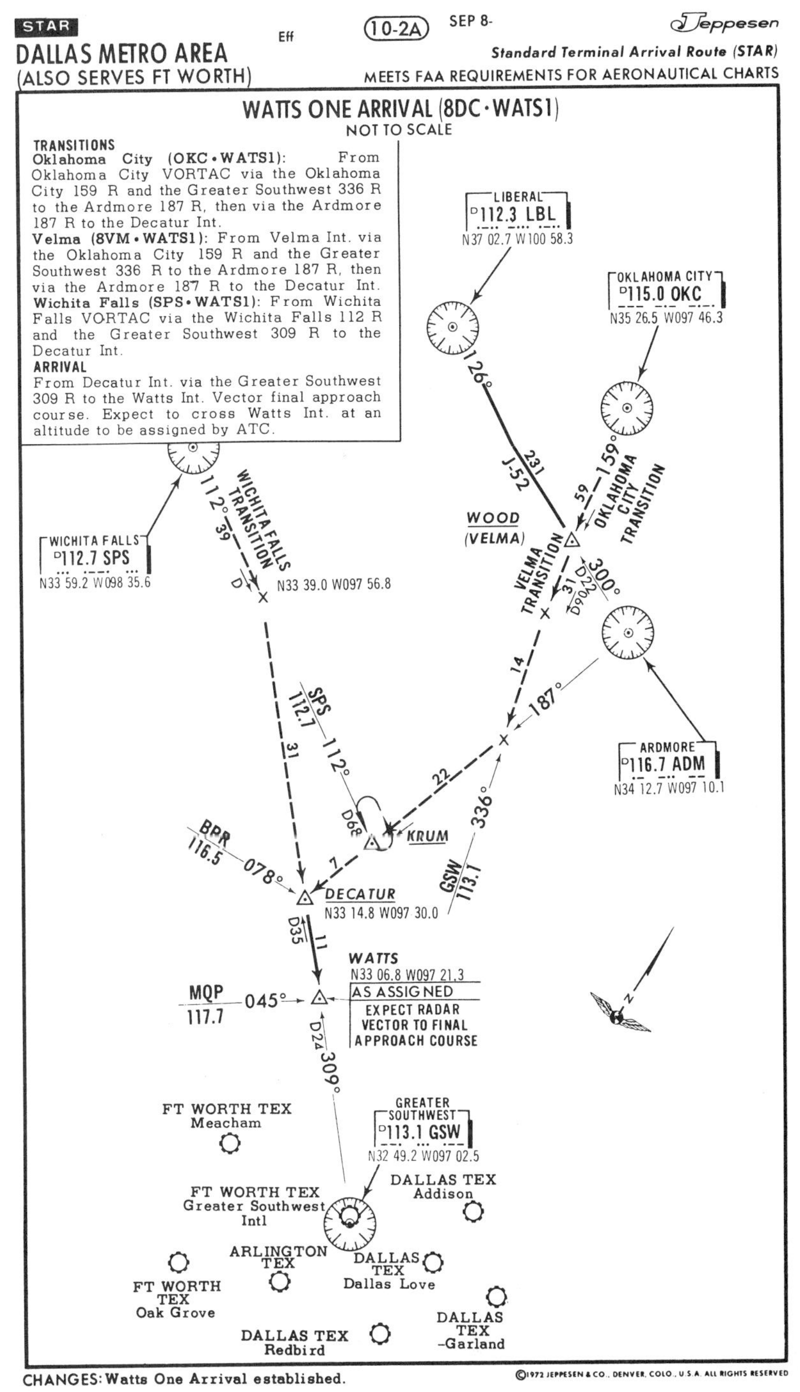

Fig. 6-33. Standard Terminal Arrival Routes (Jeppesen)

of presenting a clearance route. Through their use air traffic control more efficiently utilizes communication frequencies and the airspace in the terminal vicinity. It is important to note that clearances containing STARs and SIDs may be issued by ATC unless the pilot requests that they not be used. It is important to recognize that when a STAR is part of an air traffic control clearance and is accepted by the pilot, the STAR must be in the pilot's possession.

The symbology found on the standard terminal arrival route follows the same format as found on SID charts. As shown in figure 6-33 the STAR charts also employ a narrative and pictorial view of the procedure.

USE OF THE JEPPESEN PV–2 IFR PLOTTER

The Jeppesen IFR plotter has three major functions for use on Jeppesen IFR charts.

1. Statute and nautical mile conversions
2. Determining course direction
3. Distance measurement

STATUTE AND NAUTICAL MILE CONVERSION SCALES

On each end of the IFR plotter are conversion scales for converting statute and nautical miles. These scales show the relationship between statute and nautical miles and provide an easy means for converting one to the other.

DETERMINING COURSE DIRECTION

A pilot may determine the course direction by placing the center of the plotter compass rose over the center of the VORTAC station and positioning the plotter along the desired course line on the chart. As shown in figure 6-34, the north reference barb extending from the VORTAC symbol points directly to the magnetic course. This is the VOR radial that provides the most direct route from the VOR to the destination. In this example, the magnetic course is 281°.

DISTANCE MEASUREMENT

The IFR plotter can be used to measure distances on the Jeppesen low altitude enroute

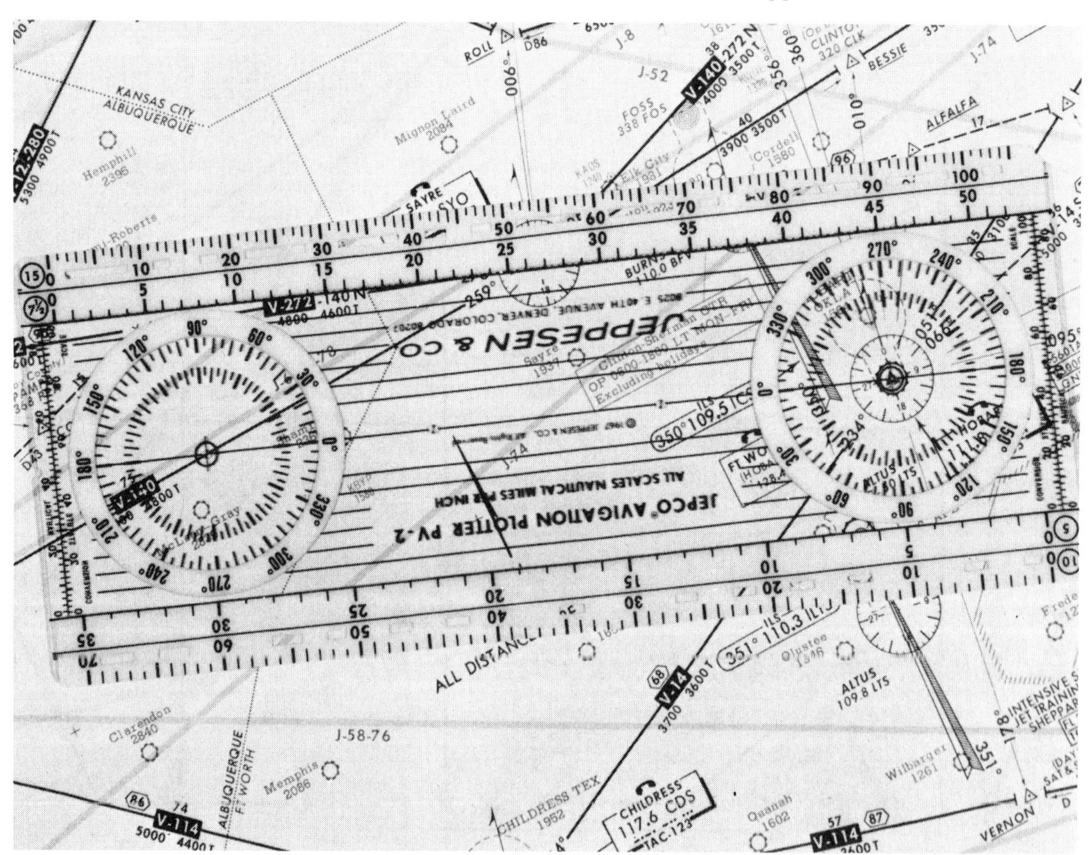

Fig. 6-34. Determining Course Direction

charts, using any one of nine different scales which correspond to the scale used on the chart. If the pilot needs to measure the distance between the Hobart VORTAC and the Shamrock Airport, he should notice that the scale on the top border of the L-12 chart is 1 inch = 15 miles. Then, using the 15 scale on the IFR plotter, the distance is easily measured between the Hobart VORTAC and the Shamrock Airport, as shown in figure 6-35. The distance in this example is 59 miles.

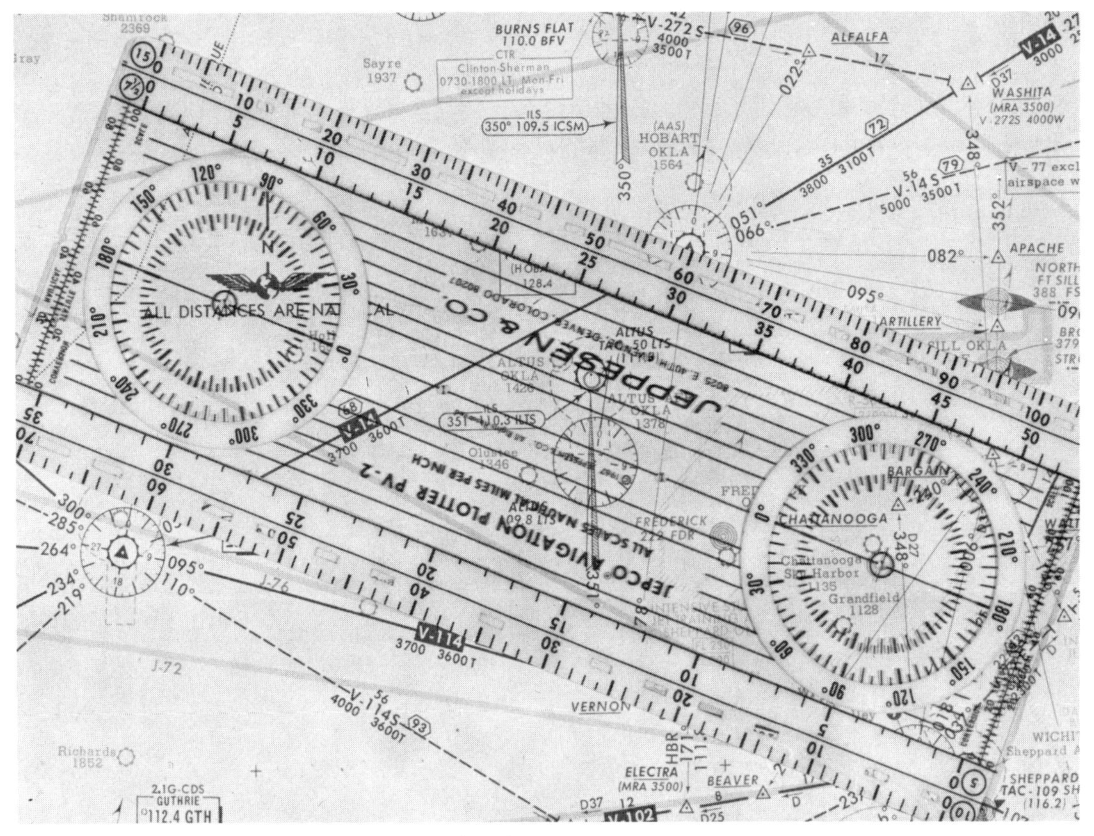

Fig. 6-35. Distance Measurement

7 INSTRUMENT APPROACHES

INSTRUMENT LANDING SYSTEM

The instrument landing system (ILS) provides a highly accurate means of navigating to the runway. It employs radio transmitting equipment on the ground plus airborne receiving equipment. When using the ILS system, the pilot determines his position on the approach by reference to instrument indications which provide the lateral and vertical guidance necessary to fly a *precision approach.*

A precision approach, by definition, is an approved descent procedure which utilizes a navigation facility aligned with a runway where electronic *glide slope* information is provided. When all components of the ILS system are available to the pilot, including an approved approach procedure, the pilot may execute a precision approach to landing.

The instrument landing system consists of the following components.

1. Localizer
2. Glide slope
3. Outer marker
4. Middle marker

The approach lights are visual aids normally *associated* with the instrument landing system, but *are not* specified as components in the Federal Aviation Regulations.

LOCALIZER

GROUND EQUIPMENT

The primary component of the ILS is the *localizer.* The localizer is a VHF radio transmitter and antenna system, which provides lateral directional guidance to the runway. The frequency of the localizer transmitter is in the same general range as VOR transmitters (between 108.10 and 111.95 MHz); however, localizer frequencies are on odd-tenths frequencies only. Additionally, the 50 kHz frequencies immediately higher than the odd-tenths frequencies are also allotted to the ILS localizer; for example, 109.50, 109.55, 109.70, and 109.75 are localizer frequencies. The transmitter and antenna are located at the opposite end of the runway from the approach threshold and on the centerline for the runway served by the facility.

LOCALIZER TRANSMITTER SIGNAL

The signal that is transmitted by the localizer consists of two signal patterns which overlap at the center, as shown in figure 7-1. They are aligned with the extended centerline of the runway served.

The right side of this pattern, with respect to an approaching aircraft, is modulated at 150 Hertz and is called the "blue" sector. The left side of the pattern is modulated at 90 Hertz and is called the "yellow" sector. As stated, these two signal patterns overlap slightly at the center and

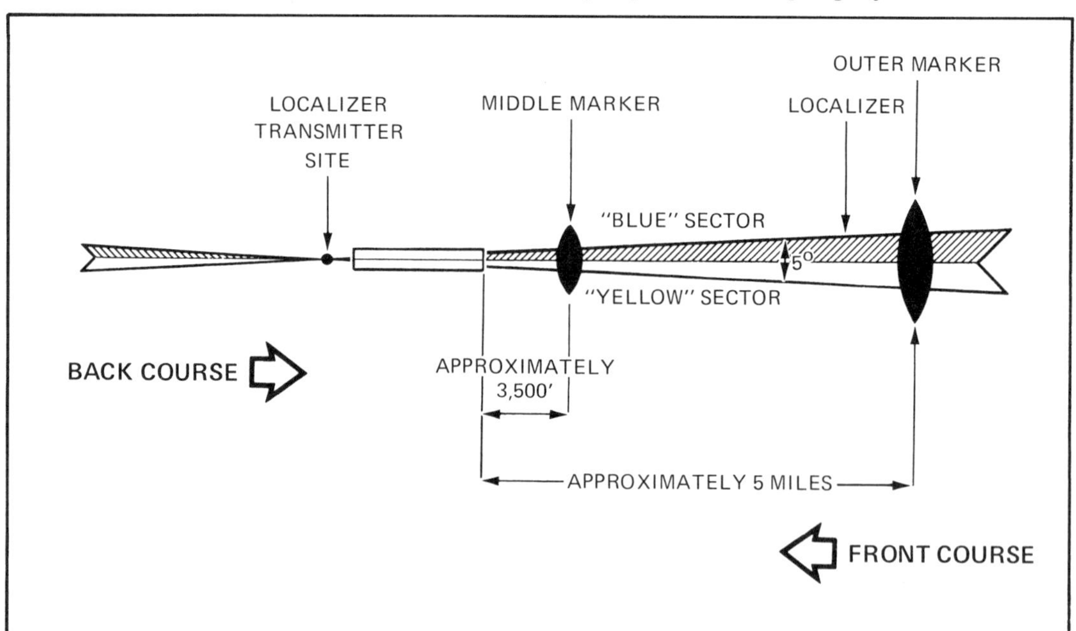

Fig. 7-1. ILS Localizer Course

align with the centerline of the runway, which forms the localizer centerline. It is this *overlap area* which provides the on-course signal.

The localizer signal provided by the transmitter is normally five degrees wide, but may vary in width from approximately three to six degrees. It is adjusted to provide a course signal which is approximately 700 feet wide at the runway threshold. The width of the beam increases as the distance from the runway increases. Thus, at 10 miles from the transmitter, the beam is approximately one mile wide.

The localizer may be identified by an audio signal which is superimposed on the navigational signal. It consists of a three-letter identification preceded by the letter "I." For example, the localizer identification at Cheyenne is "I-CYS." At some locations the ATIS may be received on the localizer frequency.

The reliable reception range of the localizer is not less than 18 nautical miles within 10° of the on-course signal. In the area from 10° to 35° of the on-course signal, the reliable reception range is not less than 10 nautical miles. This is because the primary strength of the on-course signal is aligned with the runway centerline. When operating in the area from 35° to 90° off course, erroneous reverse instrument indications may occur and should be disregarded.

AIRBORNE EQUIPMENT

The localizer signal is sensed by a receiver which is incorporated with the VOR receiver in a single unit of equipment. The two receivers share some electronic circuits and also the same frequency selector, volume control, and ON/OFF control.

The localizer signal activates the vertical needle in the navigation indicator (CDI). Assuming a final approach course aligned east and west for runway 27, an aircraft north of the extended centerline of the runway is in the area modulated at 150 Hertz. The CDI needle is deflected to the left, as shown in figure 7-2, item 1.

Conversely, an aircraft south of the runway centerline is in the area modulated at 90 Hertz, and this signal causes a right needle deflection, as shown in figure 7-2, item 2. In the overlap area, both signals apply a force to the needle causing a partial deflection in the direction of the strongest signal. Thus, if an aircraft is inbound on the front course, but slightly to the right, the CDI needle is deflected slightly to the left. This indicates that a correction to the left is necessary to place the aircraft in the precise alignment desired. At the point where the 90 Hertz and 150 Hertz signals are of equal intensity, the needle is centered, indicating that

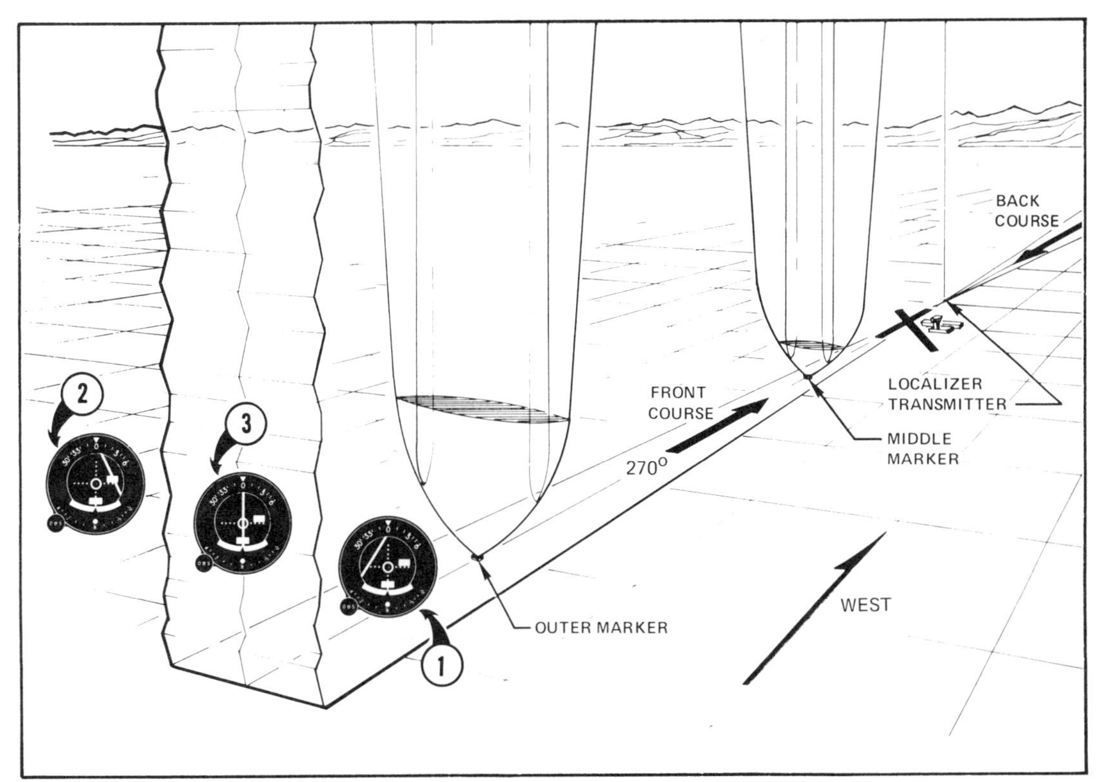

Fig. 7-2. Localizer Reception Indications

the aircraft is located precisely on the approach course, as shown in figure 7-2, item 3.

When the CDI is used for VOR navigation, a full-scale needle deflection occurs at 10° on either side of the course shown on the course selector. When this same needle is used as an ILS localizer indicator, full-scale needle deflection occurs within approximately 2.5° from the center of the localizer beam. After full-scale needle deflection, it must be realized that there will be no indication of how far the aircraft actually is to the left or right of course. The sensitivity of the CDI is approximately four times greater when used for a localizer function as when used in VOR navigation.

In contrast to VOR functions, the CDI needle in most navigation indicators is *not dependent upon a correct course selector setting;* however, it is a good operating procedure to set the course selector for the approach course as a *reminder* of the correct final approach magnetic course during the approach.

When an OFF flag appears in front of the vertical needle, it indicates that the signal strength is not sufficient and the needle indications are unreliable. Momentary OFF flag indications, CDI needle deflections, or both may occur when obstructions or other aircraft pass between the transmitting antenna and the receiving aircraft. Such indications may be disregarded providing they are only of a momentary nature.

GLIDE SLOPE

GROUND EQUIPMENT

The ILS glide slope is provided by a UHF radio transmitter and antenna system, operating within a frequency range of 329.15 to 334.85 MHz. Its function is to provide vertical guidance during the approach. The transmitter is located 750 to 1,250 feet down the runway from the threshold, offset 400 to 600 feet from the runway centerline. Like the localizer, the glide slope signal consists of two overlapping beams modulated at 90 Hertz and 150 Hertz, as illustrated in figure 7-3. Unlike the localizer, however, these signals are aligned one above the other and the signals are radiated primarily along the approach course. The thickness of the overlap area is 1.4°, or .7° above and .7° below the optimum glide slope. This glide slope signal may be adjusted between two degrees and four degrees above a horizontal plane, as shown in figure 7-3. A typical adjustment is 2.5° to 3°, depending upon such factors as obstructions along the approach path and the runway slope.

It is important to realize that false signals are generated along the glide slope, the first being approximately 12.5° above horizontal; however, there are no false signals *below the actual slope.* An aircraft flying in accordance with the pub-

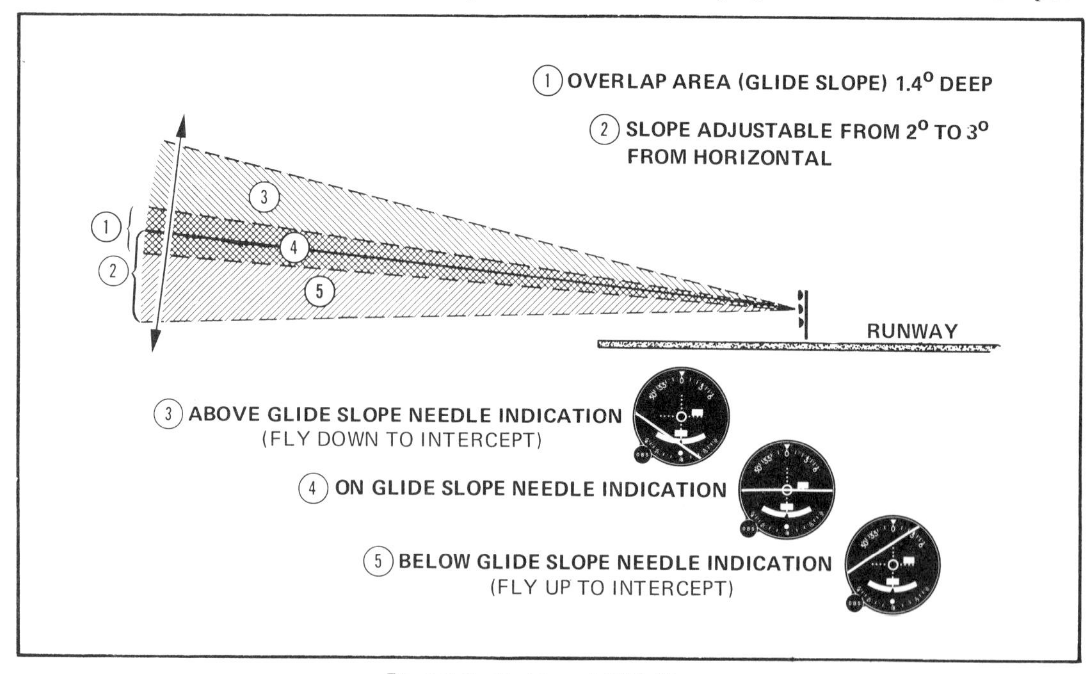

Fig. 7-3. Profile View of Glide Slope

lished approach procedure will never encounter any type of false glide slope indications.

AIRCRAFT EQUIPMENT

The glide slope signal is received by an ultrahigh frequency (UHF) receiver in the aircraft. In modern avionics installations, controls for this radio are integrated with the VOR controls so that the proper glide slope frequency will *automatically* be set when the localizer frequency is selected. In some older avionics installations, the glide slope receiver is a separate radio with its own controls and must be set to the published frequency *in addition* to setting the correct localizer frequency.

The glide slope signal activates the glide slope needle located in conjunction with the CDI, as shown in figure 7-4. There is a separate OFF flag in the navigation indicator for the glide slope needle. This flag will appear when the signal received by the glide slope receiver is not adequate for reliable use. As with the localizer, the glide slope needle will show full deflection until the aircraft reaches the point of signal overlap. At this time, the needle will show a partial deflection in the direction of the strongest signal. When both signals are equal, the needle will center horizontally, indicating that the aircraft is precisely on the glide path.

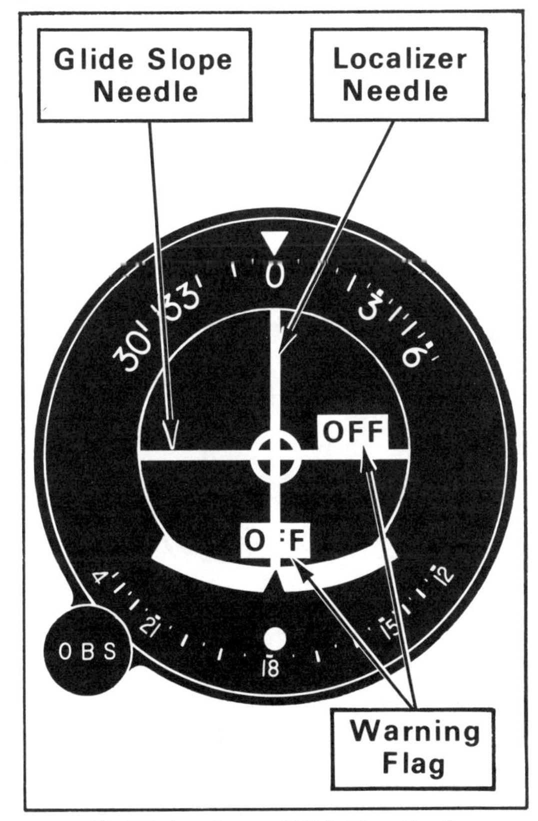

Fig. 7-4. Localizer and Glide Slope Needles

Since both vertical and lateral guidance is provided by the same navigation indicator, the pilot may determine his precise location with respect to the approach path by reference to a single instrument. The round circle in the center of the display shows where the aircraft is located in respect to the glide slope and localizer. Figure 7-5, item 1, shows both needles centered. This indicates that the airplane is located in the exact center of the approach path. An indication, such as that shown in item 2 of the same figure, tells the pilot that the aircraft is above the glide slope and to the right of course. Therefore, he must fly down and to the left to correct to the approach path. Item 3 shows the requirement to fly up and right to provide the necessary correction.

With 1.4° of beam overlap, the area of overlap will be approximately 1,500 feet thick at 10 miles, 150 feet at one mile, and less than one foot at touchdown. Thus, it may be seen that the apparent sensitivity of the instrument becomes greater as the aircraft nears the runway. Careful monitoring is necessary to keep the needle centered, since a full deflection on the needle will indicate that the aircraft is either high or low; however, *beyond* full-scale deflection, there is no indication of *how* high or low the aircraft is.

ILS MARKER BEACONS

ILS marker beacons provide the pilot with distance information from the runway during the approach by identifying predetermined points along the approach course. These beacons are low power transmitters, operating at a frequency of 75 MHz with three watts or less rated power output. They radiate an elliptical beam upward from the ground. At an altitude of 1,000 feet above ground level, the beam dimensions are 2,400 feet long and 4,200 feet wide. With further increases in altitude, the dimensions increase significantly. To decrease the sensitivity of the marker beacon receiver and, therefore, cause the marker beacon signal pattern to be effectively smaller and more accurate, the aircraft receiver sensitivity selector should be set to "low" when executing an ILS approach.

OUTER MARKER

The outer marker (OM) is located four to seven miles from the threshold, within 250 feet of an

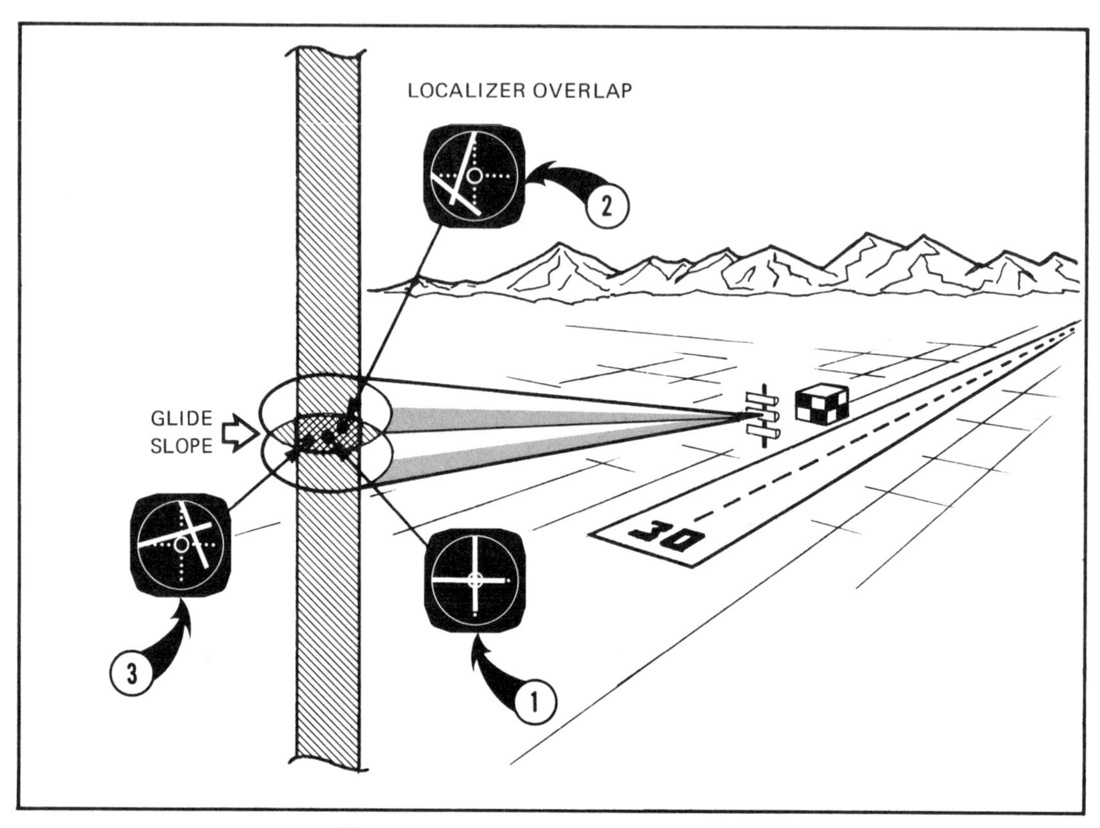

Fig. 7-5. Localizer and Glide Slope Indications

imaginary extended runway centerline. It intersects the glide slope vertically at approximately 1,400 feet above runway elevation. Also, it marks the approximate point at which *glide slope interception* normally will take place and designates the beginning of the *final approach segment* for ILS and localizer approaches. The signal is modulated at 400 Hertz, which is an audible low tone, with continuous Morse code dashes at a rate of two dashes per second. The signal is received in the aircraft by a 75 MHz marker beacon receiver. It is presented to the pilot as an audible or aural tone over the speaker or headset and by the illumination of a *blue* light which flashes in synchronization with the aural tone, as shown in figure 7-6.

MIDDLE MARKER

The middle marker (MM) is usually located approximately 3,500 feet from the threshold on the imaginary extended runway centerline. Its signal is modulated at 1,300 Hertz, an audible medium tone, with alternating Morse code dots and dashes at a rate of 95 dot-and-dash combinations per minute. The middle marker provides an aural tone and illumination of the *amber* light. (See Fig. 7-6.) The glide slope crosses the middle marker at approximately 200 feet above the touchdown zone on the runway and is near the missed approach point for the ILS approach.

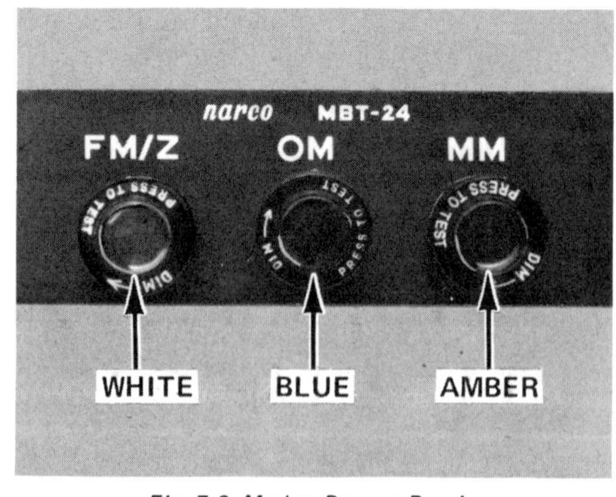

Fig. 7-6. Marker Beacon Receiver

INNER MARKER

When installed, the inner marker (IM) is located just prior to the threshold on the runway centerline. It is modulated at 3,000 Hertz, a very high aural tone, with continuous Morse code dots at the rate of six dots per second. The inner marker causes the *white* light to illuminate. (See Fig. 7-6.)

BACK COURSE MARKER

A back course marker, where installed, normally indicates the ILS back course final approach fix where the approach descent is started. The back course marker is identified by a flashing white light in the marker beacon receiver.

COMPASS LOCATOR

The compass locator is a low frequency, nondirectional radio beacon that is collocated with most outer markers and some middle markers. The transmitter, therefore, allows identification of outer and middle markers with ADF equipment. When the compass locator is situated with the outer marker, the NDB facility is called an outer compass locator.

The compass locator has a power output of less than 25 watts and a range of at least 15 miles. The transmitter has been designed to transmit Morse code identification signals consisting of the first two letters of the station identifier at the outer marker. For example, if a station is identified by the letters JMS, the LOM is identified by the letters JM.

BACK COURSE ILS

The course line along the extended centerline of the runway in the *opposite* direction from the front course is called the *back course.* The back course approach utilizes the same localizer equipment as the front course ILS approach for the reciprocal runway. The back course *normally incorporates only localizer guidance* and, in the absence of glide slope information, is considered a nonprecision approach. Also, due to terrain and other irregularities, the back course of some ILS systems cannot be utilized fully. Thus, back courses are designated as either usable or unusable. The designation "usable" indicates back courses which are suitable for ILS back course approaches, determining intersections, holding, transitions, and missed approaches. "Unusable" designates those which are unsuitable for any navigation.

Back course *glide slopes* are available at certain runways. This is provided by the addition of a second antenna array for the glide slope transmitter. The glide slope transmitter signal may be selected for either the front course or the back course at any given time, but not for both simultaneously.

Since back course marker beacons are normally not available for back course ILS approaches, other means are employed to identify the final approach fixes. Most fixes in back course approaches are based on the intersection of the localizer back course and a radial from a nearby VOR.

When flying a back course approach, the CDI needle must be read in reverse unless the equipment has reverse sensing capability. In other words, to remain on the localizer course, any necessary corrections must be made *away from* instead of toward the needle. The same situation occurs when using the front course as the departure navigation aid. Therefore, when flying *outbound* on the front course or *inbound* on the back course, reverse interpretation of the CDI is necessary.

APPROACH SEGMENTS

An instrument approach procedure may be divided into as many as five separate segments. These are the *initial*, the *intermediate*, the *final*, and the *missed approach* segments. The fifth segment of an approach is the feeder route. Approach segments usually begin and end at designated fixes; however, some may begin or end at specified points where no fixes are provided. These fixes are named to coincide with the associated segments, i.e., the *intermediate* segment begins at the *intermediate* fix and ends at the final approach fix. The concept of approach segments and the feeder route is illustrated in figures 7-7 and 7-8.

FEEDER ROUTES

Feeder routes are given in the approach plan view to provide for navigation from the enroute structure to initial approach fixes. Feeder routes are recognized by the following criteria.

1. They terminate with a prominent arrowhead.
2. Altitude, course, and distance are shown from each enroute fix to its respective initial approach fix.

INITIAL APPROACH SEGMENT

The *initial approach segment* may be made along an arc, radial, course, heading, radar vector, or a combination of any of these. Procedure turns, holding pattern descents, and high altitude jet penetrations are initial approach segments.

More than one initial approach may be established for a given approach procedure depending

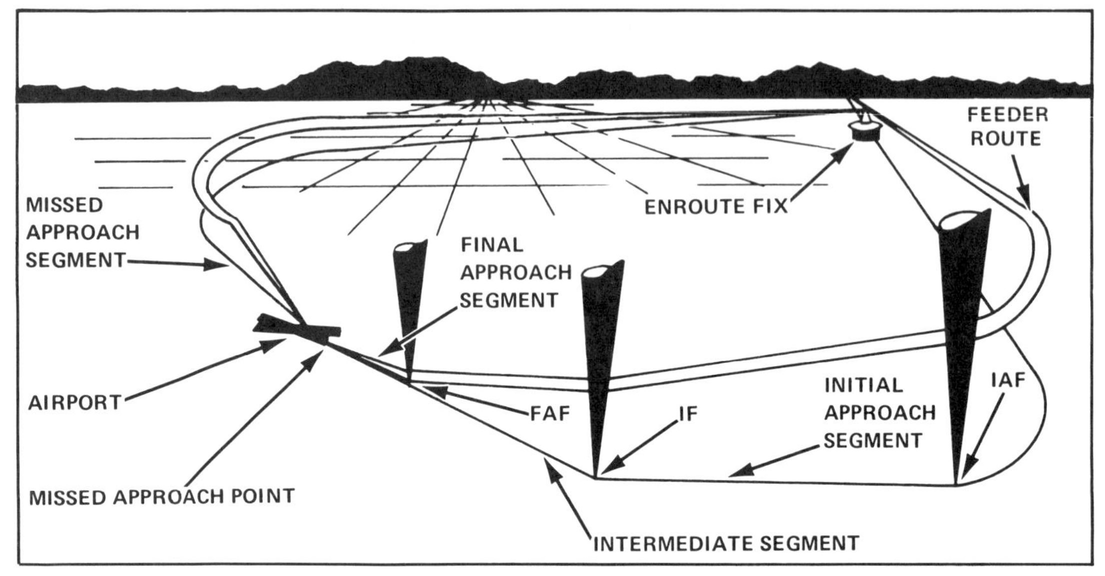

Fig. 7-7. Approach Segments Without Procedure Turn

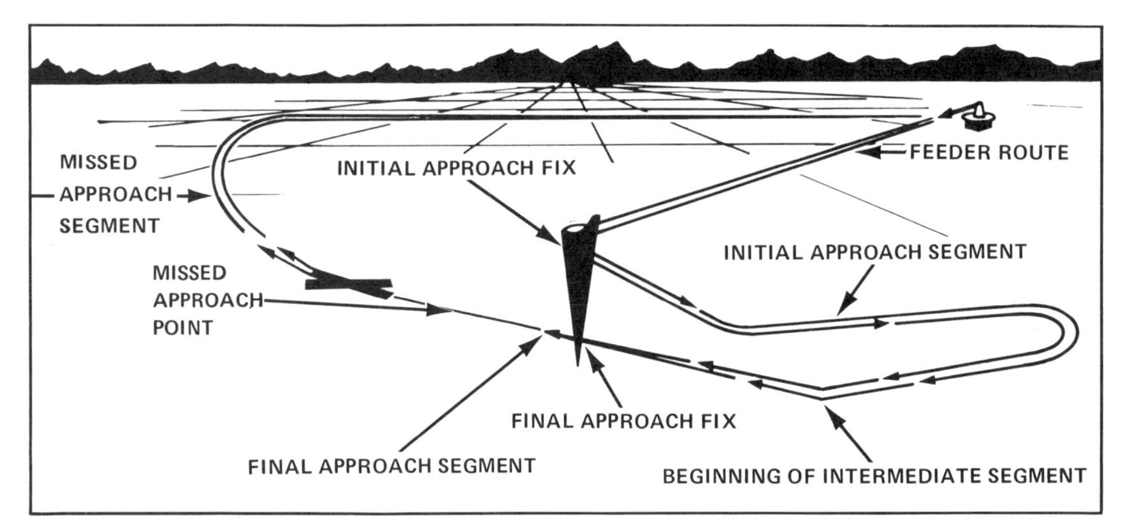

Fig. 7-8. Approach Segments With Procedure Turn

on the direction from which the aircraft is approaching the airport. Minimum altitudes may be prescribed for the initial approach segment and shown on the approach chart; however, the altitudes assigned by ATC for a given approach must be followed.

Sometimes, as in the case of an NDB approach, the initial approach fix is collocated with the final approach fix, as shown in figure 7-8. In this case, the initial approach segment begins when the aircraft passes the nondirectional beacon outbound. It continues through the procedure turn and terminates as the aircraft intercepts the approach course inbound.

INTERMEDIATE SEGMENT

The *intermediate segment* begins at an intermediate fix (IF) or point and ends at the final approach fix (FAF). It may consist of a radial, course, or arc. Its function is to connect the initial approach segment with the final approach segment. The intermediate segment course must not vary from the final approach course by more than 30°. The intermediate segment, shown in figure 7-8, begins at the completion of the procedure turn and ends at the final approach fix.

It is desirable to cross the final approach fix completely ready for the landing. Therefore, such items as adjustments in course and speed and the completion of the prelanding checklist should be accomplished during the intermediate segment.

FINAL APPROACH SEGMENT

The *final approach segment* begins at the final approach fix (FAF) and ends at the missed approach point (MAP). The final approach fix for a precision approach is identified by an outer marker, compass locator, or radar fix. The missed approach point for a precision approach is specified as the point at which an aircraft reaches a certain altitude, called decision height (DH), when descending on the ILS or radar glide slope.

The final approach fix for a nonprecision approach is usually a radio facility (i.e., VOR or NDB) or a DME fix. The missed approach point for a nonprecision approach is usually determined by timing from the final approach fix.

MISSED APPROACH SEGMENT

The *missed approach segment* begins at the missed approach point and ends at the enroute fix or initial approach fix. This segment is illustrated in figures 7-7 and 7-8.

APPROACH CHARTS

FORMAT

Approach charts are published by Jeppesen & Co. and by National Ocean Survey (formerly U.S. Coast and Geodetic Survey), an agency of the U.S. Government. Both Jeppesen and NOS charts contain the necessary data which may be used to properly execute instrument approaches. While the two types of charts contain similar information, in some cases they employ different formats and symbols. Also, the Jeppesen charts contain some additional data designed to provide the pilot with helpful information.

At the present time, only a knowledge of the NOS charts is required for the FAA written examination. However, since virtually all U.S. certificated airlines and a majority of U.S. air taxi and commercial operators use Jeppesen charts, knowledge of both types is recommended.

JEPPESEN FORMAT

The Jeppesen approach charts consist of 5-1/2" x 8-1/2" looseleaf sheets which are carried in binders. The binders are grouped by region, i.e., eastern, central, and western United States. The front side of each approach chart consists of a heading group, communications group, approach plan view, approach profile view, and landing minimums. As described in Chapter 6, an airport chart appears on the reverse side of the first approach chart for each airport. The airport chart contains a heading group, an airport plan view, additional airport data, and takeoff and alternate minimums.

NOS FORMAT

Each NOS approach chart is a 5-1/8" x 8" looseleaf sheet which is carried in a binder similar to Jeppesen binders. These are grouped into six regions, such as northeast and southeast United States.

The NOS approach chart consists of a heading group, communications group, approach plan view, approach profile view, landing minimums, notes and cautions, and an airport plan view; these are all contained on a single page.

HEADING AND BORDER DATA

Figure 7-9 shows the format for heading and border data on both the Jeppesen and NOS charts. The headings and border data are contained in the upper portion of the Jeppesen approach chart. This information is split on the NOS chart; some of the information appears at the top of the page and some appears at the bottom. The ball flag numbers which follow refer to both Jeppesen and NOS charts, as shown in figure 7-9 unless otherwise noted.

❶ Geographic Name

❷ Name of Airport

❸ Name of Approach Procedure(s)
At times, two procedures will be presented on a single Jeppesen chart. For example, a nondirectional beacon (NDB) approach procedure is often shown with an ILS procedure.

❹ Navaid Frequency and Identification
This information appears in the heading group of the Jeppesen chart. On both Jeppesen and NOS charts, the information is presented within the main body of the chart.

❺ Minimum Sector Altitudes
This information appears in the heading group on the Jeppesen chart and is contained within the plan view on NOS charts. The minimum sector altitude (MSA) pro-

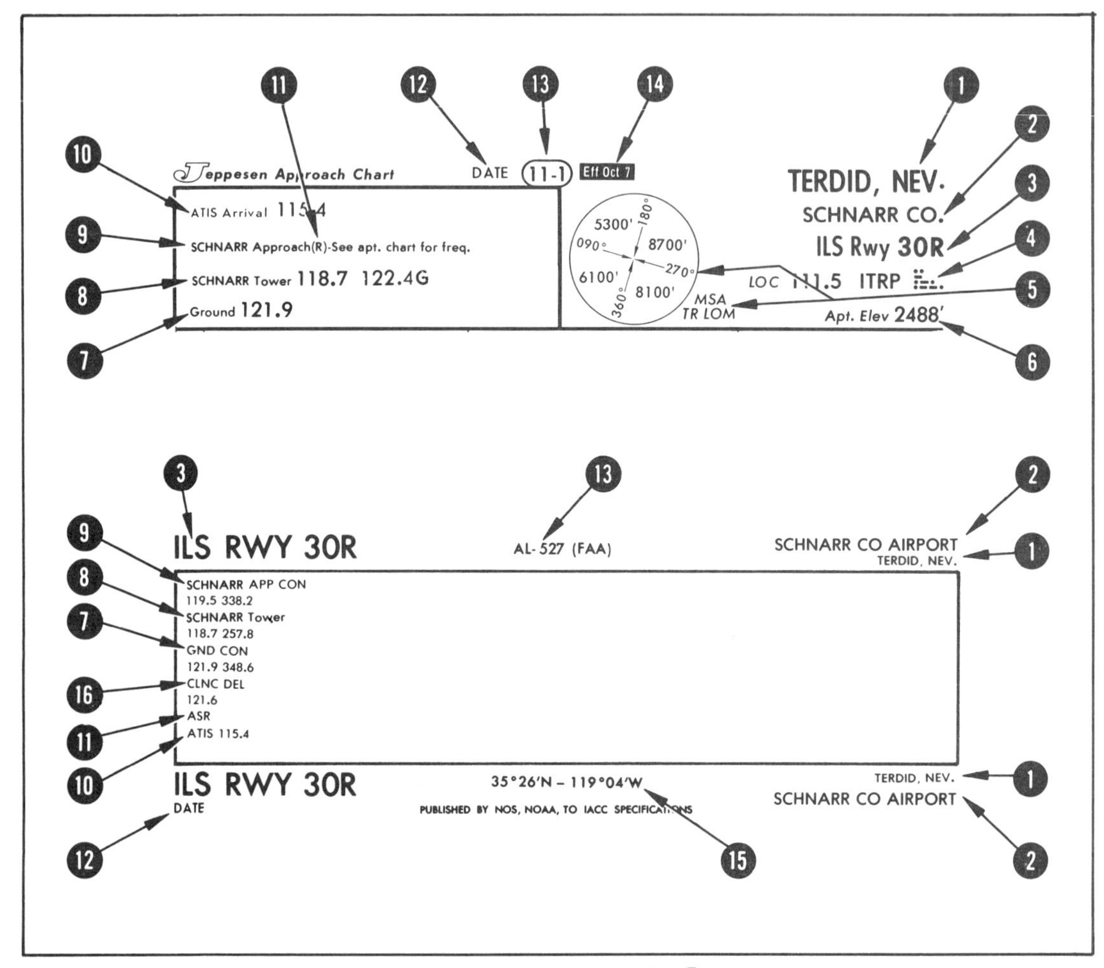

Fig. 7-9. Approach Chart Heading Format

vides for 1,000 feet of obstruction clearance within the sector indicated and within 25 nautical miles of the facility.

The MSA sectors and altitudes are depicted on Jeppesen charts. In addition, the radio facility which serves as a basis for the sectors is named.

In the example, the sectors are based on the low frequency Terps locator beacon. The arrows and associated numbers in the MSA circle indicate the *magnetic bearings to* the station that divide the sectors. When the MSA is predicated on a VOR, the sectors are based on *magnetic courses* to the station.

6 Airport Elevation
This information is presented in the heading and border data on the Jeppesen chart. On the NOS chart, the airport elevation is located at the upper left corner of the airport diagram.

7 Ground Control Frequency

8 Tower Frequency
The first frequency indicated, in this case 118.7, is the primary VHF tower frequency. The second figure shown in the Jeppesen coverage, 122.4, is followed by the letter "G." This indicates that the tower receives on, or guards, 122.4 MHz, but does not transmit on this frequency. The second frequency shown on the NOS chart (257.8) is a UHF frequency which is used primarily by the military.

Tower notes are provided by Jeppesen when tower operation is not continuous or when the altimeter setting must be obtained from another source.

9 Approach Control Frequency

Jeppesen references the airport chart (printed on reverse side of first approach chart for each airport) for approach control frequencies. This information appears on the upper portion of NOS charts and includes a UHF frequency for military use.

At some airports, different frequencies are assigned to different sectors. When the position of the aircraft inbound to the airport falls within the sectors indicated, the pilot is to use the appropriate frequency.

10 Automatic Terminal Information Service

The automatic terminal information service (ATIS) is a transcribed advisory which presents traffic information and existing weather and field conditions. The information is transmitted continuously and updated as necessary by the tower.

11 Radar Services Available

The symbol (R) located in the approach and departure control boxes on the Jeppesen charts indicates the availability of radar.

When ASR appears on the NOS chart, it indicates that both radar vectoring and ASR instrument approaches are available.

12 Date of Publication

On the Jeppesen chart, the pilot should also check **14** to see if an effective date indicates the chart should not be used until some later date.

13 Chart Code Number

The three-digit code number appearing on Jeppesen charts is intended to assist the pilot when making revisions to the approach chart coverage. The first digit represents the airport number and is an arbitrary assignment. If more than one airport exists under a single geographic name, one airport will be assigned the digit 1, the next will be assigned the digit 2, and so on. For example, Stapleton International Airport in Denver area is assigned the code number 1; Jefferson County Airport, northwest of Denver, has the code number 2. The second digit represents the following types of approach charts.

0—Area, SID, STAR, or parking diagram
1—ILS
2—PAR
3—VOR
4—Unassigned
5—Range (low frequency range)
6—NDB (ADF)
7—DF
8—ASR and other
9—Vicinity chart or RNAV

The third digit represents the filing order of the charts of the same type. For example, if an airport has more than one VOR approach, one of the approaches will be given the number 1, another will be given the number 2, etc. In the example in figure 7-9, the code number is 11-1, which indicates the following information.

1. The chart is for the first airport in the Terdid, Nevada, area.
2. The chart type is ILS.
3. The chart is the first ILS chart for the airport.

The code number which appears on the NOS chart refers to the FAA airport designation only. All of the charts for a given airport will have the same number. When making revisions, the pilot must check the airport number; in addition, he must look at the approach procedure name which appears at both the top and bottom left side of the page.

14 Effective Date

The effective date is presented on the Jeppesen chart when it is later than the date of publication.

15 Latitude and Longitude

This is the precise geographic location of the airport reference point and is located at the bottom of the NOS chart. The Jeppesen chart lists the latitude and longitude of the airport reference point in the heading of the airport information chart. This chart is located on the reverse side of the first approach chart for each airport.

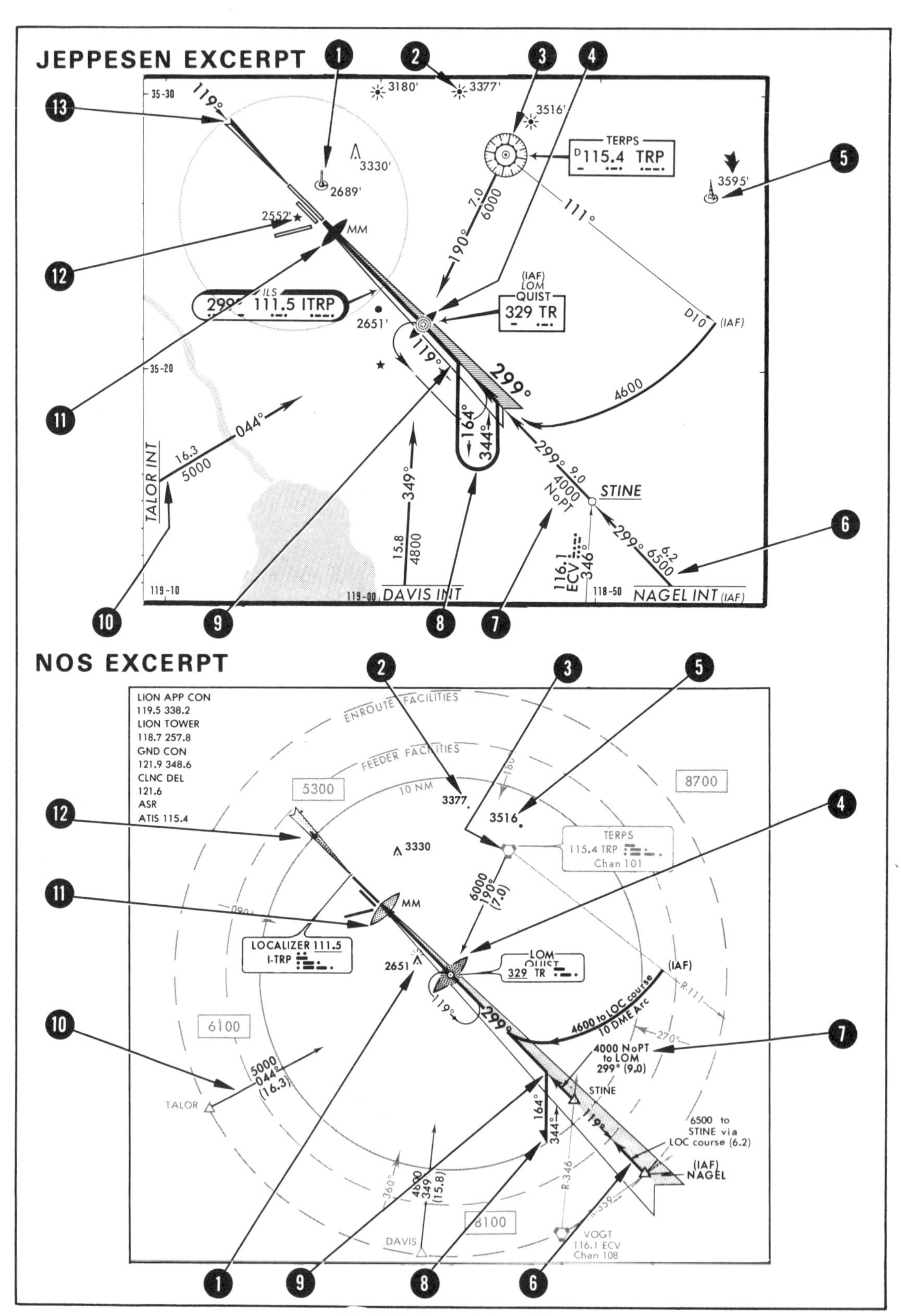

Fig. 7-10. Approach Plan Views

16 Clearance Delivery
Jeppesen charts present the clearance delivery frequency on the airport diagram side of the chart and label it "Cpt," for "Clearance pretaxi." This frequency is tabulated in the NOS format.

APPROACH PLAN VIEW

The purpose of the approach plan view is to show all segments of the approach procedure on a single chart. The ball flags in this portion of the presentation refer to figure 7-10 and apply to both the Jeppesen chart and the NOS chart, except where noted.

1 Manmade Obstruction

2 Terrain or Spot Elevation

3 VORTAC
On this approach, the VORTAC is the beginning of the feeder route. The facility provides for navigation from the enroute structure to the initial approach fix.

4 Outer Marker and Outer Compass Locator
This may be the initial approach fix in addition to being the final approach fix.
NOTE: It is the final approach fix only when the aircraft is aligned within 30° of the final approach course.

5 Highest Obstacle Charted
On the NOS chart, the highest obstacle is identified by a dot in the obstruction symbol and by printing the elevation in heavy type.

6 Terminal Route to Final Approach Course
No procedure turn is made when using these routes.

7 NoPT
This indicates no procedure turn is to be executed unless permission is issued by ATC.

8 Procedure Turn
When the procedure turn is diagramed, as in this example, the pilot may reverse course in any way he desires provided the procedure is executed on the same side of the approach course as the indicated procedure turn. Should the course reversal be diagramed as a racetrack or teardrop pattern, the indicated procedure is mandatory. The procedure turn constitutes an initial approach segment terminating at the point where the aircraft intercepts the approach course inbound, as previously shown in figure 7-8. The intermediate segment begins at this point and terminates at the final approach fix.

9 Beginning of Intermediate Segment
Interception of the approach course inbound constitutes the beginning of the intermediate segment when the procedure turn is used.

10 Feeder Route to Outer Marker Initial Approach Fix
A procedure turn is required when arriving at the outer marker from the Taylor Intersection.

11 Middle Marker
The middle marker is placed approximately 3,500 feet from the runway threshold and is depicted with a fan-shaped symbol on the plan view. The middle marker is located near the missed approach point on ILS approaches. It should be understood that the missed approach point on ILS approaches is determined by the decision height and not the middle marker.

12 Airport Beacon

13 ILS Back Course
In this illustration, the back course of the localizer constitutes the missed approach segment, and the missed approach track is nearly straight ahead. When this situation exists, no indication is made on the Jeppesen ILS approach chart. If the missed approach procedure provides for divergence from the back course ILS, the missed approach track will be shown by a dashed arrow curving in the direction specified by the procedure. NOS shows *all* missed approach tracks by a series of dashes in the direction in which the missed approach is to be made.

APPROACH PROFILE VIEW

The approach profile view shows the segments of the approach with particular emphasis on the altitudes at various points along the approach. The ball flags in the following text correspond with figure 7-11. The ball flags apply to both the Jeppesen and NOS coverages, except where noted.

1 Nonprecision Approach Path (Jeppesen charts only)

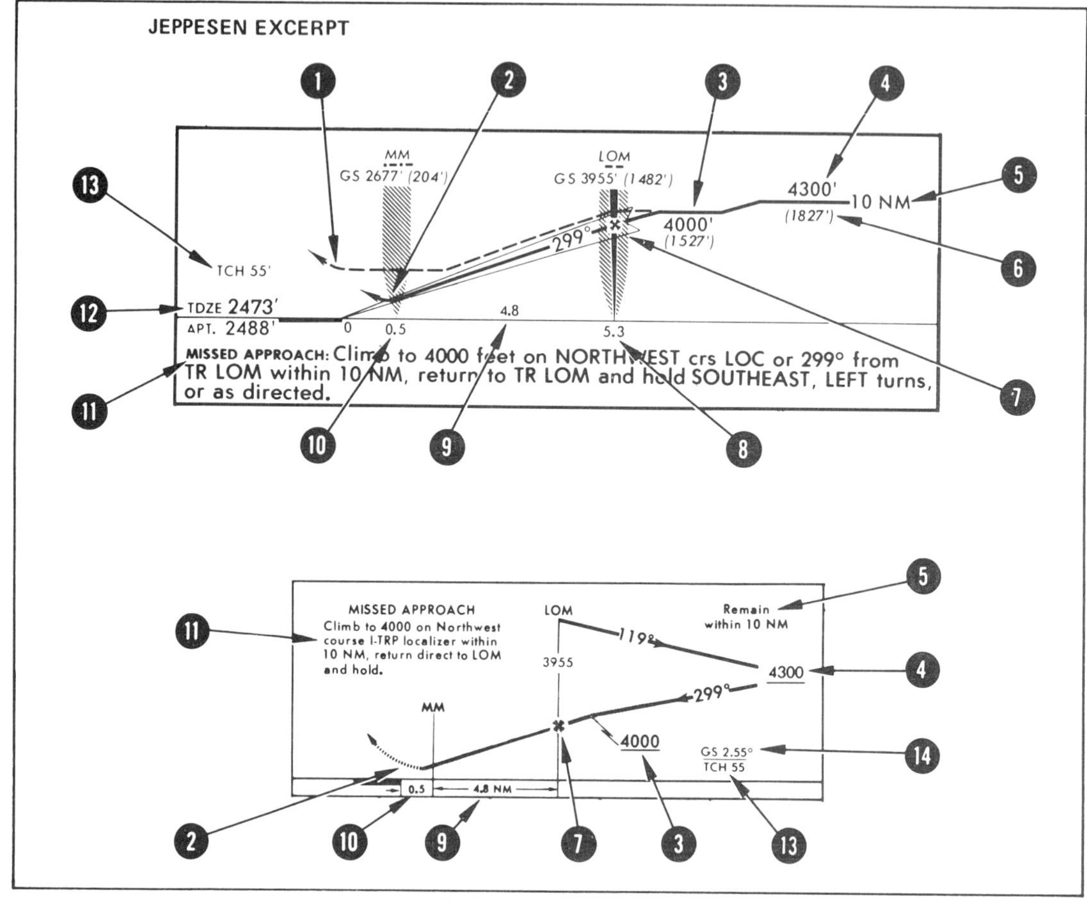

Fig. 7-11. Approach Profile View

2 Missed Approach Point for the ILS

3 Glide Slope Interception Altitude
This altitude will be different than the FAF crossing altitude when the glide slope is intercepted before reaching the FAF.

4 Procedure Turn Altitude
On Jeppesen charts, this is the minimum procedure turn altitude. The underline beneath the procedure turn altitude on the NOS chart indicates the specified altitude is a minimum altitude. A line above the altitude figure indicates a maximum altitude and a line both above and below the altitude specifies a mandatory altitude.

5 Distance
This is the distance from the approach fix within which the procedure turn must be completed.

6 Procedure Turn Height
The altitude shown within parentheses on the Jeppesen chart is the height above the touchdown zone.

7 Final Approach Fix (FAF)
When flying on the glide slope (in the example), the pilot will cross the FAF at 3,955 feet MSL. The FAF is identified by a maltese cross.

8 Distance (Jeppesen charts only)
This is measured from the outer marker to the end of the runway. Distances shown beneath the line on the Jeppesen chart are with respect to the placement of the zero. In this case, the zero is located at the runway threshold.

9 Distance
This is the distance between the outer marker and middle marker. Distances shown above the line on the Jeppesen chart are distances between facilities.

10 Middle Marker (MM)
The distance shown on both the Jeppesen and NOS charts is the distance of the middle marker from the runway threshold. The figure shown within parentheses above the MM symbol on the Jeppesen chart is the altitude of the glide slope at that point with respect to the touchdown zone.

11 Narrative of Missed Approach Procedure

12 Touchdown Zone
The touchdown zone elevation is the highest elevation along the first 3,000 feet of the straight-in runway.

13 Threshold Crossing Height (TCH)
The TCH indicates the altitude at which the glide slope crosses the runway threshold.

14 Glide Slope Angle
This angle is shown on NOS charts in the profile section and on Jeppesen charts immediately under the landing minimums section.

MINIMUMS

The instrument pilot must understand takeoff and landing minimums. The minimums, established by the FAA, vary from airport to airport, because of the existence of obstacles within the approach or missed approach path, lighting aids, and the number of radio approach aids operating at any given time.

TAKEOFF MINIMUMS

Statistics show that aircraft malfunctions are more likely to present themselves at the beginning of a flight than at other times. For this reason, the FAA has established rules which provide takeoff minimums for aircraft operated under FAR Parts 121, 123, 129, and 135. These rules provide for return to the departure airport or flight to another nearby airport, should difficulties occur immediately after takeoff. *These rules do not specifically apply to aircraft operating under FAR Part 91;* however, commonsense and good judgment suggest that all pilots comply with these minimums.

The standard takeoff minimums are, for aircraft with two engines or less, one statute mile visibility and, for aircraft with more than two engines, one-half statute mile visibility.

These takeoff minimums are provided at the bottom of each Jeppesen airport chart. On NOS charts the takeoff minimums are not printed when standard takeoff minimums apply at the airport. If other than standard minimums are established at an airport, the symbol ▽T will appear in the information box at the bottom of the chart. This symbol indicates that the pilot must consult the tabulation of IFR takeoff minimums which accompanies the NOS approach charts.

LANDING MINIMUMS

For landing, unless the pilot has *visual contact* with the *runway environment* upon arriving at the missed approach point (MAP), a missed approach is mandatory. The runway environment may be defined as the runway itself, any lighting associated with the runway, or the approach path. The minimums are shown on the front side of each Jeppesen and NOS approach chart, as shown in figure 7-12. These minimums apply only to the approach pictured.

AIRCRAFT CATEGORIES

Different approach minimums are provided for different aircraft categories. These categories are based on the aircraft approach speed which is established as 130 percent of the stall speed in the landing configuration at the maximum landing weight. Aircraft approach speed is expressed as $1.3V_{SO}$. However, some aircraft fall into the next higher category during a circling approach. The aircraft categories are shown in the table below. They also are listed in the far left column on Jeppesen charts. (See Fig. 7-12, item 1.)

APPROACH CATEGORY	SPEED (KNOTS)
A	Less than 91 knots
B	91 knots or more but less than 121 knots
C	121 knots or more but less than 141 knots
D	141 knots or more but less than 166 knots
E	166 knots or more

DESCENT LIMITS

Descent limits are minimum altitudes to which an aircraft may descend during an instrument approach while the pilot is referring solely to instrument references. For *precision approaches,* this altitude is called the *decision height* (DH) and is shown in the minimums presentation as

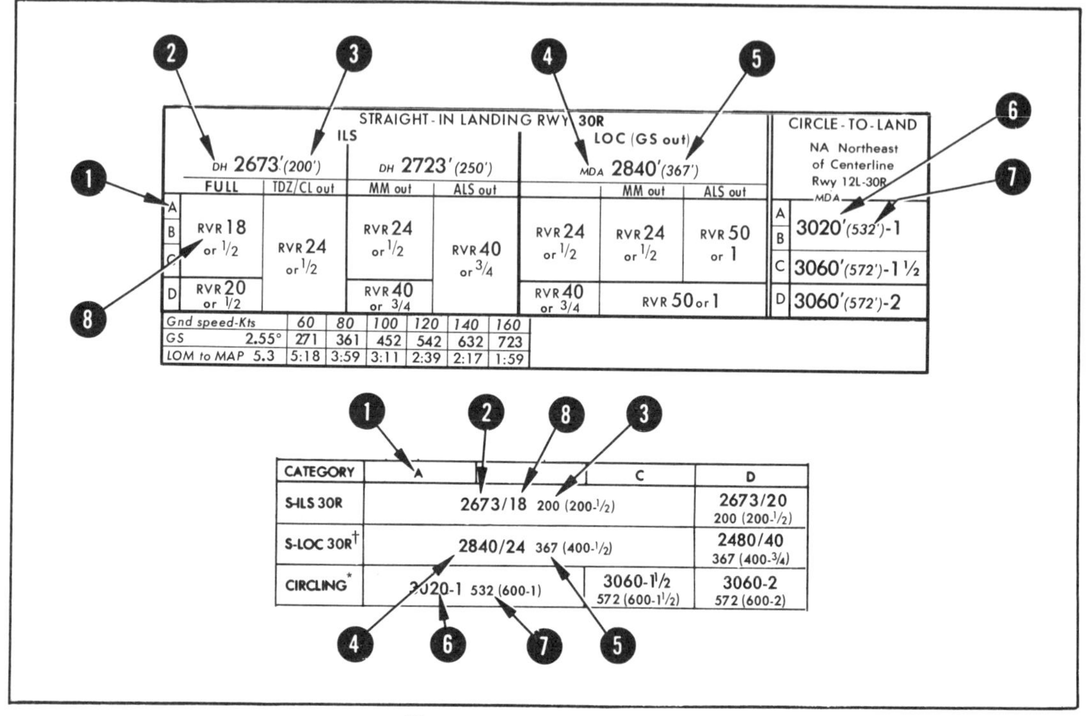

Figure 7-12. Landing Minimums

an MSL altitude in bold type. (See Fig. 7-12, item 2.) The height above touchdown (HAT) is shown in parentheses on the Jeppesen chart and to the right of the visibility on the NOS chart, as shown in item 3. When the pilot reaches DH on a precision approach, he must *decide* whether to continue the approach to a landing by reference to *external visual cues* or to execute the missed approach procedure.

For nonprecision approaches, the descent limit is the *minimum descent altitude* (MDA). It is shown in bold type (item 4) as an MSL altitude. It also is shown to the right of the MSL altitude as the height above touchdown (HAT), as shown by item 5. Circling approaches, by virtue of the fact that are conducted to an MDA rather than a DH, may be considered as on nonprecision approaches. As such, the minimum altitudes indicated by item 6 are MDAs, which are expressed as a height above airport (HAA), as shown by item 7.

When executing a nonprecision approach, the pilot should initiate a descent and begin timing the approach when he passes the final approach fix. If he arrives at the MDA prior to the expiration of approach timing, he must level off at that altitude and continue inbound until the expiration of the approach time. The missed approach point (MAP) for all non-precision procedures except VOR-DME and RNAV approaches is based upon *time* rather than a fixed point in space. During a VOR-DME approach, the MAP occurs at a specified DME mileage from a VORTAC station.

VISIBILITY REQUIREMENTS

When the missed approach point (MAP) has been reached, the pilot must either proceed visually or execute a missed approach procedure. For this reason, visibility criteria have been established which *indicate* whether it will be possible for the pilot to proceed visually from the MAP. It must be understood that the reported visibility is only a *prediction* that the pilot will have sufficient visual reference to continue the approach, since there is no way to guarantee this from a ground observation.

There are two methods used to measure visibility. One method utilizes a human observer. This measurement is reported in statute miles and fractions. The second, and more precise method, runway visual range (RVR), utilizes electronic equipment to measure the opacity of the air along a runway. The RVR measurement is reported in hundreds of feet of horizontal visibility along the runway indicated. There is a possibility that both RVR and visibility may be reported at the same time. The visibility requirements for executing an approach are shown in figure 7-12, item 8. The first number in this case is the RVR in hundreds of feet.

The second number on the Jeppesen chart represents the required visibility in fractions of miles, if the RVR is inoperative. On the NOS chart, the number in parentheses represents the ceiling and visibility requirements for military operations.

INOPERATIVE COMPONENTS AND VISUAL AIDS

On Jeppesen approach charts, the minimums column to the extreme left shows the minimums for an approach with all the ILS components and visual aids installed and operating. This condition allows the lowest possible minimums. Higher minimums appear to the right of this column. These apply when the components or visual aids shown at the top of the column are inoperative.

The pilot must refer to the table of inoperative components or visual aids (figure 7-13) when NOS charts are used. For example, the NOS minimums for a full ILS approach to runway 30R (figure 7-12) are DH 2,673 MSL, RVR 1,800, or visibility one-half mile. If the touchdown zone lights are inoperative, the decision height remains unchanged, but the visibility requirements must be increased to an RVR of 2,400 feet. This table is presented as a part of the preface material for NOS charts. This same information also can be found on the Jeppesen charts by referring to the column labeled "TDZ/CL out" and by reading the new visibility requirement of one-half mile or an RVR of 2,400 feet.

Instrument Approach Procedures (Charts)
INOPERATIVE COMPONENTS OR VISUAL AIDS TABLE

(1) ILS, MLS, and PAR

Inoperative Component or Aid	Approach Category	Increase DH	Increase Visibility
MM*	ABC	50 feet	None
MM*	D	50 feet	¼ mile
ALSF 1 & 2, MALSR, & SSALR	ABCD	None	¼ mile

*Not applicable to PAR

(2) ILS with visibility minimum of 1,800 or 2,000 RVR.

MM	ABC	50 feet	To 2400 RVR
MM	D	50 feet	To 4000 RVR
ALSF 1 & 2, MALSR, & SSALR	ABCD	None	To 4000 RVR
TDZL, RCLS	ABCD	None	To 2400 RVR
RVR	ABCD	None	To ½ mile

(3) VOR, VOR/DME, VORTAC, VOR (TAC), VOR/DME (TAC), LOC, LOC/DME, LDA, LDA/DME, SDF, SDF/DME, RNAV, and ASR

Inoperative Visual Aid	Approach Category	Increase MDA	Increase Visibility
ALSF 1 & 2, MALSR, & SSALR	ABCD	None	½ mile
SSALS, MALS & ODALS	ABC	None	¼ mile

(4) NDB

ALSF 1 & 2, MALSR, & SSALR	C	None	½ mile
	ABD	None	¼ mile
MALS, SSALS, ODALS	ABC	None	¼ mile

(REFER TO NOS FOR CURRENT TABLE)

Fig. 7-13. NOS Table of Inoperative Components

TIME AND RATE-OF-DESCENT TABLES

Both Jeppesen and NOS charts provide tables which show the time required to fly from the final approach fix (FAF) to the missed approach point (MAP) for various *groundspeeds.* For ILS approaches, this data is used when the electronic glide slope is not available. If this occurs, the approach becomes a nonprecision procedure, different minimums apply, and the pilot must determine the missed approach point by timing from the FAF.

The pilot must apply the wind factor to his approach speed in order to arrive at his groundspeed. Then he may apply his groundspeed to the table and interpolate as necessary. For example, in the upper table in figure 7-14, time from the final approach fix to the missed approach point at 80 knots groundspeed is 3 minutes, 36 seconds. At a groundspeed of 100 knots, the time is 2 minutes, 53 seconds. If the groundspeed is determined to be 90 knots, the time from the final approach fix to the missed approach point will be 3 minutes, 12 seconds.

JEPPESEN

Gnd speed-Kts	60	80	100	120	140	160
G.S. 2°41'	285	380	475	570	665	760
LOM to MAP 4.8	4:48	3:36	2:53	2:24	2:04	1:48

NOS

FAF to MAP 4.8 NM					
Knots	60	90	120	150	180
Min:Sec	4:48	3:12	2:24	1:55	1:36

Fig. 7-14. Groundspeed/Time/Rate of Descent Tables

Jeppesen also provides vertical velocity information for all precision approaches. In the upper illustration in figure 7-14, with a groundspeed of 80 knots, the vertical velocity required to maintain the glide slope is 380 feet per minute. In addition, the pilot may find it necessary to interpolate this table. If, for example, his approach *groundspeed* is 90 knots, he must maintain a rate of descent of 427 feet per minute in order to remain on the glide slope.

The rate of descent is also provided on some Jeppesen nonprecision approach charts. For these approaches, the rate of descent is predicated upon a constant rate of descent from the final approach fix to the runway.

When the approach radio facility is located at the airport, no timing figure is provided within the table for nonprecision approaches. This is because the missed approach point is designated by passing the station rather than by timing the approach.

GENERAL PROCEDURES FOR INSTRUMENT APPROACHES

While the specific procedures utilized within the various instrument approaches vary, there are certain elements common to all. For example, when flying IFR, the routing segment between the enroute structure and the approach phase must be flown in accordance with the clearance issued by ATC. In the case of those procedures for which terminal routes are provided, this clearance will utilize the terminal routes pictured on the approach procedure whenever possible. In figure 7-15, the pilot may expect to be routed via one of the terminal routes shown on the chart, or to receive a radar vector to the final approach course.

When the initial approach fix and the final approach fix utilize the same facility, the pilot will track outbound from the *initial approach fix*, reverse course by executing a procedure turn, then track inbound to the *final approach fix*, as shown on the plan view diagram. On the other hand, if the flight is being radar vectored to the final approach course, no procedure turn may be executed. Also, if the flight is cleared in such a manner it will intercept the approach course inbound by flying the 17 nautical mile DME arc or a direct course inbound from the Knock Intersection, the pilot may fly direct to the final approach fix without executing a procedure turn and commence the final approach segment.

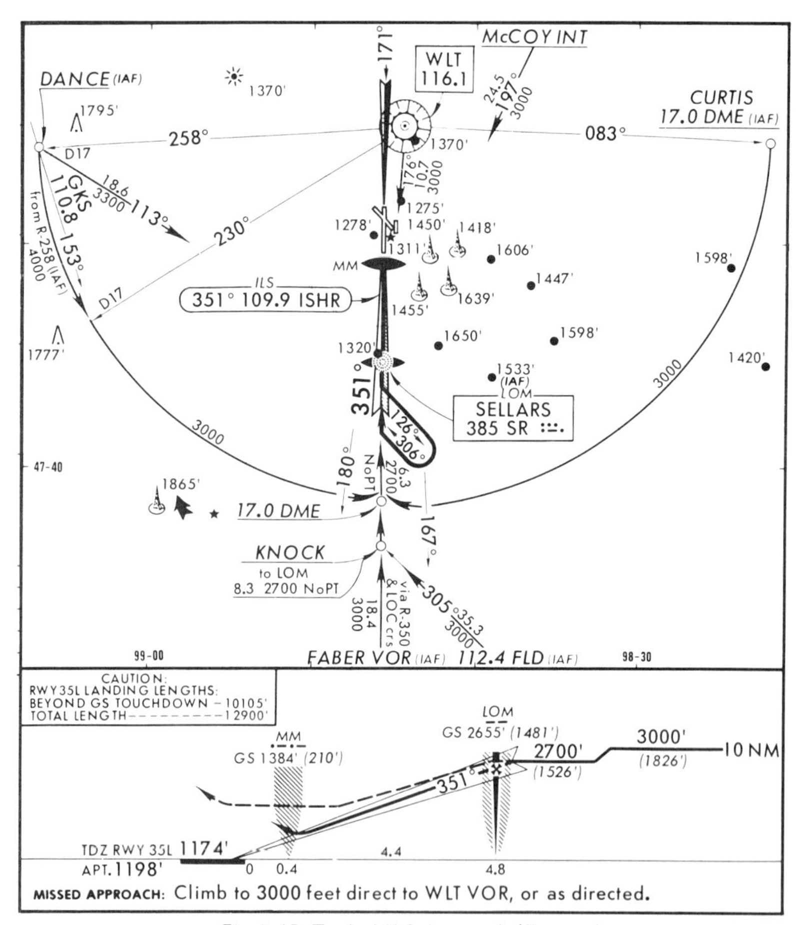

Fig. 7-15. Typical ILS Approach (Excerpt)

The altitude at which the pilot must fly will be shown on the chart for the specific segment being flown, unless a higher altitude is assigned by ATC. For example, if the pilot is cleared to fly to Knock Intersection and maintain 4,000 feet, the intersection is the clearance limit. If an approach clearance is received prior to reaching the Knock Intersection, the pilot may descend to the altitude posted for the route to that intersection, then to 2,700 feet between the intersection and the outer marker. An approach clearance constitutes a *change in altitude at the pilot's discretion.* Without an approach clearance, the pilot must maintain the last assigned altitude.

The altitude prescribed for the commencement of the final approach phase is critical. If the flight arrives at the final approach fix at an altitude significantly different from the altitude shown for initiation of the final approach phase, ATC should be notified and a revised clearance requested.

After a revised clearance is issued, the pilot can descend to the altitude prescribed for the procedure turn. For example, if the flight is cleared inbound from Knock Intersection at 4,000 feet and ATC does not issue an approach clearance or a clearance to descend, the pilot might arrive at the outer marker still at 4,000 feet. If the pilot is given approach clearance, however, and then does not have time to descend all the way to 2,700 feet, the flight might arrive at the outer marker at 3,300 feet. In these examples, the alternative procedure is to obtain an amended clearance and reduce altitude in a holding pattern. Normally, the pattern assigned will position the aircraft inbound along the final approach course prior to the FAF. Otherwise, the published procedure turn may be required for positioning prior to beginning the final approach. All aircraft maneuvering should be accomplished at the holding airspeed (if there are no speed restrictions by ATC) after passing the initial approach fix. This allows the pilot

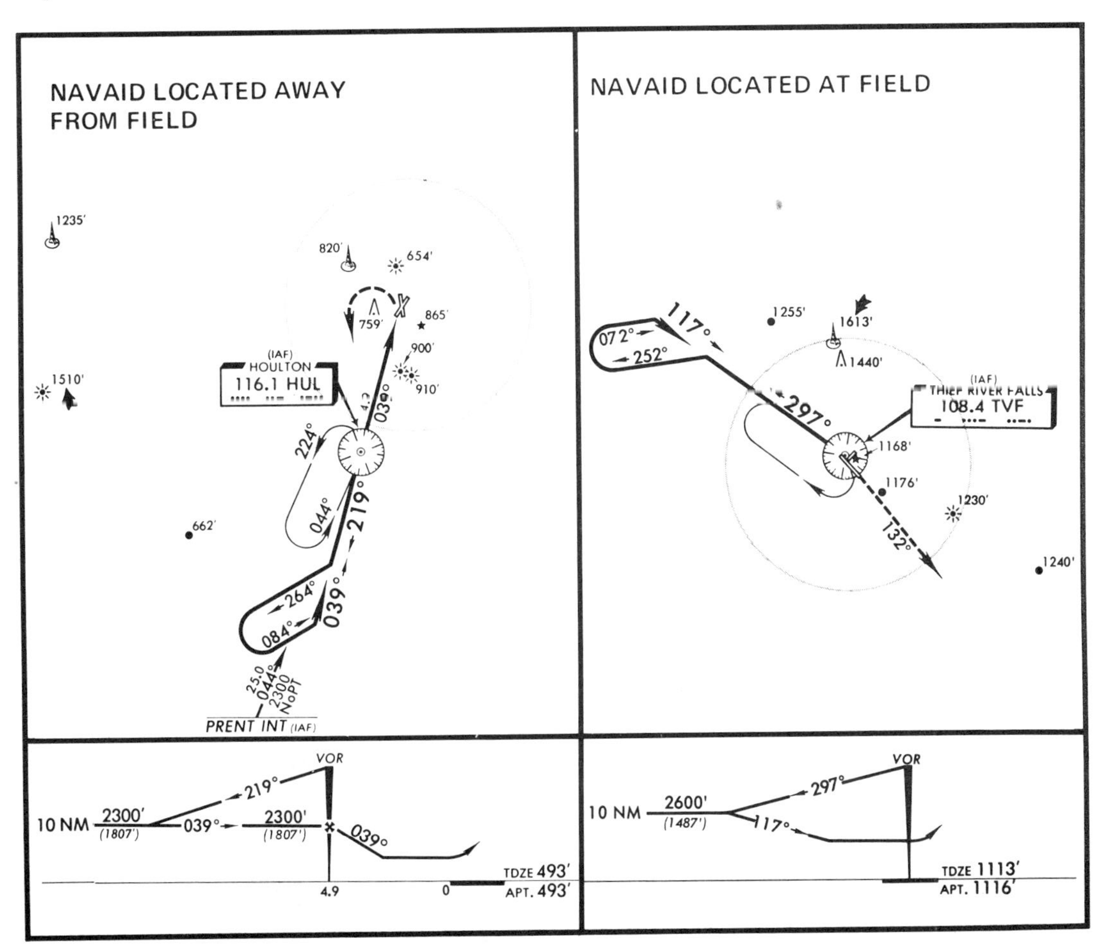

Fig. 7-16. Nonprecision Approaches

complete control over the aircraft and time to properly execute the procedure.

NONPRECISION APPROACHES

Nonprecision approaches may be grouped into two categories — those where the navaid is located away from the field, and those where the navaid is located on the field, as shown in figure 7-16. The procedures to be flown are the same for both approaches up until the time the pilot completes the procedure turn. When the navaid is located away from the field, as in the left example, the procedure turn is completed and the aircraft is established on the approach course inbound. Descent may then be made to the final approach fix crossing altitude which, in this case, is 2,300 feet. After the final approach fix is crossed inbound, further descent can be made to the *minimum descent altitude*. Timing begins upon crossing the final approach fix. The missed approach point occurs when the elapsed time for the approach ends, as determined by the table shown in the lower corner of the approach chart.

When the navaid is located at the field, as in the right example in figure 7-16, the procedure turn will be completed at the prescribed altitude. When *established* inbound on the approach course, the pilot will initiate a descent to MDA. The missed approach point for this type of approach is recognized when station passage occurs at the navaid used for the approach.

Some nonprecision approach procedures are diagramed with other than standard procedure turns. Two examples of these are shown in figure 7-17. In one such procedure, shown in the left illustration, the final approach fix (marked by the cross in the profile view) is an intersection of radials from two different navaids. In this particular case, the approach course is shown as the 268° radial of the Muncie VOR.

The final approach fix is Teeks Intersection, the point at which the 268° radial of the Muncie VOR crosses the 160° radial from Marion VOR (MZZ and 108.60 MHz.). In this case, the pilot will begin timing and descend to the MDA when Teeks Intersection is crossed. The missed approach point is determined by timing from Teeks Intersection.

Another type of approach is shown in the right illustration in figure 7-17. In this case, the course reversal is accomplished in a holding pattern. The final approach fix for this particular approach is an NDB. The missed approach point is determined by timing from the FAF.

DME is used in other ways for approach procedures, as shown in figure 7-18. One of the methods for navigating to the final approach course is flying a 10 nautical mile DME arc around the station at an altitude of 3,800 feet MSL until the inbound course is intercepted. In this case, the pilot should not execute the procedure turn, as indicated by "NoPT" on the inbound course. On the other hand, if the DME is inoperable, the pilot should navigate in accordance with clearances to the Hays VORTAC, track outbound, execute a procedure turn, then complete a normal VOR approach.

Figure 7-18 also shows that DME can be used to obtain lower landing minimums. When the four-mile DME stepdown fix is identified, the MDA is 140 feet lower than for an approach without DME.

CONTACT AND VISUAL APPROACHES

Frequently, a pilot flying on an IFR flight plan will find that his destination airport is experiencing VFR conditions. He may wish to retain his IFR clearance until touchdown, but determines that he could expedite his approach and landing by shifting to visual references, and eliminating the time required to execute a complete instrument approach. For this reason, the *contact approach* and *visual approach* have been designed. Both of these approaches relieve the pilot of the requirement to *fly the instrument approach procedure*, but they allow him to *retain* his IFR clearance. Upon accepting such a clearance, however, the pilot assumes the responsibility for avoidance of other traffic.

CONTACT APPROACH

A contact approach is made by visual reference to identifiable objects on the ground. It is *requested by the pilot* on an IFR flight plan when he is *clear of the clouds*, has at least *one*

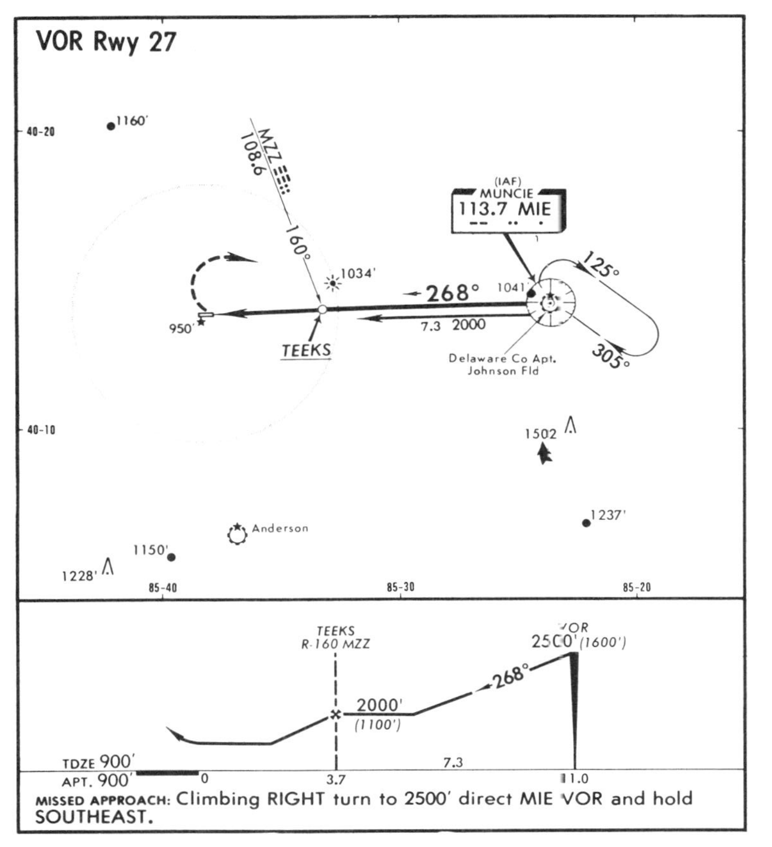

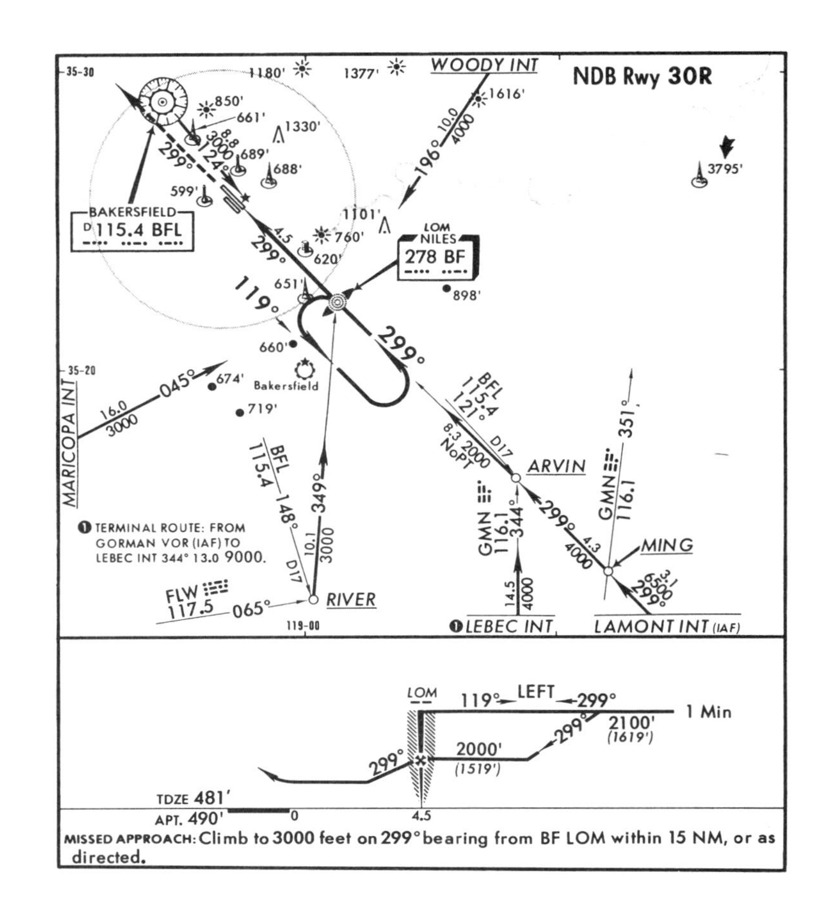

Fig. 7-17. Approaches with Non-Standard Procedure Turns

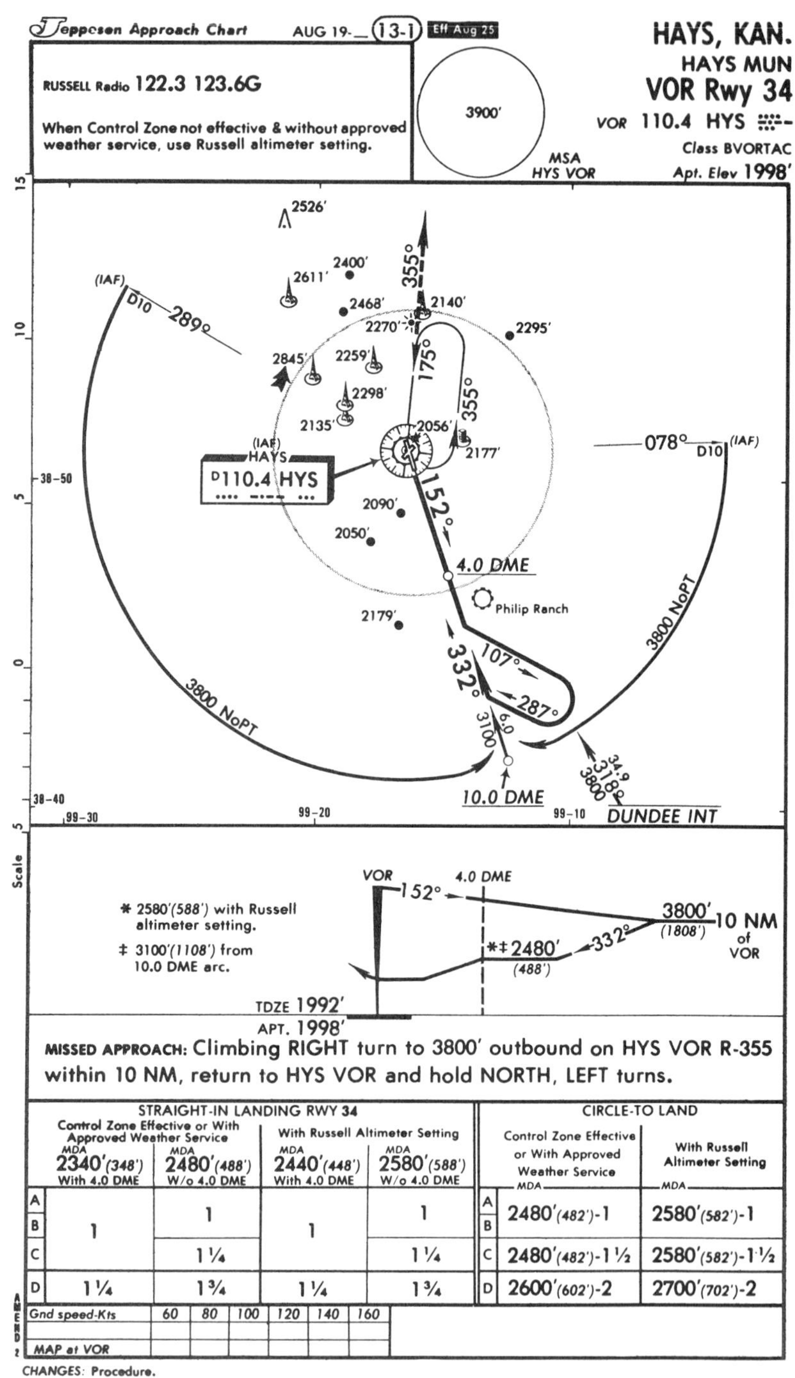

	STRAIGHT-IN LANDING RWY 34				CIRCLE-TO LAND	
	Control Zone Effective or With Approved Weather Service		With Russell Altimeter Setting		Control Zone Effective or With Approved Weather Service	With Russell Altimeter Setting
	MDA 2340'(348') With 4.0 DME	MDA 2480'(488') W/o 4.0 DME	MDA 2440'(448') With 4.0 DME	MDA 2580'(588') W/o 4.0 DME	MDA	MDA
A	1	1	1	1	2480'(482')-1	2580'(582')-1
B	1	1	1	1	2480'(482')-1	2580'(582')-1
C	1	1¼	1	1¼	2480'(482')-1½	2580'(582')-1½
D	1¼	1¾	1¼	1¾	2600'(602')-2	2700'(702')-2

Gnd speed-Kts	60	80	100	120	140	160
MAP at VOR						

CHANGES: Procedure.

Fig. 7-18. Use of DME in the VOR Approach

mile flight visibility, and can maintain continuous visual contact with the ground to complete the approach and landing. The weather requirements are the same as those for *special* VFR. The contact approach is normally used at airports which do not have radar facilities. Upon requesting and accepting a contact approach, the pilot assumes the responsibility for maintaining separation from VFR aircraft, terrain, and obstructions. Figure 7-19 shows the criteria for contact and visual approaches.

VISUAL APPROACH

Visual approaches permit a pilot on an IFR flight plan and under the control of an air traffic control facility to proceed to the destination airport in VFR conditions. After a pilot reports sighting the airport or an aircraft which is to be followed, a clearance for a visual approach can be issued.

Clearance for a visual approach may be issued by approach control on its own initiative, or upon pilot request. Normally, a pilot will be directed to contact the tower immediately after clearance for a visual approach is issued. When the pilot is told to contact the tower, radar service is *automatically* terminated. The tower will then issue traffic information, a landing sequence number, and clearance for the aircraft to land at the appropriate time.

	CONTACT	VISUAL
Weather	Ground visibility of at least one mile Aircraft must remain clear of clouds and have one mile visibility	Aircraft must maintain VFR conditions at all times.
Airport	Airport must have standard or special instrument approach procedure	Pilot must have airport or identified preceding aircraft in sight.
Pilot Action	Can only be requested by the pilot; not initiated by ATC	Initiated by ATC or requested by the pilot.

Fig. 7-19. Contact and Visual Approach Criteria

When an airport has no tower or weather reporting facility, ATC may authorize a "visual approach: if the pilot reports that descent and flight to the airport can be made in VFR conditions. The *pilot* becomes responsible for safe separation from all aircraft and obstructions upon receipt of visual approach clearance.

INSTRUMENT APPROACH TECHNIQUES

The successful execution of an instrument flight, and particularly, of an instrument approach, depends on precision flying and a thorough knowledge of the aircraft and ATC procedures. Of equal importance is the use of a well-organized, professional technique used in every flight. The following are some basic recommendations to help develop such a technique.

Every available means of navigation should be used at all times, even when on radar vectors. The pilot should always know his position, fuel status, and the current weather situation. At least one alternate course of action should be available at all times.

All charts should be arranged in the order in which they will be used. Good cockpit organization is an essential role in all flights.

The approach and missed approach procedures for the destination airport should be reviewed while enroute. The missed approach should be committed to memory. The pilot should not attempt to read a chart while executing a missed approach.

The cockpit should be cleaned up during the transition from enroute to the initial approach segment. All unnecessary charts and equipment should be put away. Depending on the type aircraft being flown, a speed reduction may be made during this portion of the flight.

The radios and navigation indicators should be set up for the approach as early as possible. Completion of these activities during the early portion of the approach reduces pilot workload during more critical segments.

During the initial approach segment, the aircraft should be slowed to approach speed and the prelanding checklist accomplished. Any final adjustments to the radios generally should be made during this portion of the approach. If flying a retractable gear aircraft, the wheels may be left up, if desired, until crossing the final approach fix.

The intermediate segment of the approach should be devoted to accurate alignment with the final approach course and to any airspeed and altitude corrections necessary. The accuracy of the heading indicator should be checked with the magnetic compass.

When crossing the final approach fix in a complex airplane, the landing gear should be lowered to establish the descent. The added drag causes a descent without the need to adjust power. In addition to using the landing gear to establish approach descents, most experienced pilots have previously determined the power settings and aircraft configurations that produce the desired rate of descent at the airspeed they normally use.

When flying a nonprecision approach, a 500 f.p.m. rate of descent should be established and maintained until reaching the minimum descent altitude (MDA). This altitude *must* be maintained until reaching the missed approach point unless visual contact is established.

If the runway environment is not in sight or the aircraft is not in a position for a normal approach at the MAP, a missed approach must be executed. Also, a missed approach must be executed any time these conditions can not be met after passing the MAP.

In the event of a missed approach, the procedures must be followed precisely. Since ATC may assign alternate procedures not published, the pilot should copy any missed approach instructions issued. If the pilot cannot comply with the published or issued missed approach procedure, he should inform ATC and request a different clearance.

VISUAL DESCENT POINT

Some nonprecision approach procedures incorporate a *visual descent point* (VDP). The VDP is a defined point on the final approach course of a nonprecision straight-in approach procedure from which a normal descent from the MDA to the runway touchdown point may commence. Descent below the MDA at this point requires visual reference and alignment with the runway of intended landing, as specified in FAR 91.117. VDPs normally are identified by DME on VOR and LOC procedures or a marker beacon on NDB procedures. VASI usually is available where a VDP is established.

To fly a VDP procedure, the pilot must maintain the MDA until he reaches the VDP *and* acquires the necessary visual references. When an aircraft lacks the required navigational equipment to identify the VDP, pilots should fly the approach as though the VDP were not established. VDPs are portrayed on both Jeppesen and NOS approach charts by a "V" symbol.

LDA AND SDF APPROACHES

The *localizer-type directional aid* (LDA) and the *simplified directional facility* (SDF) are very similar to a standard localizer. Each of these systems operates within a frequency range of 108.10 MHz to 111.95 MHz and provides course guidance to a specific runway.

The approach techniques and procedures to be used with both the LDA and SDF approaches are essentially identical to those employed while executing a standard localizer approach. Since both facilities transmit the same type of signal as a standard localizer, the pilot simply tunes the navigation radio to the appropriate frequency and monitors the navigation indicator. The course selector setting has no effect on the CDI indications.

The primary differences between the localizer, LDA, and SDF facilities are the placement of the transmitting antenna, course width, and course alignment. Due to these differences, most LDA and SDF approaches do not have usable back courses.

8 ATC PROCEDURES

ATC CLEARANCES

Aircraft flying in *controlled* airspace under IFR must file an IFR flight plan, and an ATC clearance must be obtained. ATC clearances provide separation of aircraft on IFR flights in controlled airspace. The separation is maintained vertically by assigning altitudes for each flight, longitudinally by time separation, and laterally by assigning different flight paths. A pilot flying *outside* of controlled airspace is *not* required to obtain an ATC clearance; however, other aircraft may be operating in the same area, and IFR traffic separation is not provided.

Compliance with an ATC clearance is *required* by regulations after the clearance is accepted. The pilot in command may not deviate from that clearance, unless an amended clearance is obtained, or an emergency so dictates. If a pilot is *unable* to comply with a clearance, he should refuse it and request a substitute.

The pilot may cancel an IFR clearance anytime the aircraft is operating in VFR weather below 18,000 feet MSL. It should be remembered at this point, that the flight must be conducted *strictly* in VFR conditions *from that point on*, and should IFR weather again be encountered, the pilot must *file* a new flight plan and *obtain* an IFR clearance or *adjust course* to remain VFR.

If several clearances are obtained while enroute, the last clearance received has *precedence* over all *related* items in any preceding clearances. If deviation from a clearance is required by an emergency, and ATC has given priority to the aircraft, the pilot in command *may* be requested to submit a written report within 48 hours to the chief of that ATC facility.

CLEARANCE READBACK

Pilots of airborne aircraft are expected to read back those portions of ATC clearances and instructions containing altitude assignments or vectors. There is no requirement that other ATC clearances be read back as an unsolicited or spontaneous action. However, controllers may request that a complex clearance be read back.

The pilot should read back the clearance if he feels the need for confirmation. Even though it is not specifically stated, it is generally expected that the pilot read back the enroute clearance when received from clearance delivery, ground control, or a flight service station.

CLEARANCE REQUESTS

A pilot planning an IFR cross country must first file his IFR flight plan, then receive his instrument clearance prior to departure. Depending upon the airport and facilities available, the pilot may receive his IFR clearance from clearance delivery, ground control or, if there is not a control tower in operation, the nearest flight service station.

In some situations, however, the pilot may not wish to file for a complete route. An example of this type of clearance request is when the pilot is on a VFR flight plan and needs an IFR routing clearance due to deteriorating weather conditions. Another example is when the pilot finds it necessary to obtain an instrument approach clearance to land at the destination airport. The following paragraphs describe the different clearance requests a pilot may make during IFR flight or flight planning.

CLEARANCE DELIVERY

At large airports, an additional service called *clearance delivery* is provided to relieve congestion on the ground control frequency. Some clearance delivery facilities allow a pilot to receive an IFR clearance 10 minutes prior to taxi. At locations where pretaxi clearance is available, the abbreviation "Cpt" will appear in publications along with the clearance delivery frequency. This service relieves workload and frequency congestion and provides more time for the pilot to confirm clearance details or plan for amended clearances. Specific clearance delivery frequencies are published on enroute charts, approach charts, and under the appropriate airport listing in the Airport/Facility Directory.

GROUND CONTROL

At most IFR airports equipped with a tower, but without a clearance delivery frequency, ground control has the additional function of delivering instrument flight clearances to pilots. Before departure, then, pilots must contact ground control to receive the IFR clearance and specific departure instructions.

Ground control frequencies extend from 121.60 through 121.90 MHz; however, 121.90 MHz is used most frequently. Specific frequencies are published on enroute charts, area charts, approach charts and in the Airport/Facility Directory.

The following terminology should be used when requesting an IFR routing clearance from ground control when no clearance delivery is available. *"Midway Ground, Baron N2707C, at Butler, taxi, IFR, Rockford, with Lima."*

Ground control will reply, *"Baron N2707C, taxi to runway 30, clearance on request."*

In the previous example, ATIS was available. If ATIS is not available, the clearance can be used as is, except the term "with Lima" is not applicable.

When copying an enroute clearance, the pilot should *not* be taxiing. If he is advised while taxiing that his clearance is ready, he should ask that his clearance be held until he is ready to copy it.

FLIGHT SERVICE STATIONS

When departing an airport without a control tower, but where a flight service station is in operation, the pilot may receive his IFR clearance from the FSS. In this situation, the pilot should use the same phraseology as he used with clearance delivery or ground control.

When departing IFR from an airport where no radio communications are available, it is necessary to call the flight service station after filing the flight plan and request the IFR clearance over the telephone.

Once the IFR flight is airborne, flight service stations often relay messages pertaining to IFR communications between ARTCC and pilots when communication difficulties occur on normal discrete frequencies. In some areas, a communications relay through an FSS is the only method by which the pilot may contact ATC, or vice versa. This is especially true in some geographical areas where communications facilities are not adequate for direct pilot-controller communication, or where FSSs relay IFR clearances to pilots operating from uncontrolled airports.

APPROACH CLEARANCES

If only one instrument approach procedure exists at an airport, or if ATC wishes to authorize the pilot to execute an instrument approach of his choice, the controller will issue the following clearance, *"Twin Beech N6029R, cleared for the approach."*

If more than one approach is available at an airport, and ATC wants the pilot to execute a specific approach such as the ILS runway 26 left approach, they may issue the clearance in this manner. *"Commander N6313S, cleared for the 26 left ILS approach."*

When approach control is providing radar vectors for a straight-in approach, the controller will state the distance from the final approach fix, issue a new heading to intercept the final approach course, and give the approach clearance. As an example, the controller might give the following clearance. *"Cessna N5140G, three miles from outer marker, turn right heading 240°, cleared for straight-in ILS runway 26 left approach. Contact Denver Tower 118.3 outer marker inbound."*

Note, in the previous clearance, that the controller issues the distance from the final approach fix to confirm to the pilot that he is in the intermediate segment, allowing him to descend to the final approach fix crossing altitude.

AIR ROUTE TRAFFIC CONTROL CENTER

After flying in VFR conditions from a departure airport, the pilot may find the weather along his route of flight dictates the need for an IFR clearance. In this case, it is best to call the flight service station and file an IFR flight plan while enroute. *After* the IFR flight plan is filed, the flight service station may advise the pilot to contact the ARTCC on an assigned frequency to receive the clearance directly from the air route traffic control center. In some cases, the FSS will give the clearance, followed by the appropriate center frequency. Since the IFR clearance, in this case, begins at an enroute facility rather than the departure airport, the first fix on the clearance is usually over that enroute facility. The following example is the recommended procedure to be used when requesting an IFR clearance direct from center. *"Ft. Worth Center, Commander N2705S, request IFR clearance from over Ardmore to Dallas."*

All of the clearance requests just discussed are based on the assumption that an IFR flight plan had been filed prior to the clearance request. There are times when an IFR clearance can be requested even though a flight plan has not been filed with the flight service station. The IFR clearances that are requested while enroute are called "pop up" clearances and are generally discouraged by air traffic control.

OTHER TYPES OF CLEARANCES

DETOUR CLEARANCES

While enroute, it may be necessary to detour around thunderstorms. According to the regulations, pilots are not authorized to deviate from stated clearances unless an amended clearance is received from ATC. Assuming that a large thunderstorm is straight ahead and visible to the pilot, his request should be similar to the following. *"Denver Center, Cessna 1116D, a large cumulus buildup is at our 12 o'clock position, request deviation to the south."*

If the pilot is in an area where thunderstorms have been reported, he can make the following request when in radar contact. *"Denver Center, Cessna 1116D, request vectors around the thunderstorm area."*

AMENDED CLEARANCES

For various reasons, pilots may sometimes wish to change destination airports while enroute. When the pilot requests a deviation from the flight plan and wants an amended clearance, the following sample communications should be used. *"Chicago Center, Comanche 1312M, request change of destination airport from Peoria to Springfield, Illinois."*

After the above clearance request is made, the pilot should be ready to copy his new clearance which will include the normal IFR enroute clearance information.

"VFR OVER THE TOP" CLEARANCE

A clearance to VFR over-the-top provides a simple IFR clearance to depart and climb IFR to VFR conditions over a local overcast. Once the VFR conditions are reached, the flight is conducted under visual flight rules. When requesting an IFR clearance to VFR over-the-top at a tower controlled field, it may not be necessary to file an IFR flight plan. The initial request for a VFR over-the-top clearance can be made with clearance delivery or ground control. When IFR conditions exist at an airport where an ATC facility is *not* available, a request for a clearance *to* VFR over-the-top should be made with a filed flight plan.

The following clearance is an example of the proper communications to be used when requesting an IFR clearance to VFR over-the-top when ground control, but not clearance delivery, is available. *"Great Falls Ground, Aztec 1213F, at Holman Aviation, request IFR clearance to VFR over-the-top, southbound."*

When issuing IFR clearances to "VFR over-the-top," ATC will include a nearby clearance limit, route of flight, and a request to report reaching "VFR over-the-top." The clearance limit will usually be a fix in the terminal area in the direction of flight requested by the pilot. This type of clearance should not be confused with an IFR flight plan to the destination with an altitude request of "VFR conditions on top" instead of a specific altitude request. The following terminology is an example of a VFR-conditions-on-top clearance. *"Mooney 5100V, cleared as filed, expect radar vectors, maintain VFR conditions on top, if not "VFR on top at 4,000, maintain 4,000 and advise."*

HOLDING CLEARANCES

If a delay is anticipated in the terminal area, the center or approach control normally will issue a holding clearance at least five minutes before the aircraft is estimated to reach the clearance limit. When the controller issues a holding clearance that contains only the fix and altitude of the hold, the pilot should refer to the enroute chart for the published holding pattern. If there is not a published holding pattern, the pilot should hold on the inbound radial. In the event ATC requires a holding pattern contrary to the published hold, complete instructions will be issued. In this situation, the holding pattern clearance given by ATC will include the following information.

1. Aircraft number
2. Direction of the hold from the fix
3. Holding fix
4. Radial, course, bearing, or airway on which the aircraft is to hold
5. Outbound leg length in miles, if DME or RNAV is to be used
6. Direction of holding pattern turns, if left turns are to be made
7. Altitude
8. Expect further clearance or expect approach clearance time

For example: *"Aerostar 2255E, hold north of the Jay VOR on the 350° radial, left turns, 8,000 feet, expect further clearance at one five."*

Holding Pattern Procedures

The *standard holding pattern* consists of two 180° right-hand turns connected by two straight-and-level legs as shown in figure 8-1. The

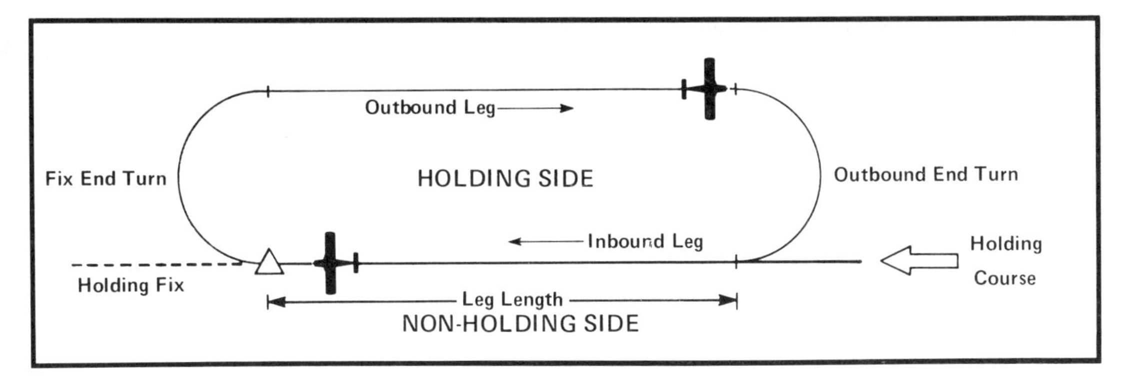

Fig. 8-1. Standard Holding Pattern

length of the *inbound leg* is one minute at holding airspeed, at or below 14,000 feet MSL. Above 14,000 feet, the inbound leg is 1-1/2 minutes in length. Turns within the holding pattern are to be made at

1. standard rate (three degrees per second), or
2. a 30° angle of bank (25° if a flight director is used), whichever requires the least angle of bank. If the holding turn is made at standard rate, the 180° turn at each end of the holding pattern will take approximately one minute.

Timing of holding legs and crosswind compensation are necessary to remain within the holding airspace. The inbound leg should be maintained at 1 or 1½ minutes, as appropriate, by shortening or lengthening the outbound leg. When a wind correction angle is needed to maintain the inbound course, the pilot should double the wind correction angle and apply it in the opposite direction on the outbound leg.

Since holding is a delaying maneuver, it is recommended that the pilot use maximum endurance airspeed. The amount of airspace reserved for holding is based on the following maximum speeds:

Propeller aircraft .... 175 knots IAS
Jet aircraft (up to and including)
6,000 feet MSL .... 200 knots IAS
6,001-14,000 feet MSL ..... 210 knots IAS
above 14,000 feet MSL ..... 230 knots IAS

These speed limits are mandatory, but may be exceeded *upon* ATC *approval* because of flight conditions or other unusual factors such as aircraft performance specifications. The pilot is required to begin a speed reduction to the mandatory indicated airspeed three minutes prior to the holding fix ETA.

Holding Pattern Entry

"The key to holding pattern entry is visualization." The pilot must visualize the holding pattern and the relationship between his *aircraft's heading* and the *inbound leg* of the pattern. The pilot may then decide which of the standard holding entry procedures applies to that relationship.

The standard holding entry procedures allow the pilot to enter the holding pattern efficiently while maneuvering within the airspace allotted for the pattern. These procedures provide for most of the maneuvering on the *holding side* of the pattern. Three separate procedures are used, depending upon the relationship between the heading of the aircraft when it crosses the holding fix the first time and the assigned holding course.

Direct Entry Procedure

When the heading of the aircraft crossing the holding fix is within a 180° arc beginning 70° toward the holding side of the assigned holding course, continuing to 110° on the nonholding side, as shown in figure 8-2, the direct entry procedure should be used. The pilot should cross the holding fix and then turn toward the *holding side* to parallel the holding course outbound. This heading should be maintained for one minute. He should then turn to intercept the holding course inbound.

Parallel Procedure

When the pilot is inbound to the holding fix and the aircraft heading falls within the *parallel* sector shown in figure 8-2, he will cross the fix and then turn to parallel the holding course outbound. This outbound heading should be maintained for one minute. The pilot should then turn toward the holding side and intercept the holding course inbound to the fix.

Teardrop Procedure

When the inbound heading to the holding fix lies within the teardrop area shown in figure 8-2, a teardrop procedure is indicated. The pilot will cross the holding fix and turn to a heading 30° divergent from the reciprocal of the holding course on the holding side, fly on this heading for *one minute*, and then turn to intercept the holding course inbound.

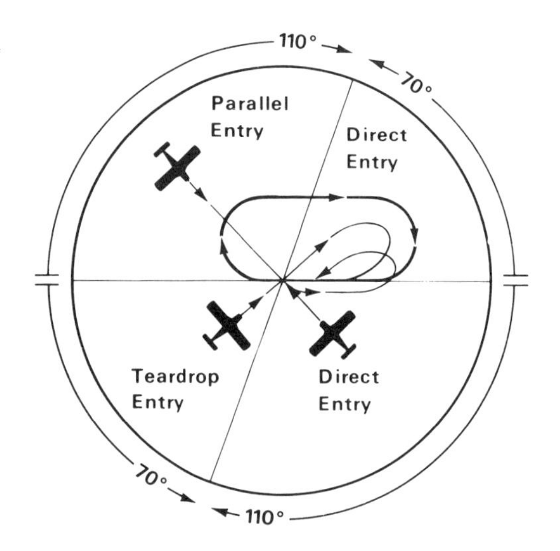

Fig. 8-2. Holding Pattern Entries

CRUISE CLEARANCE

Cruise clearances allow pilots to *climb to* and *descend from* assigned altitudes at their own discretion. Furthermore, cruise clearances are approval to proceed to, and make an approach at, the destination airport. This procedure also permits pilots to descend and land in accordance with applicable FARs governing VFR flight operations. For example, *"ATC clears Navajo 625PR to the Sipple Ranch Airport, direct, cruise 3,000."* In this clearance, the term "cruise" means the flight is cleared to the destination airport and also to climb to 3,000 feet and descend to the MEA or MOCA at the pilot's discretion.

TOWER-TO-TOWER CLEARANCE

When flying IFR between two airports, there is a possibility the pilot will not come under the jurisdiction of an air route traffic control center if the two airports are in close proximity to each other and their respective approach control areas of responsibility overlap. In this case, the pilot can request a tower-to-tower clearance without filing an IFR flight plan. The following request is an example of the recommended procedure to be used when calling ground control. *"Peterson Ground, Commander 2402B, request IFR tower-to-tower clearance to Denver."*

THROUGH CLEARANCES

When making an intermediate stop at an airport enroute to the eventual destination, a single "through" IFR flight plan can be issued at the departure airport. When filing this type IFR flight plan, a pilot can expect to receive a through clearance prior to arriving at the intermediate stop. In the through clearance, ATC will issue the approach clearance, followed by the through clearance, which includes a clearance to the destination airport. The following example illustrates a through clearance that might be issued in the Denver area. *"American 2030R, cleared for Greeley VOR approach, cleared through Greeley Airport to Denver Stapleton Airport via the 211° radial of Gill, V-81, maintain 8,000 feet. Contact Denver Center 124.1 passing through 6,000 feet."*

CONTACT APPROACH CLEARANCE

After a pilot requests a contact approach, ATC will issue a contact approach clearance if the ground visibility is at least one statute mile. In addition, the pilot must remain clear of clouds. The controller will not issue a fixed altitude on a contact approach, but may clear the aircraft to maintain an altitude at or below a specified altitude, if necessary for vertical separation. The following clearance is an example of a contact approach clearance. *"Aztec 6504M, cleared for contact approach at or below 4,000 feet."*

VISUAL APPROACH CLEARANCE

A visual approach clearance may be issued by ATC upon pilot request or ATC initiation. However, for this type of approach, the pilot must be able to descend and fly to the airport in VFR conditions and have visual contact with the field or with the aircraft immediately ahead. The following clearance is an example of a visual approach clearance. *"Mooney* 3795D, cleared for visual approach. Contact *Wichita Tower."*

CLEARANCE SHORTHAND

To operate efficiently in an IFR environment, the pilot must be able to copy and thoroughly understand IFR clearances. Copying IFR clearances becomes easy with practice. Although numerous changes have been made since the acceptance of the first shorthand used by early instrument pilots, many of the original symbols have been retained. The shorthand symbols in this chapter are considered by the Federal Aviation Administration, and experienced instrument pilots, to be the best. However, the most important consideration is not what symbol the pilot uses, but the fact that the symbols represent meaningful information after a period of time. The following symbols and contractions represent words and phrases frequently used in clearances.

WORDS AND PHRASES	SHORTHAND
ABOVE	ABV
ABOVE ("Above Six Thousand")	$\underline{60}$
ADVISE	ADV
AFTER (Passing)	< or AFT
AIRPORT	A
ALL TURNS LEFT	↰ or LT
ALL TURNS RIGHT	↱ or RT
(ALTERNATE INSTRUCTIONS)	( )
ALTITUDE 6,000–17,000	60-170
AND	&
APPROACH	APP
FINAL	F
VOR	VOR or ⊙
NONDIRECTIONAL BEACON (ADF)	ADF
SURVEILLANCE RADAR	ASR
LOCALIZER BACK COURSE	LBC
INSTRUMENT LANDING SYSTEM	ILS
LOCALIZER ONLY	LCO
PRECISION (Approach) RADAR	PAR
APPROACH CONTROL	APC
AT (Usually Omitted)	@
(ATC) ADVISES	C ADV
(ATC) CLEARS or CLEARED	C
(ATC) REQUESTS	C R
BEARING	BR
BEFORE (Reaching, Passing)	>
BELOW	BLO
BELOW ("Below Six Thousand")	$\overline{60}$
EASTBOUND	EB
INBOUND	IB
OUTBOUND	OB
CLIMB (TO)	↑
CONTACT	CT
COURSE	CRS
CROSS (Crossing)	X
CRUISE	→
DEPART (Departure)	DEP
DESCEND (TO)	↓
DIRECT	DR
DME FIX (15 DME Mile Fix)	[15]
EACH	ea
EXPECT	EX
EXPECT APPROACH CLEARANCE (Time)	EAC
EXPECT FURTHER CLEARANCE (Time or Location)	EFC
FLIGHT LEVEL	FL
FLIGHT PLANNED ROUTE	FPR
FOR FURTHER CLEARANCE	FFC
HEADING	HDG
HOLD (Direction) (Hold West)	H-W
HOLDING PATTERN	⬭
IF NOT POSSIBLE	or
ILS LOCALIZER	L
INTERSECTION	△
INTERCEPT	△ (with slash)
MAINTAIN (or Magnetic)	M
MAINTAIN VFR ON TOP	$\underline{VFR}$
MIDDLE MARKER	MM
LOW FREQUENCY BEACON LOCATED AT MIDDLE MARKER	LMM
OUTER MARKER	OM
LOW FREQUENCY BEACON LOCATED AT OUTER MARKER	LOM
OVER (Ident. Over the Line)	OKC
PROCEDURE TURN	PT
RADAR VECTOR	RV
RADIAL (092 Radial)	092R
REPORT, REPORTING	Rp
REPORT LEAVING	RL
REPORT ON COURSE	RC
REPORT OVER	RO
REPORT PASSING	RP
REPORT REACHING	RR
REPORT STARTING PROCEDURE TURN	RSPT
REVERSE COURSE	RC
RUNWAY	RW
SQUAWK	SQ
STANDBY	STBY
TAKEOFF	T-O
TOWER	Z
TRACK	TR
UNTIL	/
UNTIL ADVISED (By)	/UA
UNTIL FURTHER ADVISED	/UFA
VICTOR (Airway Number)	V-294
VOR	⊙
VORTAC	Ⓣ

IFR REPORTING PROCEDURES

When operating in an IFR environment, the pilot is required to report certain actions to ATC. Through these reports ATC is able to plan, schedule, and control aircraft movements and separation.

POSITION REPORTS

Position reports must be made enroute at *compulsory* reporting points when in a *nonradar* environment. In a *radar* environment, position reports are used primarily to *confirm* a position which the controller sees on the radar screen or to *establish* initial *radar contact.* Therefore in a *radar* environment *after* contact is *established,* position reports are not required unless requested. However, if radar contact is *lost,* normal position reporting must be resumed.

These position reports should include:

1. Ground station call sign
2. Aircraft call sign
3. Position
4. Time
5. Altitude
6. Type of flight plan (not required if an IFR position report is given directly to ARTC center or approach control)
7. ETA over next compulsory reporting point
8. The name of the next succeeding compulsory reporting point
9. Any pertinent remarks

For example, assume that a pilot is over Wichita Falls VOR enroute to Lubbock, Texas at 8,000 feet MSL. The position report is made as follows.

"*Fort Worth Center, Cessna 3324R, Wichita Falls.*" This will alert the controller that a position report is going to be made and allows him time to prepare to copy it. The controller may then tell the pilot: "*Cessna 3324R, roger.*" This will negate the requirement for making the complete position report. However, if the controller says, "*Cessna 3324R, go ahead,*" then the position report should be: "*Fort Worth Center, Cessna 3324R, Wichita Falls at 35, 8,000 feet, estimate Guthrie 13, Lubbock next.*"

TRANSFER OF CONTROL

When the aircraft is transitioning from one ARTC center to another in radar contact, the content of the report should include:

1. Center call sign
2. Aircraft call sign
3. Altitude

If the aircraft is in a climb or descent to a new altitude, this information should be stated as "*climbing/descending to (new altitude).*"

MALFUNCTION REPORTS

Any aircraft operating in controlled airspace under IFR conditions must report the loss of *any navigational* equipment, loss of an ILS *receiver,* or the impairment of any communication capability. The report of the malfunction should include the following information:

1. Aircraft identification
2. Type of malfunction
3. Effect on IFR operations
4. Any desired assistance required from ATC

REPORTS TO BE MADE WITHOUT REQUEST

It is helpful to visualize the air traffic controller's situation in order to remember the reports that are to be made "without request." Even though radar contact has been established with a specific flight, the controller may be monitoring many aircraft on the radar scope and may not immediately detect deviations from the expected altitude, route, or speed, Thus, certain reports are helpful to the controller and should be given without request. The following reports are required by regulations and must be given without request.

1. Position reports (Except while in "radar contact," only those reporting points specifically requested by ATC need be reported.)
2. Hazardous or unforecast weather
3. Anything affecting the safety of the flight
4. Failure of a navigation aid
5. Impairment of air/ground communications capability

Although *not* required by *specific* regulations, the following reports are suggested in the AIM and *should* be made without request *at the appropriate time:*

1. Leaving a previously assigned altitude for a newly assigned altitude
2. Leaving a holding fix or point
3. Crossing the final approach fix inbound
4. Executing a missed approach (and the pilot's intentions)
5. Change in altitude when the clearance specifies "VFR conditions on top"
6. Reaching a holding fix or point to which cleared
7. ETA change of plus or minus three minutes

NOTE: Items 2, 3, 6 & 7 need not be reported *when in radar contact,* unless requested.

9

AIRMAN'S INFORMATION PUBLICATIONS

The safe and efficient use of the nation's airspace is the responsibility of the Federal Aviation Administration. One method employed to fulfill this responsibility is dissemination of current aeronautical information through aeronautical charts, the national notice to airmen system, and flight information publications.

Aeronautical charts are published by Jeppesen Sanderson, Inc., a private company and the National Ocean Survey, a government agency. To the extent possible, they reflect the most current information available at the time of printing.

The notice to airmen (NOTAM) system is used to disseminate information of a "time-critical nature." This information is required for flight planning, but is not known sufficiently in advance to permit publication elsewhere.

THE AIM

The *Airman's Information Manual* (Basic Flight Information and ATC Procedures) is shown in figure 9-1. The manual is generally referred to as the AIM. It contains fundamental information required for flight in the national airspace system.

Fig. 9-1. AIM Basic Flight Information and ATC Procedures

The first portion of the AIM is the foreword which includes information on complementary publications and major changes to the current issue as shown in the excerpt in figure 9-2.

The main subjects covered in the AIM include the following.

— Navigation Aids
— Airport, Air Navigation Lighting, and Marking Aids

EXPLANATION OF MAJOR CHANGES

Para. 12 INSTRUMENT LANDING SYSTEM—Revised to include information regarding the use of DME and to include additional information concerning ILS minimums and inoperative components.

Para. 47 RADIO CONTROL OF AIRPORT LIGHTING—Revised in its entirity for clarification and to provide additional information.

Para. 132 MILITARY TRAINING ROUTES—Revised to encourage the VFR pilot to solicit current MTR route information from the FSS.

Fig. 9-2. AIM, Explanation of Major Changes Excerpt

— Airspace — Air Traffic Control
— Emergency Procedures — Safety of Flight
— Medical Facts for Pilots
— Aeronautic Charts and Related Publications
— Pilot/Controller Glossary

Navigation Aids — This section contains a detailed description of all the navaids used in the national airspace system. The discussion also addresses the components, limitations, and operational uses of each.

Airport, and Navigation Lighting, and Marking Aids — The AIM coverage of these subjects includes comprehensive illustrations of the most commonly used airport marking aids.

Airspace — This part contains pertinent information for both VFR and IFR pilots. It covers the VFR visibility and distance criteria for controlled and uncontrolled airspace, as well as the required altitudes and flight levels to be used in VFR and IFR operations.

Special use and other airspace areas also are defined. These include prohibited, restricted, warning, alert, and military operations areas. Other special airspace subjects, such as airport traffic/advisory areas, and general airspace segments and dimensions also are described.

Air Traffic Control — This is one of the most important AIM segments for all pilots, but especially commercial and instrument pilots. It covers services available to pilots, radio communications phraseology, and airport operations for controlled and uncontrolled airports.

In addition, this section lists practical steps for preflight planning and provides an insight into ATC clearance/separation practices during flight. National security and interception procedures are described at the end of this part.

Emergency Procedures — Portions dealing with emergency procedures involve topics such as common accident cause factors, the national search and rescue plan, and the rules pertaining to the aviation safety reporting program. The procedures and signals for aircraft in emergency situations also are described.

Safety of Flight — Included in this section is a discussion of weather information sources, altimetry, wake turbulence, and bird hazards. Important topics like the "do's" and "dont's" of thunderstorm flying, clear air turbulence, wind shear, and vortex avoidance procedures also are included.

Medical Facts for Pilots — This section reviews various aspects of fatigue, hypoxia, oxygen requirements, drugs/alcohol, carbon monoxide, and orientation as they relate to flight. Other subjects of interest include vision, middle ear discomfort, and vertigo.

Aeronautical Charts and Related Publications — This section lists the types of charts available and their appropriate uses. The discussion contains a detailed description of each chart series, miscellaneous auxiliary charts, and related publications.

Pilot/Controller Glossary — This glossary, located in the back of the manual, is designed to promote a common understanding of terms used in air traffic control. Because of the international nature of flying, terms used by the International Civil Aviation Organization (ICAO) are included when they differ from FAA definitions. These ICAO terms are *italicized* in the glossary, as shown in figure 9-3.

> AIRMAN'S INFORMATION MANUAL/AIM— A publication containing Basic Flight Information and ATC Procedures designed primarily as a pilot's instructional manual for use in the National Airspace System of the United States.
>
> *ICAO — AERONAUTICAL INFORMATION PUBLICATION*—A publication issued by or with the authority of a state and containing aeronautical information of a lasting character essential to air navigation.

Fig. 9-3. Glossary of Aeronautical Terms

AIRPORT/FACILITY DIRECTORY

The *Airport/Facility Directory* is published by the NOS in seven booklets — one for each of seven regional areas in the United States. Effective dates and the area of coverage for each are shown on the cover. (See Fig. 9-4.) Contents

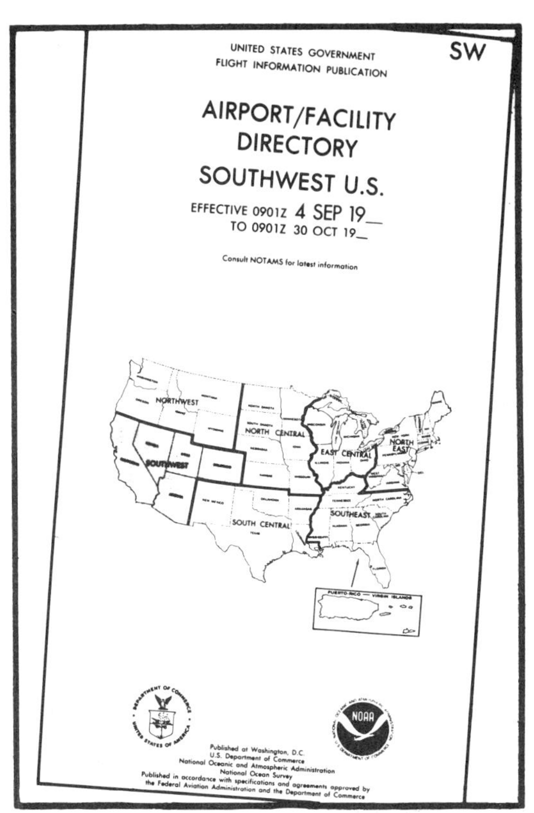

Fig. 9-4. Airport/Facility Directory

include a tabulation of all data on record with the FAA for public use civil airports, associated terminal control facilities, air route traffic control facilities, air route traffic control centers, and radio aids to navigation. Each booklet also contains additional data such as special notices, operational procedures, and preferred routes relevant to the coverage area.

Listings are alphabetical by state and cities within states. A cross reference system is used for the names of facilities. Airports are listed by associated city name and cross-referenced by airport name.

LEGEND

A comprehensive legend is printed in the first few pages of each regional directory. The format, although similar to that used in many aeronautical publications, contains several unique abbreviations and symbols. Pilots should become familiar with the directory legend. A sample legend is shown in figure 9-5.

In addition to airport information, each directory lists all national airspace system radio navigation aids within the area. An example of a listing

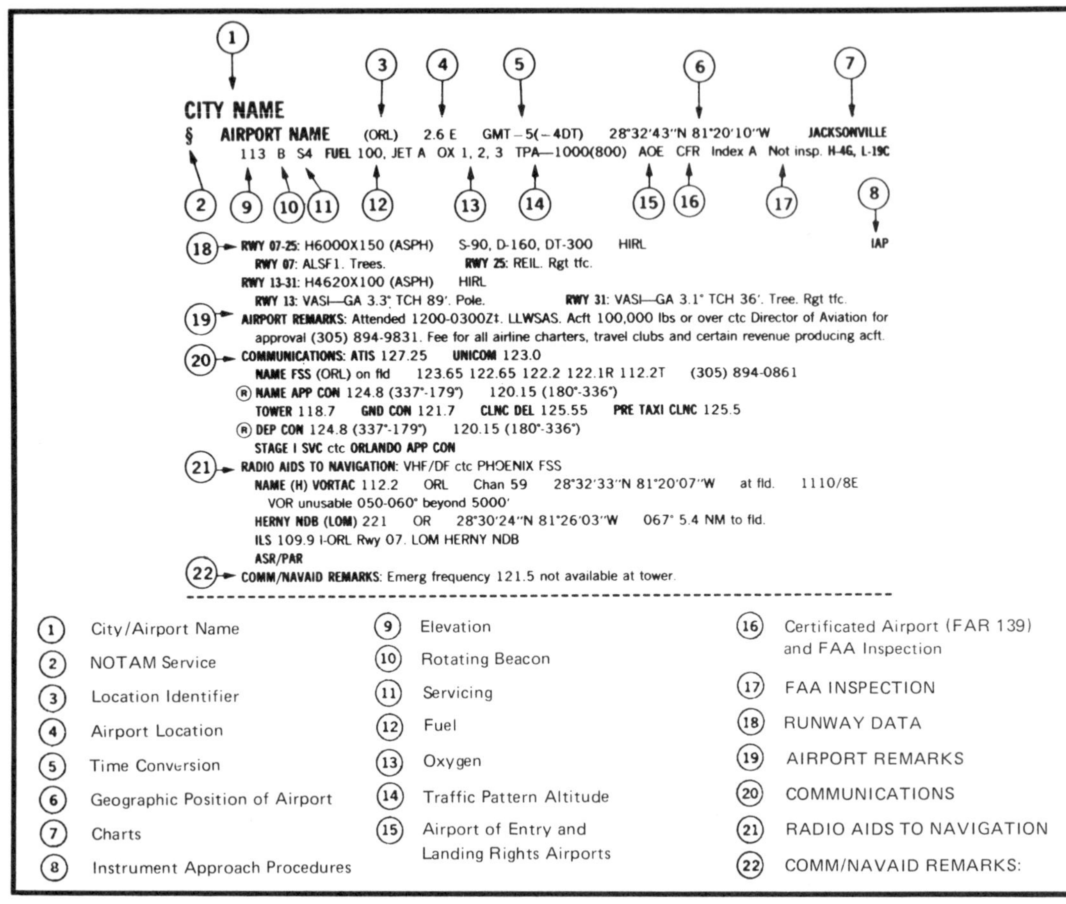

Fig. 9-5. Airport/Facility Directory (Legend Excerpt)

from the southwest directory is illustrated in figure 9-6. Radio facilities with a name different than an associated airport are listed separately under their own name and are included in the airport data.

COCHISE 32°02'00"N 109°45'27"W **PHOENIX H-2G, L-4F**
(H) VORTAC 115.8 CIE Chan 105 317° 14.8 NM to Cochise Co. 4230/13E
VORTAC unusable 005°-015° beyond 35 NM below 10,000'
015°-030° beyond 25 NM below 10,000'
030°-040° beyond 35 NM below 10,000'
190°-220° beyond 30 NM below 9000'
220°-240° beyond 25 NM below 9500'
COCHISE LRCO 122.1R 115.8T (TUCSON FSS)

Fig. 9-6. Radio Navigation Aid Listing

SPECIAL NOTICES

Most of each regional directory is devoted to the airport/facility listing, but other important information also is included. Each directory has a special notices section. This section lists "hard-to-find," significant information. For example, civil use of military fields, newly certified airports, and continuous power facilities are explained, and special procedures are identified. Special notices also contains a complete listing of useful telephone numbers. These include FSS, combined station/tower, National Weather Service, and fast file flight plan telephone numbers. All listings are alphabetical by state. An excerpt from this listing is shown in figure 9-7.

VOR RECEIVER CHECKPOINTS

VOR receiver checkpoint procedures were covered in Chapter 4 and also are discussed in the AIM. The information in the *Airport/Facility Directory* on VOR checkpoints lists each regional facility, the identifier, type of checkpoint, distance/azimuth, and a description, if necessary. Again, the listing is alphabetical by state. A typical listing of VOR receiver checkpoints is depicted in figure 9-8. VOT facilities are listed in the same section below the VOR checkpoints.

PREFERRED IFR ROUTES

Certain preferred routes have been designated by the FAA. The objective is to facilitate the flow of

FSS-CS/T AND NATIONAL WEATHER SERVICE TELEPHONE NUMBERS

Flight Service Stations (FSS) and Combined Station/Tower (CS/T) provide information on airport conditions, radio aids and other facilities, and process flight plan. CS/T personnel are not certified pilot weather briefers; however, they provide factual data from weather reports and forecasts. Airport Advisory Service is provided at the pilot's request on 123.6 by FSSs located at airports where there are no control towers in operation. (See Airman's Information Manual, Basic Flight Information, and ATC Procedures.)

"Numerous additional telephone numbers are listed under COMMUNICATIONS in the Airport/Facility Directory tabulation. If you wish to call an FSS, but do not have access to a directory listing telephone numbers, call the toll-free information number 800-555-1212. Many FAA Flight Service Station numbers may be obtained through this service." The telephone area code number is shown in parentheses. Each number given is the preferred telephone number to obtain flight weather information. Automatic answering devices are sometimes used on listed lines to give general local weather information during peak workloads. To avoid getting the recorded general weather announcement, use the selected telephone number listed.

★ Indicates Pilot's Automatic Telephone Weather Answering Service (PATWAS) or telephone connected to the Transcribed Weather Broadcast (TWEB) providing transcribed aviation weather information.
◆ Indicates a restricted number, use for aviation weather information.
■ Call FSS for "one call" FSS—Weather Service (WS) Briefing service.
Automatic Aviation Weather Service (AAWS).
§§ Indicates Fast File telephone number for pre-recorded and transcribed flight plan filing only.

Location and Identifier			Area Code	Telephone
		ALABAMA		
Anniston ANB		FSS	(205)	831-2303
Birmingham BHM		FSS	(205)	254-1387 ■
		FSS	(205)	595-2101 ★
	N.W. Route			595-5416 #
	S.E. Route			595-6452 #
	N.E. Route			595-7957 #
	S.W. Route			595-7896 #
Dothan DHN		FSS	(205)	983-3551

Fig. 9-7. FSS, CS/T & National Weather Service Telephone Numbers Excerpt

VOR RECEIVER CHECK POINTS VOR/VORTAC AND VOR TEST FACILITIES (VOT)

The use of VOR airborne and ground check points is explained in Airman's Information Manual, Basic Flight Information and ATC Procedures.

The FAA VOR test facility (VOT) transmits a test signal for VOR receivers which provides users of VOR a convenient and accurate means to determine the operational status of their receivers. The VOT is designed to provide a means of checking the accuracy of a VOR receiver while on the ground where a VOT is located. In some areas, a single VOT may provide receiver check service while on the ground or while airborne over an airport between specified altitudes.

To utilize the VOT service either on the ground or in the air, tune in the advertised VOT frequency on your VOR receiver. With the Course Deviation Indicator (CDI) centered, the omnibearing selector should read 0° with the to-from indicator being "from" or the omnibearing selector should read 180° with the to-from indication reading "to." Should the VOR receiver operate an RMI (Radio Magnetic Indicator), it will indicate 180° on any OBS setting when using the VOT. Two means of identification are used with the VOR radiated test signal. In some cases a continuous 1020 Hz tone will identify the test signal. Information concerning an individual test signal can be obtained from the local flight service station.

The airports at which airborne and/or ground VOT service is provided are tabulated in this directory in the VOR Receiver Check Points (VOR/DME, VORTAC) and VOR Test Facilities (VOT) section—See Table of Contents.

NOTE: Under columns headed "Type of Check Point" & "Type of VOT Facility" G stands for ground. A/ stands for airborne followed by figures (2300) or (1000-3000) indicating the altitudes above mean sea level at which the check should be conducted. Facilities are listed in alphabetical order, in the state where the check points or VOTs are located.

CONNECTICUT

VOR RECEIVER CHECK POINTS

Facility Name (Arpt Name)	Freq/Ident	Type Check Pt. Gnd. AB/ALT	Azimuth from Fac. Mag	Dist. from Fac. N.M.	Check Point Description
Carmel (Danbury Muni)	116.6/CMK	A/1500	050	6.7	Over apch end of Rwy 08.
Hartford (Hartford-Brainard)	114.9/HFD	G	337	7.8	On parking ramp S of terminal bldg.
Madison (Chester)	110.4/MAD	A/1500	076	9.4	Over small hangar.

Fig. 9-8. VOR Receiver Checkpoint Excerpt

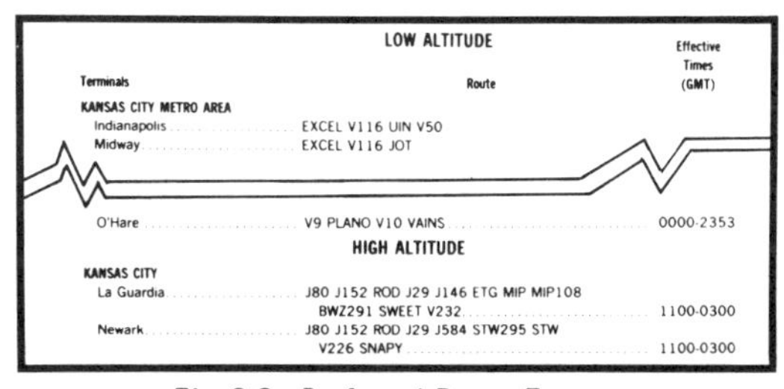

LOW ALTITUDE		
Terminals	Route	Effective Times (GMT)
KANSAS CITY METRO AREA		
Indianapolis	EXCEL V116 UIN V50	
Midway	EXCEL V116 JOT	
O'Hare	V9 PLANO V10 VAINS	0000-2353
HIGH ALTITUDE		
KANSAS CITY		
La Guardia	J80 J152 ROD J29 J146 ETG MIP MIP108 BWZ291 SWEET V232	1100-0300
Newark	J80 J152 ROD J29 J584 STW295 STW V226 SNAPY	1100-0300

Fig. 9-9. Preferred Route Excerpt

air traffic into and out of major terminal areas. The directory lists established preferred IFR routes for each region. Pilots who plan to transit a busy air terminal area during a flight should consult the preferred route section of the directory before filing an IFR flight plan. These routes pertain to both high and low altitude airspace, as shown in figure 9-9.

AERONAUTICAL CHART BULLETIN

The last section of the *Airport/Facility Directory* lists changes to aeronautical charts within the region. Generally, only changes to controlled and special use airspace which could pose a hazard to flight or impose a restriction are published in this section. Major changes to airport or radio navigation facilities are also included. When the affected aeronautical chart is republished, changes are included and the chart bulletins are removed from the directory. A chart bulletin excerpt is illustrated in figure 9-10.

AERONAUTICAL CHART BULLETIN

The purpose of this Bulletin is to provide a tabulation of the major changes in aeronautical information that have occurred since the last publication date of each Sectional Aeronautical Chart and Terminal Area Chart listed. The general policy is to include only those changes to controlled airspace and special use airspace that present a hazardous condition or impose a restriction on the pilot; major changes to airports and radio navigational facilities, thereby providing the VFR pilot with the essential data necessary to update and maintain his chart current. When the Aeronautical Chart is republished, the corrective tabulation will be removed from this Bulletin. Inasmuch as this Bulletin provides major changes only; pilots should consult the airport listing in this directory for all new information. Users of U.S. World Aeronautical Charts (WAC) should make appropriate revisions to their charts from this Bulletin. NOTE: New data which have been added to this issue are shown below the rule line under the appropriate chart.

Military Training Routes (MTRs) are shown on Sectional Aeronautical Charts and VFR Terminal Area Charts. Only the route centerline, direction of flight and the route designator are shown — route widths and altitudes are not shown. Since these routes are subject to change every 56 days and the charts are reissued every 6 months, routes with a change in the alignment of the charted route centerline will be listed in this Aeronautical Chart Bulletin below. You are advised to contact the nearest FSS for route dimensions and current status for those routes effecting your flight.

BILLINGS SECTIONAL
19th Edition, April 17, 19_

Chg. obst. elevs. 2645'MSL (515' AGL) to 3049'MSL (919' AGL) UC 46°57'29"N, 100°41'05"W. Dickinson, North Dakota control zone is established as follows: Within a 5.5 mile radius of Dickinson Municipal airport, 46°47'45"N, 102°48'00"W and within 3 miles each side of the Dickin... ...TAC 46°51'36"N, 102°46'23"W, 013° radial extending from 5.5 mile radius area to 8 miles...

Fig. 9-10. Aeronautical Chart Bulletin Excerpt

NOTICES TO AIRMEN

Time-critical aeronautical information that was not known at the time of publication of aeronautical charts or other documents receives immediate dissemination via the national notice to airmen service, a telecommunications system. Two general types of NOTAMs exist — NOTAM-D and NOTAM-L. In certain circumstances, a third category — Flight Data Center (FDC) NOTAMs — may be issued. An FDC NOTAM contains regulatory information.

NOTAM information is regularly passed to pilots by telephone or radio. The FAA also produces a separate publication, called *Notices to Airmen*, to provide further dissemination.

NOTAM-D

Information designated for NOTAM-D dissemination is the type that could affect a pilot's decision to make a flight. Events such as airport closure, interruptions in service of navigational aids, ILS, or radar service are appropriate NOTAM-D material. NOTAM-D information is provided for all navigation facilities, all IFR airports (with approved instrument approach procedures), and for VFR airports annotated by the NOTAM symbol (§) in the *Airport/Facility Directory.*

NOTAM-L

NOTAM-L information is that which is of an advisory, or nice to know, nature. It includes such information as taxiway closings, men and equipment near or crossing runways, and information on airports not annotated with the NOTAM symbol (§) in the *Airport/Facility Directory.* NOTAM-L information is maintained on file only at those local air traffic facilities concerned with the operations at these airports. However, this information can be made available upon specific request to the local FSS having responsibility for the airport concerned.

FDC NOTAM

On those occasions when it becomes necessary to disseminate information which is regulatory in nature, such as amendments to aeronautical charts, instrument approach procedures, or to effect restrictions to flight, the National FDC in Washington, D.C., issues a NOTAM containing the regulatory information as an FDC NOTAM. FDC NOTAMs are distributed through the National Communications Center in Kansas City and are transmitted to all air traffic facilities with telecommunications access. Current FDC NOTAMs are published in their entirety in the *Notices to Airmen* publication and also as part of instrument approach procedure charts.

NOTICES TO AIRMEN (CLASS II)

The *Notices to Airmen* publication contains three basic parts, or subdivisions. The first part consists of notices which meet the NOTAM-D criteria and are expected to remain in effect for an extended period. These NOTAMs are included to reduce teletype circuit congestion. NOTAM-L and other special notices may be included in the interest of flight safety. Data is republished until the information is cancelled, is no longer valid or, in the case of permanent information, is published in other documents that are revised less frequently. All notices in the first part are expected to remain in effect for at least seven days after the publication date.

Notices are arranged in alphabetical order by state and within the state by city or locality. New or revised data, in this part, is indicated by boldly italicizing the airport name. It is important to note that, unless stated otherwise, all times are local.

The second part contains all FDC NOTAMs current through specific FDC NOTAM number and date. This information is listed in the FDC legend.

The third part contains special notices that, either because they are too long or because they concern a wide or unspecified geographic area, are not suitable for inclusion in the general notices. The main criteria for inclusion of these special notices is enhancement of flight safety. Each

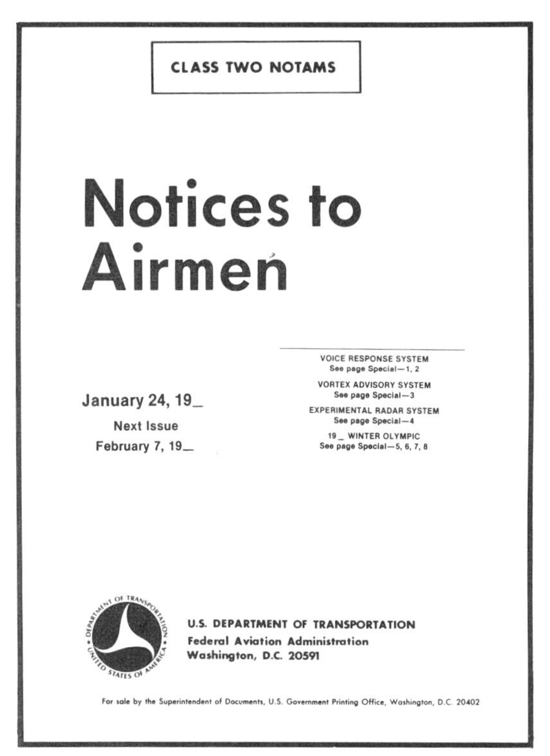

Fig. 9-11. Notices to Airmen

Fig. 9-12. Graphic Notices and Supplemental Data

biweekly publication of the *Notices to Airmen* (Class II) publication lists the special notices on the cover, as illustrated in figure 9-11.

GRAPHIC NOTICES AND SUPPLEMENTAL DATA

Graphic Notices and Supplemental Data is a publication containing aeronautical data, area notices, navigational route information which is supplemental to other operational publications and charts, and data that is not generally subject to frequent change. Information that appears in this publication, however, may be revised by inclusion in *Notices to Airmen* (Class II) which is published more frequently.

Graphic Notices and Supplemental Data is a quarterly publication that is available through subscription from the Superintendent of Documents. The effective dates of each issue are printed on the cover, as shown in figure 9-12. The major sections include area advisories, area navigation (RNAV) routes, North Atlantic Routes (NAR), Terminal Area Graphics and Terminal Radar Service Area (TRSA) graphics.

AREA ADVISORIES

This section is limited to information that does not concern a specific airport or navaid, and also does not meet the criteria for inclusion into a more permanent publication such as the *Airport/Facility Directory.* The advisories are grouped under state headings and, where appropriate, under city or geographical locations.

RNAV AND NORTH ATLANTIC ROUTES

The next section contains a listing of *area navigation* or RNAV routes. Guidelines for use of these routes within the national airspace system specify the use of approved navigation systems such as Doppler radar and inertial or courseline computers. A description of the route numbering system also is provided.

North American Routes for North Atlantic Traffic are described and listed in the next section. These routes, which generally pertain to air carrier operations, are referred to by the acronym NAR.

TERMINAL AREA GRAPHIC NOTICES

The majority of *Graphic Notices and Supplemental Data* is devoted to terminal area graphic notices and terminal radar service areas (TRSAs). Both of these sections are preceded by a map and an index of the respective graphic depictions. Listings are alphabetical by city.

Terminal area graphic notices are published for areas of concentrated IFR traffic. Normally, these areas have at least one busy air terminal with a complex air traffic flow involving a mixture of commercial, general aviation, and military aircraft. The graphics are intended to assist VFR pilots planning flights in the areas and, in many cases, recommended VFR routes or corridors are shown. The information presented is advisory only. Pilots flying within these areas are encouraged to use the graphic information and any other radar or advisory service. A typical terminal area graphic notice is shown in figure 9-13.

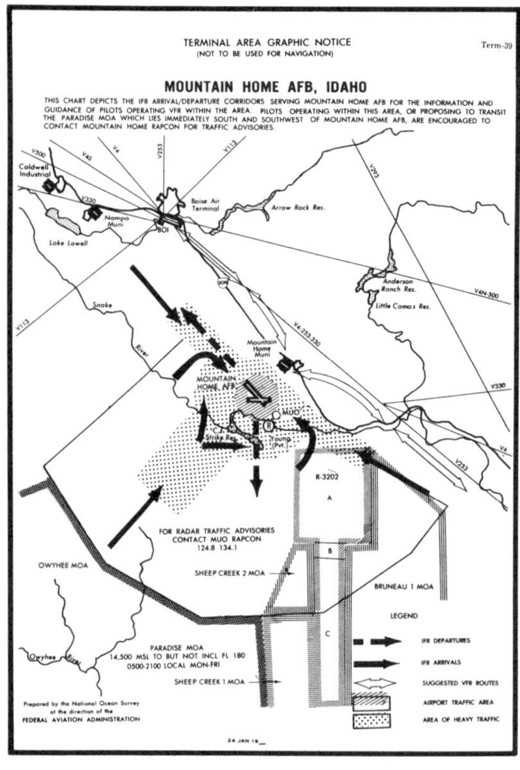

Fig. 9-13. Terminal Area Graphic Notice

TERMINAL RADAR SERVICE AREAS

As discussed earlier, the radar service programs available to VFR pilots include Stage I (advisory), Stage II (advisory and sequencing), and Stage III (sequencing and separation). Where Stage III service is implemented, the area is called a terminal radar service area, or a TRSA. Formerly all TRSAs were depicted in *Graphic Notices and Supplemental Data*, however, they will eventually be removed from this publication as they are added to Sectional Charts during normal chart revision cycles.

A typical TRSA contains frequencies, routes, checkpoints, and airspace boundaries. The procedures require IFR pilots to operate according to normal IFR; VFR pilots are expected to contact the controlling agency whenever they depart, arrive, or transit a TRSA. VFR participation is not mandatory; however, it is encouraged.

Normally, TRSA service consists of vectoring, sequencing, and separation of both VFR and IFR traffic landing at the primary airport. Separation service and advisories on unknown aircraft also are provided to all participating aircraft on a controller workload basis.

The terminal radar service areas should not be confused with terminal control areas (TCAs), which are located at very large airports with high density traffic. In the TCAs, *all* aircraft are subject to definite operating rules, specific pilot certification, and equipment requirements. Also, both VFR and IFR pilots normally must operate in accordance with an ATC clearance within the terminal control area. Areas that have a TCA are depicted on aeronautical charts. Basic information is found on the world aeronautical charts and sectional charts, while specific TCA information is shown on VFR terminal area charts.

10 FLIGHT COMPUTER

This chapter utilizes "programmed frames" to teach the step-by-step procedures necessary for computer operation. In order to use the frames most effectively, the following procedure should be employed.

1. Place a sheet of paper over the first page of frames in this text in such a manner that only Frame 1 is visible.
2. Read the programmed frame and write the answer in the blank.
3. Slide the sheet of paper down to the next frame and confirm that the correct answer to the preceding frame (shown in the right-hand column) was made. If you answered incorrectly, make a written correction to your response at once. If you answered incorrectly and do not understand the reason for the correct answer, immediately review the preceding steps.

THE JEPPESEN CR-1 COMPUTER

The Jeppesen CR-1 Computer provides fast, accurate time-speed-distance and wind problem solutions. The two sides of the computer are called the calculator side and the wind side.

CALCULATOR SIDE OF THE COMPUTER

Some of the scales on the calculator side are exactly like the scales on a slide rule, as illustrated in figure 10-1. There are three scales around the computer — the outer scale (sometimes called the mile scale), the middle scale (minute scale), and the inner scale (hour scale). The inner and middle scales are on one disc which is movable in relation to the base disc on which the outer scale is printed.

A given figure on the scales can stand for any number containing the same digits. That is, the point marked "50" can stand for 0.5, 5, 50, 500, etc. The correct position of the decimal point must be determined from a given problem.

The divisions between the numbers on the scales are not always the same. For example, there are 10 divisions between the numbers 13 and 14 on the mile scale; therefore, each division represents one unit. The first mark past 13 can be given a value of 13.1, 131, etc. Next, note the marks between 25 and 30 on the mile scale. There are only five divisions between 25 and 26, so each division has a value of two. The second mark past 26 can have a value of 26.4, 264, etc. The longer marks represent the figures 26, 27, 28, and 29. The value of these increments must be determined carefully to arrive at the correct solution for each problem.

The calculator side of the computer is constructed so any relationship between two numbers (one on the stationary scale and one on the movable scale) holds true for all other numbers on the two scales. If the two 10's are placed opposite each other, all other numbers are identical around the whole scale. If the movable

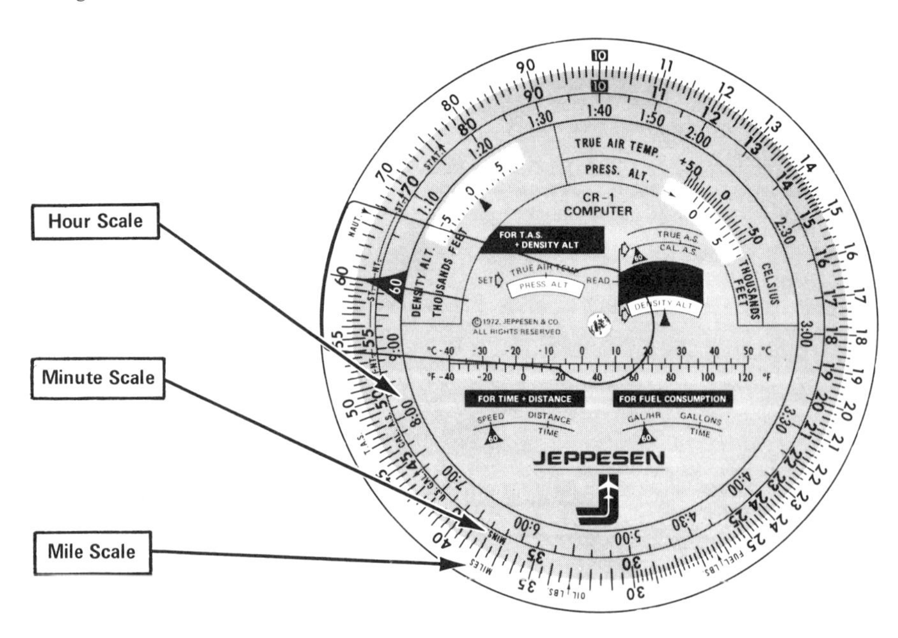

Fig. 10-1. The Calculator Side

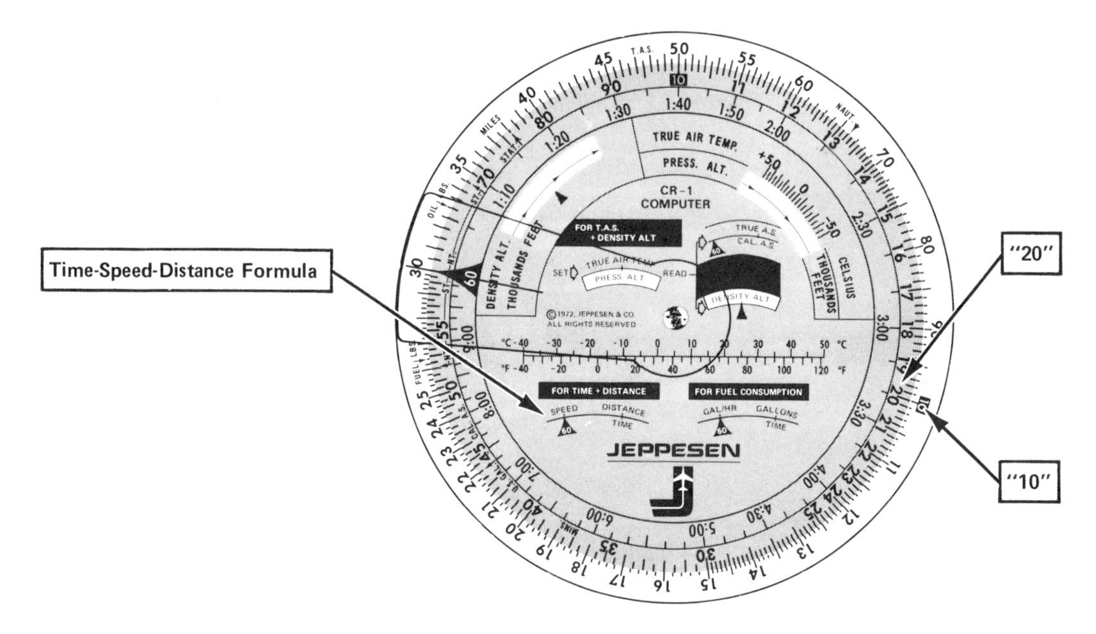

Fig. 10-2. Proportions

20 is placed under the stationary 10, all inner numbers will be twice the value of the outer numbers. This is illustrated in figure 10-2.

Time-speed-distance problems can be solved using this proportion method. In most problems, the answer is desired in terms of knots, miles or gallons per hour. The computer has a speed index "60" at 60 on the minute scale to give the answer directly without dividing by 60.

An easy way to remember these proportions is to learn the relationships between speed and distance as illustrated in figure 10-3. In the lower left quadrant is the speed index, and in the lower right quadrant is the value representing time. Above the index, read either speed or gallons per hour, depending on the problem being solved. Above the time value, read total distance or total fuel. The two halves of the square may be widely separated on the computer but the relationship never varies. Two values in a problem are always known, thus the question revolves around finding the third value.

EXAMPLE:

Given: Speed . . . . 174 m.p.h.
Time . . . . 40 min.

Find: Distance

Using the time-speed-distance relationship shown in figure 10-3, place the speed index under miles per hour (174) and read the distance traveled over the time (40). The answer is 116 miles. This problem is set up on the computer shown in figure 10-4.

The solution is equally accurate when nautical miles are being used; however, all values of the problem must be in the same units. That is, if speed is given in knots, distance must be given in terms of nautical miles.

EXAMPLE:

Given: Speed . . . . 113 kts.
Distance . . . . 180n.m.

Find: Time

First, set the speed index under the speed (113 kts.), then, read the time under the distance (180 nautical miles). The answer is 95½ minutes or 1:35½ as shown in figure 10-5.

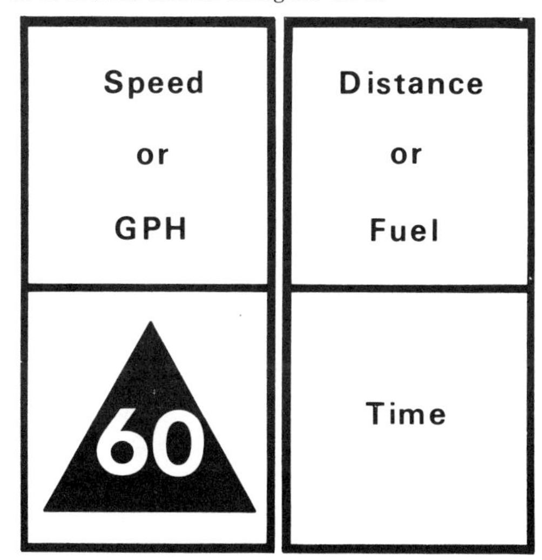

Fig. 10-3. Proportion on the Computer

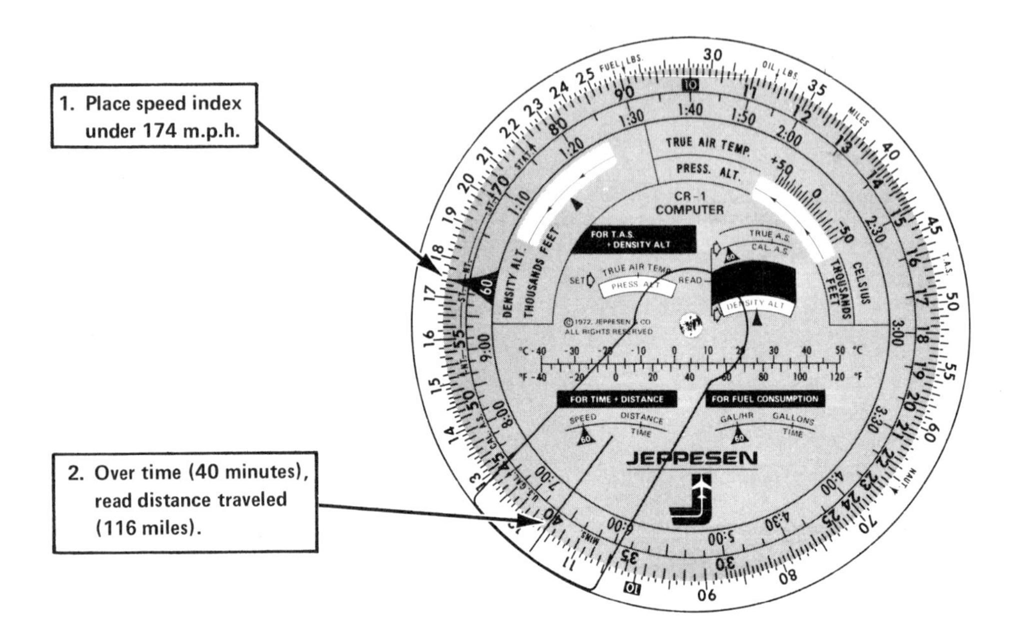

Fig. 10-4. Solving for Distance

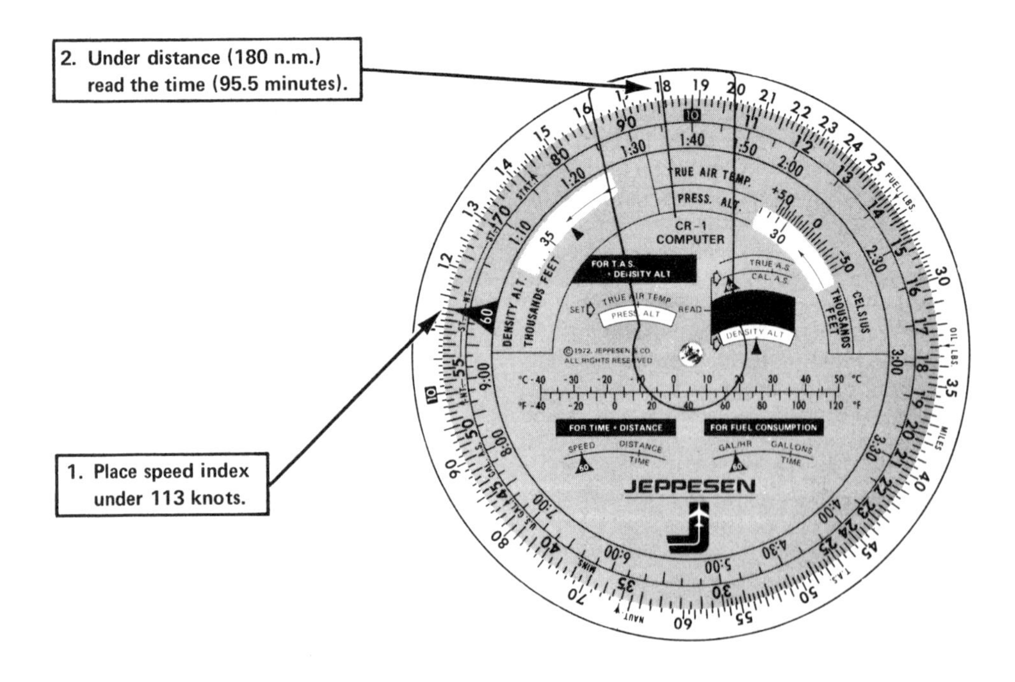

Fig. 10-5. Solving for Time

Speed can be found when time and distance are given.

EXAMPLE:

Given: Distance . . . . .266 s.m.

Time . . . . . . .1 hr., 25 min. (85 min.)

Find: Speed

Set the time (1:25) under the distance (266) and read speed over the speed index. The answer in this case is 188 m.p.h., as shown in figure 10-6.

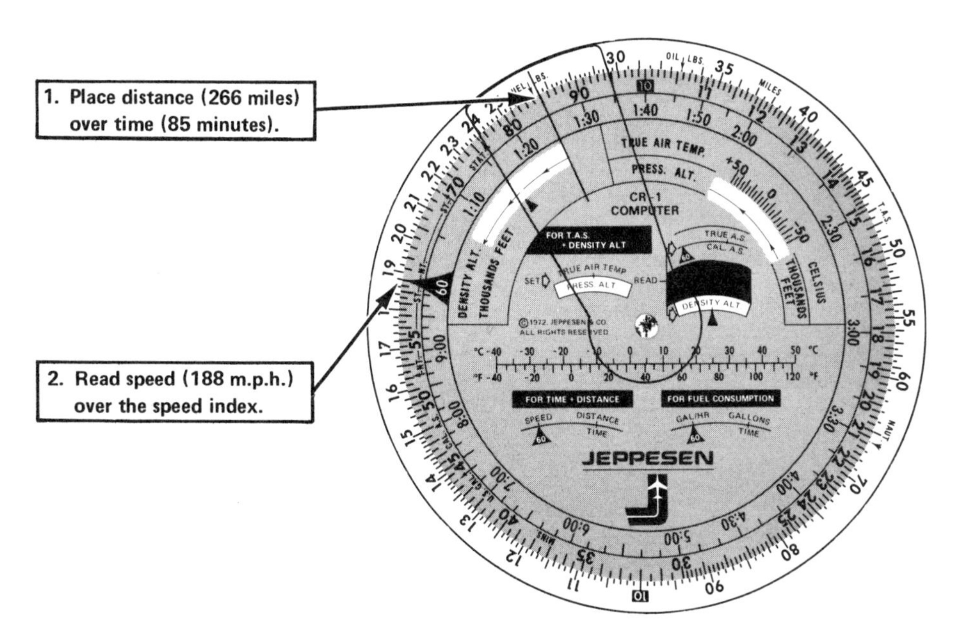

Fig. 10-6. Solving for speed

1. The first mark past 15 on the mile scale can be equal to __________(151,152).	
2. The first mark past 30 on the mile scale can be equal to __________(30.1, 30.5).	*152*
3. The third mark past 60 on the mile scale can be equal to __________(63, 60.3).	*30.5*
4. To find the distance when given speed and time, place the __________ __________ under the speed and read the distance above __________.	*63*

5. To find the distance when given a speed of 160 m.p.h., place the speed index under 160. To find the distance traveled in 30 minutes, read the figure above 30 on the time scale as ______ miles.	*Speed index time*
6. The distance traveled in 30 minutes at a speed of 170 m.p.h. is ______ miles.	*80*
7. At a speed of 140 m.p.h., the distance traveled in 2:15 is ______ miles.	*85*
8. At a speed of 150 m.p.h., the distance traveled in two hours is ______ miles.	*315*
9. Assuming a speed of 120 knots and a distance traveled of 50 nautical miles, place the speed index under 12. Below 50 on the mile scale, the correct response can be seen as ______ minutes.	*300*
10. At a speed of 150 m.p.h., it takes ______ minutes to fly 200 miles.	*25*
11. At a speed of 130 knots, it takes ______ to fly 300 nautical miles.	*80*
12. Read speed above the speed index when time and distance are given. If 60 miles are flown in 30 minutes, the speed is ______.	*2:18.5*
13. If 80 miles are flown in 50 minutes, the speed is ______.	*120 m.p.h.*
14. If 400 miles are flown in 2:30, the speed is ______.	*96 m.p.h.*
	160 m.p.h.

CONVERSIONS

The computer has a Celsius — Fahrenheit temperature conversion scale. Figure 10-7 shows that temperature conversions can be read directly from this scale. For example, 10° Celsius is equal to 50° Fahrenheit.

STATUTE MILES — NAUTICAL MILES

Statute miles can be converted to nautical miles or vice versa by using the time-speed-distance scales. Position the nautical arrow on the mile scale opposite the statute arrow on the minute scale, then read any nautical mile value on the mile scale opposite the identical distance in terms of statute miles shown on the minute scale. This procedure is illustrated in figure 10-7.

The CR-1 Computer incorporates an additional method of converting from statute to nautical miles or vice versa. As a convenience to the pilot, there is a statute-nautical conversion method incorporated on either side of the black triangular speed index. As an exercise in the use of this method, set the speed index directly under the value 12, as illustrated in figure 10-8.

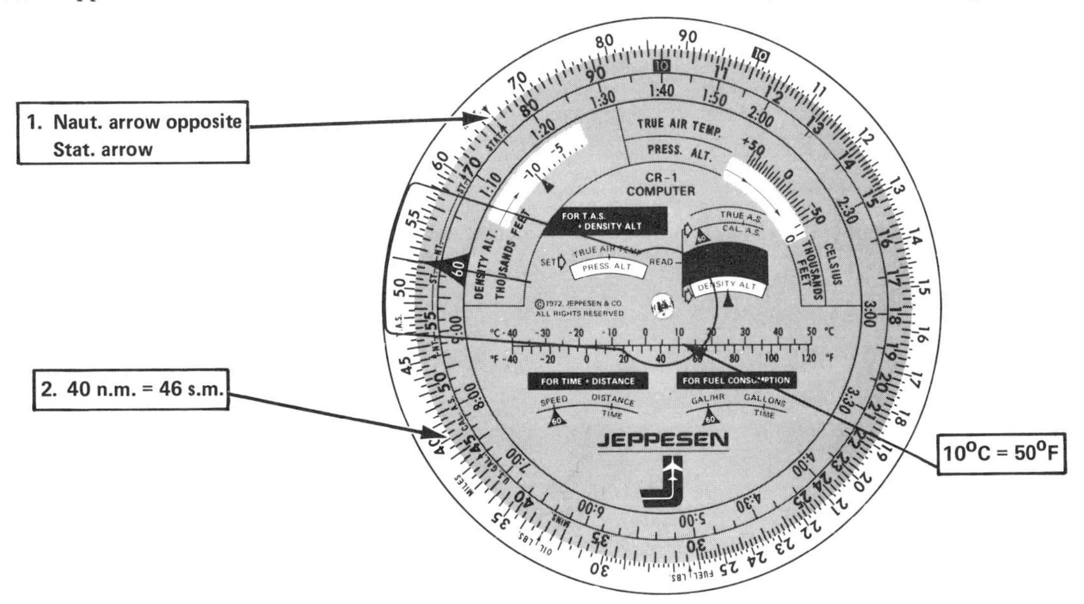

Fig. 10-7. Temperature Conversion

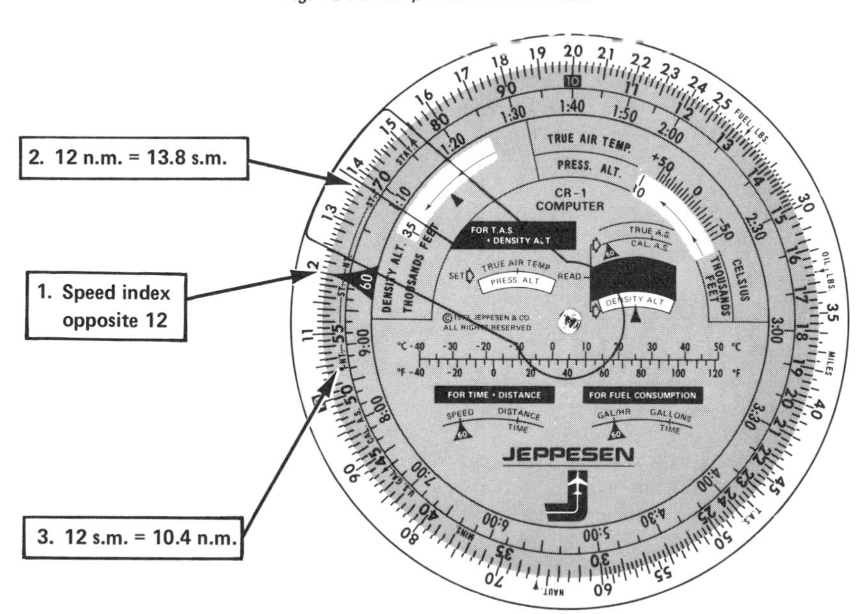

Fig. 10-8. Nautical to Statute Miles

Assume this figure represents 12 nautical miles. To convert this value to statute miles, proceed to the right to the arrow labeled ST (statute). Note that the statute arrow shows that 12 nautical miles is equal to exactly 13.8 statute miles, as read on the mile scale.

On the other hand, if the value 12 is given in terms of statute miles, to convert the value to nautical miles, proceed left from the speed index to the arrow labeled NT. (nautical). Note that 12 statute miles is equal to 10.4 nautical miles. Both of the conversions discussed above are equally effective in making conversions between statute and nautical miles, and miles per hour and knots.

1. A temperature of 32°F equals ______ °C.	
2. A temperature of 30°C equals ______ °F.	*0*
3. A temperature of -22°F equals ______ °C.	*+86*
4. After setting the nautical arrow on the mile scale opposite the statute arrow on the minute scale, it can be determined that 16 nautical miles is equal to ______ statute miles and the 145 m.p.h. is equal to ______ knots.	*-30*
5. Using the method described in the previous frame, determine that 250 knots is equal to ______ m.p.h.	*18.4* *126*
6. After setting the speed index at 127 knots, determine that this speed is equal to ______ m.p.h.	*288*
7. After setting the speed index at 90 m.p.h., determine that this speed is equal to ______ knots.	*146*
	78

GALLONS – POUNDS

The CR-1 Computer provides a fast, efficient method of converting from gallons to pounds and vice versa. For example, to convert gallons of aviation fuel to pounds, position the FUEL LBS. arrow on the mile scale directly over the U.S. GAL. arrow on the minute scale. With this relationship established, all values on the mile scale represent fuel in pounds and the opposite value on the minute scale represents the same amount of fuel in gallons. This relationship is illustrated in figure 10-9. Note that one gallon of fuel weighs six pounds and that 25 gallons weighs 150 pounds.

A similar procedure is used to convert gallons of oil to pounds of oil. Simply set the arrow labeled OIL LBS. on the mile scale opposite the U.S. GAL. arrow on the minute scale. When this relationship is set on the computer, any value on the mile scale is equal to the value on the minute scale when speaking in terms of quantity of oil. Note that one gallon of oil weighs 7.5 pounds and 120 pounds of oil can be equated to 16 gallons as illustrated in figure 10-10.

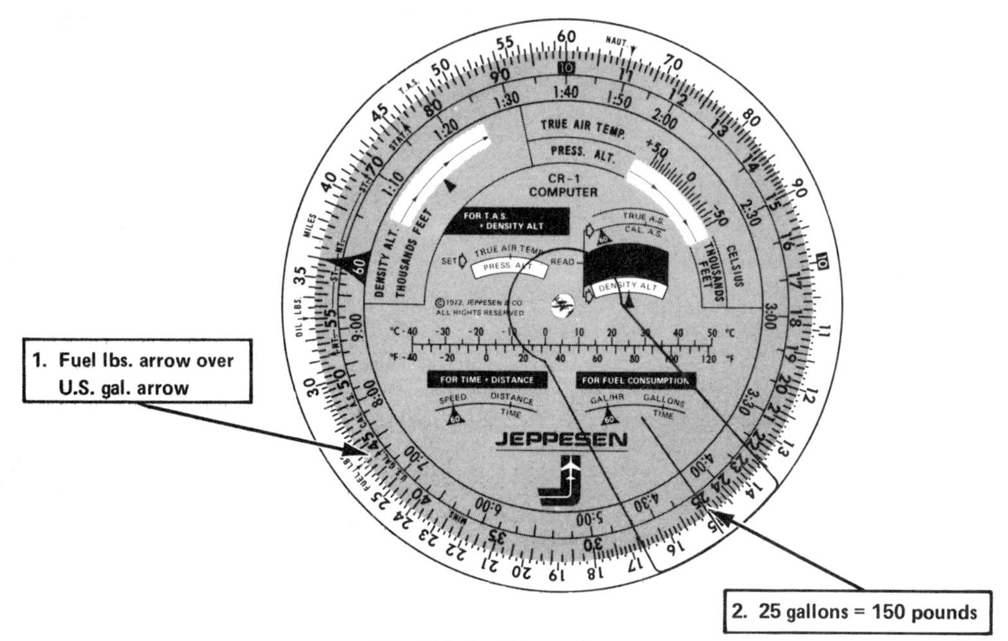

Fig. 10-9. Fuel Conversions

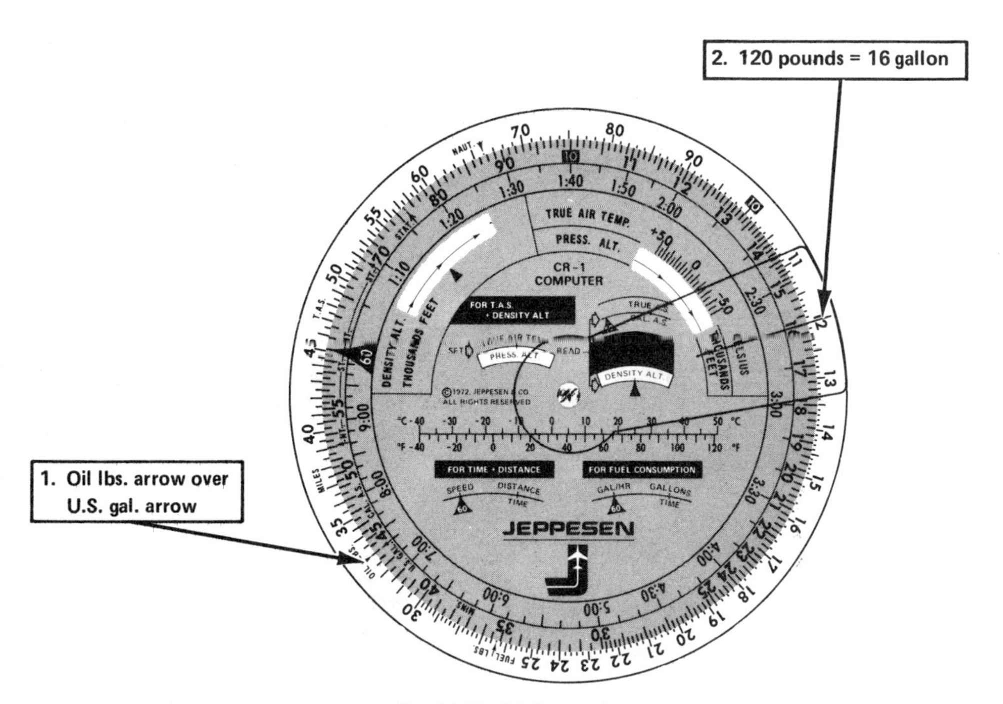

Fig. 10-10. Oil Conversions

1. Nineteen gallons of aviation fuel weigh ________ pounds.

2. Fifty gallons of fuel weigh ____________ pounds.	*114*
3. Eight gallons of oil weigh ____________ pounds.	*300*
4. Sixteen gallons of oil weigh __________, pounds.	*60*
5. Ninety pounds of oil is equal to ____________ gallons.	*120*
	12

FUEL CONSUMPTION

Fuel consumption problems are solved in exactly the same manner as problems involving speed, except that the rate is expressed in gallons per hour instead of miles per hour or knots. Note that the proportion formula for solving fuel consumption problems is printed on the lower portion of the movable disc.

EXAMPLE:

Given: Total fuel consumed . . . . . . 19 gal.
Flight time 1:35

Find: Rate of fuel consumption (g.p.h.)

Solution: Figure 10-11 shows the time (1:35) under the fuel consumed (19 gal.). Read the rate of fuel consumption (12 g.p.h.) over the speed index. Good operating practice requires that, in planning a flight, fuel on board at takeoff is at least the normal amount required for the flight, with an additional 45 minute reserve.

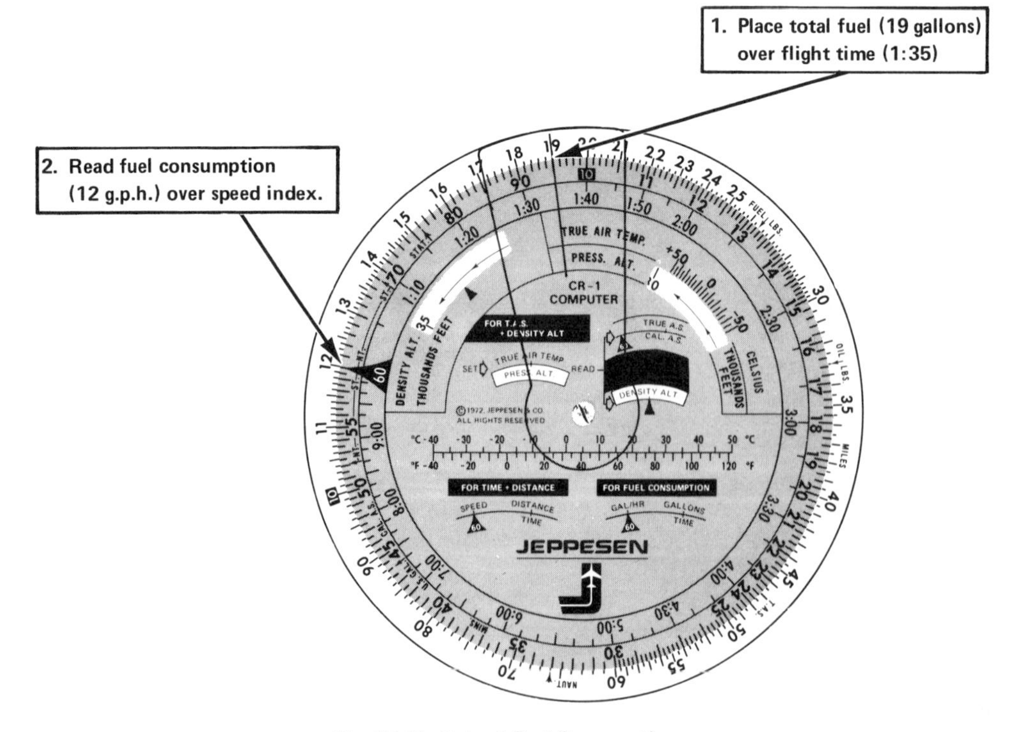

Fig. 10-11. Rate of Fuel Consumption

EXAMPLE:

Given: Fuel consumption . . . . . 14 g.p.h.
Flight time 2:30

Find: Fuel required

Solution: As shown in figure 10-12, set the speed index under the g.p.h. (14) and read the total fuel (35 gallons) over the time (2:30). Without moving the computer, read the fuel required for a 45-minute reserve (10.5 gallons) over the 45. The fuel requirement then, with fuel reserve is 35 gallons plus 10.5 gallons, or 45.5 gallons.

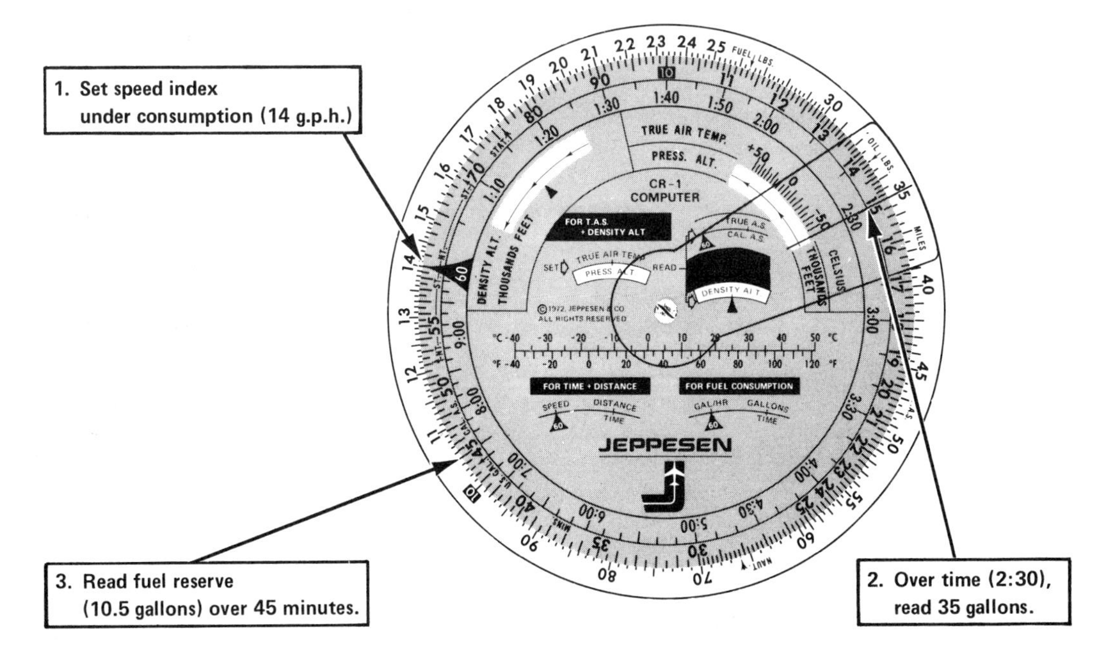

Fig. 10-12. Fuel Required

1 If 20 gallons of fuel are consumed in 2:00, the rate of fuel consumption is__________ g.p.h.	
2. If 10 gallons of fuel are consumed in 1:10, the rate consumption is __________ g.p.h.	*10*
3. If an airplane consumes 10 g.p.h., a flight of 2:00 with a 45-minute reserve requires a fuel load of __________ gallons.	*8.6*
4. To have a 45-minute reserve for a 1:30 flight with a consumption rate of 11 g.p.h., the fuel load must be __________ gallons.	*27.5*
	24.8

AIRSPEED AND DENSITY ALTITUDE COMPUTATION

The airspeed indicator in an airplane is constructed to operate correctly at sea level under standard atmospheric conditions, which include a pressure of 29.92 inches of mercury and a temperature of 15°C. (59°F). However, standard atmospheric conditions rarely exist, so the pilot must correct the readings of his airspeed indicator for altitude and temperature variations in order to determine the true airspeed. The CR-1 Computer incorporates a scale which enables the pilot to make the necessary corrections. This scale, shown in figure 10-13, is located above and to the right of the instructions labeled FOR TAS AND DENSITY ALT. COMPUTATIONS.

True airspeed can be obtained in one setting of the computer.

EXAMPLE:

Given: True air temperature . . . . . . -10°C
Pressure altitude . . . . . . 7,000 ft.
Calibrated airspeed . . . . 110 m.p.h.

Find: True airspeed
Density altitude

Solution: With reference to figure 10-13, set the true air temperature (-10°C) opposite the pressure altitude (7,000 feet) in the proper window. Opposite calibrated airspeed (110 m.p.h.) on the minute scale, read the true airspeed (119.5 m.p.h.) on the mile scale. Read the density altitude (5,800 feet) in the density altitude window.

NOTE: In most cases where light, training airplanes are involved, indicated airspeed may be used in lieu of calibrated airspeed without excessive error. Calibrated airspeed can be defined as the airspeed indicator reading (indicated airspeed) corrected for position and instrument error. Essentially, calibrated airspeed differs from indicated airspeed due to the errors caused by the pitot-static tube and static port installations, plus inherent instrument errors.

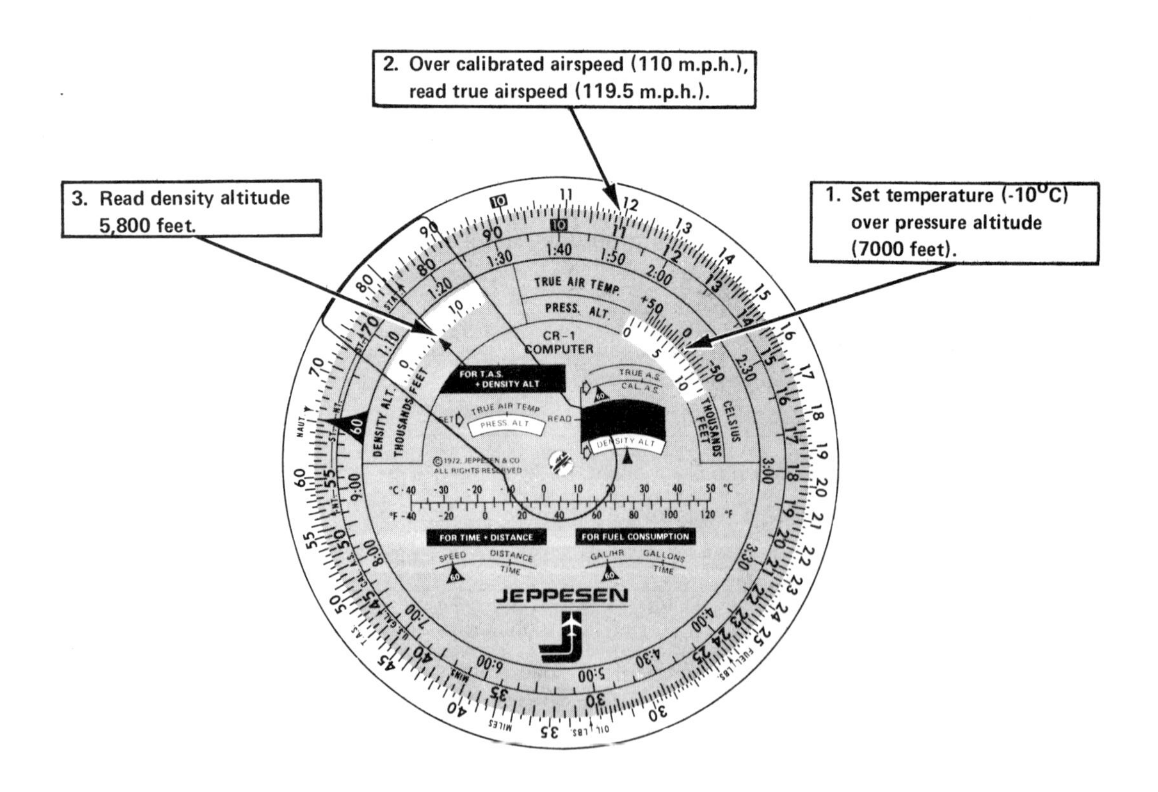

Fig. 10-13. True Airspeed and Density Altitude

1. Assume a temperature of 9°C at a pressure altitude of 5,000 feet. With a calibrated airspeed of 150 m.p.h. the true airspeed is ________ m.p.h.	
2. Use the following conditions to calculate the true airspeed. Temperature . . . . +5°C Pressure altitude . . . . 5,000 ft. Calibrated airspeed . . . . 130 m.p.h. The true airspeed (TAS) is ________ m.p.h.	*162*
3. Temperature . . . . -20°C Pressure altitude . . . . 9,000 ft. Calibrated airspeed . . . . 150 kts. In this instance, the true airspeed is ________ knots.	*140*
4. Using the values given in the previous frame, calculate the density altitude as ________ feet.	*166*
	7,000

WIND SIDE OF THE COMPUTER

Every wind problem contains six variables — wind direction, wind velocity, true course, true heading, true airspeed, and groundspeed. By knowing any four of the quantities, it is possible to find the other two on the CR-1 Computer.

COURSE AND HEADING

True course (TC) is the intended flight path over the ground, drawn relative to true north on the chart. Wind, variation, and deviation may be applied to the true course, each necessitating a change in terminology.

Wind applied to true course changes the word "course" to "heading." True course, plus or minus wind correction angle is *true heading* (TH). Because the wind, except as given by the tower for takeoff or landing, is always given in terms of true north, wind problems should be worked in terms of true, rather than magnetic, direction.

Before solving any wind problem, visualize the effect that the wind will have on the airplane. To maintain a given course, an airplane must correct into the wind. Thus, when the wind is from the right, add the wind correction angle to the true course to get true heading. If the wind is from the left, the wind correction must be subtracted from the true course.

VARIATION AND DEVIATION

Variation is the difference between true north and magnetic north at the geographic locality of the observer. When variation is applied to a course or heading, the word "true" changes to "magnetic." Thus, true course with variation applied is *magnetic course* (MC) and true heading with variation applied equals *magnetic heading* (MH).

Deviation is the error due to compass installation and the magnetism of various airplane components. Deviation may be obtained from a compass card located near the magnetic compass in the airplane and varies with the direction the airplane is heading. Deviation is applied after variation and changes the wording from "magnetic" to "compass." Thus, magnetic course, plus or minus deviation, equals *compass course* (CC), and magnetic heading, plus or minus deviation, equals *compass heading* (CH).

TRUE COURSE AND MAGNETIC COURSE ON A CR-1 COMPUTER

When converting from true course to magnetic course, subtract easterly variation and add westerly variation. However, the rule is reversed when converting from magnetic course to true course. In this case, add easterly variation and subtract westerly variation.

There can be no confusion when making this conversion using the CR-1 Computer since the computer applies variation automatically on the black scale on either side of the TC index. (See Fig. 10-14.) The true course should be directly opposite the TC index and the magnetic course should be opposite the corresponding variation on the black scale.

EXAMPLE:

Given: Magnetic course . . . . .090°
Variation . . . . .11°E

Find: True course

Solution: Note that easterly variation is on the black scale to the left of the TC index and westerly variation is on the right. Locate 11° on the left (easterly side). Over 11 on the black scale, place 090° on the green scale of the top disc. Read the true course of 101° over the TC index as shown in figure 10-14.

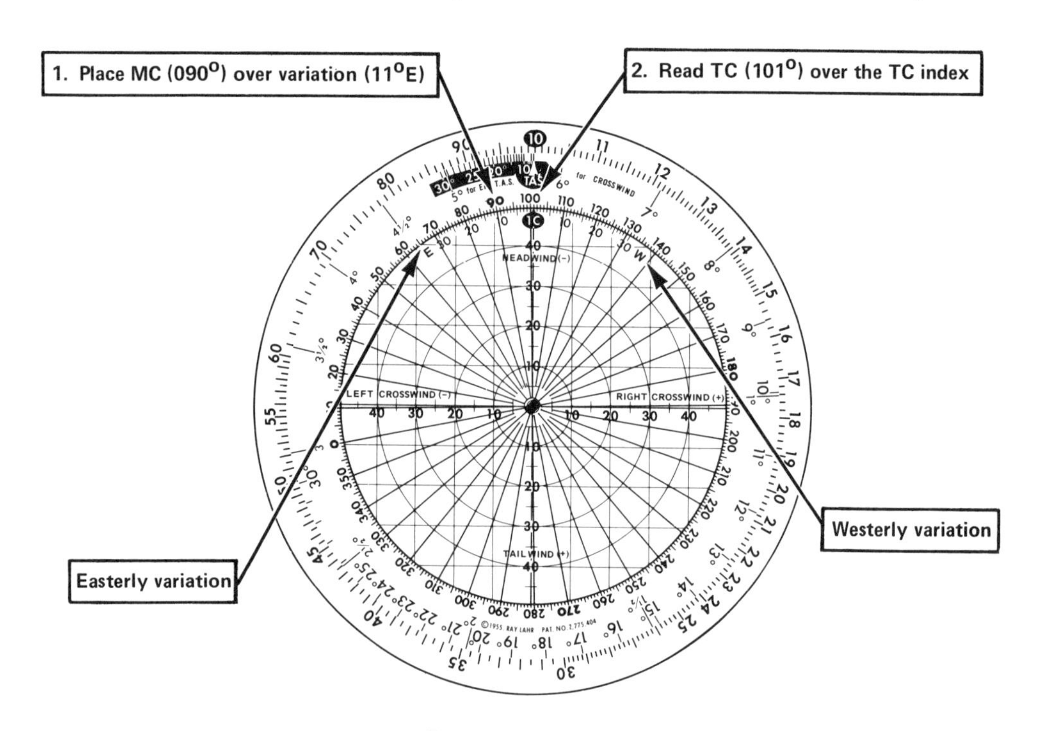

Fig. 10-14. True Course

1. When wind is applied to true course, the result is true ____________.	
2. If the wind in a given flight situation is from the left, the true heading will be less than the ____________ ____________.	*heading*

3. When variation is applied to the true course, the result is ________ ________.	*true course*
4. When converting true course to magnetic course, westerly variation is ________.	*magnetic course*
5. When converting magnetic heading to true heading, easterly variation is ________.	*added*
6. When deviation is applied to magnetic heading, the result is ________ ________.	*added*
7. When making good a magnetic course of 170° in a location where the variation is 11° west, the true course is ________.	*compass heading*
	159°

THE WIND TRIANGLE

The wind triangle in Part A of figure 10-15 illustrates the basis for any wind solution. When the job of drawing such a triangle for each wind problem is considered, it becomes readily apparent that a computer adds simplicity and brevity to the wind solution.

Part B of figure 10-15 illustrates a line drawn from the end of the TH-TAS line perpendicular to the TC-GS line. This, of course, forms a smaller triangle at the top of the original wind triangle. Part C of the same figure illustrates the fact that it is the smaller triangle which actually fits on the CR-1 Computer. Note that the three sides of this small triangle represent the tailwind

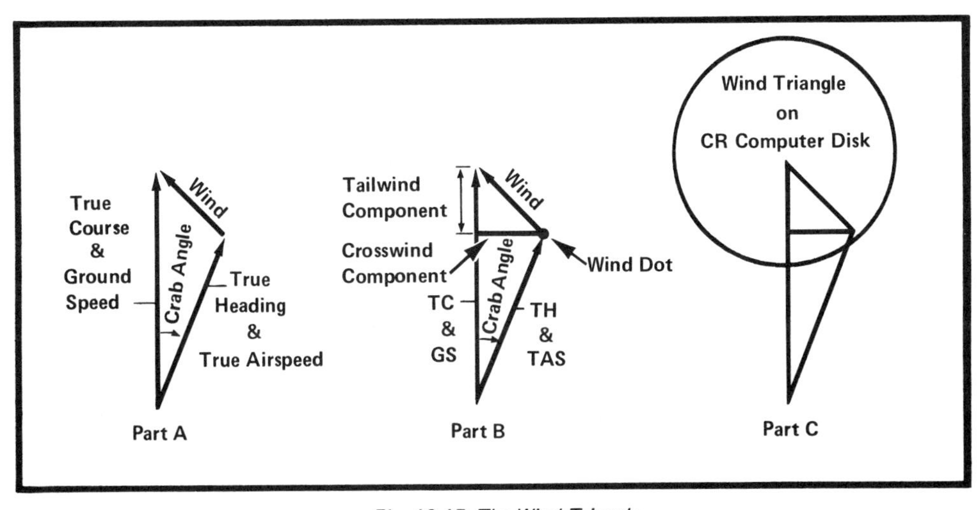

Fig. 10-15. The Wind Triangle

or headwind component, crosswind component, and wind. Each of these are of the utmost importance in any wind solution.

As is noted by reading subsequent instructions in this section, it is not necessary to draw the actual wind vectors on the computer. It is only essential to correctly locate the wind dot which represents the tail of the wind vector as is illustrated in the upper portion of Part B.

MARKING THE WIND VECTOR ON THE COMPUTER

Although the wind vector may be marked on the computer at any of several stages during the problem, it is usually more convenient to place the wind first. To designate 30 knots blowing from 040°, first locate the 040° line on the transparent disc. Then, measure outward from the center to the point where the circle marked 30 intersects the 040° line as illustrated in figure 10-16. Place a small dot accurately with a sharp pencil at this intersection and circle the dot to facilitate finding the mark during the problem.

Wind may be marked on the computer in terms of miles per hour or knots. However, all speed designations such as TAS, GS, and wind must be given in the same units for a given problem.

Remember that a small dot, correctly placed, will provide an accurate answer. In the case of a wind direction that lies between two lines, the following procedure will aid in accurate placement of a wind dot.

EXAMPLE:

Given: Wind . . . . 25 knots at 037°

Find: Position of the wind dot

Solution: Move the transparent disc until the 037° line is directly over one of the horizontal or vertical lines on the black and white disc. Place the wind dot at the halfway point between 20 and 30 on the black line as illustrated in figure 10-17.

Since there are four possible lines for use in measuring, it is unnecessary to move the top disc more than 45° to position the wind dot in this manner.

SETTING THE PROBLEM ON THE COMPUTER

Two computer motions are necessary to set a wind problem on the CR-1 Computer. The true airspeed must be placed above the TAS index and the true course must be placed above the TC index. If the true airspeed and the true course

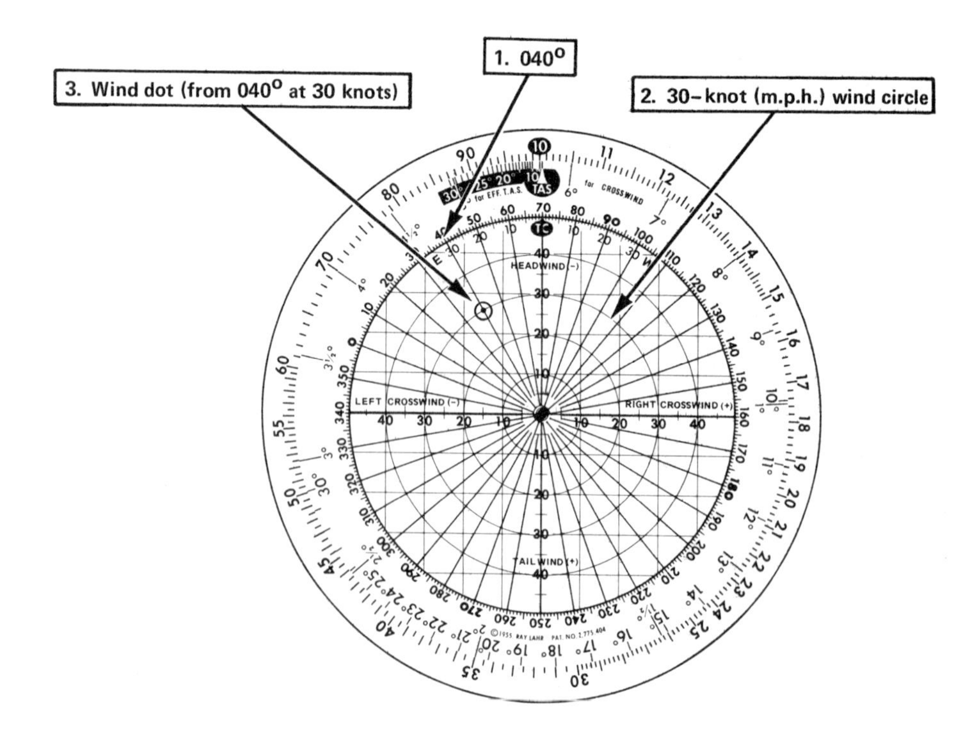

Fig. 10-16. Wind Dot

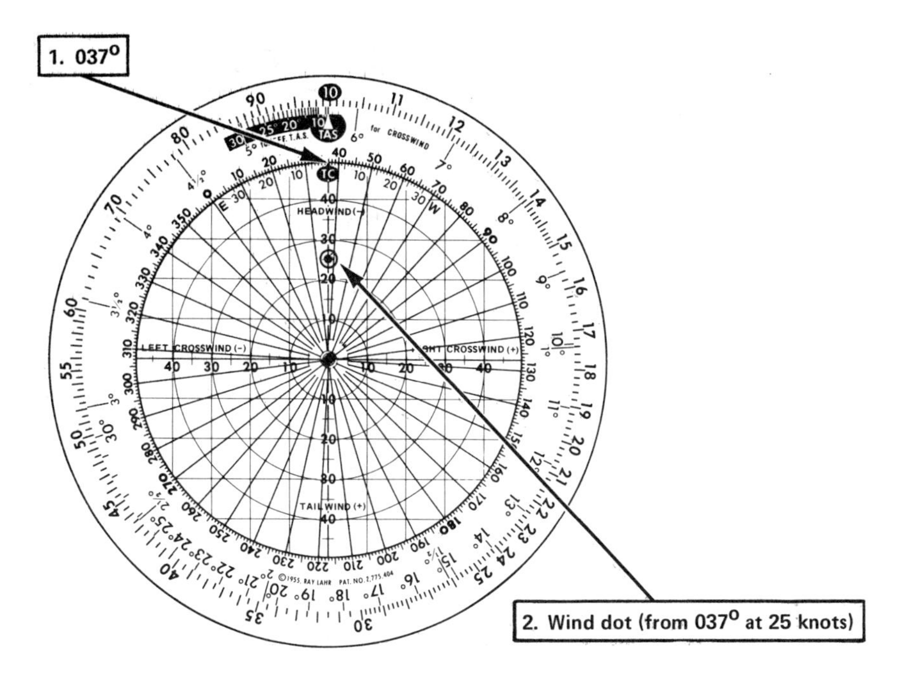

Fig. 10-17. Positioning the Wind Dot.

are given in the problem, place these quantities above the proper index and consider the computer fixed for that particular problem.

When true heading is given in lieu of true course in a problem, it is necessary to make a temporary setting of the true heading opposite the TC index and then make one or more computer adjustments before finding the true course. Remember, though, that until the true airspeed and true course are set opposite the proper indexes, the computer is not completely set for reading the solution.

FINDING THE TRUE HEADING AND GROUNDSPEED

Solving for true heading and groundspeed is the most common type of wind problem, because GS and TH are normally the two values missing in a flight planning problem.

EXAMPLE:

Given: Wind . . . . 20 knots from 010°
True airspeed . . . . 170 knots
True course . . . . 050°

Find: Groundspeed, wind correction angle, and true heading.

Solution:
1. Locate the wind dot by finding the 010° line on the transparent disc and placing the dot on the intersection of this line and the 20-knot circle as illustrated in figure 10-18.
2. Rotate the base disc until the true airspeed (170) is opposite the TAS index.
3. Rotate the transparent disc until the true course (050°) is positioned over the TC index, as illustrated in figure 10-18. Be sure that the true airspeed setting made in step 2 is not disturbed. Since both true airspeed and true course are now in place, the computer is fixed for the problem, and no further movement of the discs is necessary.
4. On the black horizontal scale, locate the point directly beneath the wind dot as shown in figure 10-19. This point represents a value of 12. Since the point is to the left of the computer center, there is a 12-knot left crosswind component signifying that the wind correction angle is to the left.
5. Locate the crosswind component on the outside scale of the computer by finding the figure 12 on the base disc. Approximately opposite this 12-knot figure, read the wind correction angle (WCA) of 4°.

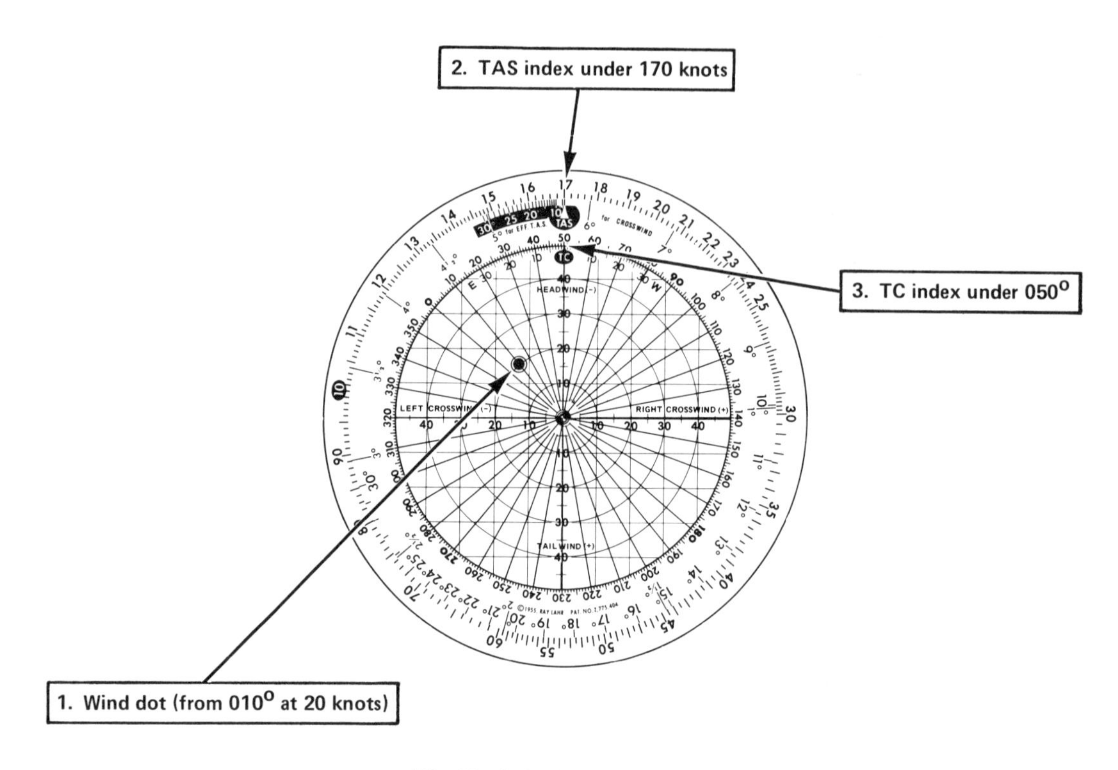

Fig. 10-18. Setting the Computer

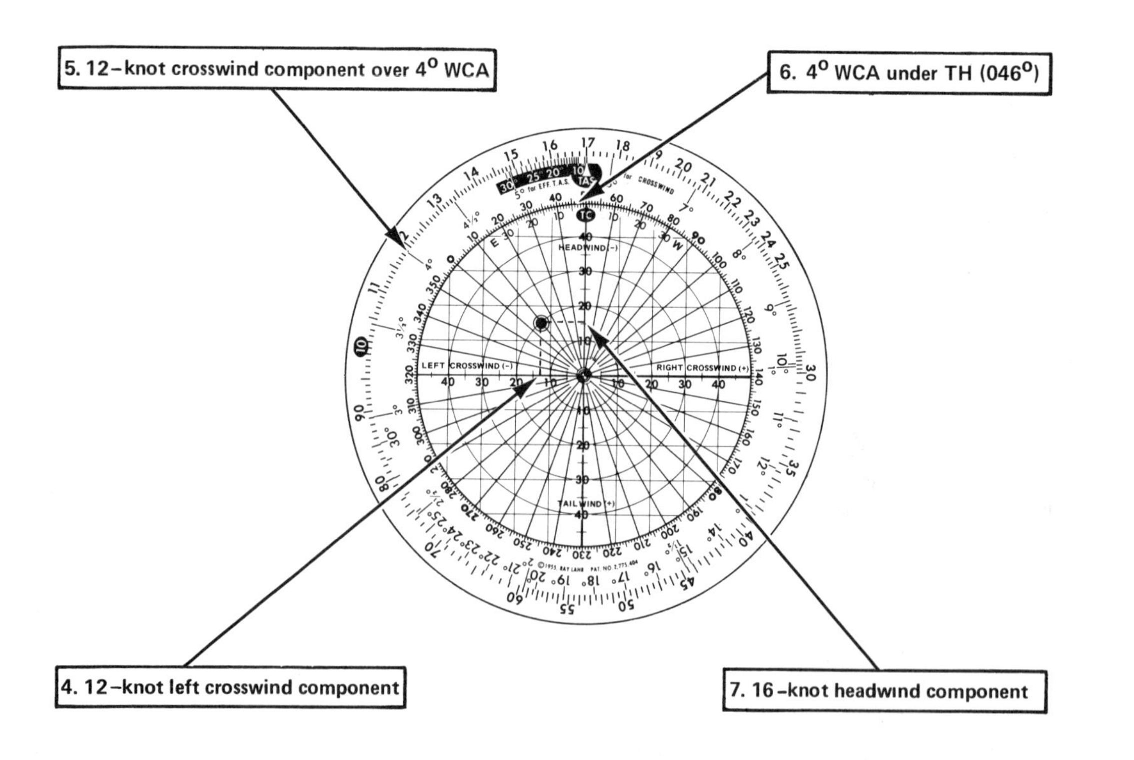

Fig. 10-19. Groundspeed, WCA, True Heading

6. Since the crosswind is from the left, the wind correction angle must be subtracted from the true course to find the true heading. This is easily done by proceeding 4° to the left of the true course index and reading the true heading (046°) on the transparent disc.
7. Next, the groundspeed must be calculated. To the right of the wind dot, interpolate on the black vertical scale to find the value 16. Since the point is above the computer center, on the headwind part of the scale, there is a 16-knot headwind component. If the point had been below center, there would have been a tailwind component. Subtract the 16-knot headwind component from the true airspeed (170 knots) to get the groundspeed of 154 knots.

Answers: Groundspeed — 154 knots; wind correction angle — 4°; true heading — 046°.

A slightly different solution is employed when the wind correction angle is 5° or greater.

EXAMPLE:

Given: Wind . . . . .25 kts. from 225°
True airspeed . . . . 120 kts.
True course . . . . .108°

Find: True heading and groundspeed

When the wind correction angle is 5° or greater, it is necessary to find the *effective true airspeed.* The headwind or tailwind component must be applied to the effective true airspeed, rather than true airspeed, in order to get the groundspeed. This procedure is explained in the following frames. The first three steps in this problem are illustrated in figure 10-20.

1. Place the wind dot on the transparent disc on the intersection of the 225° line and the ________ knot circle.	
2. Over the TAS index, place the true airspeed of ________ knots.	*25*
3. Over the TC index, place the true course of ________°. Since both true course and true airspeed are set, the computer is now in position for the problem. No further computer adjustments should be made.)	*120*
4. Measure upward from the wind dot to find a right crosswind component of ________ knots as illustrated in figure 10-21.	*108*
5. Approximately opposite 22 on the outside scale, read the crosswind correction of ________°.	*22*
6. Since the crosswind is from the right, read the true heading 11° to the right of the TC index. The true heading is ________°.	*11*

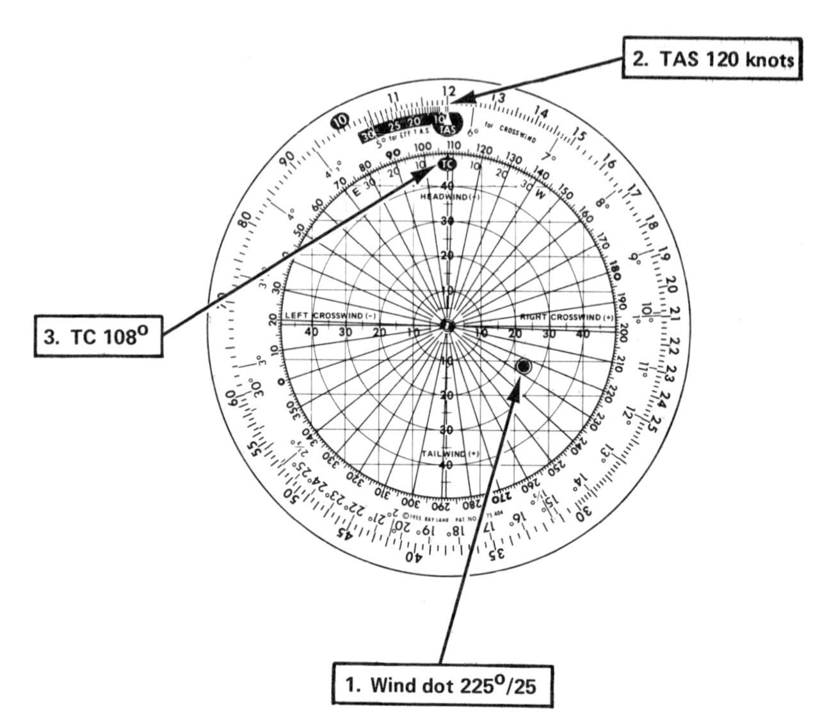

Fig. 10-20. Setting the Computer

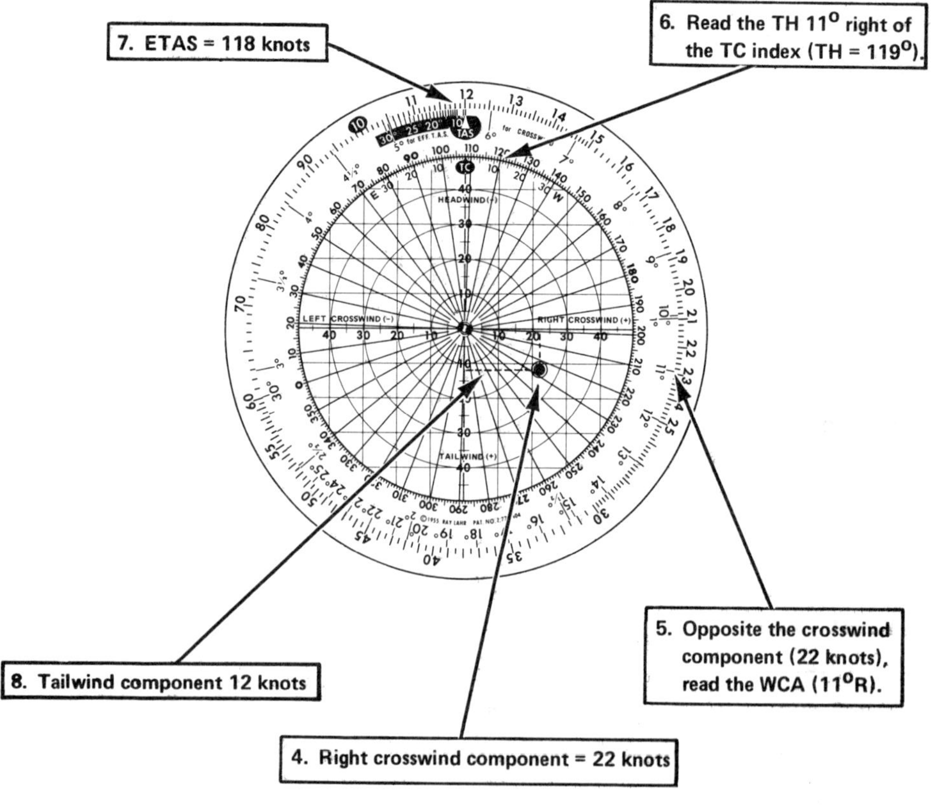

Fig. 10-21. True Heading and Groundspeed

7. Since the wind correction angle is greater than 5°, it is necessary to use effective true airspeed instead of true airspeed in finding the groundspeed. On the black scale to the left of the TAS index, locate the 11° point. Above this, read the effective true airspeed of ____________ knots.	*119*
8. When the wind correction angle is 5° or greater and an effective true airspeed is used, the headwind or tailwind component is applied to the effective true airspeed instead of the true airspeed to find groundspeed. Directly to the left of the wind dot, measure the tailwind component of ____________ knots.	*118*
9. Since the 12 is on the lower or tailwind side of the center, add 12 to the effective true ____________.	*12*
10. When the tailwind component (12) is added to the effective true airspeed (118 knots) the groundspeed is found to be ____________ knots.	*airspeed*
	130

EFFECTIVE TRUE AIRSPEED

The effective true airspeed scale should always be used when the wind correction angle is 5° or more. A small mark just to the left of the TAS index, between the index and the 10° mark, indicates a 5° wind correction angle. Interpolate between 5° and 10° for wind correction angles between these values.

When the wind correction angle is less than 10°, the error caused by ignoring effective true airspeed is less than 1½ percent. At a 5° wind correction angle, the error is less than one-half of one percent. The answers which are given in this manual are based on an effective true airspeed only when the wind correction angle (WCA) is 5° or more.

READING THE ANGLE SCALE

When reading the wind correction angle on the middle disc opposite the crosswind component on the outside scale, note that when going from 10° clockwise to approximately 30° on the middle disc, there are two scales — the outside one with the large angles 10°, 11°, 12°, etc., and the inside scale with the small angles from 1° through 1½°, 2°, 2½°, etc. Common sense will dictate whether the large or small angle should be used. However, one useful rule is that if the crosswind component is less than 10% of the true airspeed, the crab angle should be read on the inside angle scale and will be less than 6°. If a crosswind component is greater than 10% of the true airspeed, the wind correction angle should be read on the outside angle scale and will be 6° or greater.

FINDING TRUE COURSE AND GROUNDSPEED

EXAMPLE:

Given: Wind . . . . 20 knots from 090°
True airspeed . . . . .110 knots
True heading . . . . .050°

Find: True course and groundspeed

It is necessary to set the true course and true airspeed opposite the proper indexes on the computer in order to position the disc for a given wind problem. Since the true course is not available at the start of this problem, obviously the computer disc cannot be placed immediately in the final setting. It is necessary to place the TC index temporarily under the true heading in order to get an approximate crosswind component. Then, the top disc must be moved one or more times to try out various true courses until the wind correction angle opposite the crosswind component corresponds to the difference between the true course and the true heading. The solution to this problem is provided in the following frames.

NOTE: It is necessary to check the crosswind component and drift angle again to see whether this is the proper true course for the problem.

1. Place the wind dot on the transparent disc at the point where the 20-knot circle intersects the __________ line.	
2. Set the true airspeed (110 knots) over the TAS __________.	*090°*
3. Over the true course index, set the true heading of __________°. (Figure 10-22 shows the first setting.)	*index*
4. Under the wind dot, on the black horizontal scale, read the estimated crosswind component of __________ knots. (See Fig. 10-22.)	*050*

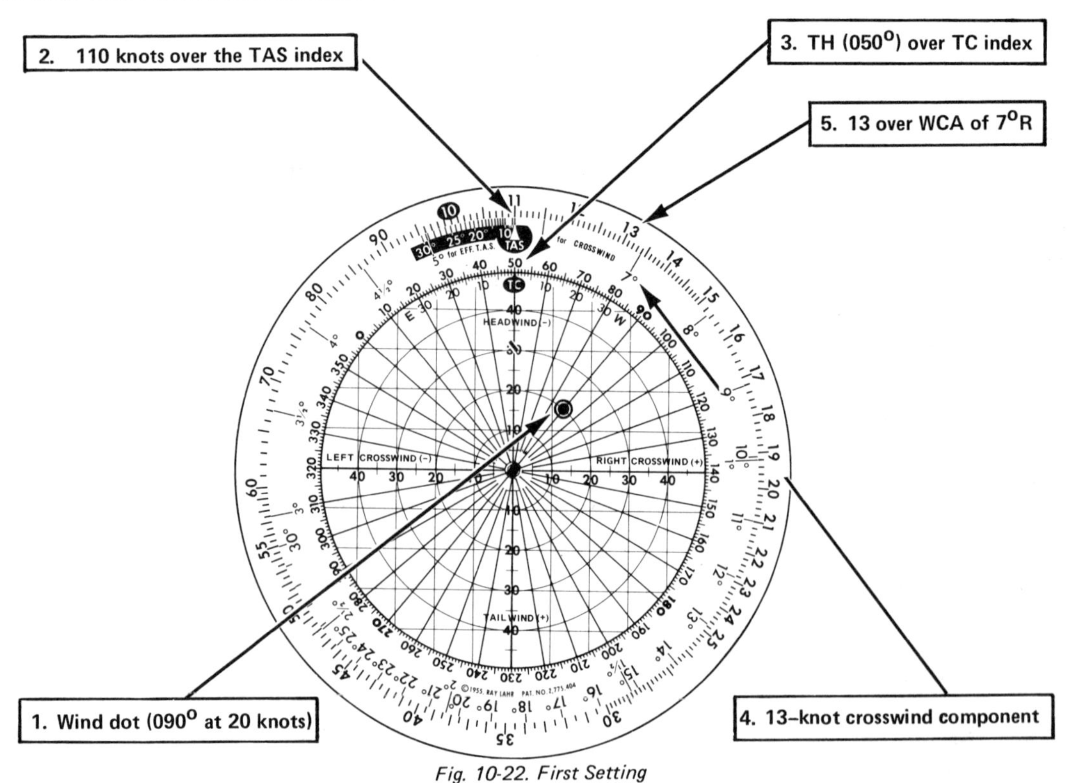

Fig. 10-22. First Setting

5. Since the top disc will be moved at least once when finding the true course, this 13-knot value is not necessarily the final crosswind component. Opposite 13 on the outer scale of the computer, read the approximate wind correction angle of ________°.	13
6. The crosswind is from the right and thus the aircraft must be corrected to the right to counteract the effect of wind. Hence, the top disc must be rotated clockwise so that the true heading is 7° to the right of the ________ index.	7
7. This action may be accomplished by rotating the transparent disc until the 050° true heading lies over the point marking 7° on the black scale. Note that the TC index now points to ________°. (See Fig. 10-23.)	TC

NOTE: It is necessary to check the crosswind component and drift angle again to see whether this is the proper true course for the problem.

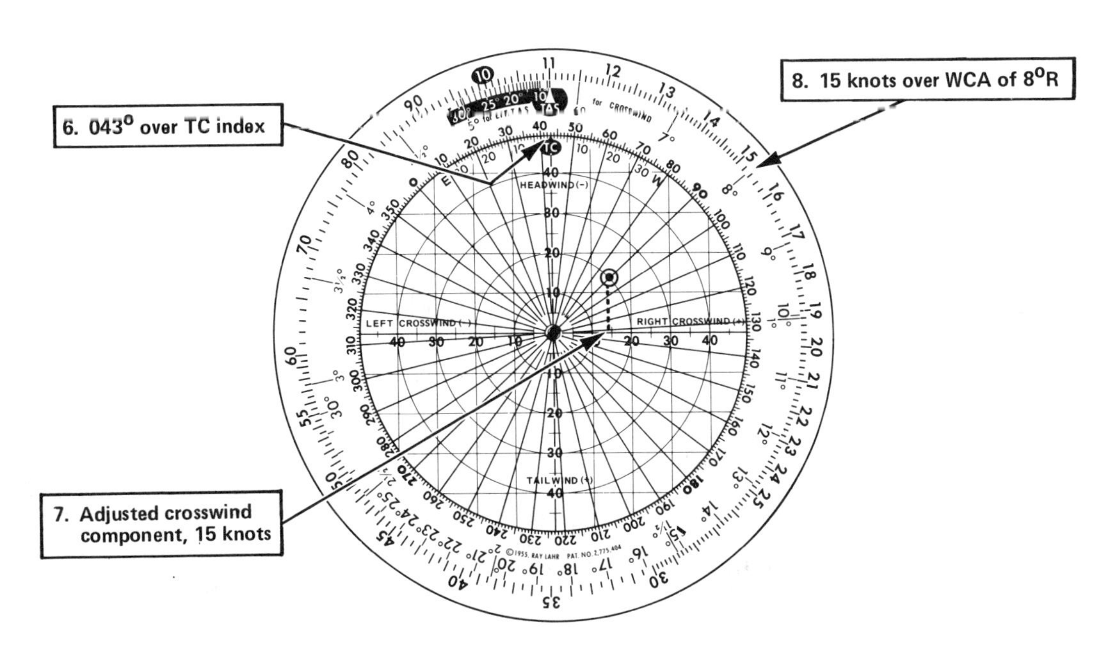

Fig. 10-23. Second Setting

8. After the transparent disc adjustment in the previous frame, the wind dot is now in a different position. Reading down from the dot to the horizontal scale, locate a new estimated crosswind component of ________ knots.	*043*
9. On the outside computer scale, the 15 knot crosswind component lies closest to the wind correction angle of ________°.	*15*
10. The 8° wind correction angle calculated in the last frame must also be reflected at the TC index. This is accomplished by positioning the TH 8° to the right of the ________ ________ as illustrated in figure 10-24.	*8*
11. The true course can be established as ________°.	*TC index*

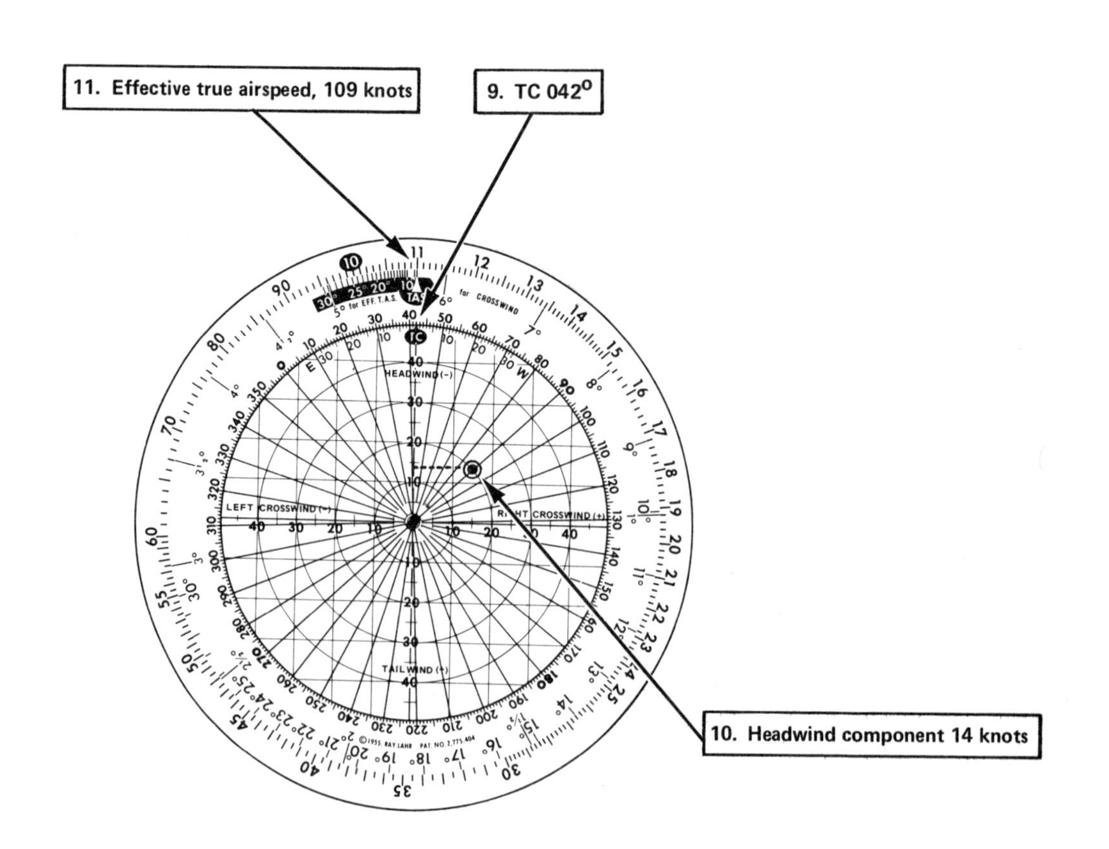

Fig. 10-24. True Course and Effective True Airspeed

12. On the vertical scale, to the left of the wind dot, read the headwind compoment of ________ knots.	*042*
13. With an 8° wind correction angle, the effective true airspeed is more nearly ________ knots.	*14*
14. After subtracting the headwind compoment (14 knots) from the effective true airspeed (109 knots), the groundspeed is ________ knots.	*109*
	95

FINDING THE WIND IN FLIGHT

Previously in this section, it was stated that all wind problems have six components and given any four, the other two may be solved. In order to find the wind in flight, all of the problem components except wind velocity and direction must be given. Since true course and true airspeed are given, it is possible to position the computer disc immediately by setting those values opposite the proper indexes. Then, after the crosswind component and the headwind or tailwind component are known, it is a simple matter to locate the two coordinates required to fix the wind dot.

EXAMPLE:

Given: True course . . . . .216°
True airspeed . . . . 150 kts.
True heading . . . . .220°
Groundspeed . . . . 130 kts.

Find: Wind direction and velocity.

The solution to this problem is provided in the following frames

1. Set the true airspeed (150 knots) at the TAS index. The number representing 150 knots on the outer scale is ________.	
2. Set the true course index at ________°.	*15*
3. Subtract the true course from the true heading to get the wind correction angle (220° - 216° = ________°).	*216*
4. Next, locate the 4° wind correction angle on the ________ ________.	*4*
5. Opposite 4° on the middle disc, the crosswind component of ________ knots is found.	*middle disc*
6. Since the true heading is greater than the true course, there ia a ________ crosswind.	*10.5*

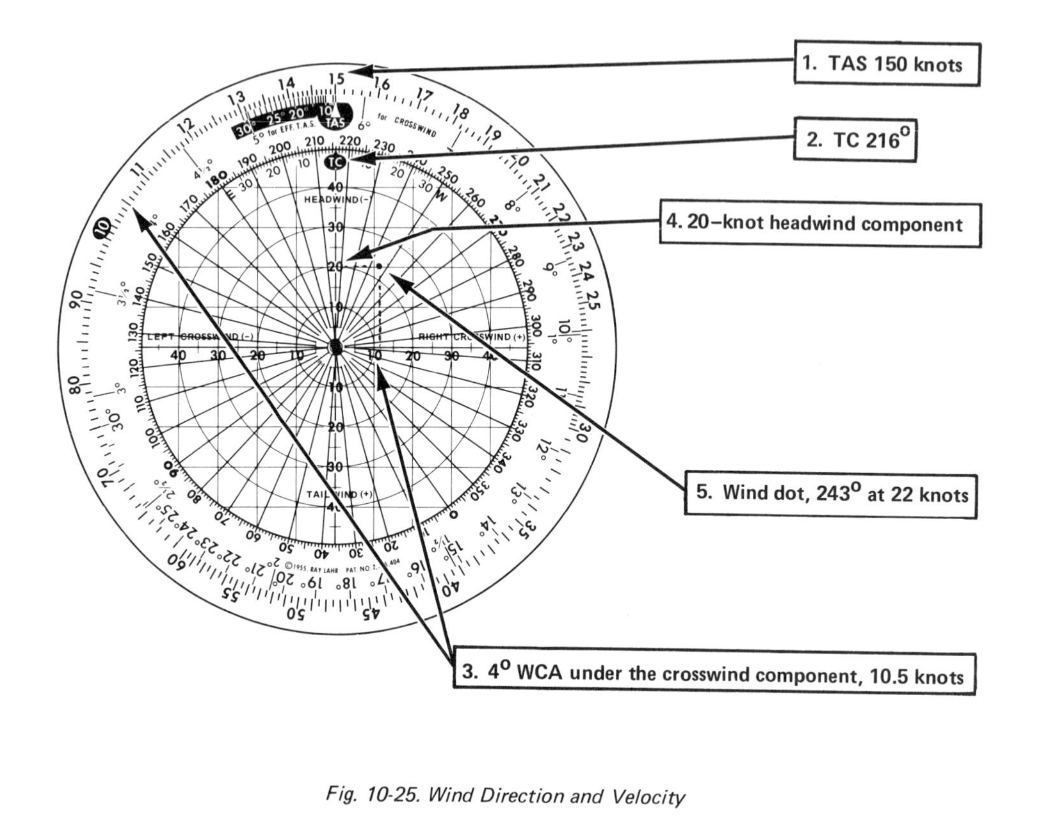

Fig. 10-25. Wind Direction and Velocity

7. Since the groundspeed is less than the true airspeed, there is a __________.	*right*
8. Next, measure to the right from the grommet along the horizontal scale to a total crosswind component of __________ knots.	*headwind*
9. Subtract the groundspeed (130 knots) from the true airspeed (150 knots) to arrive at the headwind component of __________ knots.	*10.5*
10. From a 20-knot headwind component measured above the grommet, proceed right until intersecting the line representing a 10-knot __________ component.	*20*
11. At the intersection of the lines representing these two components, place the __________ __________.	*crosswind*

12. The wind may be determined by noting that the wind dot falls on the 243° line of the transparent disc and by measuring the distance from the wind dot to the grommet as ______________ knots.

wind dot

22

NOTE: Had the wind correction angle exceeded 5°, effective true airspeed would have had to be taken into consideration. In this case, the wind correction angle would have had to be applied to the effective true airspeed scale. Then, the headwind component could be calculated by subtracting the groundspeed from the effective true airspeed.

THE CR-1 COMPUTER AS A SLIDE RULE

Most flight computers incorporate logarithmic time-distance scales, but do not have a movable indicator which is a necessity for successive multiplications and divisions in the same problem. The movable cursor on the CR-1 makes it a convenient circular slide rule. Businessmen, engineers, and pilots with a frequent use for figures can use the CR-1 as a handy pocket slide rule.

In using the computer for multiplication or division, care must be taken to treat the speed index as 6 or 60, and not to confuse it with the unit index 10. Also, remember that slide rule operations give only the successive digits in a number, thus it is often necessary to place the decimal point or to add zeros to the number. In order to get the correct answer, an estimate of the answer must be made. If an answer has the digits 294, for example, an estimate will determine whether it should be .294, 2.94, 2940, or another variation of the digits in that order.

MULTIPLICATION

In order to multiply two numbers, place the unit index of the minute scale opposite one of the numbers to be multiplied. Pressing the disc with the thumb and forefinger to prevent it from rotating, move the cursor until the hairline intersects, on the minute scale, the second number to be multiplied. Then read the answer beneath the hairline on the mile scale.

Use the following steps, with reference to figure 10-26, to multiply 2 x 3.

1. Place the unit index on the minute scale opposite the figure 20 on the mile scale.
2. Move the cursor until the hairline intersects 30 on the minute scale.
3. Opposite the figure 30 on the minute scale, read 60 on the mile scale and interpret the answer as 6.

ESTIMATING

Using the method just described, multiply 14 x 95. The answer consists of the digits 133. In order to estimate the answer, note that 95 is close to 100, and 100 times 14 equal 1400. Hence the correct answer must be 1330. Always estimate each answer to determine the location of the decimal point of final zeros, and to check the logic of the answer.

DIVISION

To divide one number by another on the CR-1, simply put the number to be divided on the mile scale and place the divisor beneath it. Read the answer on the mile scale opposite the minute scale unit index. Use the following steps to divide 244 by 67.

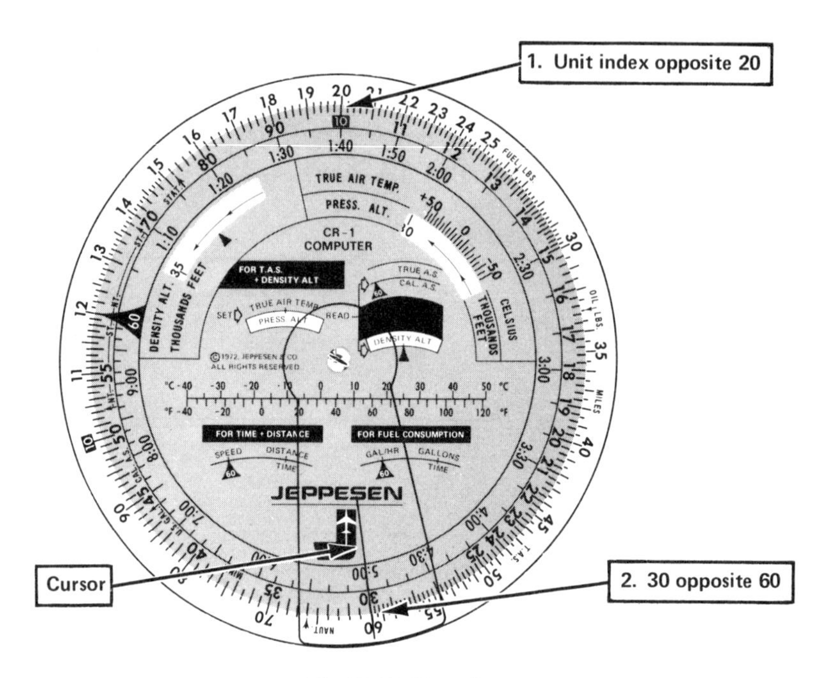

Fig. 10-26. Multiplication

1. Place 67 on the minute scale under 244 on the mile scale as shown in figure 10-27.
2. Read the answer (364) on the mile scale opposite the unit index on the minute scale.
3. Estimate the answer by noting that 244 is close to 240 and 67 is close to 60. 240 divided by 60 equals 4, therefore the answer must be 3.64 (not 36.4 or 364).

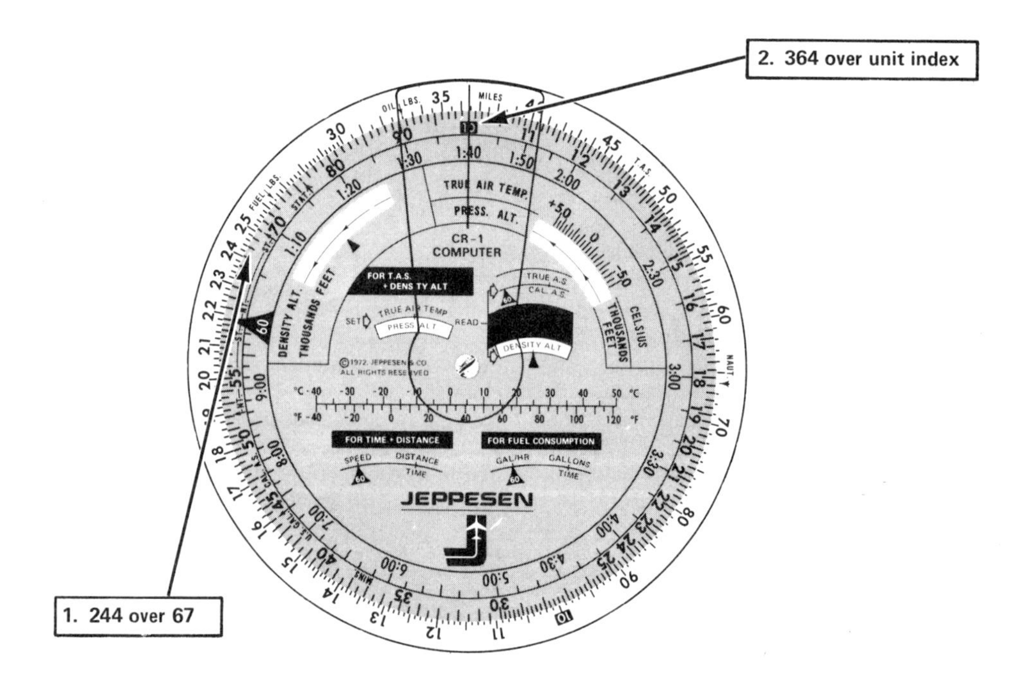

Fig. 10-27. Division

11 FLIGHT PLANNING CONSIDERATIONS

WEATHER CONSIDERATIONS

The first consideration in adequate flight planning is a preliminary check of existing and forecast weather at the departure airport, enroute, and at the destination airport. The pilot conducting an IFR flight should receive a broad, general scan of the total enroute weather from the weather charts, followed by a more specific briefing, including the weather reports and forecasts.

The pilot should consult surface analysis charts, weather depiction charts, low-level significant weather prognostic charts, and radar summary charts. The weather reports and forecasts that give more detailed information are the hourly aviation weather reports, terminal forecasts, area forecasts, winds aloft forecasts, and pilot reports (PIREPs).

The method of first gaining a general weather scan and then narrowing down to specific enroute information is a recommended procedure. However, each pilot may find that alternate methods of obtaining weather briefings are suitable, depending upon the aircraft's capabilities, terrain, and distance of the flight.

AIRCRAFT PERFORMANCE

Aircraft performance refers to how well the airplane will react under a given set of circumstances. Generally, the pilot can accurately predict this performance by use of graphs and charts. The aircraft owner's manual, or flight manual, contains those charts and graphs.

Before any presentation on the use of specific performance data, it is necessary to understand the principles of a graph, and how it is used. Graphs and charts accurately display the relationship between two or more variables. Pertaining to aircraft performance, graphs and charts are used for computing range, fuel consumption, takeoff distance, available power, and rate of climb. For example, a power setting selected from the range charts is more efficient than a random setting. These charts provide accurate fuel flow settings from which close estimates of fuel consumption can be made. The information in the charts and graphs has been compiled from actual flight tests (using average piloting techniques) with the aircraft and engines in good condition.

Aircraft owner's manuals or flight manuals may use both straight line or curved line graphs to present essential performance data. By using a combination of straight line and curved line graphs, as many as five variables may be presented.

1. Pressure altitude
2. Outside air temperature
3. Gross weight
4. Headwind
5. Distance

By compensating for these five variables, the pilot can determine the accelerate-stop distance, normal takeoff distance, and additional performance items.

The charts will be presented and discussed in the general order of their occurrence in the flight profile — accelerate-stop, takeoff, climb, cruise, and landing. These charts and graphs are typical of those found in the flight manuals or owner's manuals for multi-engine aircraft; however, similar types of charts and graphs are used for light, single-engine aircraft as well.

ACCELERATE-STOP DISTANCE

Prior to flight, Federal Aviation Regulations require pilots to determine that the runway lengths are sufficient at airports they intend to use. Also, if a multi-engine aircraft is used, safe operating practice dictates that the pilot determine if the runway is long enough for the airplane to accelerate to safe single-engine liftoff speed, then stop while there is still sufficient runway remaining.

If an engine should fail on the takeoff roll, prior to reaching takeoff speed, the aircraft must be stopped regardless of available runway length remaining. However, it is vital that the pilot know, prior to the takeoff, the distance required to accelerate the aircraft to its normal liftoff speed, discontinue the takeoff, and stop the aircraft. This distance is termed the accelerate-stop distance, and may be determined by the use of a chart, as shown in figure 11-1.

To allow for the variables that affect accelerate-stop distance, some criteria must be adopted to attain maximum safety without optimum pilot technique. This is accomplished by considering the same variables that are present when computing normal takeoff distance. These variables are temperature, pressure altitude, gross weight, and headwind.

In the following example, the determination of the accelerate-stop distance is predicated on the following variables.

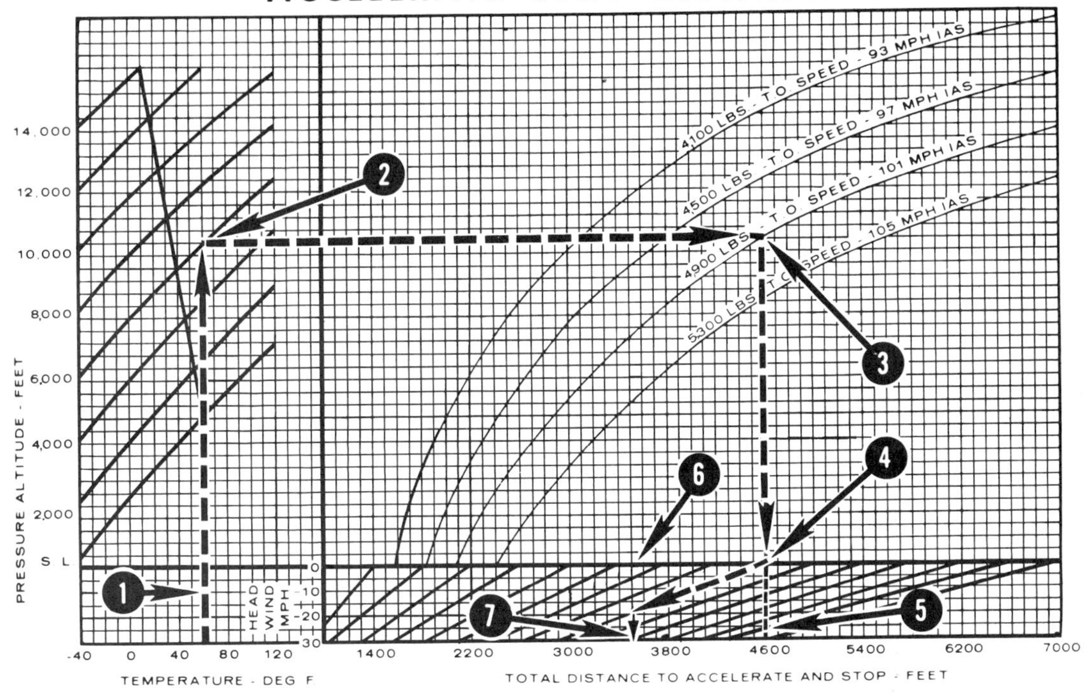

1. Temperature 60°F.
2. Pressure Altitude 6,000 ft.
3. 4,900 lbs. Gross Weight
4. No-Wind Distance
5. Distance 4,600 ft.
6. To 20 m.p.h. Wind Line
7. Distance = 3,500

Fig. 11-1. Typical Accelerate-Stop Distance Graph

1. Hard-surfaced runway
2. Wing flaps up
3. Full power before releasing the brakes
4. Mixture at recommended fuel flow
5. Engine failure at takeoff speed
6. Heavy braking after engine failure

Referring to the chart in figure 11-1, the accelerate-stop distance under the following conditions should be found.

Temperature . . . . .60°F
Pressure altitude . . . . .6,000 ft.
Takeoff weight . . . . .4,900 lb.
Headwind . . . . 20 m.p.h.

The steps in solving for accelerate-stop distances are as follows.

1. Locate the temperature of 60° Fahrenheit (item 1) on the referenced illustration.
2. Move up to the 6,000-foot pressure altitude line (item 2).
3. Move right horizontally to the 4,900-pound gross weight curved line (item 3).
4. Move down to the no-wind line (item 4). To find the no-wind distance of 4,600 feet, continue down to item 5.
5. In this example, move from item 4 diagonally left to the 20 m.p.h. headwind line at item 6, then, move vertically down to item 7 to find the accelerate-stop distance of 3,500 feet.

TAKEOFF PERFORMANCE

Takeoff performance is frequently a critical consideration in flight planning. The takeoff itself limits the total load of the aircraft since every airplane can handle a considerably heavier load in flight than on takeoff. Because of variations in atmospheric conditions, available power, and pilot technique, it is not a simple matter to accurately predict takeoff performance without the aid of a reliable chart or graph.

Normal takeoff procedures and considerations are standard in that computations are predicated on the basis of a hard surfaced runway, full power and, loaded weight of the aircraft. Although gross weight is the primary factor determining the length of the takeoff run, there are adjustable factors and additional factors.

1. Flap setting
2. Type of runway
3. Atmospheric pressure
4. Air temperature
5. Wind direction and velocity

The takeoff distance chart in figure 11-2, is used for calculating normal twin-engine takeoff distance. For example, referring to the chart shown, the pilot can determine the total ground run under the following conditions:

Temperature	90°F
Pressure altitude	2,000 ft.
Gross weight	3,600 lb.
Headwind	10 m.p.h.

The example problem is solved in six steps.

1. Locate the temperature of 90° Fahrenheit at the bottom of the chart.

2. Move up vertically to the diagonal line representing 2,000 feet of pressure altitude (item 1).

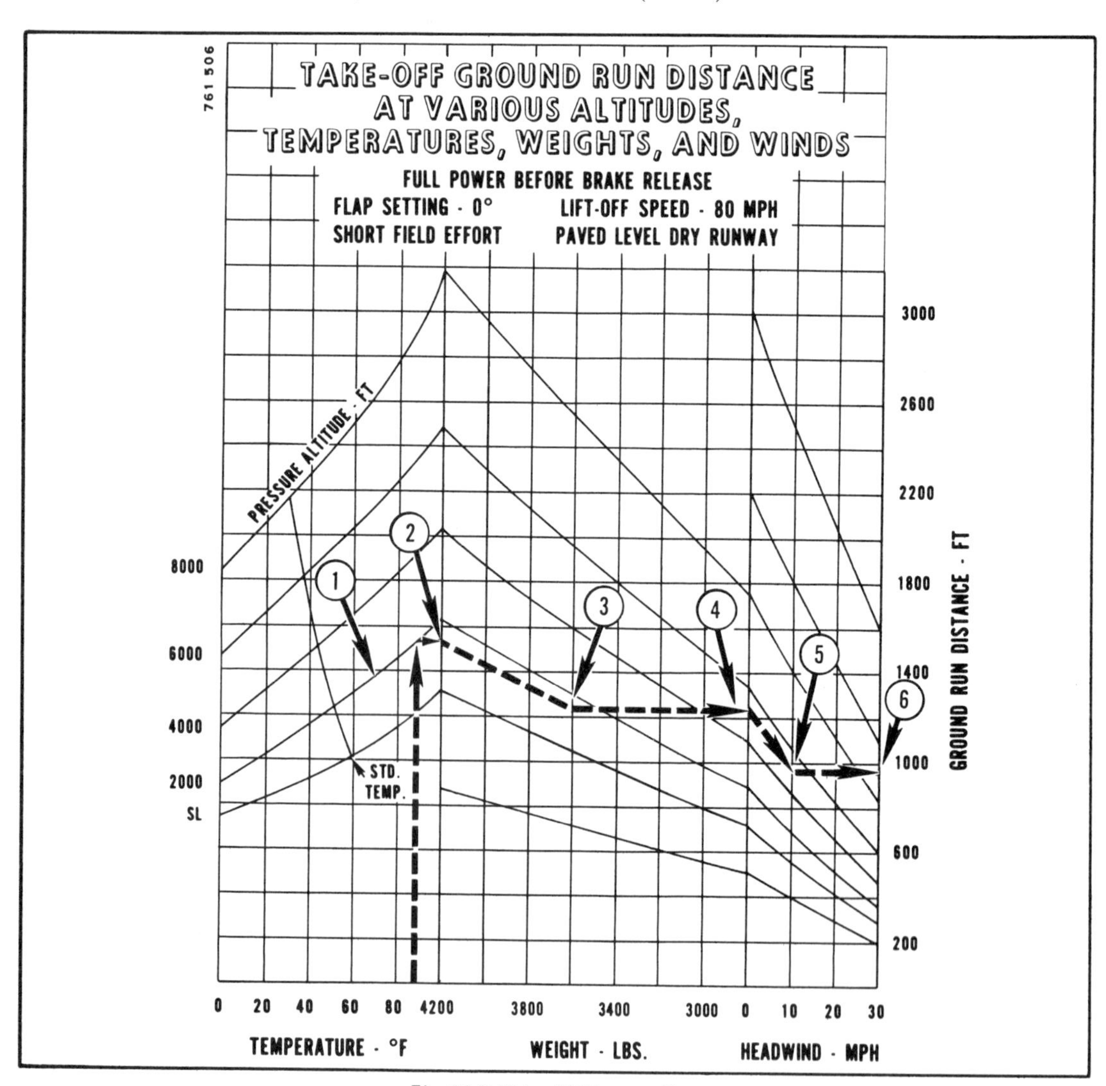

Fig. 11-2. Takeoff Distance Chart

3. Move horizontally (right) to the gross weight line (item 2).

4. Move parallel to the gross weight lines and proceed to the line that represents 3,600 pounds (item 3).

5. Proceed horizontally to the zero headwind line (item 4).

6. Proceed parallel to the headwind lines until reaching the headwind component of 10 m.p.h. (item 5), then move horizontally (right) to find the total takeoff distance of 950 feet.

NOTE: *When using this chart, the pilot will need to convert the windspeed from knots to miles per hour.*

It is important for the pilot to realize, however, that by consulting the takeoff distance charts and computing the distance based on known or estimated atmospheric conditions, it is possible, under extreme circumstances, to find that a takeoff is impossible within the confines of the available runway. Therefore, minor factors not allowed for in the performance charts should be considered. When marginal conditions appear, such as humidity, runway gradient, turbulence, or the age and condition of the airplane, the pilot may wisely postpone takeoff until conditions are more favorable.

CLIMB

The climb segment of the flight profile can best be defined as complex; that is, there is usually a broad variety of speeds associated with the ascent to altitude. The different types of climbs the pilot should understand are best angle-of-climb, best rate of climb, and cruise climb.

BEST ANGLE OF CLIMB

The best angle-of-climb speed will provide the greatest gain in altitude over a *given distance.* This speed is most often used immediately after liftoff in order to clear obstacles that may be at the end of the runway. Since the aircraft is at a high angle of attack at this speed, there is a wasteful increase in drag which, of course, requires more power and reduces the rate of climb. Therefore, once obstacles are cleared, the pilot should reduce the angle of attack so the aircraft may accelerate to a more efficient climb speed.

BEST RATE OF CLIMB

As the nose of the aircraft is lowered from the best angle-of-climb position, the speed increases and the rate-of-climb increases. At the best rate-of-climb airspeed, the aircraft will gain the greatest altitude in a given unit of time. The best rate-of-climb is often used to reach an assigned altitude in the least amount of time. Terrain, winds, and altitude are considerations which may influence the pilot to strive to obtain the greatest altitude gain in a given amount of time. The speed which accomplishes this tends to decrease as the altitude increases. (See Fig. 11-3.) The amount of decrease is so slight, however, that it can be disregarded for most operations.

CRUISE CLIMB

The airplane owner's manual usually will specify a normal or cruise climb airspeed which will result in adequate climb yet will save time and fuel for the overall flight. This speed may be used after suitable altitude is gained after takeoff.

As shown in figure 11-3, an aircraft at sea level, at a gross weight of 5,300 pounds, will climb at

MULTI-ENGINE CLIMB DATA AT 5300 POUNDS

NOTE: DECREASE RATE OF CLIMB 20 FT/MIN FOR EACH 10°F ABOVE STANDARD TEMPERATURE FOR A PARTICULAR ALTITUDE.

MAXIMUM CLIMB

SEA LEVEL 59°F			5000 FT. 41°F			10,000 FT. 23°F			15,000 FT. 5°F			20,000 FT. -12°F		
Best Climb IAS MPH	Rate of Climb Ft/Min	Lbs of Fuel Used	Best Climb IAS MPH	Rate of Climb Ft/Min	From S.L. Fuel Used	Best Climb IAS MPH	Rate of Climb Ft/Min	From S.L. Fuel Used	Best Climb IAS MPH	Rate of Climb Ft/Min	From S.L. Fuel Used	Best Climb IAS MPH	Rate of Climb Ft/Min	From S.L. Fuel Used
123	1495	24	122	1136	39	121	782	56	120	427	80	119	66	136

NOTE: FULL THROTTLE, 2625 RPM, MIXTURE AT RECOMMENDED FUEL FLOW, FLAPS AND GEAR UP. FUEL USED INCLUDES WARM-UP AND TAKEOFF ALLOWANCE AND IS EXPRESSED IN POUNDS.

Fig. 11-3. Rate-of-Climb Chart

1,495 feet per minute if the temperature is standard (59°F). If the temperature, for example, were 99°F, the rate-of-climb would be decreased by 80 feet per minute. This computation is the result of decreasing the climb 20 feet per minute for each 10° above the standard temperature. Therefore, if the outside air temperature is 99°F, the rate of climb is 1,415 feet per minute.

CRUISE PERFORMANCE

When planning the cruise segment of the flight profile, the pilot should be economy conscious; however, the attainment of economy depends largely on his ability to read and interpret cruise performance charts. The cruise charts provided in the airplane owner's manual indicate the true airspeed, range, endurance, and fuel consumption that can be expected for a given power setting at various altitudes.

To accurately project flight endurance, compensation must be made for data not included in the chart figures. Fuel consumed during taxi, warm-up, takeoff, and climb is not allowed for in the charts. Headwind and altitude factors are such variables that it is impractical to chart their effects; hence, the pilot must allow for this factor in addition to mixture leaning techniques in order to have an adequate fuel reserve.

The cruise performance charts for the aircraft, illustrated in figure 11-4, are based on gross weight. It can be seen that 2,450 r.p.m. and 19 inches of manifold pressure are required to achieve 60 percent power at 10,000 feet. This power setting should deliver a true airspeed of 206 m.p.h. with a total fuel consumption rate of 22.8 gallons per hour. Any desired flight condition can be determined by referring to the chart appropriate to the altitude.

If the planned flight altitude falls between the altitudes covered by the charts, the values sought may be found by a process known as "interpolation." When it is known that the reading desired will fall between two values given on the charts, the difference between these two values should be divided into 10 equal parts. Then the point at which the reading is desired can be said to be so many tenths from either value and added or subtracted to obtain the desired figure.

LANDING PERFORMANCE

Performance information for the landing phase is also included in the airplane owner's manual, or aircraft flight manual. Landing and stopping distances are plotted on the landing performance chart shown in figure 11-5. Wing flaps have little direct effect on aircraft stopping distances, but *indirectly they have an appreciable effect*, since a high wing flap angle will materially reduce landing speed.

The information presented in the landing chart, figure 11-5, is based on maximum braking, 40° flap extension, and a hard-surfaced runway. Theoretically, maximum braking is realized just before the point at which skidding occurs. The chart directly makes allowances for temperature, pressure altitude, wind, and aircraft weight.

The steps listed are used in determining the total landing distance under the following conditions:

CRUISE PERFORMANCE WITH NORMAL LEAN MIXTURE AT 10,000 FT.

RPM	MP	% BHP	TAS	Total Gal/Hr.	Endurance 100 Gal.	Range 100 Gal.	Endurance 130 Gal.	Range 130 Gal.
2450	20	64	212	24.2	4.1	875	5.4	1140
	19	60	206	22.8	4.4	905	5.7	1175
	18	56	198	21.2	4.7	935	6.1	1215
	17	52	191	19.7	5.1	970	6.6	1260
2300	20	59	203	22.1	4.5	920	5.9	1195
	19	55	197	20.9	4.8	940	6.2	1225
	18	51	191	19.7	5.1	970	6.6	1260
	17	47	182	18.3	5.5	990	7.1	1290
2200	20	54	196	20.6	4.8	950	6.3	1235
	19	51	189	19.5	5.1	970	6.7	1260
	18	48	182	18.4	5.4	990	7.1	1285
	17	44	175	17.3	5.8	1010	7.5	1310
2100	20	50	188	19.2	5.2	980	6.8	1275
	19	47	181	18.2	5.5	995	7.1	1295
	18	44	174	17.2	5.8	1010	7.6	1310
	17	40	164	16.2	6.2	1010	8.0	1315
	16	37	147	15.3	6.5	960	8.5	1245

CRUISE PERFORMANCE IS BASED ON STANDARD CONDITIONS, ZERO WIND, NORMAL LEAN MIXTURE, 100 AND 130 GALLONS OF FUEL (NO RESERVE), AND 5100 POUNDS GROSS WEIGHT.

Fig. 11-4. Cruise Performance Chart

Temperature . . . . .60°F
Pressure altitude . . . . 4,000 ft.
Gross weight . . . . .3,800 lb.
Headwind . . . . 15 m.p.h.

1. Locate the temperature of 60° F at the bottom of the chart.
2. Move up vertically to the diagonal line representing 4,000 feet pressure altitude (item 1).
3. Move horizontally (right) to the gross weight line (item 2).
4. Move parallel to the gross weight lines and proceed to the area that represents 3,800 pounds (item 3).
5. Proceed horizontally to the zero headwind line (item 4).
6. Proceed parallel to the headwind lines until reaching the headwind component of 15 m.p.h. (item 5); then move horizontally (right) to find total landing distance of 525 feet.

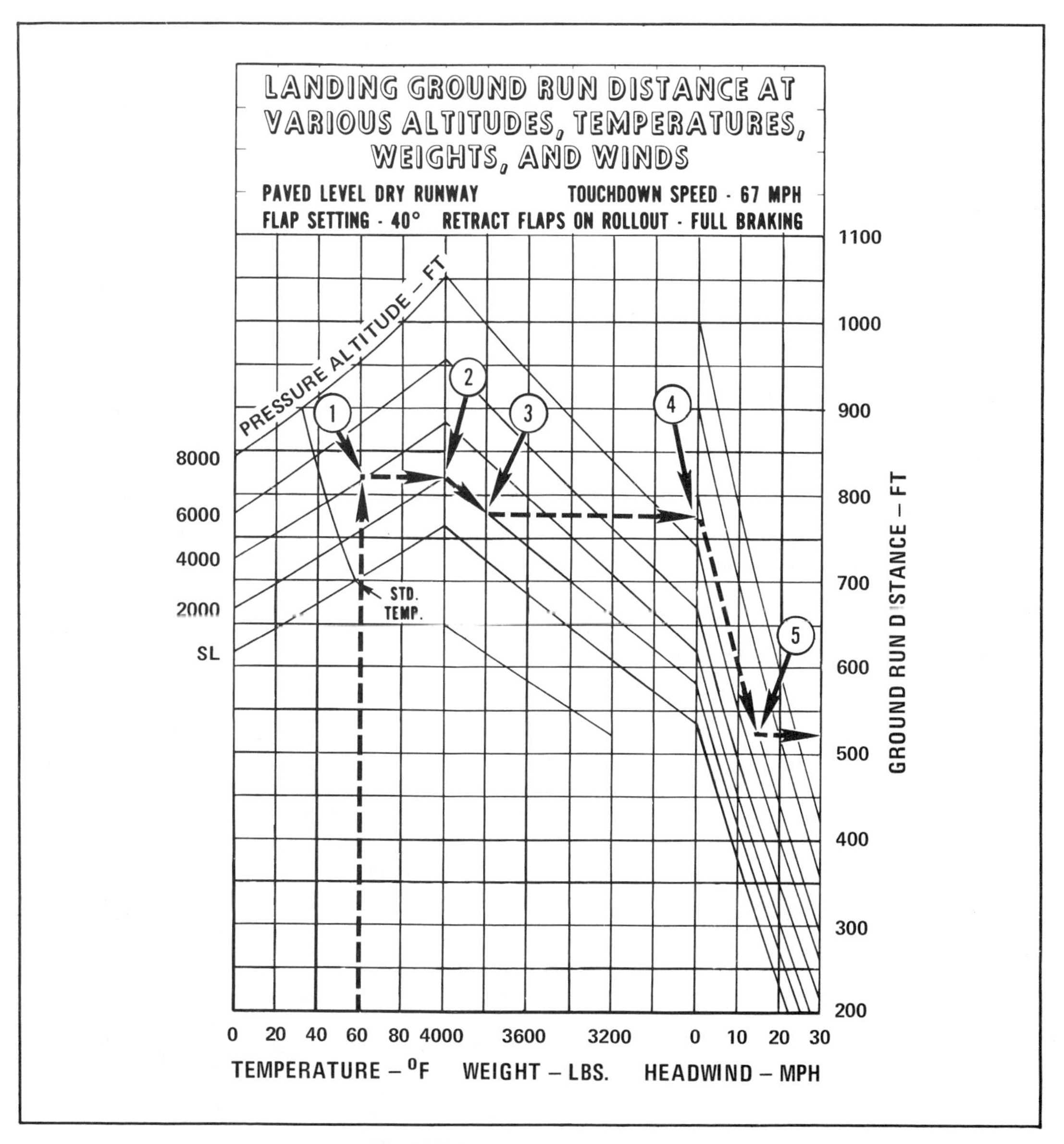

Fig. 11-5. Landing Performance Graph

OXYGEN CONSUMPTION

An oxygen supply system in the airplane permits flight operations at altitudes where the atmospheric density is so low that the pilot and passengers otherwise would not receive sufficient oxygen. By having oxygen available, the pilot can realize greater fuel economy as well as take advantage of favorable winds at higher altitudes. Oxygen is stored under a pressure of 1,800 p.s.i. in the oxygen cylinder. Before being routed to the continuous flow couplings provided for the pilot and each passenger, it is reduced to a breathing pressure by a preset automatic pressure regulator.

The system is designed to simultaneously supply adequate oxygen to the pilot and passengers and conserve the oxygen supply. This is accomplished by providing the pilot with an oxygen mask which supplies more oxygen than a standard passenger mask. In order to readily identify the masks of different capacities, a color coding system is used. Generally, the pilot's oxygen mask hose assembly is color coded red.

Assuming the conditions given in figure 11-6 for a flight planned at 20,000 feet with pilot and three passengers, how long will the fully charged 76.6 cubic foot system supply the flight with oxygen?

OXYGEN CONSUMPTION RATE CHART

CYLINDER CAPACITY CUBIC FEET	76.6		48.3	
ALTITUDE RANGE FEET	10,000 22,000	22,000 30,000	10,000 22,000	22,000 30,000
HOSE ASSEMBLY COLOR	ORANGE	RED	ORANGE	RED
CONSUMPTION PSI/HR	(1) 125	195 (2)	197	308

OXYGEN DURATION CALCULATION:

Total Oxygen Duration (Hours) = oxygen pressure indicator reading [oxygen consumption (PSI/HR) x number of passengers + pilot consumption rate]

EXAMPLE (76.6 cu. ft. capacity) (1800 psi, oxygen pressure indicator reading)

1. Planned Flight-Pilot and 3 passengers at 20,000 feet
2. From Chart — At 20,000 feet altitude, passenger flow rate is 125 PSI/HR. and the pilot flow rate is 195 PSI/HR.
3. Oxygen Duration = 1800 ÷ (3 x 125 + 195) = 3.16 hours

Fig. 11-6. Oxygen Consumption Rate Chart

This problem can be solved using the four steps listed.

1. Find the flow rate per passenger (item 1) and note that this rate is 125 p.s.i. per hour.
2. Multiply this value times the number of passengers. (3x125=375 p.s.i. per hour).
3. Find the flow rate for the pilot (item 2) and add this figure to the value for the passengers (375 + 195 = 570 p.s.i. per hour).
4. The oxygen duration is found by dividing the available supply (1,800 p.s.i.) by 570 p.s.i. per hour (1,800 ÷ 570 = 3.16 hours).

WEIGHT AND BALANCE

The pilot is responsible at all times for observing the operating limitations of the aircraft. One important limitation is the aircraft's weight and balance limitation. Each particular aircraft type is only capable of carrying a certain maximum load — exceeding this load, or weight, harms the stability and control characteristics of the aircraft and may even result in structural damage or the inability of the aircraft to take off in some conditions. The other main consideration is whether or not the aircraft is balanced properly; that is, whether the loads are properly distributed throughout the airplane. Improper weight distribution also adversely affects the aircraft stability and control.

If either the weight or the balance of the aircraft exceeds the limitations established by the manufacturer, then, the pilot is attempting to operate the aircraft under conditions demonstrated as unsatisfactory during its certification and approval. At best, the aircraft performance will suffer; and at worst, the aircraft may have very unfavorable handling characteristics that may not be known to the pilot. On an instrument flight this is particularly dangerous because of the added workload on the pilot. An unstable aircraft is extremely difficult to manage under IFR weather conditions.

MAXIMUM ALLOWABLE LOADS

There are three terms used in the computations involved in the loading of any kind of aircraft in general aviation.

1. *Basic empty weight* — the actual or computed weight provided by the manufac-

turer for new aircraft or by certificated mechanics for modified aircraft. It includes unusable fuel, full operating fluids, and full oil.

2. *Useful load* — the weight of the contents of the aircraft, including fuel, passengers, cargo, baggage and crew. However, in the data provided by some manufacturers, the weight of the engine oil may or may not be included in the useful load.

3. *Gross weight* — the sum of the empty weight and useful load.

Federal Aviation Regulations require that the pilot have access to the empty weight and the maximum allowable (maximum certificated) gross weight data for every individual aircraft. Gross weight must be determined by adding the weights of all planned loads and then adding that figure to the aircraft empty weight. If the maximum certificated gross weight has been exceeded, then the pilot must determine the best way to decrease the load to make it fall within allowable limits.

LOAD DISTRIBUTION

Even an aircraft that is operated at less than its maximum certificated gross weight may be unsafe if the loads have not been correctly distributed in the airplane. Usually the larger the aircraft, the greater the possibilities for incorrect loading. Several different methods are used to compute load distribution but they all have in common the following considerations.

All of the weight of the empty aircraft can be considered to be concentrated at a single point called the center of gravity. If the aircraft were placed in any attitude and suspended from or balanced on the center of gravity the attitude of the aircraft would not change. However, as loads are placed in the aircraft, they have the effect of changing the pitch attitude of the aircraft, depending on whether the loads are placed ahead or aft of the original center of gravity. These pitching or rotating forces are called *moments.* The moment is proportional both to the amount of weight in a load and to the distance of the load from the center of gravity. Increasing either the weight or the distance increases the moment.

Since the CG of an airplane changes with each loading situation, it cannot be used as a reference point. To provide a constant reference point for use in computations, a "datum line" is established. A datum line is an arbitrarily chosen reference line about which all moments are computed. Moments are referred to as positive (plus) if they are aft of the datum line and negative (minus) if they are forward. As seen in Figure 11-7, moments to the right of the datum line are positive and moments to the left are negative.

In many airplanes, the datum line is located at the firewall, as shown in figure 11-7. In this case, all of the moments ahead of the firewall will be negative and all those behind the firewall will be positive. To avoid use of negative moments, many manufacturers have placed the datum line at, or ahead of, the nose of the airplane. It does not matter where the datum line is located; however, once the datum line is established by the manufacturer it must be the basis from which measurements are calculated.

The distance of a load from the datum line is called the arm. All aircraft weight and balance calculations depend on the principle that *weight times arm equals moment.* Thus a 10-pound

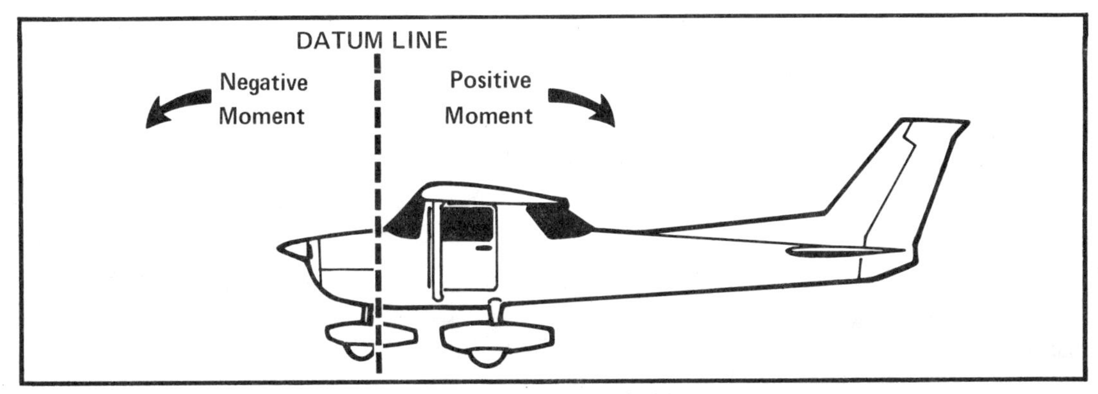

Fig. 11-7. Datum Line Position

load placed 104 inches aft of the datum line would have a moment of 1040 pound-inches. Placing the same 10-pound weight only 50 inches aft of the datum line would give it a moment of only 500 pound-inches.

As each load is put into an aircraft its moment affects the balance of the aircraft, and so the pilot must add together all the moments for the loads he intends to carry. Even the unloaded aircraft has a moment of its own, which is the aircraft empty weight multiplied by the arm (distance of the center of gravity from the datum line). The moment of each load is added to the moment of the empty aircraft and the proper loading of the airplane can then be determined from the resulting total moment.

The balance limitations of an aircraft loaded to its maximum certificated gross weight can be described in terms of either arm limits or center of gravity (CG) limits. In other words, if the aircraft is thought to be loaded too far forward, the pilot can determine whether the arm is greater or less than the prescribed minimum value. The aircraft would be seen to be loaded too far aft if the arm exceeds a certain maximum value. For certain types of aircraft, calculations of this kind are done with minimum and maximum moment limitations rather than arm limitations, but the safety of any combination of loads can be determined by either method chosen by the manufacturer.

EFFECTS OF IMPROPER LOADING

Although the takeoff and climb performance of an overloaded airplane may be drastically reduced, it might be possible under some circumstances to undertake a flight with a seriously over-loaded condition. However, the airplane would require more power to overcome high induced drag while cruising. The drag would be caused by the high angle of attack necessary to produce enough lift to carry the increased load. The structure of the aircraft would also fail more readily in extreme turbulence or abrupt maneuvers, and an overloaded aircraft would stall at a higher-than-normal airspeed.

Placing a load too far aft in an airplane (carrying too much baggage, for example) could cause undesirable flight characteristics such as instability and loss of control. With the CG too far aft, the pilot might be unable to lower the nose to recover from a stall — control forces could be insufficient to overcome an extreme tail-heavy condition.

With the center of gravity too far forward, it could be difficult to raise the nose in some maneuvers. The fuel consumption, range and speed would be adversely affected. Stability about all three axes of the aircraft might also be greatly decreased. To avoid such situations, it is essential the preflight planning of the pilot include careful attention to weight and balance considerations.

SAMPLE LOADING PROBLEM

Figure 11-8 shows a sample loading chart with spaces to be filled in as the weights and moments are determined. The basic empty weight and moment of the airplane are given since they are constant figures. The moments given are in thousands of pound-inches.

SAMPLE LOADING PROBLEM	WEIGHT (LBS.)	MOMENT (/1000)
Basic Empty Weight	1860.5	63.0
Pilot & Front Passenger		
Fuel (55 gal. at 6 lbs. per gal.)		
Rear Passengers		
Baggage		
TOTAL WEIGHT AND MOMENT		

Fig. 11-8. Sample Problem

The pilot and front seat passenger weigh a total of 360 pounds. This weight should now be entered in the weight column. Referring to the loading graph in figure 11-9, the pilot finds 360 pounds on the left scale (item 1). He then draws a line horizontally across the chart until it

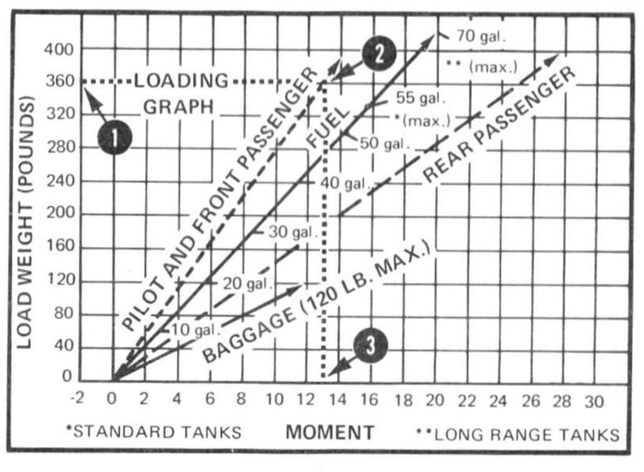

Fig. 11-9. Sample Loading Graph

intersects the line labeled "pilot and front passenger" (item 2). From this point a line is drawn straight down to the moment scale. The moment of 13 (item 3) is now entered in the moment column of the loading chart.

Since fuel weighs 6 pounds per gallon and the aircraft holds 55 gallons, the fuel weight is determined to be 330 pounds. Using the loading graph, the pilot determines the moment of the fuel to be 15.8.

The two rear passengers weigh a total of 300 pounds. Referring to the loading graph the pilot computes their moment as 21.0.

Forty pounds of baggage will be carried on this trip. The moment for the baggage is determined to be 4.0.

With many older airplanes, the weight and moment of engine oil must be added. In this case, oil weight is calculated at 7.5 pounds per gallon. If the oil tank is located forward of the datum line, its moment is negative and must be subtracted.

When the weight column is added the total gross weight of the aircraft is found to be 2,890.5 pounds. Since this is below the maximum gross weight of 2,900 pounds, the aircraft is within weight limits. Next, the moment column is added to arrive at a total moment of 116.8.

The pilot now should refer to the center of gravity moment envelope shown in figure 11-10. A horizontal line representing the loaded aircraft weight of 2,890.5 pounds is projected from the left scale and the vertical line representing the moment of 116.8 is projected from the lower scale. Since these two lines intersect *within the moment envelope*, the aircraft is properly loaded.

The previously described system of determining weight and balance is referred to as the graph method. However, the pilot should also understand the table and computation methods.

The table method provides the moments for various weights of items such as passengers, fuel, and baggage. Interpolation may be necessary to determine some moments when using the table method. With the computation method, the moment of each item is determined by multiplying its weight by the arm instead of referring to a graph or table. Next, the weight of

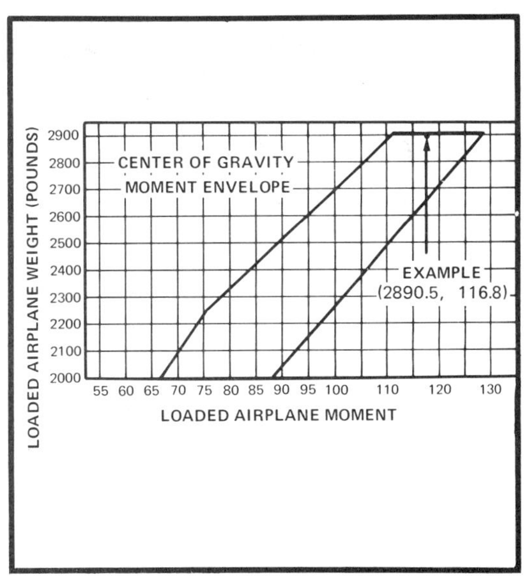

Fig. 11-10. Sample Moment Envelope

each item is added to determine the gross weight and the moments are added to find the total aircraft moment. Then, the total moment is divided by the gross weight to find the center of gravity position. Both the graph and table methods eliminate the need for the multiplication and division functions necessary in the computation method.

Some aircraft manufacturers identify locations in the fuselage by station. A station is identified by the number of inches it is located from the datum line. Thus the term *station* is synonymous with the weight and balance term *arm*.

CHARTS AND FORMS USED IN FLIGHT PLANNING

A Jeppesen low altitude chart may be used to determine the route. Entries are then made in the flight log form, including checkpoints, frequencies, and airways, as shown in figure 11-11. Communications frequencies needed both enroute and at the destination should also be listed.

When the initial flight log entries are complete, the flight plan is then completed. The pilot should obtain a final weather briefing to be sure of having latest available weather information for the destination, alternate, and enroute emergency airports, then file the flight plan, as illustrated in figure 11-12.

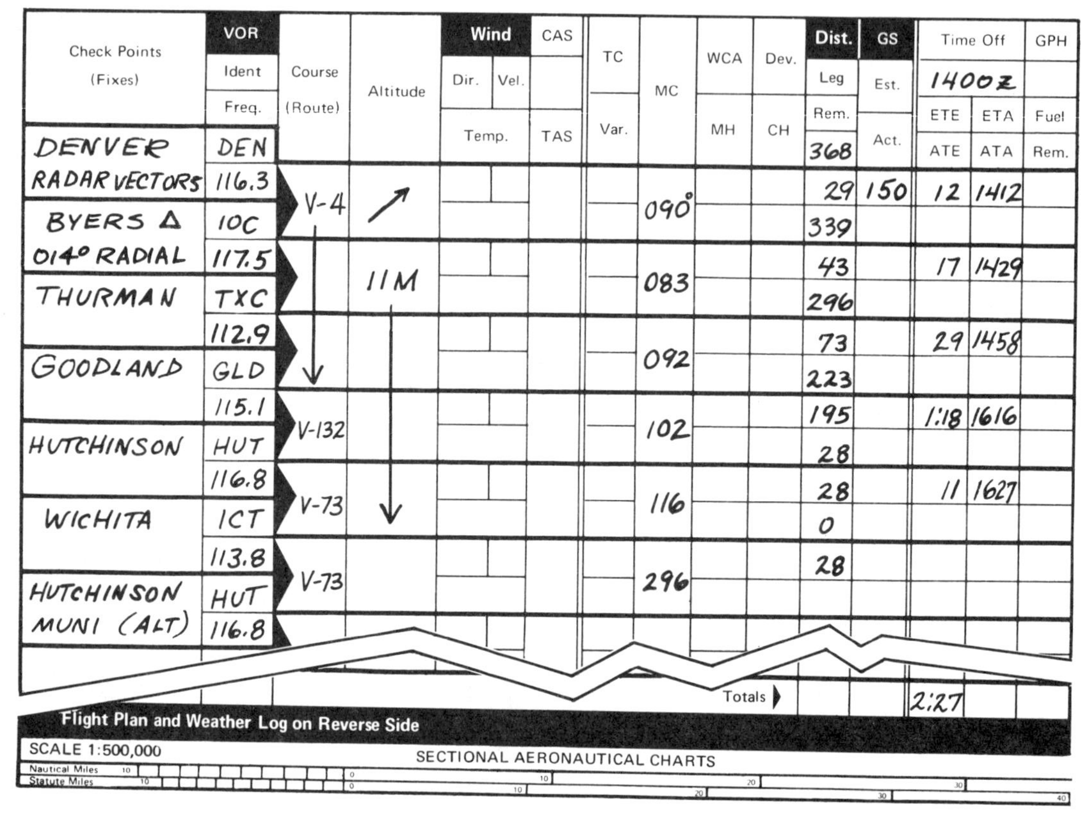

Check Points (Fixes)	VOR Ident / Freq.	Course (Route)	Altitude	Wind Dir. / Vel. / Temp.	CAS / TAS	TC / Var.	MC	WCA / MH	Dev. / CH	Dist. Leg / Rem.	GS Est. / Act.	Time Off 1400Z: ETE / ATE	ETA / ATA	GPH / Fuel Rem.
DENVER	DEN									368				
RADAR VECTORS	116.3	V-4	↗				090°			29	150	12	1412	
BYERS Δ	IOC	↓								339				
014° RADIAL	117.5		11M				083			43		17	1429	
THURMAN	TXC		↓							296				
	112.9						092			73		29	1458	
GOODLAND	GLD									223				
	115.1	V-132					102			195		1:18	1616	
HUTCHINSON	HUT									28				
	116.8	V-73					116			28		11	1627	
WICHITA	ICT									0				
	113.8	V-73					296			28				
HUTCHINSON MUNI (ALT)	HUT 116.8													
										Totals ▶		2:27		

Flight Plan and Weather Log on Reverse Side

SCALE 1:500,000 SECTIONAL AERONAUTICAL CHARTS

Nautical Miles 10 0 10 20 30

Statute Miles 10 0 10 20 30 40

Fig. 11-11. Flight Log Form

DEPARTMENT OF TRANSPORTATION— FEDERAL AVIATION ADMINISTRATION

FLIGHT PLAN

Form Approved OMB No. 04-R0072

1. TYPE	2. AIRCRAFT IDENTIFICATION	3. AIRCRAFT TYPE/ SPECIAL EQUIPMENT	4. TRUE AIRSPEED	5. DEPARTURE POINT	6. DEPARTURE TIME PROPOSED (Z)	ACTUAL (Z)	7. CRUISING ALTITUDE
VFR ☐ / IFR ✓ / DVFR ☐	1008 Q	PA-39/U	160 KTS	DEN	1400		11,000

8. ROUTE OF FLIGHT

DEN V-4 GLD V-132 HUT V-73 ICT

9. DESTINATION (Name of airport and city)	10. EST. TIME ENROUTE HOURS	MINUTES	11. REMARKS
ICT	2	27	

12. FUEL ON BOARD HOURS	MINUTES	13. ALTERNATE AIRPORT (S)	14. PILOT'S NAME, ADDRESS & TELEPHONE NUMBER & AIRCRAFT HOME BASE	15. NUMBER ABOARD
5	00	HUT	J.D. CAMPBELL, 7408 N. HAVENLOCK WAY, FARGO, N. DAKOTA, 452-7310, HECTOR FIELD (FAR)	3

16. COLOR OF AIRCRAFT: RED / WHITE

CLOSE VFR FLIGHT PLAN WITH ____________ FSS ON ARRIVAL

FAA Form 7233-1

Fig. 11-12. Flight Plan Form

ALTERNATE LANDING AIRPORT REQUIREMENTS

FAR 91.83 specifies rigid rules for selection of a landing alternate to an IFR destination airport. These rules are based on weather forecasts at the first airport of intended landing, as well as the alternate airport, when required. An alternate airport is *not required* if there is a standard instrument approach procedure for the first airport of intended landing and, for at least one hour before and one hour after the estimated time of arrival, the weather reports or forecasts or any combination of them, indicate that the ceiling will be at least 2,000 feet above the airport elevation and the visibility will be at least three miles.

QUALIFYING AN ALTERNATE AIRPORT

For an airport with a published instrument approach procedure to qualify as an alternate airport, the weather forecast at the ETA at that airport must not be less than 600 feet and two statute miles visibility for a precision approach or 800 feet and two statute miles for a nonprecision approach. However, if the flight must proceed to the alternate, the published instrument approach minimums apply.

NOS instrument approach charts do not show standard alternate minimums (600 feet or 800 feet and two statute miles visibility). However, if the **A** symbol appears in the information box, the airport has nonstandard minimums and the pilot must then consult a separate NOS tabulation of "**A** IFR Alternate Minimums." This tabulation is contained in the approach chart binders. An excerpt of the NOS tabulation for the alternate in this example flight is shown in figure 11-13. An airport without a published instrument approach procedure also may be used as an alternate airport; however, forecast weather conditions at the estimated arrival time must allow a descent, approach, and landing from the MEA in VFR conditions.

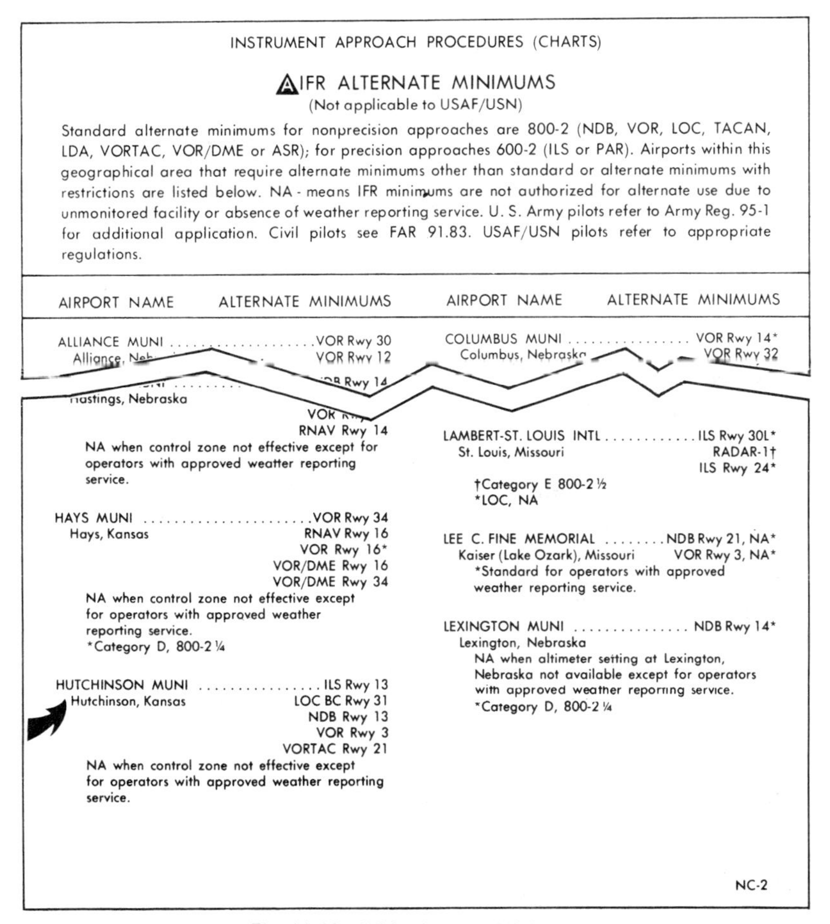

INSTRUMENT APPROACH PROCEDURES (CHARTS)

A IFR ALTERNATE MINIMUMS

(Not applicable to USAF/USN)

Standard alternate minimums for nonprecision approaches are 800-2 (NDB, VOR, LOC, TACAN, LDA, VORTAC, VOR/DME or ASR); for precision approaches 600-2 (ILS or PAR). Airports within this geographical area that require alternate minimums other than standard or alternate minimums with restrictions are listed below. NA - means IFR minimums are not authorized for alternate use due to unmonitored facility or absence of weather reporting service. U. S. Army pilots refer to Army Reg. 95-1 for additional application. Civil pilots see FAR 91.83. USAF/USN pilots refer to appropriate regulations.

AIRPORT NAME	ALTERNATE MINIMUMS	AIRPORT NAME	ALTERNATE MINIMUMS
ALLIANCE MUNI, Alliance, Neb...	VOR Rwy 30 VOR Rwy 12	COLUMBUS MUNI, Columbus, Nebraska	VOR Rwy 14* VOR Rwy 32
... Hastings, Nebraska	...B Rwy 14 VOR R... RNAV Rwy 14 NA when control zone not effective except for operators with approved weatter reporting service.	LAMBERT-ST. LOUIS INTL, St. Louis, Missouri	ILS Rwy 30L* RADAR-1† ILS Rwy 24* †Category E 800-2 ½ *LOC, NA
HAYS MUNI, Hays, Kansas	VOR Rwy 34 RNAV Rwy 16 VOR Rwy 16* VOR/DME Rwy 16 VOR/DME Rwy 34 NA when control zone not effective except for operators with approved weather reporting service. *Category D, 800-2 ¼	LEE C. FINE MEMORIAL, Kaiser (Lake Ozark), Missouri	NDB Rwy 21, NA* VOR Rwy 3, NA* *Standard for operators with approved weather reporting service.
HUTCHINSON MUNI, Hutchinson, Kansas	ILS Rwy 13 LOC BC Rwy 31 NDB Rwy 13 VOR Rwy 3 VORTAC Rwy 21 NA when control zone not effective except for operators with approved weather reporting service.	LEXINGTON MUNI, Lexington, Nebraska	NDB Rwy 14* NA when altimeter setting at Lexington, Nebraska not available except for operators with approved weather reporting service. *Category D, 800-2 ¼

NC-2

Fig. 11-13. NOS Alternate Minimums

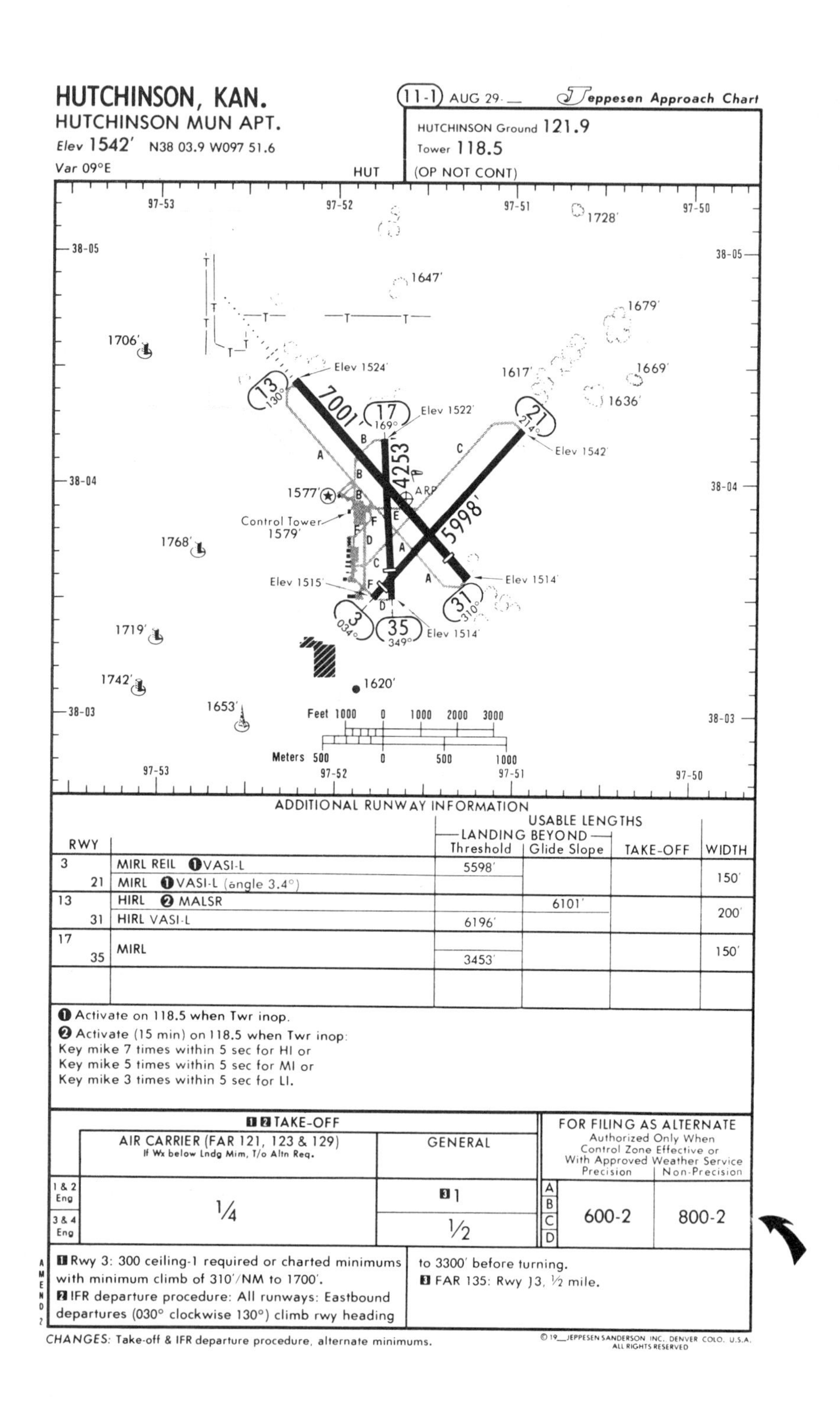

HUTCHINSON, KAN.
HUTCHINSON MUN APT.
Elev 1542′ N38 03.9 W097 51.6
Var 09°E HUT

(11-1) AUG 29 — Jeppesen Approach Chart

HUTCHINSON Ground 121.9
Tower 118.5
(OP NOT CONT)

ADDITIONAL RUNWAY INFORMATION

RWY		USABLE LENGTHS LANDING BEYOND Threshold	Glide Slope	TAKE-OFF	WIDTH
3	MIRL REIL ❶VASI-L	5598′			150′
21	MIRL ❶VASI-L (angle 3.4°)				
13	HIRL ❷ MALSR		6101′		200′
31	HIRL VASI-L	6196′			
17					150′
35	MIRL	3453′			

❶ Activate on 118.5 when Twr inop.
❷ Activate (15 min) on 118.5 when Twr inop:
Key mike 7 times within 5 sec for HI or
Key mike 5 times within 5 sec for MI or
Key mike 3 times within 5 sec for LI.

	❶❷ TAKE-OFF AIR CARRIER (FAR 121, 123 & 129) If Wx below Lndg Mim, T/o Altn Req.	GENERAL		FOR FILING AS ALTERNATE Authorized Only When Control Zone Effective or With Approved Weather Service Precision	Non-Precision
1 & 2 Eng	1/4	❸ 1	A B C D	600-2	800-2
3 & 4 Eng		1/2			

❶ Rwy 3: 300 ceiling-1 required or charted minimums with minimum climb of 310′/NM to 1700′.
❷ IFR departure procedure: All runways: Eastbound departures (030° clockwise 130°) climb rwy heading to 3300′ before turning.
❸ FAR 135: Rwy J3, ½ mile.

AMEND 2

CHANGES: Take-off & IFR departure procedure, alternate minimums.

Fig. 11-14. Jeppesen Airport Chart

Jeppesen approach charts, on the other hand, list minimums in the same section with the approach depiction. On the reverse side of the first approach sheet, the detailed airport diagram (described in section 6) contains additional runway information, including runway lighting, pertinent remarks, takeoff minimums, and alternate minimums. This information for the alternate Hutchinson (HUT) is illustrated in figure 11-14.

IFR FUEL REQUIREMENTS

The pilot in command is responsible for ensuring that sufficient fuel is carried on an IFR flight. When operating under IFR, the fuel supply must be sufficient to enable the flight to reach the airport of intended landing (including the landing approach and known traffic delays), then proceed to the designated alternate (if one is required), and fly thereafter for 45 minutes, at *normal cruise speed.* If an alternate airport is *not* required, the fuel supply for flight in instrument conditions must be sufficient to fly to the airport of intended landing, and thereafter for 45 minutes at normal cruise speed.

When planning *any* cross-country flight, it is *good operating practice* to provide a minimum of 45 minutes fuel reserve, in addition to any auxiliary fuel already provided for traffic delays, wind shifts, etc. For an IFR flight, such a fuel reserve is *mandatory*, and the FAA, in the examination questions regarding minimum fuel requirements, expects the pilot applicant to automatically include the 45-minute fuel reserve in his computations.

12 EMERGENCY IFR OPERATIONS

COMMUNICATIONS LOSS

INITIAL ACTION

When a pilot suspects that he has experienced a loss of two-way radio communications, his first action should be to contact the controlling agency and request a communications check.

Once the pilot determines that a communications failure has occurred, his next logical step is to define the problem by answering these questions.

1. Is the aircraft transmitter inoperative?
2. Is the aircraft receiver inoperative?
3. Has the controlling agency lost communications?

There are several steps the pilot can take to answer these questions. First, he should switch to a previously assigned ATC frequency. If this is not successful, he should tune to a flight service station frequency and attempt to make radio contact. Although not required by regulation, commonsense dictates that, *as a last resort*, an attempt should be made on the emergency frequency of 121.50 MHz. Each of these steps should be performed in an unhurried manner. The pilot must remain on the frequency long enough to determine positively whether his transmission has been received. In addition, while on a given frequency, the pilot should listen closely to see whether he receives any transmission from other sources, such as an aircraft relaying information for an ARTCC. This will aid him in the determination of whether or not his aircraft receiver is operating.

RESOLUTION

If it has been determined that the aircraft transmitter is inoperative, there are several steps the pilot can take in attempting to resolve his problem. First, he should ensure that the microphone is plugged in and functioning. If the aircraft is equipped with a secondary or backup transmitter, the resolution of his problem is very simple. He can utilize the secondary transmitter to replace the failed transmitter.

If both the transmitter and the receiver are inoperative, it is possible that there has been a loss of electrical power to the radio. This problem may be resolved by checking the fuses or circuit breakers and then replacing the fuse or resetting the circuit breaker, as appropriate.

Some aircraft have separate battery and alternator switches. In these aircraft, it is possible the pilot failed to turn on the alternator prior to takeoff. The radio equipment will operate on battery power for some period of time, and then, begin to fail as the battery voltage drops. If this situation occurs, the radio equipment should be turned *off* before turning on the alternator. This procedure will protect the equipment from the electrical surge which occurs as the alternator attempts to recharge the depleted battery. When the voltage and load have stabilized, the pilot should return the radios to operation and attempt to regain radio contact.

If the aircraft is equipped with a transponder, it can be used to reply to ATC transmissions. When the controller has noted that an aircraft is no longer replying to his radio transmissions, he may direct the pilot to "*Squawk IDENT*," or to change his transponder code to establish that there is, in fact, a communications problem. Once the controller understands the loss in transmitter capability and the aircraft has receiver capability, the pilot may avoid the necessity of resorting to IFR lost communications procedures.

With certain types of failure within the transmitter, it is possible that voice modulation will not be transmitted; however, each time the pilot depresses the mike button, the radio will transmit a *click*. In this case, the controller who hears the click *may* recognize that the pilot has had a microphone failure or some other type of transmitter failure.

The controller may direct the pilot to "*key*" his mike one or more times in response to questions or directions.

If it has been determined that the aircraft *receiver* is inoperative, the pilot most likely will not be able to determine whether his transmitter is operating. If a secondary, or backup, receiver is available, the pilot should use that equipment. He also should check the radio fuses and circuit breakers and the output of the generator or alternator.

If the pilot is not able to communicate utilizing any of the preceding suggestions, he should listen carefully for instructions from air traffic control, *comply* with the instructions, *if it is possible* for him to do so, and squawk 7600 on his transponder. If the pilot's navigational equipment is still in operation, he should *increase the volume* on the navigational receivers. Upon noting the communication failure code from the transponder, air traffic control will begin transmitting instructions on any frequency the pilot is likely to be monitoring. An additional benefit can be received from this procedure — there have been cases when a pilot has presumed communications failure only to find that his headset was not plugged in properly. If a pilot receives apparently normal indications from the navigational equipment and cannot receive the audio identification signal for that navaid, it *may* indicate that his speaker is inoperative and that he should switch to his headphones in order to bypass the speaker circuitry.

LOST COMMUNICATIONS REGULATIONS

If the pilot finds that he cannot regain communications, he must comply with FAR 91.127. If the communications failure occurs while the aircraft is in VFR *conditions*, or if VFR conditions are encountered following the failure, the pilot should *remain* in VFR conditions, if possible, and land as soon as practicable. It should be noted that the pilot is not *required* to land at the first level ground he encounters *unless*, in his judgment, this is the best course of action. The key word is *practicable.* The regulations cannot possibly cover every conceivable problem a pilot might encounter during a lost communications situation. When his situation is not specifically covered by the regulations, the pilot is expected to take whatever action is necessary to safeguard the flight in accordance with his emergency authority.

ROUTE

When a pilot experiences a radio failure on an IFR clearance, he is expected to proceed via the *last route* assigned by air traffic control. He may have received a modified clearance after takeoff which specifies a route of flight different from that issued in the original clearance.

Assume the pilot has filed a flight plan from the departure airport via checkpoints A, B, C, and D, to the destination airport, as shown in figure 12-1, which is referred to throughout this discussion. He receives a clearance stating, "*cleared as filed.*" Shortly after takeoff from the departure airport and prior to reaching checkpoint A, he is given a modified clearance to proceed direct to checkpoint B. If a communications failure occurs while on the direct flight, he should proceed in accordance with his *last* clearance (direct to checkpoint B). Then, since this was the only modification received, he should proceed to checkpoints C, D, and on to his destination.

If a failure occurs while the pilot is being radar vectored, he is expected to proceed via a direct route from the point of radio failure to the fix, route, or airway specified in the vector clearance. In the example, the pilot has been cleared via radar vectors from checkpoint B to intersection X between checkpoints B and C in order to avoid traffic. The vector clearance specifies that the pilot is to proceed to the intersection, then, from the intersection direct to checkpoint C. If a failure should occur halfway between checkpoint B and the intersection, the pilot is expected to *continue* to the intersection via his own navigation and then proceed direct to point C, as specified in the clearance.

However, a pilot may be given a radar vector with no additional information. For example, upon reaching checkpoint B, the pilot is directed to turn left 30° for weather avoidance. If a communications failure should occur after receiving such a clearance, the pilot is expected to proceed direct to his next flight-planned checkpoint. For example, if the communications failure occurs three minutes after receiving the vector, the pilot should turn right and proceed direct to checkpoint C. Upon arriving at checkpoint C, he should resume the route of flight as stated in his previous clearance, proceed to checkpoint D, then to his destination.

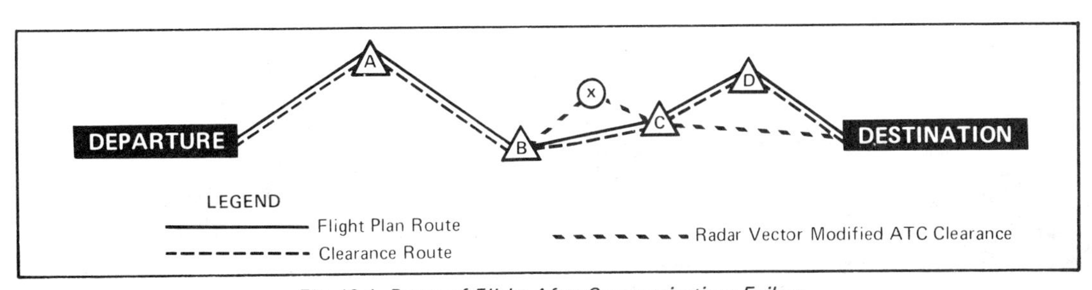

Fig. 12-1. Route of Flight After Communications Failure

In the absence of an *assigned* route, the pilot is expected to proceed via the route air traffic control has advised may be expected in a further clearance. For example, the pilot files from the departure airport to the destination via checkpoints A, B, C, and D and is cleared to fly via the flight planned route. He takes off and complies with that clearance. While enroute, however, he is advised to hold at checkpoint C until 1515 Zulu and to expect further clearance at that time to point D and his destination. Prior to arriving at point C, the pilot experiences communications failure. He is expected to hold at point C until 1515 Zulu, the time specified in the amended clearance, and then proceed via the route which has been advised.

In the event no further clearance is given to the pilot under these circumstances (for example, if he is told simply to hold at checkpoint C), he is expected to proceed via the route filed in the flight plan. Since a certain element of confusion can exist as the result of such a clearance, a pilot will be well advised *not* to accept a clearance to a point short of the destination airport without receiving an *expect further clearance time.*

A possible exception might be when a pilot is cleared direct to the navaid which serves the destination airport via a radar vector. For example, in figure 12-1 assume the pilot has flown his filed route to checkpoint C. Then, he is given a radar vector direct to the navaid serving the destination airport. In this case, the pilot should ask for an *expect approach clearance time* if it is not given as part of the clearance.

ALTITUDE

In the event of a communications failure, the pilot is expected to fly at the *highest* of the following altitudes or flight levels *for the route segment being flown.*

1. Altitude assigned in the last clearance received
2. Minimum altitude for IFR operations for the route segment (may be an MEA, MCA, or MOCA, depending upon the situation)
3. Altitude air traffic control has advised may be expected in a further clearance

In figure 12-2 the pilot is cleared to his destination airport via checkpoints A, B, C, and D, and cleared to climb to and maintain 6,000 feet. If a communications failure occurs at position 1, before he enters IFR conditions, the pilot is expected to remain in VFR conditions and land as soon as practicable. In this case, he will probably return to the departure airport. If communications failure occurs at position 2, while the pilot is in IFR conditions, he is expected to climb after passing checkpoint B to the new MEA of 10,000 feet. For purposes of this illustration, assume that the VFR area between checkpoint B and checkpoint C does not exist. Upon reaching checkpoint C, the pilot should descend to 6,000 feet, the last assigned altitude which, in this case, is higher than the new MEA of 4,000 feet.

When the pilot reaches checkpoint D, he normally expects to be cleared for an approach and to descend to the initial approach altitude, MEA, or MOCA. However, since he is required to fly at the *highest* of the various altitudes which apply to his flight, he must remain at the *last assigned altitude* of 6,000 feet. If he arrives at the navaid from which the approach begins before the ETA, he should enter a standard holding pattern at his *assigned* altitude of 6,000 feet.

If a communications failure occurs at position 3, the pilot should maintain the MEA for that leg. In this illustration, the pilot encounters VFR conditions shortly after position 3. When this occurs, the pilot is expected to remain VFR and land as soon as practicable while remaining in VFR conditions. In this example, the pilot could proceed to the destination airport while maintaining VFR conditions.

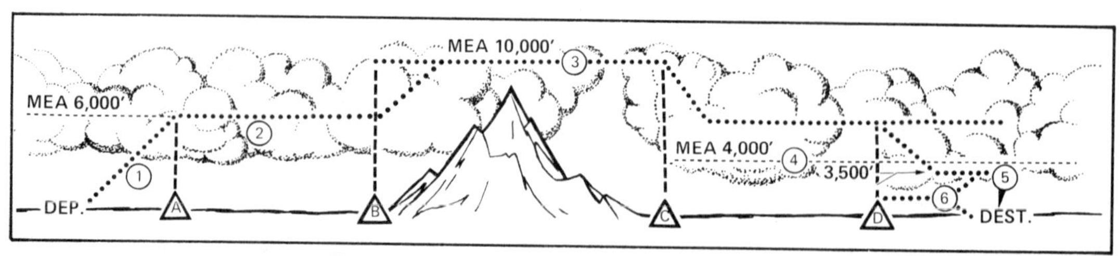

Fig. 12-2. Altitudes After Communications Failure

Assume the flight proceeds normally past checkpoint C, and the pilot is given a clearance to descend to the MEA. If a communications failure occurs at position 4, he will be enroute toward D at the MEA of 4,000 feet. After passing checkpoint D, the MEA remains 4,000 feet but the MOCA is 3,500 feet. Presuming the pilot *has not yet received* an approach clearance when the communications failure occurs, he must remain at his assigned altitude of 4,000 feet and proceed to the navaid which serves as an approach facility for the destination airport. If the airport has more than one approach facility, the pilot may select whichever one he wishes. His altitude will be cleared by air traffic control at all of the approach facilities for the destination airport.

If the pilot has received approach clearance prior to the communications failure, he is expected to continue with the approach as if no failure had occurred. He should conform to the routes, altitudes, and procedures depicted in the approach chart and commence the approach upon arrival. If the pilot loses communications at position 6 while in VFR conditions, he is expected to remain in VFR conditions, proceed to his destination, and land.

TIME

There are three *clearance* times that a pilot must be concerned with in the event of a communications failure. The first of these is the ETA. The ETA for the destination airport will be determined by both the pilot and controller by adding the estimated time enroute shown on the flight plan to the actual takeoff time from the departure airport. The ETA may be revised, as necessary, during the flight. The last ETA revision is the one which will apply in the event communications are lost. Second, the EFC is the time the pilot has been advised he may *expect further clearance.* This time applies to an enroute or holding clearance. Third, the EAC is the time the pilot may *expect approach clearance.*

There is a fine distinction between the application of ETA, EFC, and EAC in lost communications procedures. For example, as shown in figure 12-3, if a communications failure occurs while the pilot is holding at an enroute fix (navaid 1), he is expected to depart from that fix *at the time specified in the holding clearance,* or the time the pilot has been told to *expect further clearance* (EFC).

If air traffic control inadvertently omits the EFC or EAC when the clearance limit is at a location other than the destination airport, the pilot must *request* an expect further clearance time. However, if he has been given an *expect approach clearance* (EAC) time, he is expected to depart the holding fix (navaid 1) to *arrive* at the fix from which the approach begins (navaid 2) at the *expect approach clearance* time. In most cases, the approach begins at the initial approach fix (IAF). In figure 12-3, navaid 2 is the IAF since a procedure turn must be expected if the pilot arrives at the angle illustrated in the diagram.

When a communications failure occurs after the pilot is *"cleared as filed"* to the destination airport, he is expected to commence his approach from the fix at which the approach begins *at the ETA*, as shown on the flight plan. If the pilot should arrive at that fix prior to the ETA, he is expected to enter a holding pattern until the ETA. If he should arrive at that fix *at the ETA or later*, he should immediately descend, if necessary, in a holding pattern to the altitude specified for crossing the final approach fix and then execute the approach. If the approach is initiated from a holding pattern, it is not necessary for the pilot to then complete the procedure turn unless this is indicated by a special note on the approach profile view. In each case, the pilot should remain at his assigned cruising altitude until arrival over the fix from which the approach begins.

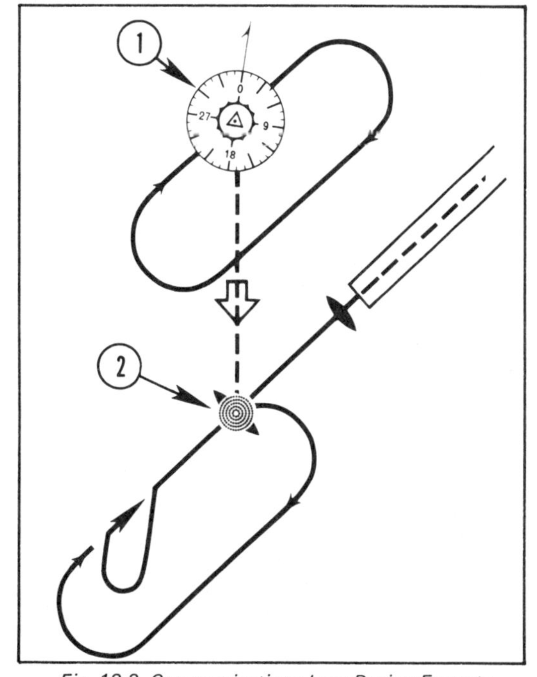

Fig. 12-3. Communications Loss During Enroute Holding Pattern

Because of the possible confusion which might exist in the pilot's mind regarding this time, once again it is pointed out that a pilot should not accept a clearance to hold without an accompanying EFC or EAC time, *unless* the pilot knows that he will be able to proceed from that holding point to his destination airport in VFR conditions.

DISTRESS COMMUNICATIONS

If a pilot operating in IFR conditions finds that he is not able to comply with the preceding lost communications procedures because of a *compound difficulty, such as a loss of navigation systems as well as communications,* he may attempt to alert civil or military radar systems by the use of the transponder on code 7700. This provides the pilot with the best and most reliable method of alerting ATC to the fact that his aircraft is in distress.

FLIGHT INSTRUMENT FAILURE

GYRO INSTRUMENTS

When a pilot notices irregularities in the operation of the gyro instruments, three aspects must be reviewed to determine the nature of the problem.

1. Is a vacuum pump failure rendering the attitude indicator and heading indicator inoperative?
2. Is one instrument malfunctioning, creating the effect that the remaining gyro instruments are erroneous?
3. Is the turn-and-slip indicator faulty, or is it simply not receiving electrical power?

The first step in determining the nature of a suspected gyro insturment failure is to check the suction gauge. For the newer type instruments, the suction gauge should read 4.6 to 5.4 inches of mercury. If the suction is less than 4.6 inches, the heading indicator and attitude indicator will not function properly and will produce erroneous indications. A total vacuum pump failure will be indicated by a zero reading on the suction gauge.

If, for example, the suction gauge is indicating normal suction, and the attitude indicator is suspected as faulty, it should be cross-checked with the supporting instruments. The turn-and-slip indicator and heading indicator should be used to verify bank information, and the airspeed indicator to verify pitch. In this example, if the attitude indicator is showing a 15° angle of bank, but the heading indicator and turn-and-slip indicator do not reflect a turn, the attitude indicator is probably in error.

To confirm there is a malfunction, the wings should be leveled according to the attitude indicator. If, in fact, the instrument has failed, the heading indicator will show a heading change, and the turn-and-slip indicator will also show a turn with the wings-level indications. From this information, the pilot can determine that the attitude indicator has failed, and it should be covered, or ignored, to prevent the possibility of creating vertigo.

As a second example, if the pilot experiences difficulty with the turn-and-slip indicator, he should first check the appropriate circuit breaker or fuse. Thus, since most modern turn-and-slip indicators are electric, the problem may be corrected quickly. In the event the electrical source is not faulty, verification of instrument failure may be accomplished by cross-checking other instruments. Once an instrument is determined to be inoperative or inaccurate, it should be covered to prevent inadvertantly including it in the instrument scan and ultimately producing vertigo.

PITOT-STATIC INSTRUMENTS

In review, the pitot-static instruments are the airspeed indicator, altimeter, and vertical velocity indicator. To determine a failure of any or all of these instruments, a cross-check with the gyro instruments must be accomplished. For example, if the pilot notices the airspeed is slowly decreasing, he may interpret this indication to mean the aircraft is in a nose-high attitude, climbing, and perhaps approaching a stall. In this situation, the pilot should immediately cross-check the attitude indicator for pitch information. The vertical velocity indicator and altimeter also must be checked to verify the nose-high attitude. If these three instruments are showing normal indications, the airspeed indicator should be covered and disregarded in the normal instrument scan. Regardless of the type of instrument failure, either gyro or pitot-static, the pilot must be prudent and precise in his decisions, and the malfunction reported to ATC immediately. Also, he should include a request for flight to possible VFR conditions.

13

SAFETY OF FLIGHT

FEDERAL AVIATION REGULATIONS PERTAINING TO SAFETY OF FLIGHT

The Federal Aviation Regulations (FARs) are constructed upon principles that promote a safe, expeditious flow of air traffic. This chapter discusses particular regulations that pertain to safety aspects, rather than procedural matters.

AUTHORITY OF PILOT IN COMMAND

FAR 91.3 states that "the pilot in command of an aircraft is directly responsible for, and is the final authority as to, the operation of that aircraft." It should be noted that the words "responsible" and "authority" are placed together within this quotation. The pilot in command has the ultimate authority with regard to the *operation* of his aircraft and the *safety* of the aircraft, its passengers, crew, and cargo. Along with this authority, he also accepts full responsibility for the flight.

FAR 91.3 also states, "In an emergency requiring immediate action, the pilot in command may deviate from any rule (in Part 91) . . . . to the extent required to meet that emergency." Emphasis should be placed on the words "*immediate*" and "*extent required.*" Therefore, in the event an emergency occurs, and a reasonable cause exists for deviating from the rules, the pilot is allowed and *expected* to do whatever he believes necessary to conduct his flight operation in a safe and proper manner.

There are, however, provisions made in FAR Part 91 for guidance in general emergencies, and these provisions must be adhered to in applicable situations. The pilot may, under certain circumstances, be required to send a written report of any deviation, and he should be prepared to defend his actions.

PILOT IN COMMAND RECENT FLIGHT EXPERIENCE

Federal Aviation Regulation, Part 61.57 (c) (e) states the basic recent flight experience requirements for pilot in command operations. These regulations are established as a matter of safety to insure the pilot's competency for the flight operation he conducts.

VFR DAY

The regulation states that before a person may act as pilot in command of an aircraft carrying passengers during the day, he must execute three takeoffs and three landings as the sole manipulator of the controls in an aircraft of the same category, class, and if required, type. If the aircraft is a tailwheel aircraft, the landings must be made to a full stop.

VFR NIGHT

If the pilot in command determines he meets the recent flight experience for VFR day, he must consider whether any portion of his flight will be conducted at night. In the event the pilot anticipates a night flight carrying passengers, he must execute three takeoffs and three landings *to a full stop* during the period beginning one hour after sunset, and ending one hour before sunrise.

IFR

To fulfill the recent IFR experience requirements of FAR 61.57(e), the pilot in command of an IFR flight must, within the preceding six months, log at least six hours of instrument flight time (under actual or simulated IFR conditions) and six instrument approaches. A pilot also may meet the recent flight experience requirement by taking an instrument competency check conducted by an FAA inspector, an FAA-approved check pilot, or a certificated instrument flight instructor. The competency check is mandatory if 12 months have elapsed since the pilot in command last met the recent IFR experience requirements or had an instrument competency check.

PREFLIGHT ACTION

As contained in the regulations, the pilot in command must familiarize himself with *all available information affecting the flight.* Prior to all flights, available runway lengths and estimated takeoff and landing distances must be determined. Additionally, on IFR flights or flights not in the vicinity of an airport, weather reports and forecasts, fuel requirements, alternate routes of flight, and known traffic delays (advised by ATC) must be included in the planning. "Alternate courses of action" should include reviewing the alternate routes of flight and airports that are available for use, and considering the navigation and airport facilities, weather, and fuel range available for that flight.

Preflight attention must also be given to the aircraft. If, while enroute, the pilot in command determines that an aircraft does not meet the requirements for safe flight, he is required by regulations to discontinue the flight operation. Also, an aircraft must be airworthy *prior* to flight. The pilot in command is the final authority in determining if the aircraft is in an airworthy condition. This decision must be based on knowledge of regulations, examination of pertinent aircraft documentation, thorough preflight inspection, and good commonsense.

EMERGENCY LOCATOR TRANSMITTERS

U.S. civil aircraft must be equipped with an approved type of emergency locator transmitter (ELT). Frequencies used by these transmitters are the world-wide VHF and UHF emergency frequencies of 121.50 MHz and 243.00 MHz. These units can operate from 24 to 48 hours on their selfcontained batteries. ELTs are used for locating downed aircraft by providing a DF homing signal.

Since accidental activation of the ELT is a possibility, the pilot should check the ON/OFF switch for the proper position prior to flight. It also is recommended that the communications radio be tuned to 121.50 MHz to check for an inadvertent locator beacon signal as part of the preflight check.

SUPPLEMENTAL OXYGEN

It is possible to experience the effects of hypoxia, or oxygen deficiency, at altitudes of *less than 14,000 feet MSL.* IFR flight procedures and techniques place an increased mental load on the pilot, requiring that he consume more oxygen than might otherwise be demanded. In order to insure that the flight crew is not overcome by the subtle onset of hypoxia, the minimum flight crew is required to use supplemental oxygen above 12,500 feet MSL.

A special provision is made in FAR 91.32, to provide for short duration climbs and descents above 12,500 feet MSL. If the short duration climb does not exceed 14,000 feet MSL, and is of less than 30 minutes duration, the pilot is not required to use supplemental oxygen. If, however, the flight continues above 14,000 feet MSL, or is above 12,500 feet MSL for more than 30 minutes, oxygen *must be used* by the minimum required flight crew. Although supplemental oxygen may not be *required* for flight below 14,000 feet MSL, it is *recommended* for IFR flight (and VFR flights) at or above 10,000 feet MSL. At altitudes above 15,000 feet MSL, supplemental oxygen must be provided for each occupant of the aircraft.

The altitudes referred to in the regulations governing supplemental oxygen are stated in terms of *cabin pressure altitudes.* Pilots flying in pressurized aircraft below FL 250 (25,000 feet MSL) are not required to have supplemental oxygen available, as long as the cabin pressure altitude is below 12,500 feet MSL, even though the aircraft's true altitude may be well above 12,500 feet. While certain pressurized aircraft are not *required* to have supplemental oxygen available, it is recommended that a sufficient supply of oxygen be available to meet the unpressurized aircraft requirements in the event the aircraft becomes depressurized and immediate descent to 12,500 feet or lower is not possible.

ALTIMETER AND STATIC SYSTEM INSPECTION

At least once in every 24 calendar month period, the static system and altimeter used in IFR flight must be tested in accordance with procedures in FAR Part 43. Accurate altimetry is particularly critical to the safety of an IFR flight at lower altitudes in IFR weather conditions. At higher altitudes, regardless of weather, traffic separation depends heavily on maintaining the proper altitude or flight level. The pilot cannot legally fly at an altitude higher than the altitude to which his aircraft's altimeter system has been tested, and he cannot accept a clearance from air traffic control to climb to a higher altitude.

VOR EQUIPMENT CHECK FOR IFR FLIGHT

Accurate navigation is vital to the safety of an IFR flight. Modern VOR receivers are less susceptible to developing significant bearing errors, but they must be checked and found to be within the accuracy limits of Part 91.25. The VOR system must have been checked and found to be accurate within the preceding 30 days prior to an IFR flight.

The preferred method of checking the indicated bearing accuracy is the VOT (very high frequency omnidirectional test facility) which is located at selected airports. This facility utilizes one of the VHF navigational frequencies normally assigned to a VOR station. A regular VOR station radiates signals in all directions that correspond to the 360° of the magnetic compass. The VOT, however, utilizes the 360° radial signal only, and transmits that signal in all directions. The reception range of a VOT usually is very limited. The VOT causes the course deviation indicator to center with a FROM indication when the course selector is set to 360° and to center with a TO indication when the course selector is set to 180° *regardless of the aircraft's position relative to the VOT antenna.* With either the TO or FROM indication and the course deviation indicator centered, the maximum allowable indication error is *plus or minus four degrees.*

If a VOT signal is not available, the next most desirable means of checking VOR system accuracy is to use a designated ground checkpoint. While the aircraft is parked over the designated checkpoint, the VOR receiver is tuned to the station that serves the airport. With the deviation indicator centered, the indicated radial is compared to the published radial for that checkpoint. The maximum allowable error is, again, *plus or minus four degrees.*

If *dual* VOR systems are installed in the aircraft, the pilot may check the indications of one VOR system against the other. In this case, the maximum permissible *variation* between the two indications is *four degrees.* One VOR system may be entirely accurate, the other four degrees in error. It is possible, however, for one system to be in error by more than four degrees and the other system to be in even greater error, with the *variation between the two* being less than four degrees. Although such a situation is remote, the pilot must consider this source of potential navigation error when making a VOR check using this method.

Another acceptable VOR check is to note the VOR indication over a prominent landmark that lies along the centerline of an established VOR airway. However, since the actual position of an airway may differ from the position shown on sectional or WAC charts due to cartographic errors and electromagnetic disturbances, designated airborne checkpoints have been established. These checkpoints are used in the same manner as ground checkpoints, except the allowable error is plus or minus six degrees. Both ground and airborne checkpoints are published in the Airport/Facility Directory.

After the VOR check is made, the pilot should sign the aircraft's logbook or other permanent record and enter the date, place, and bearing error. The personnel of a certificated and appropriately rated radio repair station may use a test signal radiated by that station to make the required VOR check. In this case, they will make the necessary logbook entry.

MINIMUM IFR ALTITUDES

Except for takeoff and landing, and while under radar control, an aircraft may not operate under IFR below the altitudes shown on the enroute charts. On direct routes (off airways), the pilot in command is responsible for staying 2,000 feet above the highest obstacles within *five* statute miles of course over designated mountainous areas. In areas *other* than designated mountainous areas, the MEA is 1,000 feet above the highest obstacle within *five* statute miles of the course. The designated mountainous areas are shown in figure 13-1.

When both a minimum enroute altitude (MEA) and a minimum obstruction clearance altitude (MOCA) are prescribed for a route, the pilot may descend to the MOCA when within 25 statute miles ((22 nautical miles) of the VOR concerned. When flying at the MEA or MOCA, the pilot must begin a climb to a higher minimum altitude when passing the point for which the new altitude applies. However, when obstacles are present, the pilot must climb to the higher minimum enroute altitude so that the point will be crossed at or above the minimum crossing altitude (MCA).

GOOD OPERATING PRACTICES

To conduct safe, efficient IFR operations, the pilot must adopt certain good operating practices and be aware of the factors affecting flight safety. These factors are human, mechanical, and environmental. Although each of these factors relate to good operating practices and safety of flight, it must be recognized that the pilot has the opportunity to control their effects.

HUMAN FACTORS

Safety of flight depends primarily upon the pilot. Although many people — the aircraft owners, mechanics, ATC controllers, flight service and National Weather Service personnel — have a bearing on each flight, it is the pilot who ultimately must bear the final responsibility for safety.

LIMITATION

A successful and safe instrument pilot can be defined as *one who understands his own limitations and always remains within them.* The instrument pilot, whether newly rated or an "old pro," should analyze his capabilities, training, experience, and mental preparedness, then set certain limits within which he will fly. A short proficiency flight with an instrument

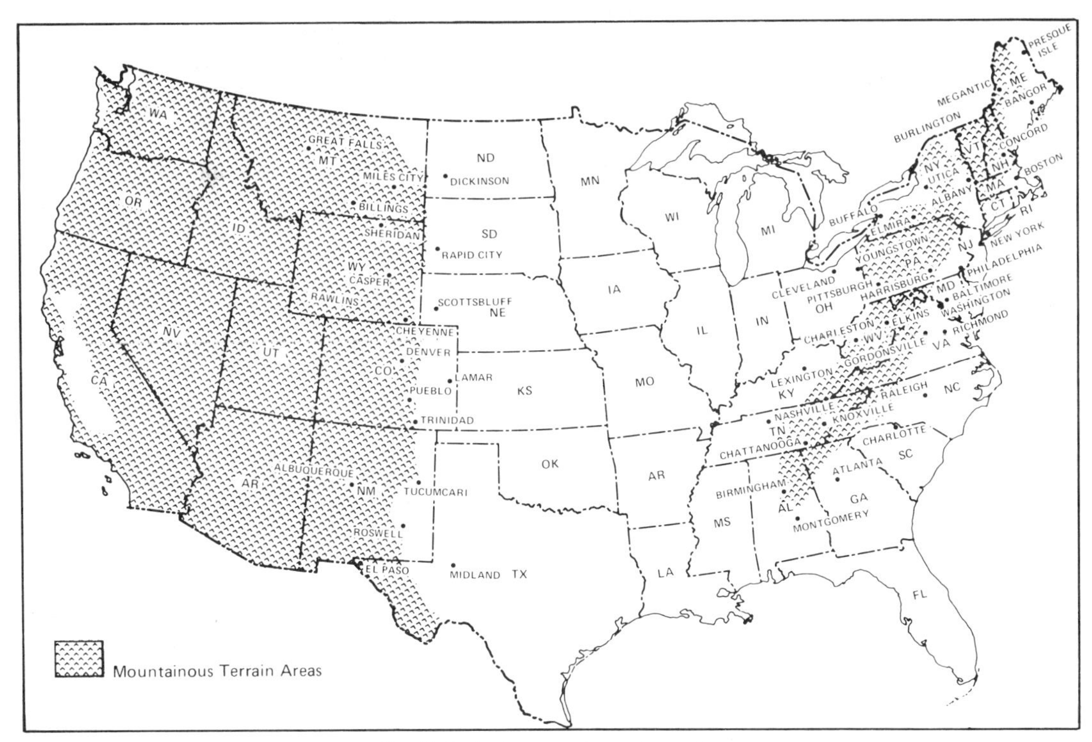

Fig. 13-1. Designated Mountainous Areas

instructor at regular intervals (6 or 12 months) will be helpful in aiding the pilot to establish these limitations for himself and to regain proficiency.

GAINING EXPERIENCE

After earning his instrument rating, the pilot must establish a program to gain the experience and knowledge of an "old pro." Through experience and practice, the pilot develops instrument proficiency and the ability to be at ease in the IFR environment.

It is suggested that the pilot gain experience initially by flying a few short IFR cross-country flights. As confidence and ability develop, he should extend his cross-country flights into unfamiliar areas. This experience is highly recommended prior to making IFR flights into the busier air terminals of the country.

Development of skill in instrument flying is influenced to a large extent by the type of weather encountered. The novice instrument pilot will gain more *useful experience* and confidence by breaking out of the clouds with the airport in view at 1,000 feet AGL or more, than he will by "sweating it out" down to the approach minimums. The enroute weather should also be chosen carefully on the first few flights. Weather with little or no turbulence and no chance of icing is preferred. By gaining and extending experience in a step-by-step process, the pilot can become the "old pro" from whom new pilots seek advice.

STATE OF MIND

An important part of preparation for each instrument flight is the mental state of the pilot. Anything that causes the pilot to be in the wrong frame of mind is detrimental to both safety and the pilot's outlook on instrument flying. Probably, the *single most important* consideration in developing a relaxed and tranquil state of mind is realistic self-confidence. The pilot who lacks confidence will probably be edgy, worried, and ill-prepared for a flight. Conversely, the pilot who has confidence in his ability, knowledge, and aircraft equipment will be able to maintain positive control of the aircraft.

There are two major steps the pilot can take to develop the proper confidence. First is *preparation for the flight.* Self-confidence, is increased by the degree that the pilot prepares for each flight. With proper planning, he will be prepared and mentally alert. For example, the pilot should plan the flight so the types of navigation, clearances, and approaches he will receive can be anticipated. He should use preferred routes, SIDs, STARs, and current weather information and try to be prepared for any contingency.

The second step in developing confidence is the *upgrading of pilot abilities and qualifications.* This can be accomplished by studying the areas of aviation subjects pertaining to instrument flight and staying abreast of new developments in aviation. The pilot can also benefit by flying with and observing qualified instrument rated pilots.

INSTRUMENT FLIGHT CURRENCY

To remain legally proficient, the pilot must have a minimum amount of recent experience. The Federal Aviation Regulations require that a pilot have at least six hours of instrument flight and six instrument approaches within the preceding six months to remain current. The pilot should remember that this is an absolute *minimum.*

All too often, a pilot will simply practice instrument maneuvers with a safety pilot aboard the aircraft; however, to become really proficient, the pilot must not only have recent instrument flying, but interaction with air traffic control as well. Copying clearances, communicating, following instructions, and executing different types of approaches will aid in developing the greatest amount of proficiency. The pilot who is thoroughly familiar with ATC procedures and can fly his aircraft under difficult conditions while complying with complex instructions will be best qualified, most proficient, and prepared for his next IFR operation.

KEEPING INFORMED

In keeping with the definition of a safe instrument pilot, one who understands his limitations and remains within them, the next factor to be considered is the national airspace system. This system is constantly being refined and updated with an emphasis on the safe, expedient movement of air traffic. The *safe* instrument pilot will remain current in his knowledge of the total system and any changes, or limiting factors, contained therein.

New information, procedures, and techniques are introduced in the *Airman's Information Manual.* In addition, Exam-O-Grams, advisory circulars, aviation magazines, and other publications are of great help to the pilot in remaining current in his knowledge. Through the dissemination and usage of this information, each pilot in the system will improve in efficiency.

MECHANICAL FACTORS

Mechanical factors, as well as human factors, relate to flight safety. Since any type of mechanical failure presents a greater problem during instrument flight conditions than during visual flight conditions, the instrument pilot should check his aircraft closely before each flight. During this check, emphasis should be placed on the essential IFR equipment, including navigation/communication radios and flight instruments.

Also, it must be realized than an aircraft with a current airworthiness inspection (annual or 100-hour) is not necessarily airworthy. Although airworthy when inspected, it may not meet the airworthiness criteria at the present time. Therefore, when making the preflight check of a plane used for instrument flying, the pilot should make a thorough inspection of aircraft surfaces, controls, engine, equipment, and other components. It generally is recommended that even if a pilot lands only for a fuel stop, he should make at least a walk-around inspection before the next takeoff.

AIRCRAFT OPERATING LIMITATIONS

To obtain maximum utilization from the aircraft, the pilot must be familiar with the aircraft operating limitations. This information is available in FAA approved flight manuals and/or owner's manuals. The limitations also are displayed by various placards, listings, instrument markings, or a combination of these. The method of presenting this information is left to the manufacturer, but it is the pilot's responsibility to familiarize himself with aircraft operating limitations.

ENVIRONMENTAL FACTORS

Another important area of flight safety and good operating practice is an awareness of the diverse and changing flight environment. While flying within the United States, the pilot will encounter all types of terrain features and weather conditions. At one time, the pilot may find himself flying from an airport near sea level to a high elevation airport in the mountains. At another time, the pilot may take his aircraft from one type of weather to its extreme — from the hot, dry areas of the southwest to the cold, moist areas in the north, or the hot subtropical areas of Florida. Therefore, the pilot should become familiar with each area in which he flies, or anticipates flying, and consider the total environment as an important set of factors with which he must reckon.

WEATHER

Weather is the most important facet of the environmental factor. Even with an instrument rating, the pilot will find that weather will impose certain flight limitations. For example, flight within a thunderstorm is extremely dangerous with or without an instrument rating.

While in a radar environment, the pilot on an instrument flight normally will be vectored around severe weather. However, it is best for the pilot to delay the flight or take another route rather than plan to fly in or near an area of severe weather. Through avoidance of these hazardous conditions, the flight not only will be more comfortable and enjoyable, but safer.

Widespread areas of very low ceilings and visibilities can cause problems in selection of an alternate airport. If there is a chance (based on forecasts and known conditions) that the alternate may also go below IFR minimums, the pilot must either delay the flight or choose an alternate airport in an area of better weather.

Structural icing problems exist during instrument flight in areas of visible moisture and freezing temperatures. A pilot of light aircraft without de-icing equipment must plan flights of short duration through these areas. However, it is possible for the pilot to climb safely or descend through *mild* icing conditions. To reduce the possibility of ice accumulation during these flights, the pilot can spray the leading edges of the wings, stabilizers, and propellers with an anti-ice chemical, which is available commercially in aerosol cans. The chemical will reduce the possibility of ice readily adhering to the aircraft surfaces, but should not be used as total ice protection.

Even pilots of aircraft with de-icing equipment must be cognizant of icing conditions. A comparison should be made between the de-icing capabilities and the extent and intensity of the icing which may be encountered. The information required for this comparison can be found in the aircraft owner's manual. Figure 13-2 illustrates typical de-icing equipment available for general aviation aircraft.

TERRAIN

Another environmental factor the pilot should consider is the terrain over which the flight will be conducted. A direct route through the western part of the country may require an MEA as high as 16,000 or 17,000 feet. This will necessitate oxygen equipment or cabin pressurization. In addition to a high MEA, a departure toward high mountains may require a rate of climb that is beyond the capability of the aircraft. In either case, it is possible that a slight variance in the route of flight will provide a lower MEA or the additional time to reach

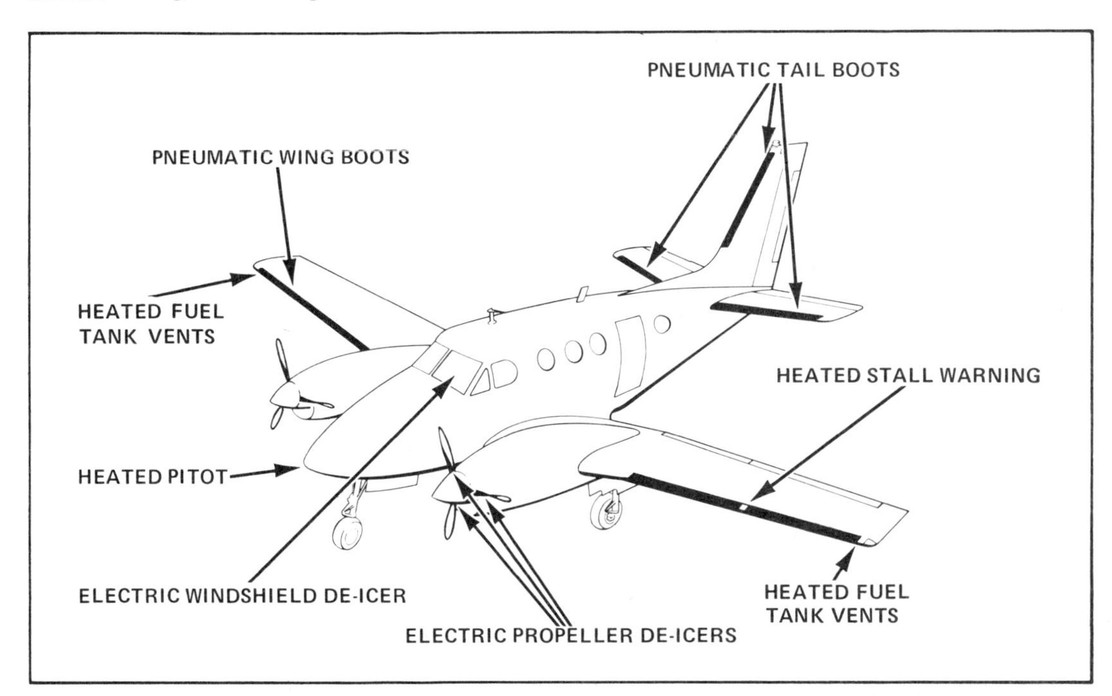

Fig. 13-2. Typical De-icing Equipment

cruise altitude before crossing the higher terrain. Terrain also should be considered in anticipation of emergency operations. Should an engine or navigation aid fail, it is comforting to know that an unobstructed area could be reached for landing. Many times a flight over more desirable terrain would only add a few miles to the total distance of the trip.

UNFAMILIAR AREAS

One of the best and easiest ways for the pilot to learn and develop his abilities is to benefit from the experience of others; as such, a pilot flying through an unfamiliar area will be aided by the advice of pilots familiar with the area and the inherent problems. A good rule of thumb pertaining to such flight is to seek advice and analyze the situation.

MEDICAL FACTORS

OXYGEN REQUIREMENTS AND AVAILABILITY

The amount of oxygen consumed by the body during the respiratory cycle depends primarily upon the degree of physical or mental activity of the individual. A person walking at a brisk pace will consume about four times as much oxygen as when he is resting. In the course of an average day, a normal adult male will consume about 2½ pounds of oxygen. This is approximately equivalent to the weight of food consumed daily. An oxygen supply which might be adequate for a person at rest would be inadequate for the same individual when flying under severe weather conditions or when under mental stress.

Oxygen becomes more difficult to obtain as the altitude increases, because the air becomes less and less dense, and the total pressure decreases. As the total pressure decreases, so does the pressure of oxygen even though the percentage of oxygen in the air remains constant. For example, at 40,000 feet, the total pressure is only 2.7 p.s.i., and the oxygen pressure is 0.56 p.s.i.

As altitude increases and the pressure of oxygen is reduced, the amount of oxygen transfer in the lung air sacs is reduced resulting in a decrease in the percentage of oxygen saturation in the blood. This causes a deficiency of oxygen throughout the body. For this reason, supplemental oxygen, pressurized cabins, and/or pressurized suits are required if the body is to receive adequate oxygen.

The total effect on an oxygen-deprived person is the result of both altitude and time of exposure. Every cell in the body is affected by the lack of oxygen, but the primary effects are on the brain and the nervous system. Above 10,000 feet, deterioration of physical and mental performance is a progressive condition. The impairment becomes more severe with increased altitude as well as prolonged exposure.

HYPOXIA

An insufficient supply of oxygen to the body is called *hypoxia.* The following are common symptoms of hypoxia.

1. Increased breathing rate
2. Lightheaded or dizzy sensation
3. Tingling or warm sensation
4. Sweating
5. Reduced visual field
6. Sleepiness
7. Blue coloring of skin, fingernails, and lips
8. Behavior changes

Subtle hypoxic effects may be noticeable at 5,000 feet at night. In the average individual, night vision will be blurred and narrowed. Also, dark adaptation will be affected, and at 8,000 feet, night vision is reduced as much as 25 percent without supplemental oxygen. Little or no effects will be noticed during the daylight at these altitudes.

At 10,000 feet, the oxygen pressure in the atmosphere is approximately two p.s.i. Accounting for the dilution effect of water vapor and carbon dioxide in the lung air sacs, this is not enough to deliver a normal supply of oxygen into the lungs, but is enough to keep the blood about 90 percent saturated.

This mild deficiency is ordinarily of no great consequence. However, the pilot may experience difficulty in concentrating, reasoning, solving problems, and making precise adjustments under prolonged flight conditions at this altitude.

It is true that susceptibility to hypoxia varies, and there are some who can tolerate altitudes well above 10,000 feet without any significant effects. It is equally true that there are those who develop hypoxic effects below 10,000 feet. As a general rule, individuals who do not exercise regularly or who are not in good physical condition will have less resistance to oxygen deficiency.

Also, those persons who have recently overindulged in alcohol, who are moderate to heavy smokers, or who take certain drugs will be more susceptible to hypoxia. Susceptibility can also vary in the same individual from day to day, or from morning to evening.

The most hazardous feature of hypoxia, as it may be encountered in general aviation aircraft, is its gradual and insidious onset. The production of a false feeling of well-being, called euphoria, is particularly dangerous. Since it obscures a person's ability and desire to be critical of himself, he generally does not recognize the symptoms. The hypoxic individual commonly believes things are getting progressively better as he nears total collapse.

TIME OF USEFUL CONSCIOUSNESS

The term "time of useful consciousness" refers to the maximum length of time an individual has to perform the purposeful tasks necessary for his survival such as putting on an oxygen mask. Although the hypoxic individual may remain conscious for a longer period, he has a *limited* time in which his brain receives sufficient oxygen to make decisions and perform useful acts.

Figure 13-3 shows the times of useful consciousness for various altitudes. The times shown represent the average times for flying personnel. Note that the times vary from 10 minutes at 22,000 feet to 1 minute at 30,000 feet. For example, if the pressurization equipment fails in an aircraft flying at 28,000 feet, the pilot and passengers have only 2½ to 3 minutes to get their oxygen masks on before they exceed their time of useful consciousness.

RECOVERY FROM HYPOXIA

Recovery from hypoxia is rapid, usually within 15 seconds after oxygen is administered. Transient dizziness may occur during the recovery. The severely hypoxic individual recovering from moderate or severe hypoxia is usually quite fatigued and may suffer measurable deficiency in mental and physical performance for many hours.

PREVENTION OF HYPOXIA

The best protection from hypoxia is to be constantly aware of the problem and use the altimeter as the primary guide for the use of oxygen. It is recommended that oxygen be used by pilots when they fly at altitudes over 10,000 feet during the day and altitudes of over 5,000 feet at night. Night vision is reduced and dark adaptation is affected above 5,000 feet. Passengers are urged to use supplemental oxygen above 10,000 feet if the cabin is not pressurized.

The chart in figure 13-4 shows the percentage of oxygen needed to maintain sea level and 10,000-foot equivalents. Flying at 15,000 feet, 39 percent oxygen is needed to maintain a sea level atmosphere. If it is desired to maintain a 10,000-foot atmosphere while flying at 15,000 feet, only 26 percent oxygen is required. Notice that above 33,000 feet, even 100 percent oxygen will no longer provide inhaled air with oxygen equal to that in sea level air. At 40,000 feet, 100 percent oxygen can maintain only a 10,000-foot atmosphere.

HYPERVENTILATION

Hyperventilation is defined as excessive ventilation of the lungs from breathing too rapidly and too deeply with resulting loss of too much carbon dioxide from the body. This excessive loss of carbon dioxide causes the blood to become more alkaline.

ALTITUDE	TIME OF USEFUL CONSCIOUSNESS
22,000 FT.	5 TO 10 MINUTES
25,000 FT.	3 TO 5 MINUTES
28,000 FT.	2½ TO 3 MINUTES
30,000 FT.	1 TO 2 MINUTES

Fig. 13-3. Time of Useful Consciousness versus Altitude

INHALED AIR --- % OXYGEN REQUIRED

I ALTITUDE	II S.L. EQUIV.	III 10,000 FT. EQUIV.
40,000 FT.		100
36,000 FT.		81
33,000 FT.	100	67
30,000 FT.	84	56
25,000 FT.	63	42
20,000 FT.	49	33
15,000 FT.	39	26
10,000 FT.	31	20.9
5,000 FT.	25	
S.L.	20.9	

Fig. 13-4. Percent of Oxygen Required versus Altitude

Overventilation of the lungs of pilots or passengers is usually caused by anxiety or apprehension. When this condition occurs, the respiratory center normally compensates by adjusting the breathing rate. However, persons who are anxious frequently override the respiratory signal, continue the forced breathing, and aggravate the carbon dioxide washout.

The resulting chemical imbalance may produce symptoms that are often mistaken for hypoxia. The following symptoms are associated with hyperventilation.

1. Dizziness
2. Tingling of the fingers and toes
3. Muscle spasms
4. Increased sensation of body heat
5. Nausea
6. Rapid heart rate
7. Blurred vision
8. Finally, loss of consciousness

A person may induce such symptoms experimentally by simply breathing deeply and rapidly over a period of time. Caution is advised, however, and such breathing should be limited to no more than two minutes. In this short length of time, a number of the symptoms will be experienced by the individual.

If hyperventilation progresses until unconsciousness occurs, the body's respiratory signals will take over, and the breathing rate will be exceedingly low until carbon dioxide is increased to a normal level. Hyperventilation symptoms can be relieved by voluntarily reducing rate and depth of breathing, allowing the body to build up a normal carbon dioxide level. Pilots are sometimes advised to rebreathe from a paper bag to hasten the return to normal carbon dioxide levels. This procedure artificially increases the percentage of carbon dioxide in the available air supply.

VERTIGO

During day-to-day activities, the ability of the human body to function correctly and precisely depends partly upon a person's ability to correctly determine his position relative to the earth. This ability to correctly orient himself is especially important to the pilot. The sensations from several sources are used to maintain balance.

1. Eyes
2. Nerve endings in the muscles and about the joints
3. Certain tiny balance organs which are part of the inner ear structure

A pilot who has entered clouds and can no longer perceive ground references literally will not be able to tell "which end is up" from his balance organs' signals. This happens because these nerve endings and balance organs depend, for the most part, upon a correct orientation to the pull of gravity and yet, these forces may not be oriented in the same direction as gravity. The brain struggles to decipher signals sent from the senses, but without the clue normally supplied by vision, incorrect or conflicting interpretations may result.

The results of such sensory confusion is a dizzy, whirling sensation termed *vertigo*, sometimes referred to as *spatial disorientation.* Vertigo may take a variety of forms and may be produced by a number of different flight situations in IFR *and* VFR conditions.

For example, a pilot conducting a night flight above an inclined cloud layer, which is well-illuminated by moonlight, might experience sensory confusion. The reason for this confusion is that most people naturally assume a cloud layer to be parallel to the surface of the earth. However, if the cloud layer is inclined, the pilot might attempt to align the aircraft with the cloud layer, thereby creating a wing-low attitude or gradually increasing rate of climb or descent. (See Fig. 13-5.)

Fig. 13-5. Flying on Top of a Sloping Cloud Layer

In turbulence, if an aircraft is rolled abruptly right or left, then, slowly resumes straight-and-level flight, the pilot may be aware of the roll, but not the recovery. On the other hand, if the aircraft rolls very slowly to the left, the pilot might still believe the aircraft to be in straight-and-level flight. A pitch change of 20° from the horizontal, if done slowly and without visual reference, may go unnoticed.

Also, in a banked turn, centrifugal force tends to press the body into the seat, much the same as when the aircraft is entering a climb or leveling off from a descent. Without visual reference, a banked turn may be interpreted as a climb. When completing a turn, the reduction in pressure gives the pilot the same sensation as going into a dive and may be interpreted this way.

INSTRUMENT CONDITIONS

A very serious, but common type of sensory illusion may occur when a noninstrument rated pilot continues flight into instrument conditions. When this occurs, the pilot has lost outside visual reference.

Often the aircraft will enter a very slight bank at a rate undetectable to the sense of balance. This bank will generally increase until there is a noticeable loss of altitude. The pilot, noting the decrease in altitude and still believing that he is in level flight, may apply elevator back pressure and perhaps add power in an attempt to gain back the lost altitude. This maneuver only serves to tighten the spiral, unless he has the presence of mind to first correct the bank attitude of the aircraft.

Once the spiral has started, the pilot will suffer an illusion of turning in the opposite direction if he tries to stop the turning motion of the aircraft. Under these circumstances, it is unlikely that he will take the appropriate corrective action. Rather, he will continue tightening the spiral until a possible dangerous situation develops. This unfortunate situation all too often occurs to pilots without instrument training who mistakenly believe that they can maintain their orientation in clouds.

STARS AND GROUND LIGHTS

A common problem associated with night flying is the confusion of ground lights with stars. (See Fig. 13-6.) Many incidents are recorded in which pilots have put their aircraft into very unusual attitudes in order to keep ground lights above them, having mistaken them for stars. Some pilots have misinterpreted the lights along the seashore as the horizon, and have maneuvered their aircraft dangerously close to the sea while under the impression that they were flying straight and level.

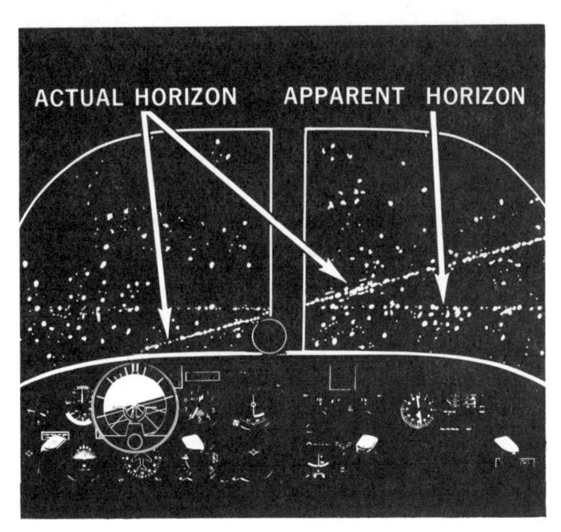

Fig. 13-6. Confusion of Ground Lights with Stars

FLICKER VERTIGO

There are other special causes of vertigo in addition to spatial disorientation (balance organ confusion). among them is the effect of a flickering light or shadow at a constant frequency. *Flicker vertigo* can result from a light flickering 14 to 20 times a second and may produce unpleasant and dangerous reactions in some persons. These reactions include dizziness, nausea, disorientation, and unconsciousness. They are especially insidious because the subject is often not aware of the cause of distress.

In a single-engine airplane flying toward the sun, the propeller can cause a vertigo-producing flickering effect, especially when the engine is throttled for a landing approach. The flickering shadows of helicopter blades have been known to cause flicker vertigo, as has been the bounce-back from rotating beacons or strobe lights when operated in or near the clouds at night. Pilots operating under instrument conditions are advised to turn off the rotating beacon or strobe lights in order to avoid this effect. Slight changes in propeller or rotor r.p.m. usually will produce relief when the effect cannot be avoided otherwise.

DRUGS

There are dozens of drug prescriptions and medicines available for every day common ailments and conditions. Many persons think that most commonly used remedies are harmless items which can be used routinely to add to daily well-being. This may not be so, and pilots should know the possible pitfalls and their tolerance to some common medications.

DRUG SIDE EFFECTS

Almost all drugs have side effects detrimental to the body, and the severity of the effects is often increased by flying. Major side effects of common medications can include drowsiness, mental depression, decreased coordination, reduced sharpness of vision, diminished function of the balance organs, increased nervousness, decreased depth perception, and impaired judgment. Any medical condition for which drugs are necessary is sufficient cause for the pilot not to fly at all.

Probably the best general recommendation for flying personnel, and others directly associated with flight control, is abstinence from all drugs. However, some illnesses and symptoms may not preclude flying or ground traffic control work, but may be benefited by appropriate drugs. The question, therefore, arises as to whether use of the required drugs in such instances may be safe and, in fact, advantageous. This question led to the preparation of *Drug Hazards in Aviation Medicine* for aviation medical examiners. It lists drugs, the toxic effects relevant to aviation that they may produce, and a conservative estimate of their allowable use. This book, shown in figure 13-7, may be purchased from the U.S. Government Printing Office.

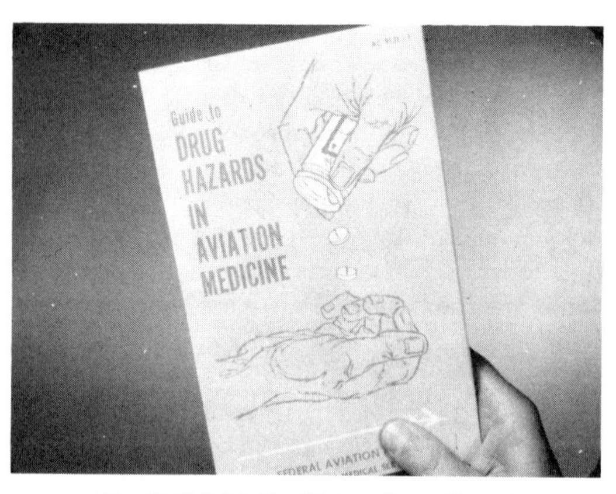

Fig. 13-7. FAA Booklet on Drug Hazards

ALCOHOL

Alcohol has effects similar to tranquilizers and sleeping tablets and may remain circulating in the blood for a considerable length of time, especially if consumed with food. The FARs require that a pilot must allow eight hours between the consumption of alcoholic beverages and flying an aircraft. Commonsense dictates that large amounts require an even longer recovery period. A good rule is to allow 24 hours between the last drink and takeoff time.

The following physiological and psychological effects of alcohol have been substantiated by research and are considered to be extremely detrimental to piloting an aircraft.

1. Dulling of critical judgment
2. Decreased sense of responsibility
3. Diminished skill reactions and coordination
4. Decreased speed and strength of muscular reflexes (even after one ounce of alcohol)
5. Efficiency of eye movements decreased 20 percent during reading (after one ounce of alcohol)
6. Significantly increased frequency of errors (after one ounce of alcohol)
7. Constriction of visual field
8. Decreased ability to see under dim illumination
9. Loss of efficiency of sense of touch
10. Decrease of memory and reasoning ability
11. Increased susceptibility to fatigue and decreased attention span
12. Decreased relevance of responses in free association test, with an increase in nonsensical reactions
13. Increased self-confidence with decreased insight into immediate capabilities and mental and physical status

PSYCHOLOGICAL CONSIDERATIONS

The psychological effect of desire may drive a person to be overmotivated to achieve a goal, perhaps to the extent judgment is impaired. A pilot may become so intent on getting home, he may attempt to fly through weather that demands skill beyond his experience.

Ironically, some pilots who are in professions requiring a high degree of knowledge sometimes place themselves in undesirable situations because of an overpowering desire to fulfill an ambition or goal. An ambitious or aggressive person who lives daily with challenge continually must examine his motives carefully and avoid flying beyond his experience and ability levels.

ANXIETY AND STRESS

Since a pilot, in his aviation environment, is required to evaluate information continually, perform complex tasks, and make decisions, aviation psychology deals with a pilot's performance during various emotional states. Two of these states — stress and anxiety — may have adverse effects on a pilot's performance. The causes of emotional stress are many and varied. They can be divided into two categories. The first includes those situations not specifically related to flying, such as family problems, financial or business considerations, or the demands of a pressing schedule.

The second category includes situations directly related to flight; for instance, apprehension about adverse weather conditions, malfunctioning equipment, or lack of confidence on the part of the pilot.

Fear is a normal, protective emotion which can build from stress or anxiety. Fear progressing to panic, however, certainly is undesirable. Panic can be avoided or overcome by forcibly maintaining or reestablishing self-control. A person who understands a situation completely can maintain control of his emotions, think more clearly, and reason properly. Once reason and logic are applied to the facts, the proper decision can be made and appropriate action taken.

If a pilot becomes lost or disoriented and panics, for instance, he may wander around aimlessly and compound his problem. If, however, he maintains self-control, he can follow a logical means of evaluating his situation, reorienting himself, and taking proper corrective action.

As man has created more complex, faster, and higher flying aircraft, his physiological and psychological characteristics have become increasingly important. Although these characteristics create some limitations, man's adaptability and his capacity to design a protective environment have permitted aeronautical exploration undreamed of a few short years ago. It is expected that aviation physiology will become an even greater aspect of the total flight environment of the future.

14 JOB PERFORMANCE

This *Job Performance* chapter employs programmed frames to provide an immediate, correct response for each step of the trip planning. In this manner, every step is completed properly before proceding to the next. The frames are to be used as they were in Chapter 10.

EXPLANATION OF JOB PERFORMANCE SECTION

This section summarizes all of the material covered in the *Instrument Rating Manual* and, at the same time, provides a series of problems that might be encountered during the planning and enroute phases of an instrument flight.

The flight is scheduled to leave Oklahoma City (Will Rogers Airport) at 0700 CST for Denver, Colorado, via Gage, Garden City, Goodland, and Thurman. The references needed to complete the following programmed frames are found at the end of this chapter. The appropriate approach charts, SIDs, and STARs are included in the appendix. Complete the blank flight plan form (Reference 1), assuming a cruising altitude of 8,000 feet. The OKC winds should be used as far as GCK, and the GCK winds should be used for the remaining distance to DEN. Keep the log current as the flight progresses. An altitude of 10,000 feet and the DEN winds should be used for the alternate leg from Denver to Pueblo on V-81.

1. A review of the enroute and destination weather situations at the Will Rogers Weather Service Office (See Ref. 5.) indicates the Denver area forecast was issued on the ____________day of the month at____________Zulu.	
2. Area forecasts are issued every____________hours for an ____________-hour period.	*11th* *1240*
3. The area forecast is valid for the period starting at ____________Tuesday and ending at____________Wednesday.	*12* *18*
4. According to the area forecast, this flight____________ (should, should not) encounter icing.	*1300Z* *0700Z*
5. Turbulence in northern Colorado will be severe near the ____________.	*should not*
6. The surface weather map in reference 7 indicates a cold front 90 nautical miles northwest of Denver, moving at a speed of 30 knots. The cold front should arrive in Denver at approximately____________.	*thunderstorms*
7. The warm front is 110 n.m. east of Denver moving at a speed of 15 knots. It will pass Goodland at approximately ____________.	*1500Z*
8. Referring to the station model on the surface weather map, the middle cloud type at Denver is thin layered ____________.	*1420Z*

9. According to the station model, the barometric pressure at Goodland has risen __________ millibars in the last three hours.	*altostratus*
10. The surface aviation weather reports show that between 0300MST and 0500MST the ceiling at Denver rose __________ feet.	*.5*
11. The pilot decides to use Pueblo as the alternate airport. If the control tower is operating, the alternate minimums for an ILS approach are __________ feet and __________ miles.	*2,000*
12. The winds aloft forecast in reference 4 is to be used from __________ Zulu to __________ Zulu.	*700* *2 ½*
13. The winds aloft forecast for Oklahoma City (OKC) at 8,000 feet indicates the wind will be from __________° at __________ knots.	*0600* *1500*
14. At a fuel consumption rate of 24 gallons per hour, (65 percent power) the flight from Oklahoma City to Denver and Pueblo (plus a 45-minute reserve) will require at least __________ gallons of fuel.	*040* *20*
15. The terminal forecast for Stapleton indicates that, upon arrival, the ceiling should be __________ feet.	*108.0*
16. Does any information in the NOTAM publication preclude a landing at Stapleton Airport? __________	*2,500*
17. After completing the flight planning and checking NOTAMs, insure that the current approach charts for both Stapleton and Pueblo are available, including the VOR, NDB, and __________ approach charts.	*No*
18. VASI is available at Stapleton Airport on runways __________, __________, and __________.	*ILS*
19. The primary control tower frequency for landing from the east at Stapleton is __________ MHz.	*8R, 17R, 26R*
20. Oxygen service __________ (is, is not) available at Stapleton.	*118.30*

21. The frequency for the VOR test signal at Stapleton is ____________MHz.	*is*
22. Using the no-wind takeoff chart, a temperature of 50° F, and a field elevation of 1,280 feet, the takeoff distance at Will Rogers will be____________feet.	*111.00*
23. Using the time-to-climb chart, the time required to climb to 8,000 feet will be____________minutes.	*1,900*
24. Referring to the low altitude enroute chart, it is determined that the ATIS frequency for Will Rogers World Airport is ____________MHz.	*5.2*
25. ATC issues the following clearance: "*N3562Y, cleared as filed — maintain 8,000 — maintain runway heading for departure vectors, departure control will be 124.6, squawk 1000.*" The limit of this clearance is____________.	*125.60*
26. After the tower issues takeoff clearance, instructions will be given to contact departure control. Departure control amends the clearance and provides a vector to Kingfisher VORTAC and a direct clearance to Gage. The frequency of the Kingfisher VORTAC is____________MHz.	*Stapleton Airport*
27. Arrival over Kingfisher occurs at 0711 and the flight continues directly to Gage. ETA at Gage is____________.	*114.70*
28. After passing Kingfisher, this flight is handed off to an ARTCC; the controlling center (approximately 12 nautical miles west of Kingfisher) is ____________ ____________.	*:42*
29. The initial call to Kansas City Center should include the aircraft identification and____________.	*Kansas City*
30. At 0719, Kansas City Center requests N3562Y to contact them on 126.95 when 30 miles east of Gage. The best point on the direct route between Kingfisher (IFI) and Gage (GAG) for a groundspeed check is abeam the____________ Intersection.	*altitude*
31. To make a groundspeed check at Camar, tune and identify the ____________ VORTAC.	*Camar*

32. With the Sayre VORTAC tuned and identified, rotate the course selector to________°.	*Sayre*
33. The VOR will indicate the airplane is nearly abeam the Camar Intersection when the CDI needle is ________.	*011*
34. If the CDI centers at 0733, the groundspeed is ________knots.	*centered*
35. With a groundspeed of 178 knots, N3562Y should arrive at Gage at________.	*178*
36. At Camar a new Gage ETA of :43 is computed. According to the *Airman's Information Manual*, ATC must be notified when an ETA changes as much as ________ minutes.	*:43*
37. Prior to reaching Gage, compute the ETA for Garden City. Use the new groundspeed (from frame 34) to make the computation. The ETA at Garden City is ________.	*:03*
38. The VOR indicates station passage at Gage as :43. Turn the course selector to________° and, since radar contact has been lost, make the position report to the Kansas City ________.	*:18*
39. Write the position report as it should be given to Kansas City Center. "*Kansas City Center,* 1. ________ 4. ________ 2. ________ 5. ________ 3. ________ 6. ________."	*327* *Center*
40. Flack Intersection is established by the ________ and ________ VORTACs.	*N3562Y,* *Gage,* *43,* *8,000 feet,* *Garden City 18,* *Goodland*

41. Passing Flack Intersection the MEA raises to 4,700 feet. Since this area is designated as nonmountainous terrain, the MEA provides reception of a reliable navigation signal and a terrain clearance of ______ feet.	*Garden City (GCK)* *Liberal (LBL)*
42. N3562Y arrives at Garden City at 0818 and is unable to contact Kansas City Center to make the position report. Kansas City Center is remoted at Garden City on ______ MHz.	*1,000*
43. If unable to contact Kansas City Center on 125.20, try to contact ______ ______ FSS.	*125.20*
44. To contact Garden City Flight Service Station, use the simplex frequencies of 122.20, ______, or ______ MHz.	*Garden City*
45. If a position report is made with the FSS, the initial call should include ______ and ______.	*122.45* *123.60*
46. After crossing the Garden City VORTAC, the outside air temperature is +5° Celsius. Using a CAS of 177 knots, the TAS will be ______ knots.	*identification* *position*
47. With a TAS of 202 knots and a wind of 350°/35, the new groundspeed will be ______ knots.	*202*
48. The Holcomb NDB (8 n.m. northwest of Garden City) transmits weather continuously, which is a service called ______ ______ ______ (TWEB).	*169*
49. Five minutes north of Garden City, thunderstorm buildups are sighted. For permission to alter left of course to avoid the thunderstorms, attempt to contact ______ ______ Center or ______ FSS.	*transcribed* *weather* *broadcast*
50. ATC amends the clearance, by directing N3562Y to proceed to the Goodland VOR via the 307° radial GCK and the 155° radial GLD. Turn left and select the ______° radial of Garden City VOR.	*Kansas City* *Garden City*

51. With the 307° radial selected, the TO/FROM indicator will read____________.	*307*
52. N3562Y will be on the 307° radial when the CDI ____________.	*FROM*
53. After establishing the aircraft on the 307° radial from Garden City, tune and____________the____________ VORTAC with the number two VOR receiver.	*centers*
54. When Goodland is positively identified, the course selector is adjusted to____________°.	*identify* *Goodland*
55. Until established on the selected course, the CDI will be ____________of center.	*335* *(reciprocal of 155°)*
56. Between Garden City and Goodland, expect a handoff to Denver Center. The new frequency for Denver Center will be ____________MHz.	*left*
57. To use V-216 as a cross-check on your position, Hill City is tuned and identified and 231° is set in the course selector. The CDI is fully deflected to the left. V-216 is located to the ____________of your present position.	*132.50*
58. The minimum altitude at which Modoc Intersection can be identified is____________feet.	*south*
59. The minimum enroute altitude along V-4 after passing Goodland is____________feet MSL.	*7,000*
60. For navigation purposes, change from Goodland VORTAC to Thurman VORTAC at____________nautical miles from Goodland.	*7,000*
61. Between Goodland and Thurman, moderate to severe turbulence is encountered. The aircraft's speed should be reduced to the recommended____________speed.	*36*
62. Maneuvering speed is the maximum speed at which abrupt ____________movements should be made, and the speed recommended for____________air.	*maneuvering*

63. To aid other pilots that may be flying the same route, make a PIREP of the turbulent condition to any ______________ ______________ station.	*control* *turbulent or rough*
64. Before changing frequencies to make the PIREP, the pilot of N3562Y ______________ (is, is not) required to notify Denver Center.	*flight service*
65. The "T" ends of V-4 on each side of Byers Intersection indicate a change in ______________.	*is*
66. Denver Center instructs N3562Y to climb to 10,000 feet and contact Denver Approach Control at Byers Intersection. Since icing was reported at 10,000 feet, acknowledge the approach control contact but ______________ a lower altitude.	*MEA*
67. The arrow with the letter "D" just west of Byers Intersection indicates the intersection can be established with a single VOR and DME. The distance to DEN VOR is ______________ nautical miles.	*request*
68. The ATIS reports the following weather information, "*measured ceiling 600 feet, visibility two miles in light snow.*" An RVR of ______________ feet is required for a 26L ILS approach at Denver.	*29*
69. ATC clears N3562Y to hold at Watki Intersection and expect approach clearance for an ILS approach at :39. If the flight reaches Watki at :31, ______________ complete holding patterns with one-minute legs must be made.	*2400*
70. The minimum holding pattern altitude for the ILS approach is ______________ feet MSL.	*2*
71. The distance from Watki Intersection to the ILS outer marker is ______________ nautical miles.	*9,000*
72. The ILS outer marker is ______________ from the end of the runway.	*8.4*

73. The inbound course on the 26L ILS is ________ °.	*5.5 n.m.*
74. Upon arrival at the outer marker, approach control reports the ceiling at Stapleton has dropped to 100 feet with an RVR of 2,000. The decision height for the approach to Stapleton is ________ .	*257*
75. The ILS missed approach procedure entails a climb to ________ feet MSL, then a climbing ________ turn to 9,000 feet MSL direct to the Denver VORTAC.	*5,531*
76. N3562Y informs the tower that the flight will go to the alternate airport (Pueblo). Departure control gives the vector to Pueblo, and then assigns the altitude to fly. Using the DEN winds, the flying time to Pueblo VOR from the Denver VOR via V-81 will be ________ minutes.	*7,000* *right*
77. At the Franktown Intersection, 69 gallons of fuel remain; upon arrival at Pueblo VORTAC, ________ gallons of fuel remain.	*31*
78. Once an ETA at the PUB VOR has been given to ATC, a descent should not be made before the expect approach clearance time (if received) or the ________, in the event of communication failure.	*60.0*
79. If any portion of this flight is conducted above 18,000 feet, ________ ________ airspace will have been entered.	*ETA*
80. In that case, certain requirements must be met. The aircraft must have a transponder, appropriate communications capability, and must operate under ________ at a ________ ________ assigned by ATC.	*positive control*
81. When operating below the positive control area, the pilot should know the VFR-on-top requirements. Assuming proper cloud clearance is possible, the lowest altitude for a VFR-on-top flight between COS and PUB is ________ feet.	*IFR* *flight level*

82. To operate under IFR in controlled airspace, a pilot should be sure that the altimeter has been checked within the preceding__________ calendar months.	*11,500*
83. According to the low altitude enroute chart, the frequency of the Metro LOM at Pueblo is__________kHz.	*24*
84. If no wind exists as the PUB VOR is reached on V-81, the relative bearing to the LOM should be about__________°.	*302*
85. The time during which the airplane has been flown by instrument reference may be logged as__________time.	*101*
	instrument

REFERENCES

The references on the following pages pertain to the aircraft flight manual, aircraft specifications, loading, and weather. The appropriate SIDs, STARs, and approach procedure charts are located in the appendix. The Jeppesen low altitude US (LO) 11/12 chart is included with the manual. Reference 13 contains the completed flight plan form. This reference can be used to check the flight plan form completed as the programmed frames were answered.

CLEARANCES	
	TIME OFF

FIXES	ROUTE	MC	MH	GS	DIST	ETE	ETA	ATA
Will Rogers	D→ ↗					7		
OKC	8,000 V-17							
GAG	↓							
GCK	↓							
GLD	8,000 V-4							
TXC	↓							
DEN	approach					7		
Stapleton	↗					7		
DEN	10,000 V-81							
COS	10,000 V-81							
PUB	approach					7		
TRIP TOTALS								

FUEL	
TAXI/TO	2.0
CLIMB	3.0
CRUISE	
APPROACH	3.0
ALTERNATE	
RESERVE	18.0
TOTAL	

TC → ± VAR = MC → ± DEV = CC
± WCA ± WCA
TH → ± VAR = MH → ± DEV = CH

BLOCK IN	
BLOCK OUT	
LOG TIME	

TIME FLIGHT PLAN EXPIRES

POSITION REPORT

Acft. Ident.	Position	Time	Alt.	IFR/VFR	Est. Next Fix	Name Next Fix

Reference 1 — Flight Time Analysis Form

Airplane	N3562Y, four-place, twin-engine
Gross Weight	4,880 pounds
Basic Empty Weight	2,730 pounds
Useful Load (Pounds)	2,150 pounds
Baggage Capacity	200 pounds
Fuel	Four 36-gallon fuel cells located outboard of engines (Total fuel -- 144 gallons of grade 100 gasoline)
Engine	Two six-cylinder, 250 H.P. @ 2,575 r.p.m. 65 percent power -- 2,400 r.p.m.
Fuel Consumption	65 percent power — 12.0 g.p.h. per engine 75 percent power — 14.0 g.p.h. per engine
Speeds	Cruise 65 percent power — 180 Kts. TAS Cruise 75 percent power — 188 Kts. TAS V_{SO} — 68 Kts. IAS V_Y — 105 Kts. IAS V_A — 146 Kts. IAS
Radio Equipment	Dual VHF transmitter/receivers with 360-channel simplex and VOR, ILS, ADF, marker beacon receiver, and 4096-code transponder

Reference 2 — Airplane Specifications

```
    HOURLY WEATHER 11 1Ø12
DEN M15 OVC 3R- 156/49/45/25Ø8/ØØ2
GLD M8 OVC 5F 181/37/35/16Ø8/ØØ4
GCK 2Ø SCT M35 OVC 7 189/39/32/2215/ØØ7
ICT CLR 1GF 224/38/37/16Ø2/Ø14
LBF M12 OVC 1Ø 16Ø/35/3Ø/151Ø/ØØ3

    HOURLY WEATHER 11 1112
DEN M2Ø OVC 4R- 15Ø/5Ø/46/2412/ØØØ
GLD M1Ø OVC 8 178/38/34/2314/ØØ3
GCK 2Ø SCT M25 OVC 8 186/4Ø/33/2218/ØØ6
ICT CLR 6H 221/38/33/16Ø4/Ø13
LBF M9 OVC 6S- 156/36/34/Ø91/ØØ2

    HOURLY WEATHER 11 1212
DEN M35 BKN 7Ø OVC 7R- 144/52/49/221Ø/998
GLD M1Ø BKN 12 172/39/36/2215/ØØ1
GCK 25 SCT M5Ø OVC 1Ø 18Ø/42/35/232Ø/ØØ4
ICT CLR 15 218/38/28/161Ø/Ø12
GAG 5Ø SCT 2ØØ/56/46/1812/ØØ6
OKC 4Ø SCT 226/45/42/171Ø/Ø12
LBF M5 OVC 2S- 15Ø/38/37/Ø81Ø/999
```

DEN—Denver, Colorado
GAG—Gage, Oklahoma
GCK—Garden City, Kansas
GLD—Goodland, Kansas
HLC—Hill City, Kansas
ICT—Wichita, Kansas
LBF—North Platte, Nebraska

Reference 3 — Surface Aviation Weather Report Excerpts

```
FDUS1 KWBC 110545
DATA BASED ON 0000Z
VALID 11 1200Z   FOR USE 0600-1500Z.  TEMPS NEG ABV 24000
```

FT	3000	6000	9000	12000	18000	24000	30000	34000	39000
ALS			3219	3320-05	3435-19	3444-31	345248	345355	344557
BFF		3222	3328-02	3329-09	3437-21	3445-34	355254	355254	354156
DEN			2328+00	3329-08	3440-20	3449-33	355948	356055	344857
GCK		3531+01	3537-08	3540-12	0140-24	0140-28	014030	035138	055545
HLC		3321+00	3431-06	3431-11	3432-23	3439-35	354348	343451	322152
ICT	3111	3314+00	3217-05	3217-10	3120-22	3024-34	292747	282549	262351
OKC	0112	0216+12	0522+09	0626+03	0629-04	0640-09	074118	065126	055639

Reference 4 — Winds Aloft Forecast Excerpt

```
MKC FA 111240
13Z TUE-07Z WED
OTLK 07Z WED-19Z WED

WY CO KS NE SD ND...
HGTS ASL UNLESS NOTED...

SYNS...HI PRESS WITH CNTR OVR GRT LKS RGN DMNTS AREA. SYM
MOVG SLWLY EWD WITH SLY FLO RTRNG TO WRN PTNS AREA BY 21Z AND
OVR ALL OF AREA BY END OF PD. LO PRES TROF AND FNT TO THE W
OF AREA MVG EWD TO ERN WY WRN CO BY 19Z WED.

SGCLD AND WX...

CO...
EXTRM NE AGL CIG OCNL AOB 10 AND VSBY AOB 3 MIS IN R-/L-F IPVG
TO CIG 15 BKN 20 OVC AFT 21Z. ELSW 120 SCT 300 BKN AND OCNL
120 BKN 300 OVC. BCMG 120 OVC NWRN AND N CNTRL PTNS AFT 18Z
AND CHC SCT SW-/RW- DVLPG THOSE AREAS AFT 22Z WITH PTN NRN MTNS
BCMG OBSCD ABV 80-110 AFT 23Z. OTLK... VFR.

ICG... OCNL MDT ICGICIP FRZ LVL LWRG TO SFC WRN WY AND WRN
CO SLPG TO 100 WRN KS.

--------------------------------------------------------------

AIRMET BRAVO 1.  OVR NERN COLO ERN WYO OCNL MDT TURBC BLO 180
SVR NEAR TSTMS.
```

Reference 5 — Area Forecast Excerpt

```
FT CO 110448

DEN 110505 C15 OVC 6R- 2212. 13Z C20 OVC 6S- 3620. 15Z C25 OVC 6S- 3625..
PUB 110505 C25 OVC 2510. 14Z 30 SCT C80 OVC 3625. 16Z 30 SCT C80 BKN 3625..
```

Reference 6 — Terminal Forecast Excerpt

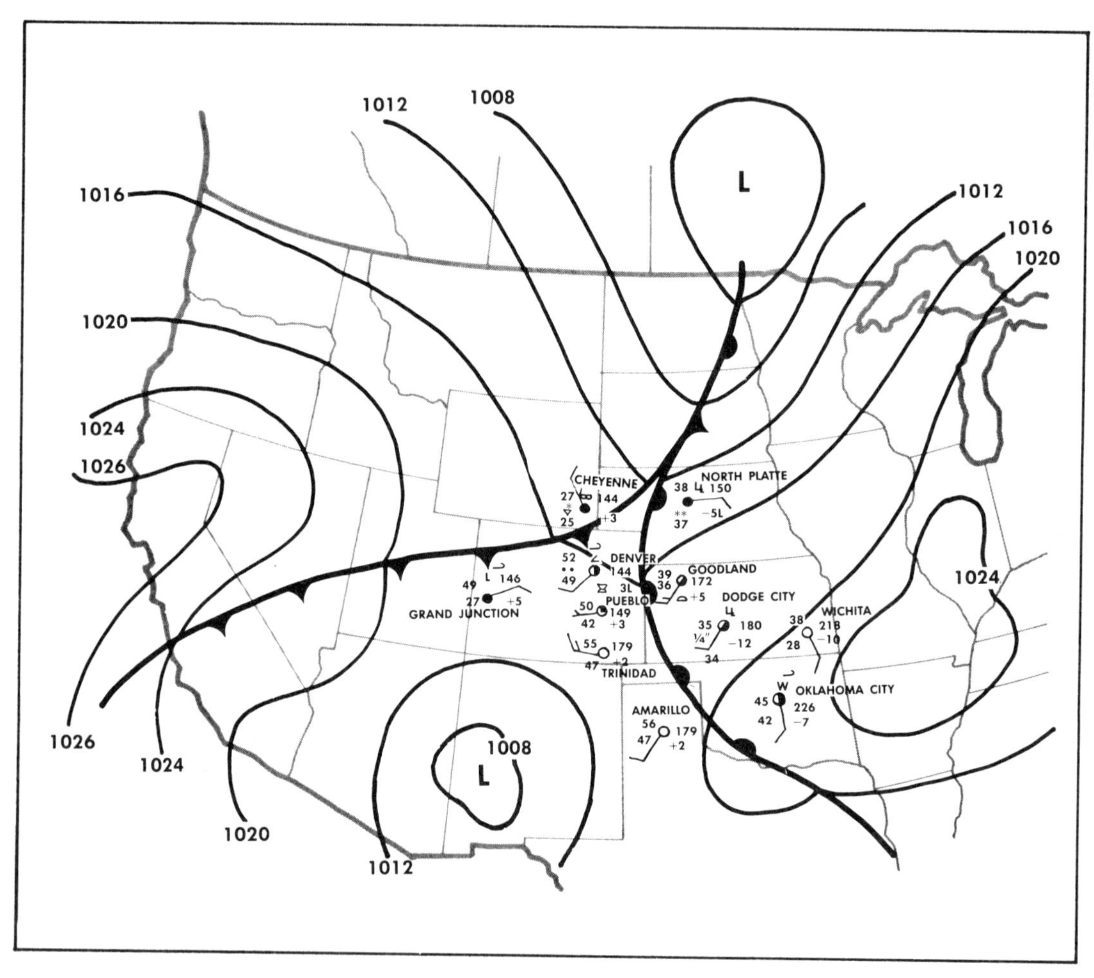

Reference 7 – 1200Z Surface Weather Map

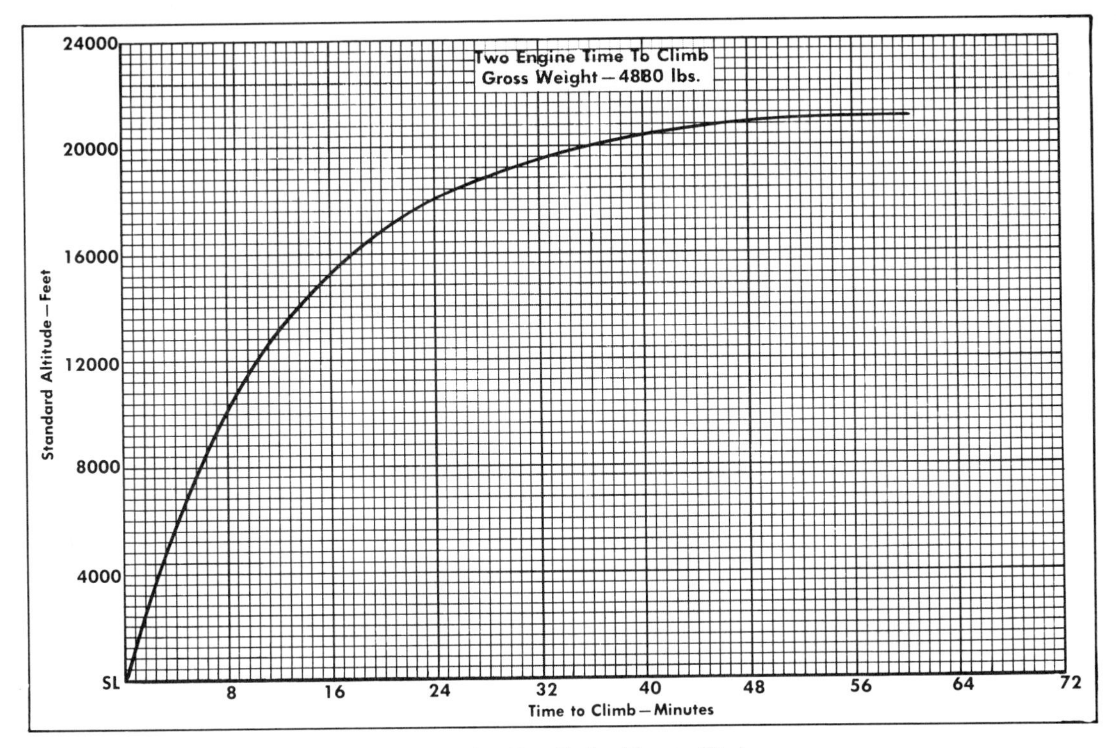

Reference 8 — Two-Engine Time to Climb

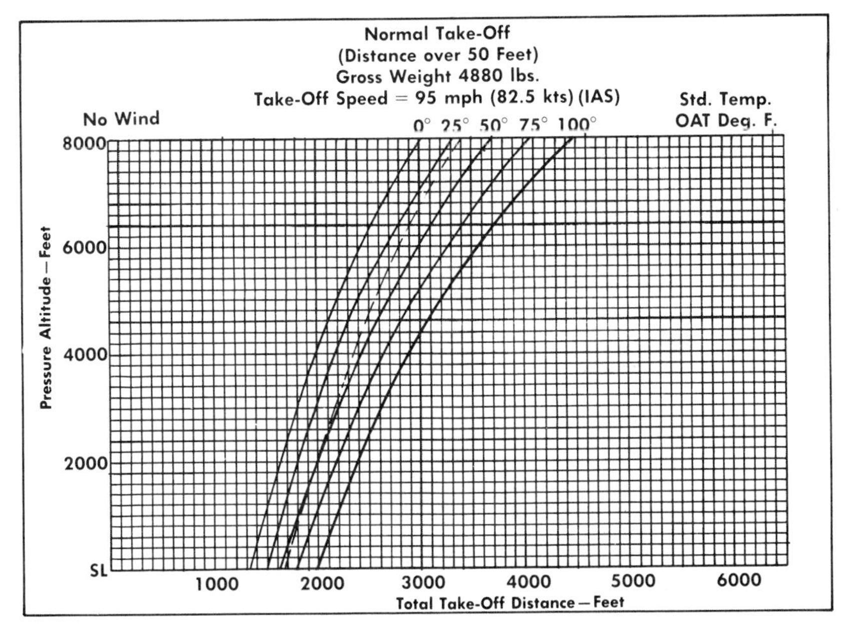

Reference 9 — Takeoff Distance Chart

EXPLANATION OF WEATHER STATION MODEL AND SYMBOLS

At Weather Bureau offices, maps showing conditions at the earth's surface are drawn 4 times daily or oftener. The location of the reporting station is printed on the map as a small circle. A definite arrangement of the data around the station circle, called the station model, is used. The station model is based on international agreements. Thru such **standardized** use of numerals and symbols, a meteorologist of one country can use the weather maps of another country even though he does not understand the language. An abridged description of the symbols is presented below.

WEATHER SYMBOLS

AREA SYMBOLS

FRONTAL SYMBOLS

Type	Color
Cold	Solid Blue
Warm	Solid Red
Occluded	Solid Purple
Upper Cold	Dashed Blue
Upper Warm	Dashed Red
Stationary	Alt. Blue and Red

Many of the elements in the STATION MODEL are entered in values which can be interpreted directly. Some, however, require reference to coded tables and these STATION MODEL entries are described in the tables below:

C_L	Description (Abridged From W. M. O. Code)
	Cu of fair weather, little vertical development and seemingly flattened.
	Cu of considerable development, generally towering, with or without other Cu or Sc bases all at same level.
	Cb with tops lacking clear-cut outlines, but distinctly not cirriform or anvil-shaped; with or without Cu, Sc, or St.
	Sc formed by spreading out of Cu; Cu often present also.
	Sc not formed by spreading out of Cu.
	St or Fs or both, but no Fs of bad weather.
	Fs and/or Fc of bad weather (scud).
	Cu and Sc (not formed by spreading out of Cu) with bases at different levels.
	Cb having a clearly fibrous (cirriform) top, often anvil-shaped, with or without Cu, Sc, St, or scud.

C_M	Description (Abridged From W. M. O. Code)
	Thin As (most of cloud layer semi-transparent).
	Thick As, greater part sufficiently dense to hide sun (or moon), or Ns.
	Thin Ac, mostly semi-transparent; cloud elements not changing much and at a single level.
	Thin Ac in patches; cloud elements continually changing and/or occurring at more than one level.
	Thin Ac in bands or in a layer gradually spreading over sky and usually thickening as a whole.
	Ac formed by the spreading out of Cu.
	Double-layered Ac, or a thick layer of Ac, not increasing; or Ac with As and/or Ns.
	Ac in the form of Cu-shaped tufts or Ac with turrets.
	Ac of a chaotic sky, usually at different levels; patches of dense Ci are usually present also.

C_H	Description (Abridged From W. M. O. Code)
	Filaments of Ci, or "mares tails", scattered and not increasing.
	Dense Ci in patches or twisted sheaves, usually not increasing, sometime like remains of Cb; or towers or tufts.
	Dense Ci, often anvil-shaped, derived from or associated with Cb.
	Ci, often hook-shaped, gradually spreading over the sky and usually thickening as a whole.
	Ci and Cs, often in converging bands, or Cs alone; generally overspreading and growing denser; the continuous layer not reaching 45° altitude.
	Ci and Cs, often in converging bands, or Cs alone; generally overspreading and growing denser; the continuous layer exceeding 45° altitude.
	Veil of Cs covering the entire sky.
	Cs not increasing and not covering entire sky.
	Cc alone or Cc with some Ci or Cs, but the Cc being the main cirriform cloud.

Cloud Abbreviation	R_t	Time of Precipitation	h	Height in Feet (Rounded Off)	N	N_h	Sky Coverage
St or Fs-Stratus or Fractostratus	0	No Precipitation.	0	0 - 149		0	No clouds.
Ci-Cirrus	1	Less than 1 hour ago	1	150 - 299		1	Less than one-tenth or one-tenth
Cs-Cirrostratus	2	1 to 2 hours ago	2	300 - 599		2	Two and three-tenths
Cc-Cirrocumulus	3	2 to 3 hours ago	3	600 - 999		3	Four-tenths
Ac-Altocumulus	4	3 to 4 hours ago	4	1,000 - 1,999		4	Five-tenths
As-Altostratus	5	4 to 5 hours ago	5	2,000 - 3,499		5	Six-tenths
Sc-Stratocumulus	6	5 to 6 hours ago	6	3,500 - 4,999		6	Seven and eight-tenths
Ns-Nimbostratus	7	6 to 12 hours ago	7	5,000 - 6,499		7	Nine-tenths or overcast with openings.
Cu or Fc-Cumulus or Fractocumulus	8	More than 12 hours ago	8	6,500 - 7,999		8	Completely overcast
Cb-Cumulonimbus	9	Unknown	9	At or above 8,000, or no clouds		9	Sky obscurred

Reference 10 — Surface Analysis Chart Legend

NOTICES TO AIRMEN

INFORMATION CURRENT AS OF NOVEMBER 17, 19__

THIS SECTION CONTAINS NOTICES TO AIRMEN THAT ARE EXPECTED TO REMAIN IN EFFECT FOR AT LEAST SEVEN DAYS.
NOTE: NOTICES ARE ARRANGED IN ALPHABETICAL ORDER BY STATE (AND WITHIN STATE BY CITY OR LOCALITY). NEW OR REVISED DATA: NEW OR REVISED DATA ARE INDICATED BY BOLD ITALICIZING THE AIRPORT NAME.
NOTE: ALL TIMES ARE LOCAL UNLESS OTHERWISE INDICATED.

ALABAMA

ALBERTVILLE MUNI: A/C freq now 118.25. (12/)
BIRMINGHAM MUNI ARPT: ATCT ground control freq 121.7 cmsnd, freq 121.9 dcmsnd. Rwy 5–23 closed 2300–0700 daily until Dec 15. ALS ... and rw... thru ...
15. ILS ...v 5 and L...
15. (9/
GADSD... wy 14–... sed thru ... 23,
GRE... wy for non turbojet acft 12500 lbs
... days ...y. (9/)
...A, ROHERVILLE ARPT: VASI rwy 11 cmsnd.
–2)
FRANKLIN FIELD: TPA 800 ft. (11/ –2)
FULLERTON MUNI ARPT: RAIL rwy 24 OTS. (3/)
GROVELAND, PINE MOUNTAIN LAKE ARPT: Right tfc rwy 27. (10/ –2)
HAWTHORNE MUNI ARPT: ATCT freq 122.5R dcmsnd. (9/ –3)
HESPERIA, AIR LEDGE ARPT: Threshold rwy 21 dsplcd 390 ft. Rotating beacon OTS. (10/ –2)
LIVERMORE MUNI ARPT: VASI rwy 25 cmsnd. (10/ –3)
LONE PINE ARPT: Rwy 13–31 2535 ft x 100 ft gravel. Threshold rwy 13 dsplcd 1100 ft emergency use only by prior arrangement. (12/ –2)
LOS ANGELES FSS: Freq 122.5 cmsnd, freq 122.7 dcmsnd. (10/ –3)
MONTAGUE, SISKIYOU COUNTY ARPT: ALS rwy 35 now simplified short apch light system with RAIL. For rwy lights contact FSS 0615–2145 thereafter key 123.0 5 times and for ALS key 123.0 3 additional. (12/ –2)
NOVATO, GNOSS FLD: UNICOM now freq 123.0. (11/ –2)
OAKLAND, METROPOLITAN OAKLAND INTL ARPT: Rwy lights rwy 15–33 OTS. (8/)
ORLAND, HAIGH ARPT: A/C and departure control svc provided by Oakland ARTCC freq 120.4. Rwy 12–30 closed permly. (10/ –2)
OXNARD: VOR "OAF" OTS thru Feb 20. (12/)
PACOIMA: NDB "PAI" 370 kHz cmsnd. (9/ –3)
QUARTZ HILL ARPT: Right tfc rwy 23. (10/ –2)
RIALTON MUNI/MIRO FLD: TPA 800 ft. Rwy 34 open full length. (10/ –2)
SACRAMENTO EXECUTIVE ARPT: ALS rwy 2 OTS. (10/ –3)
SALINAS MUNI ARPT: VASI rwy 8 and rwy 13 cmsnd. VASI rwy 31 dcmsnd. (12/ –3)
SAN FRANCISCO: VOR/DME unusable 030–070 degrees beyond 20 NM below 6500 ft, 150–190 degrees beyond 25 NM below 4500 ft, 190–260 degrees beyond 10 NM below 4500 ft, 260–295 degrees beyond 35 NM below 3000 ft, and 295–330 degrees beyond 20 NM below 4000 ft. (11/ –3)
SAN JUAN CAPISTRANO, CAPISTRANO ARPT: Rwy 5–23 wt brg capacity S –5000 lbs. Threshold rwy 5 dsplcd 60 ft. (10/ –2)
SANTA ANA (MCAS): CTLZ hours 0800–1800 Mon, 0800–2200 Tues-Thur, 0800–1700 Fri, not in effect Sat-Sun and holidays. (11/ –3)

HUNT GREEN VALLEY ARPT: Arpt closed. (11/)
MARBLE CANYON CLIFF DWELLERS LODGE ARPT: Rwy 4–22 3800 ft x 85 ft dirt. (11/ –2)
NEW RIVER, LAKE PLEASANT ARPT: Arpt closed permly. (10/ –3)
PAGE MUNI ARPT: Rwy 7–25 now 2380 ft. (11/ –2)
PARKER MUNI ARPT: MIRL rwy 1–19 cmsnd. (11/ –2)
PHOENIX, FARM AERO ARPT: Rwy lights rwy 17–35 dcmsnd. (11/ –2)
PRESCOTT MUNI ARPT: UNIC... dcm... (1...
SCOTTSDALE ... ARPT: ...
...000 lbs, ... (10/ ...
...nge ... ...9. (11/ –3)

COLORADO

ALAMOSA MUNI ARPT: UNICOM freq 122.8 cmsnd. (11/ –2)
BOULDER MUNI ARPT: Rwy 8–26 now rwy 8R-26L. Rwy 8L-26R 4100 ft x 75 ft dirt cmsnd. Gliders use only rwy 8L-26R. (10/ –2)
CENTER LEACH ARPT: Threshold rwy 12 dsplcd 105 ft. Threshold rwy 30 dsplcd 1400 ft. (10/ –2)
DENVER, ARAPAHOE COUNTY ARPT: ATIS hours 0600–2200. (11/ –3)
DENVER, STAPLETON INTL ARPT: Profile Descent Procedured. A specified profile descent clearance must be issued by ATC before a pilot flies the profile descent procedure into Stapleton Intl Arpt. Clearance to fly the associated STAR does not constitute clearance to fly the charted profile descent procedured. (9/)
DENVER STAPLETON INTL ARPT: RVR, RVRT, RVRR rwy 35L OTS until Jan 4. LOC BC rwy 17R OTS until Jan 4. (10/)
DURANGO: VOR "DRO", DME cmsnd. Bondz fm ident now "R". (12/)
HALE, BONNY DAM ARPT: TPA 700 ft. Threshold rwy 33 dsplcd 200 ft. Rwy 15–33 now 3300 ft. Threshold rwy 33 dsplcd 200 ft. (9/ –2)
HAYDEN YAMPA VALLEY ARPT: Arpt closed to unscheduled CAB certificated air carrier except by prior permission. (8/)
LAJUNTA: RCAG freq 134.05 cmsnd. (12/)
LONGMONT MUNI ARPT: UNICOM now freq 123.0. (12/ –2)
WATKINS, SKYLINE ARPT: Arpt closed permly. (11/ –3)

CONNECTICUT

MADISON GRISWOLD ARPT: TPA 1000 ft. (12/ –2)
MARLBOROUGH, LESNEWSKI FLD: Arpt closed. (5/)
SIMSBURY TRI-TOWN ARPT: For rwy lights key freq 122.8 with voice transmission within 10 NM or by prior arrangement. Rwy 14–32 now 1000 ft. Rwy 3–21 now 2205 ft. (12/ –2)
WATERBURY AIRPORT INC ARPT: Rotating beacon cmsnd. Rwy 2–20 now 1660 ft. Rwy 17–35 now 2037 ft. (12/ –2)
WILLIMANTIC, WINDHAM ARPT: Numerous Sea Gulls frequently on or in vicinity of arpt. (10/ –2)
WINDSOR LOCKS BRADLEY INTL ARPT: Stage III radar service provided within 15 NM. (11/ –4)

Reference 11 — Excerpt-For Training Purposes Only

DENVER
$ STAPLETON INTL (DEN) 5E GMT-7(-6DT) 39°46′26″N 104°52′39″W — DENVER
5331 B S4 **FUEL** 100 LL. Jet A OX 1, 3 LRA CFR index D — H-1C, L-6E, 8G, A-2G
RWY 07-25: H5020 X 75 (CONC) S-12.5
RWY 08L-26R: H7926 X 150 (CONC) MI RWY LGTS S-120, D-150, DT-260
RWY 08L: THRESHOLD DISPLCD 1227′ RWY 26R: VASI, RGT TFC
RWY 08R-26L: H10004 X 150 HI RWY LGTS, S-200, D-200, DT-360
RWY 08R: VASI RWY 26L: SSALR, RGT TFC
RWY 17C-35C: H6480 X 100 (CONC) S-12.5
RWY 17L-35R: H12000 X 200 (CONC) C/L HI RWY LGTS S-200+, D-200+, DT-360+
RWY 17R-35L: H11500 X 150 (CONC) HI RWY LGTS S-200+, D-200+, DT-360+
RWY 17R: REIL VASI RWY 35L: MALS
AIRPORT REMARKS: 747 & JUMBO ACFT ON RWY 8R-26L MAX WGT 800,000 lbs.
COMMUNICATIONS: ATIS 124.45 (DEP) 125.6 (ARR) UNICOM 123.0
DENVER FSS (DEN) on fld 122.1R, 122.2, 122.6, 123.65 (303) 321-0031
DENVER APCH CTL - 120.5 (NORTH) 125.3 132.05 (SOUTH) 128.05
DENVER DEP CTL - 119.3 (080-218) 123.85(219-349) 126.9(350-079)
DENVER TOWER: 118.3(EAST WEST) 119.5 (NORTH SOUTH) GND CON 121.9
CLNC DEL 127.6 PRE-TAXI CLNC 121.6
RADIO AIDS TO NAVIGATION VOT 111.0
DENVER (H) ABVORTAC 116.3 DEN Chan 110 39°51′36″N, 104°45′6″W 218° 7.5 NM to fld
ILS RWY: 35L(108.1/I-SPO), 08R(110.3/I-DEN), 35R(109.3/I-RRV)
ILS/DME RWY: 26L(110.3/I–DEN)
ASR

Reference 12

CLEARANCES								
					TIME OFF			
FIXES	ROUTE	MC	MH	GS	DIST	ETE	ETA	ATA
Will Rogers	D→ ↗					7		
OKC	8,000 V-17	289	294	182	116	38		
GAG	↓	327	333	169	103	37		
GCK	↓	322	326	147	99	40		
GLD	8,000 V-4	273	283	162	73	27		
TXC	↓	267	278	165	72	26		
DEN	approach					7		
Stapleton	↗					7		
DEN	10,000 V-81	161	171	178	55	18		
COS	10,000 V-81	153	162	182	40	13		
PUB	approach					7		
TRIP TOTALS								

FUEL	
TAXI/TO	2.0
CLIMB	3.0
CRUISE	67.0
APPROACH	3.0
ALTERNATE	15.0
RESERVE	18.0
TOTAL	108.0

TC → ± VAR = MC → ± DEV = CC
↓ ± WCA ↓ ± WCA
TH → ± VAR = MH → ± DEV = CH

BLOCK IN	
BLOCK OUT	
LOG TIME	

TIME FLIGHT PLAN EXPIRES

POSITION REPORT

Acft. Ident.	Position	Time	Alt.	IFR/VFR	Est. Next Fix	Name Next Fix

Reference 13 — Completed Flight Log

INSTRUMENT APPROACH PROCEDURES (CHARTS)
WEST CENTRAL UNITED STATES

▽ IFR TAKE-OFF MINIMUMS AND DEPARTURE PROCEDURES

FAR 91.116(c) prescribes take-off rules and establishes standard take-off minimums as follows:

(1) Aircraft having two engines or less – one statute mile.

(2) Aircraft having more than two engines – one-half statute mile.

Aerodromes within this geographical area with IFR take-off minimums other than standard are listed below alphabetically by aerodrome name. Departure procedures and/or ceiling visibility minimums are established to assist pilots conducting IFR flight in avoiding obstructions during climb to the minimum enroute altitude.

Take-off minimums and departure procedures apply to all runways unless otherwise specified.

AERODROME NAME — TAKE-OFF MINIMUMS

ABERDEEN REGIONAL .... Rwy 31, ½ mile
Aberdeen, South Dakota

AITKIN MUNI .... Rwys 16, 26, 300-1
Aitkin, Minnesota
IFR DEPARTURE PROCEDURE: Rwys 8, 16, 26, when weather is below 500-1, climb runway heading to 1700 before turning.

ALAMOSA MUNI
Alamosa, Colorado
Climb direct Alamosa VORTAC, continue climb in ALS VOR holding pattern NW, right turn 142° inbound to MCA for direction of flight: N-bound J-13, 10,500; NE-bound J-64, 10,500; S-bound J-13, V-83 8300.

ALEXANDRIA MUNI .... Rwy 4, 200-1
Alexandria, Minnesota
IFR DEPARTURE PROCEDURE:
Rwys 4, 13 and 22, when weather is below 200-1, climb runway heading to 1800 before turning.

ALGONA MUNI .... Rwy 12, 500-1
Algona, Iowa

AMES MUNI
Ames, Iowa
Rwys 1, 19, 13, 31, when weather is below 700-2, climb on runway heading to 1700 before proceeding W bound.

ANOKA COUNTY-BLAINE AIRPORT (JANES FIELD)
Minneapolis, Minnesota
All runways: SE bound departures climb to 2380′ on runway heading before proceeding on course.

ARROWHEAD .... Rwys 15, 20, 400-1
So. Louis, Missouri

ATKINSON MUNI
Pittsburg, Kansas
Rwys 3, 10, 16, 21, 28 and 34 climb runway heading to 1200 before turning on course.

ATLANTIC MUNI .... Rwy 8, 500-1
Atlantic, Iowa
When weather is below 500-1; Rwy 12: Maintain runway heading to 1600 before proceeding on course; Rwys 17, 26, 30, 35: When planned route of flight is E bound, maintain runway heading to 1600 before proceeding on course.

AERODROME NAME — TAKE-OFF MINIMUMS

AUDUBON MUNI .... Rwy 32, 300-1
Audubon, Iowa
Rwy 14: When planned route of flight is N bound and weather is below 300-1, climb runway heading to 1600 before turning.

AUGUSTA MUNI
Augusta, Kansas
IFR DEPARTURE PROCEDURE:
Rwys 18 and 36, when weather is below 500-1, climb runway heading to 1800 before turning.

BEATRICE MUNI .... Rwy 21, 400-1
Beatrice, Nebraska
IFR DEPARTURE PROCEDURE: When weather is below 400-1, departure on Rwys 3, 13, 17 climb runway heading to 2200 before turning.

BEECH FACTORY .... Rwy 18, 200-1
Wichita, Kansas — Rwy 31, 500-1
IFR DEPARTURE PROCEDURE:
When weather is below 500-1 all W bound departures 180° to 360° climb runway heading to 1800 before turning.

BELLEVILLE MUNI
Belleville, Kansas
Rwy 13, 31, 35 climb on runway heading to 1900 before turning.

BENSON MUNI .... Rwy 5, 200-1
Benson, Minnesota
When weather is below 500-2: Rwy 5: Climb to 2000 on runway heading before proceeding on course. Rwy 14: Right climbing turn to 2000 on 180° bearing from RBn before proceeding on course.

BERT MOONEY SILVER BOW COUNTY
Butte, Montana
Rwys 2, 11, 15, 20, 29, 33, 1500-2
Climb visually over the airport to 7000 climb direct to BTM VORTAC on R-096, continue climb on R-343 within 10 NM to cross BTM VORTAC at or above: E bound V86, 8800; SE bound V257, 9000; N bound V257, 8800. NE bound V113, 8000; SW bound V113, 8600.
SE Bound Departures V257: When a ceiling of 4000-2 exists, aircraft may be cleared to climb visually over the airport to 9500, continue climb via the 11 DME arc BTM VORTAC CW to intercept BTM R-151 at or above 11,000.

(Continued on page 2)

PUBLISHED BY NOS, NOAA, TO IACC SPECIFICATIONS

Fig. A-1

INSTRUMENT APPROACH PROCEDURES (CHARTS)
WEST CENTRAL UNITED STATES

△ IFR ALTERNATE MINIMUMS

(Not applicable to USAF USN)

Standard alternate minimums for nonprecision approaches are 800-2 (NDB, VOR, LOC, TACAN, LDA, VORTAC, VOR DME or ASR); for precision approaches 600-2 (ILS or PAR). Aerodromes within this geographical area that require alternate minimums other than standard or alternate minimums with restrictions are listed below. NA - means IFR minimums are not authorized for alternate use due to unmonitored facility or absence of weather reporting service. U. S. Army pilots refer to Army Reg. 95-1 for additional application. Civil pilots see FAR 91.83. USAF/USN pilots refer to appropriate regulations.

AERODROME NAME	ALTERNATE MINIMUMS
LIBERAL MUNI Liberal, Kansas	VOR Rwy 17*† VOR Rwy 35* RNAV Rwy 12*
	*NA when control zone not effective except for operators with approved weather reporting service. †1100-2, Non DME.
MANHATTAN MUNI Manhattan, Kansas	VOR/NDB Rwy 3* VOR-H
	*NA when control zone not effective, except for operators with approved weather reporting service.
MANKATO MUNI Mankato, Minnesota	VOR Rwy 15 VOR Rwy 33
	NA when control zone not effective, except for operators with approved weather reporting service.
MARSHALL MUNI RYAN FIELD Marshall, Minnesota	VOR Rwy 12 NA*
	*Standard for operators with approved weather reporting service.
MC COOK MUNI Mc Cook, Nebraska	VOR Rwy 12* VOR Rwy 30* RNAV Rwy 12*
	*NA when control zone not effective except for operators with approved weather reporting service.
MINNEAPOLIS-ST. PAUL INTL (WOLD CHAMBERLAIN) Minneapolis, Minnesota	ILS BC Rwy 11R ILS BC Rwy 22
	ILS, LOC, 800-2
MINOT INTL Minot, North Dakota	ILS Rwy 31*
	*ILS CAT D, 700-2
MISSION FIELD Livingston, Montana	VOR-A, 1700-2
Montrose, Colorado	VOR-2 Rwy 12 1000-2*†
	*Category D †NA when control zone not effective, except for operators with approved weather reporting service.
NATRONA COUNTY INTL Casper, Wyoming	VOR Rwy 21, 1000-3 ILS Rwy 3*†
	*ILS, 800-2 †ILS; LOC Category D 800-2 ½
PARK RAPIDS MUNI Park Rapids, Minnesota	VOR Rwy 13 NA* VOR Rwy 31 NA†
	*Standard for operators with approved weather reporting service. Cat. A, B, C; Cat. D, 800-3 †Standard for operators with approved weather reporting service.
PHILLIP Phillip, South Dakota	VOR-A, 900-2*†
	*NA when Phillip, S.D. altimeter setting not available. †Cat. D, 1000-2
PHILIP BILLARD MUNI Topeka, Kansas	LOC BC Rwy 31* ILS Rwy 13*
	*NA when control tower not in operation
PIERRE MUNI Pierre, South Dakota	LOC BC Rwy 13*
	*LOC, 1100-2
PUEBLO MEMORIAL Pueblo, Colorado	VOR Rwy 25R (TAC)* LOC BC Rwy 25R*† NDB Rwy 7L, 1000-2 ¾ NDB Rwy 25R, 800-2 ½ ILS Rwy 7L%† ILS Rwy 25R%† RADAR-1, 800-3**
	*Cat D, 800-3; Cat E, 1000-3 †NA when control tower not in operation. %ILS 700-2 ½, Cat. E 1000-3; LOC 800-2 ½, Cat. E 1000-3. **Cat E, 900-3
RAPID CITY REGIONAL Rapid City, South Dakota	ILS Rwy 32*
	*ILS Category D and E, 700-2

15 DEC 19__

PUBLISHED BY NOS NOAA TO IACC SPECIFICATIONS

Fig. A-2

50

STAPLETON INTL
DENVER, COLORADO

KIOWA ONE ARRIVAL (HGO.IOC1)

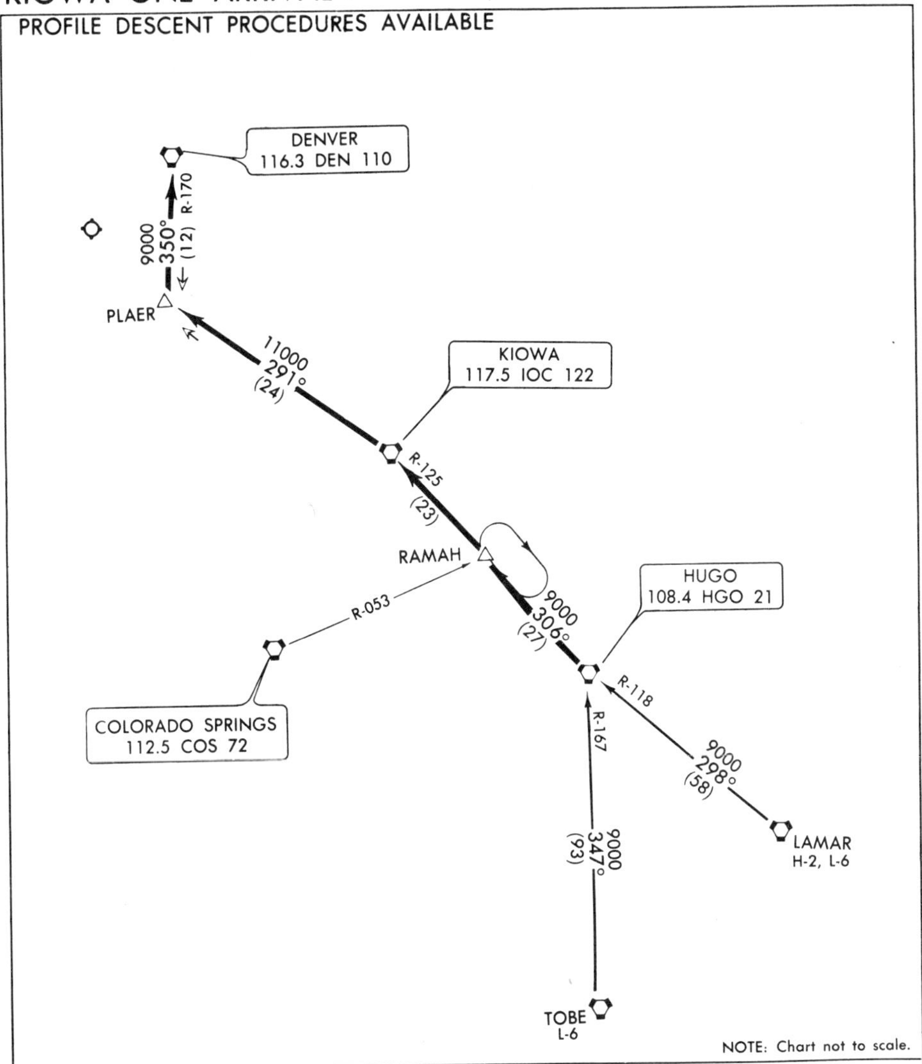

<u>LAMAR TRANSITION (LAA.IOC1)</u>: From over LAMAR VORTAC via LAMAR R-298 and HGO R-118 to HGO VORTAC. Thence . . . .
<u>TOBE TRANSITION (TBE.IOC1)</u>: From over TOBE VORTAC via TOBE R-347 and HGO R-167 to HGO VORTAC. Thence . . . .
. . . . From over HGO VORTAC via HGO R-306 to IOC R-125 to IOC VORTAC. Thence IOC R-291 to DEN R-170 to DEN VORTAC.

KIOWA ONE ARRIVAL (HGO.IOC1)
50

DENVER, COLORADO
STAPLETON INTL

Fig. A-3

DENVER COMMON TWO DEPARTURE (DEN2.DEN)

STAPLETON INTL
DENVER, COLORADO

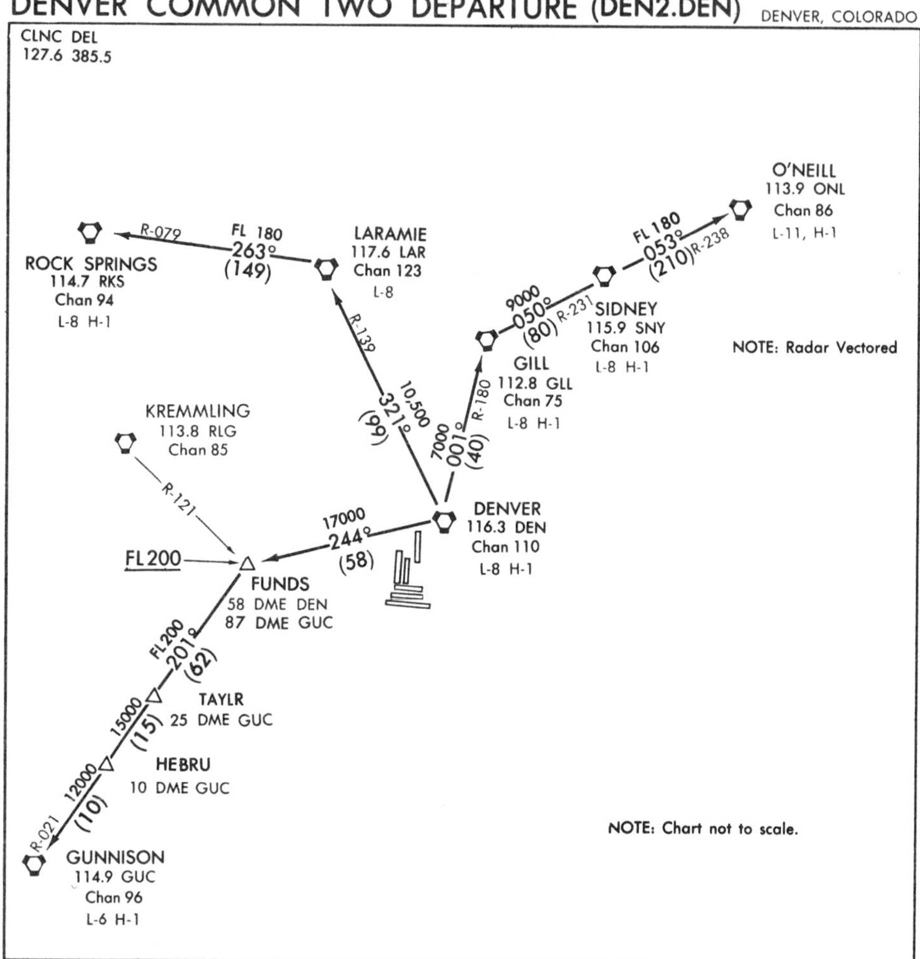

DEPARTURE ROUTE DESCRIPTION

All aircraft expect radar vectors to appropriate airway, route, or transition. Fly runway heading for initial radar vector. Maintain 10,000′ or assigned lower altitude. Expect clearance to Flight level 200 (or filed altitude if below FL 200) 15 NM from Stapleton Intl Airport.
LOST COMMUNICATIONS PROCEDURES: If no transmissions are received for one minute after departure; Gunnison transition climb direct to DEN VORTAC via R-270 or R-190. Depart DEN VORTAC at or above 10,000. All other departures climb direct to DEN VORTAC via R-270 or R-190 and proceed on course.

(Continued on next page)

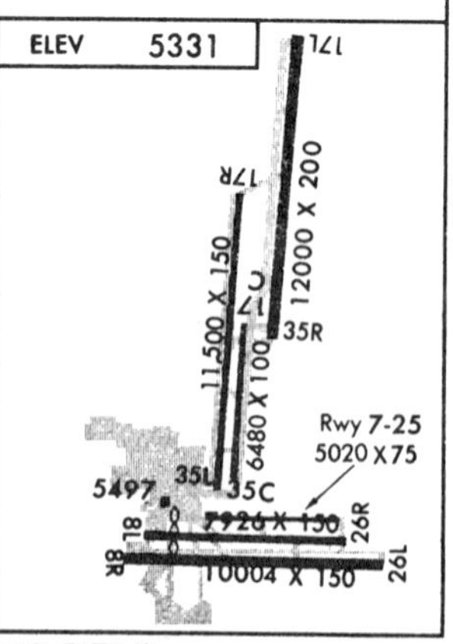

DENVER COMMON TWO DEPARTURE (DEN2.DEN)
DENVER, COLORADO
STAPLETON INTL
153

Fig. A-4

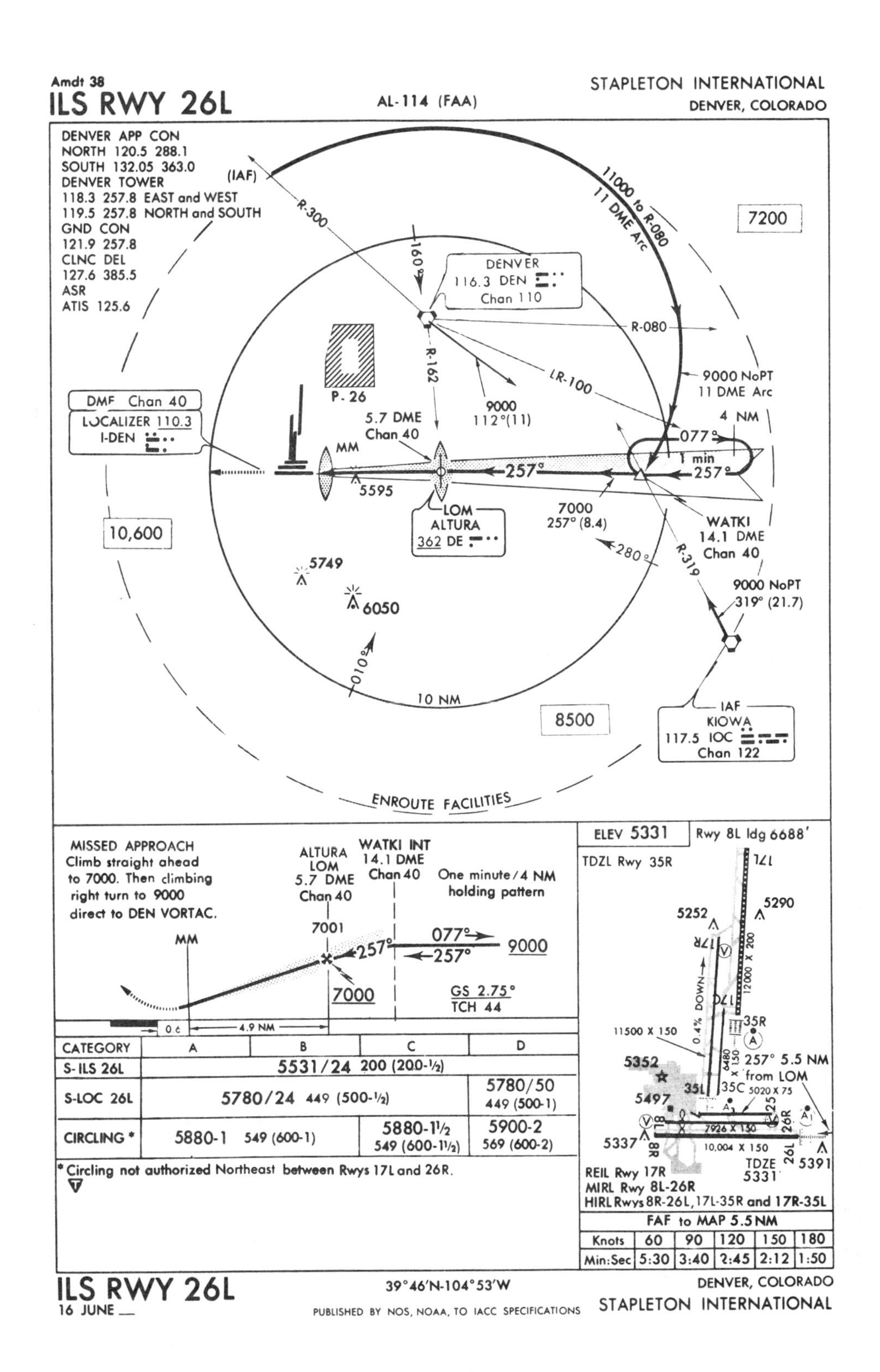

CATEGORY	A	B	C	D
S-ILS 26L	5531/24 200 (200-½)			
S-LOC 26L	5780/24 449 (500-½)			5780/50 449 (500-1)
CIRCLING *	5880-1 549 (600-1)		5880-1½ 549 (600-1½)	5900-2 569 (600-2)

* Circling not authorized Northeast between Rwys 17L and 26R.

FAF to MAP 5.5 NM					
Knots	60	90	120	150	180
Min:Sec	5:30	3:40	2:45	2:12	1:50

ILS RWY 26L
16 JUNE __
39°46'N-104°53'W
PUBLISHED BY NOS, NOAA, TO IACC SPECIFICATIONS
DENVER, COLORADO
STAPLETON INTERNATIONAL

Fig. A-5

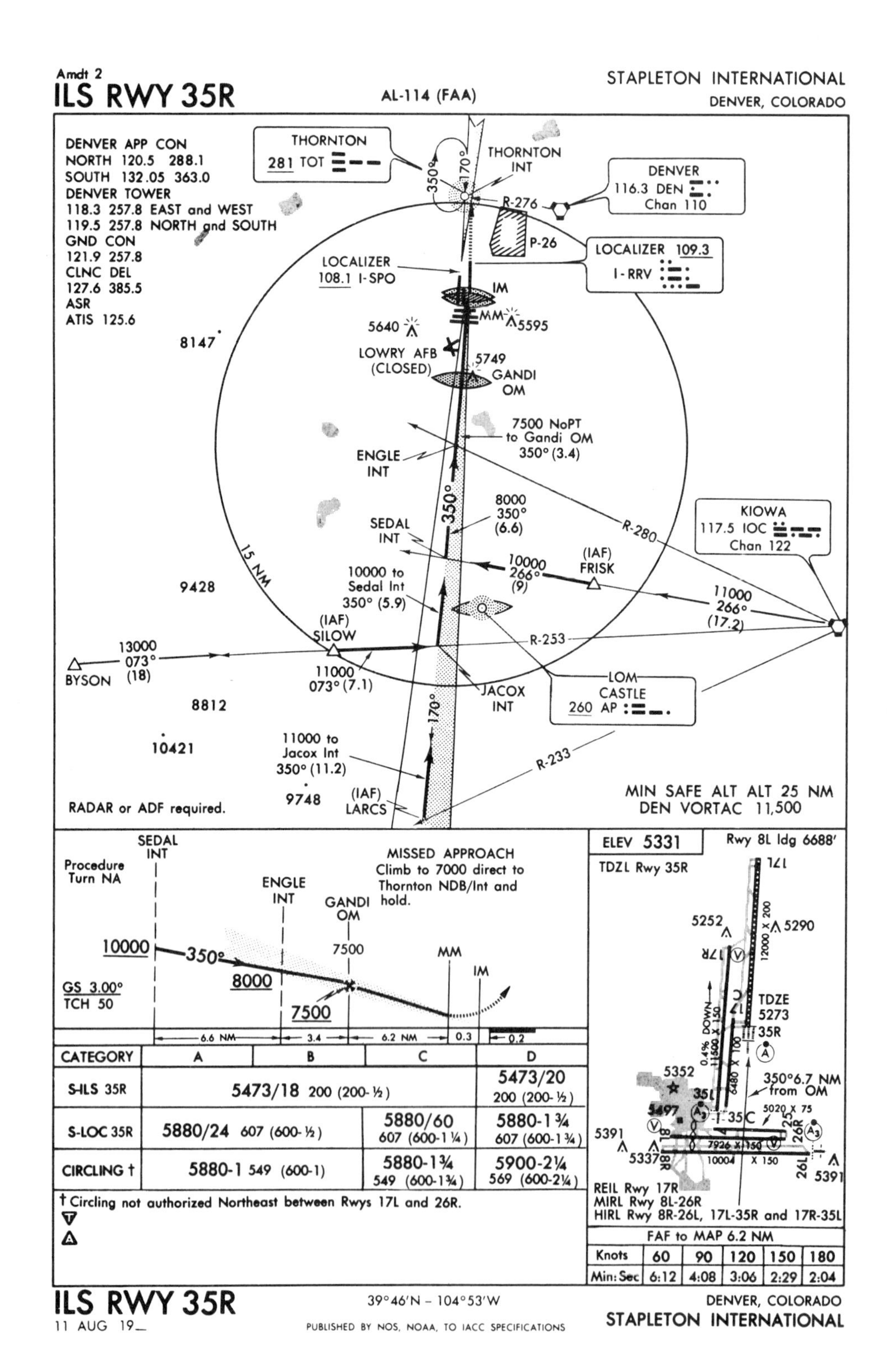

CATEGORY	A	B	C	D
S-ILS 35R	5473/18 200 (200-½)			5473/20 200 (200-½)
S-LOC 35R	5880/24 607 (600-½)		5880/60 607 (600-1¼)	5880-1¾ 607 (600-1¾)
CIRCLING †	5880-1 549 (600-1)		5880-1¾ 549 (600-1¾)	5900-2¼ 569 (600-2¼)

† Circling not authorized Northeast between Rwys 17L and 26R.

FAF to MAP 6.2 NM

Knots	60	90	120	150	180
Min:Sec	6:12	4:08	3:06	2:29	2:04

ILS RWY 35R
11 AUG 19__
39°46'N – 104°53'W
PUBLISHED BY NOS, NOAA, TO IACC SPECIFICATIONS
DENVER, COLORADO
STAPLETON INTERNATIONAL

Fig. A-6

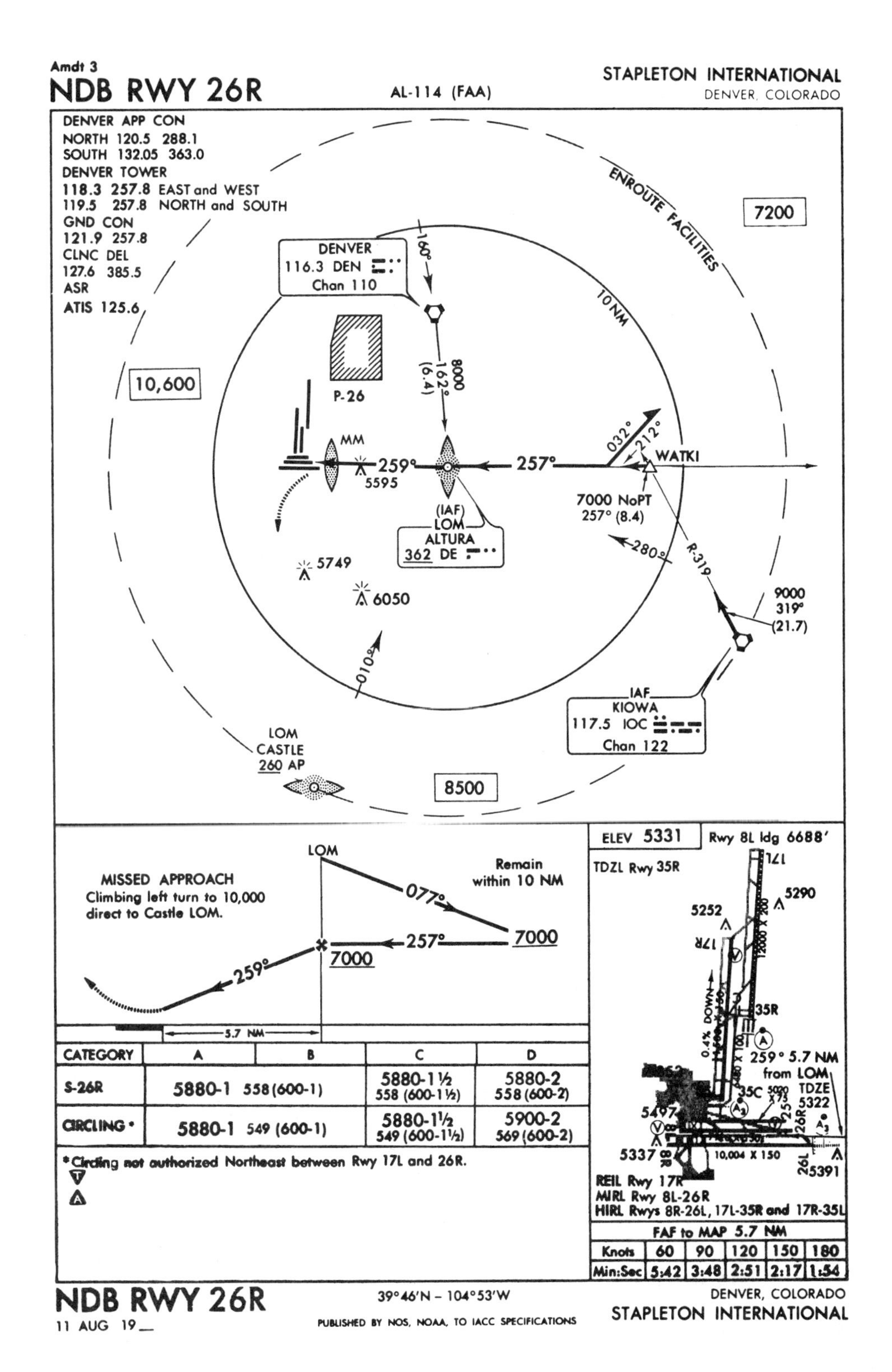

CATEGORY	A	B	C	D
S-26R	5880-1 558 (600-1)		5880-1½ 558 (600-1½)	5880-2 558 (600-2)
CIRCLING*	5880-1 549 (600-1)		5880-1½ 549 (600-1½)	5900-2 569 (600-2)

FAF to MAP 5.7 NM

Knots	60	90	120	150	180
Min:Sec	5:42	3:48	2:51	2:17	1:54

Fig. A-7

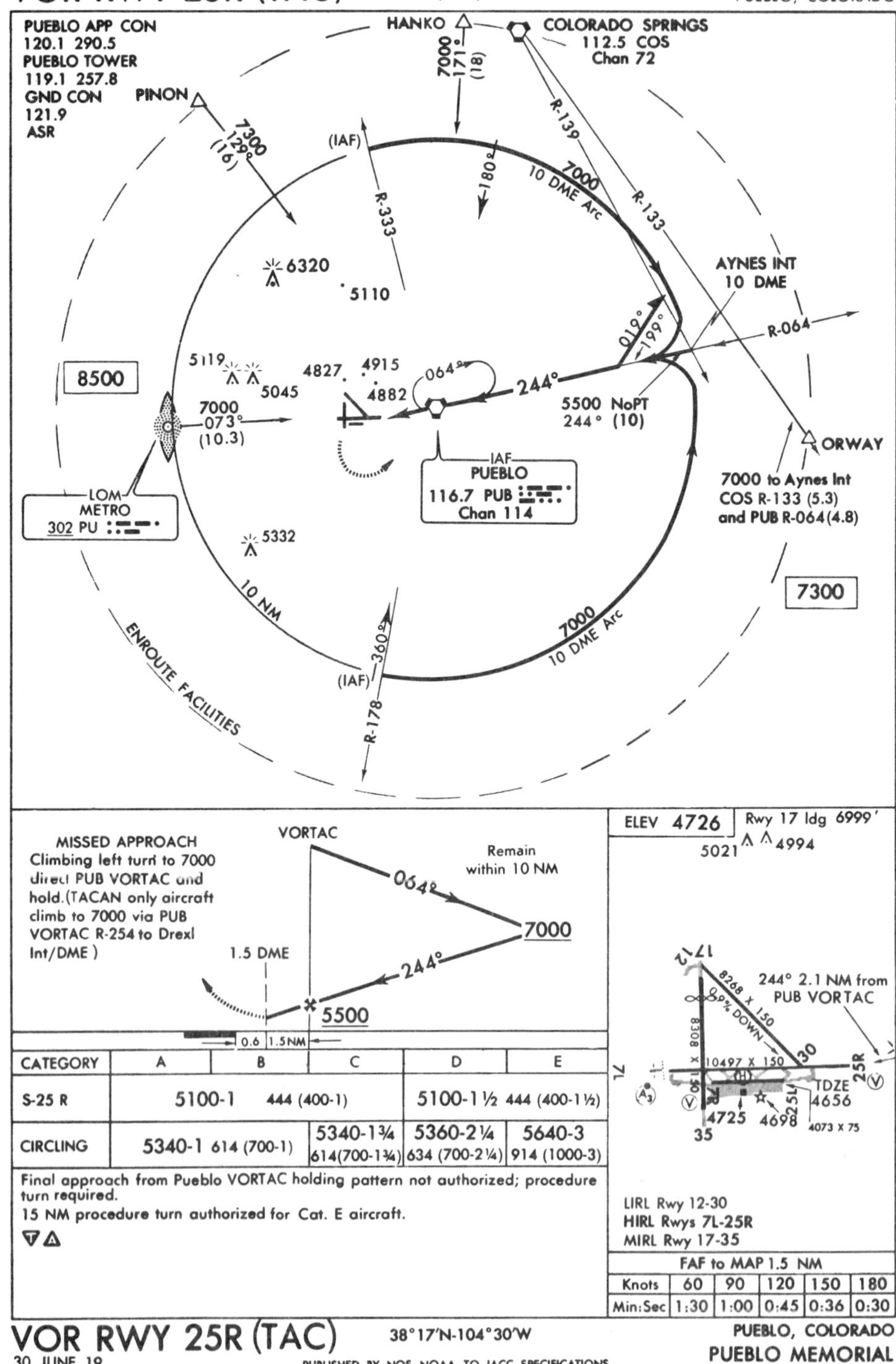

Amdt 18
VOR RWY 25R (TAC)
AL-334 (FAA)
PUEBLO MEMORIAL
PUEBLO, COLORADO
PUEBLO APP CON
120.1 290.5
PUEBLO TOWER
119.1 257.8
GND CON
121.9
ASR
HANKO
COLORADO SPRINGS
112.5 COS
Chan 72
7000
171°
(18)
PINON
7300
129°
(16)
(IAF)
R-139
R-133
R-333
180°
7000
10 DME Arc
6320
5110
AYNES INT
10 DME
R-064
019°
199°
8500
5119
5045
4827
4915
4882
064°
244°
5500 NoPT
244° (10)
7000
073°
(10.3)
IAF
PUEBLO
116.7 PUB
Chan 114
ORWAY
7000 to Aynes Int
COS R-133 (5.3)
and PUB R-064 (4.8)
LOM
METRO
302 PU
5332
7300
10 NM
7000
10 DME Arc
ENROUTE FACILITIES
(IAF)
360°
R-178
MISSED APPROACH
Climbing left turn to 7000 direct PUB VORTAC and hold. (TACAN only aircraft climb to 7000 via PUB VORTAC R-254 to Drexl Int/DME)
VORTAC
Remain within 10 NM
064°
7000
244°
1.5 DME
5500
0.6
1.5NM
ELEV 4726
Rwy 17 ldg 6999'
5021
4994
244° 2.1 NM from PUB VORTAC
8268 X 150
.9% DOWN
8308 X 150
10497 X 150
TDZE 4656
4725
4698
4073 X 75
CATEGORY | A | B | C | D | E
S-25 R | 5100-1 444 (400-1) | 5100-1½ 444 (400-1½)
CIRCLING | 5340-1 614 (700-1) | 5340-1¾ 614(700-1¾) | 5360-2¼ 634 (700-2¼) | 5640-3 914 (1000-3)
Final approach from Pueblo VORTAC holding pattern not authorized; procedure turn required.
15 NM procedure turn authorized for Cat. E aircraft.
LIRL Rwy 12-30
HIRL Rwys 7L-25R
MIRL Rwy 17-35
FAF to MAP 1.5 NM
Knots | 60 | 90 | 120 | 150 | 180
Min:Sec | 1:30 | 1:00 | 0:45 | 0:36 | 0:30
VOR RWY 25R (TAC)
38°17'N-104°30'W
PUEBLO, COLORADO
PUEBLO MEMORIAL
30 JUNE 19__
PUBLISHED BY NOS, NOAA, TO IACC SPECIFICATIONS

Fig. A-8

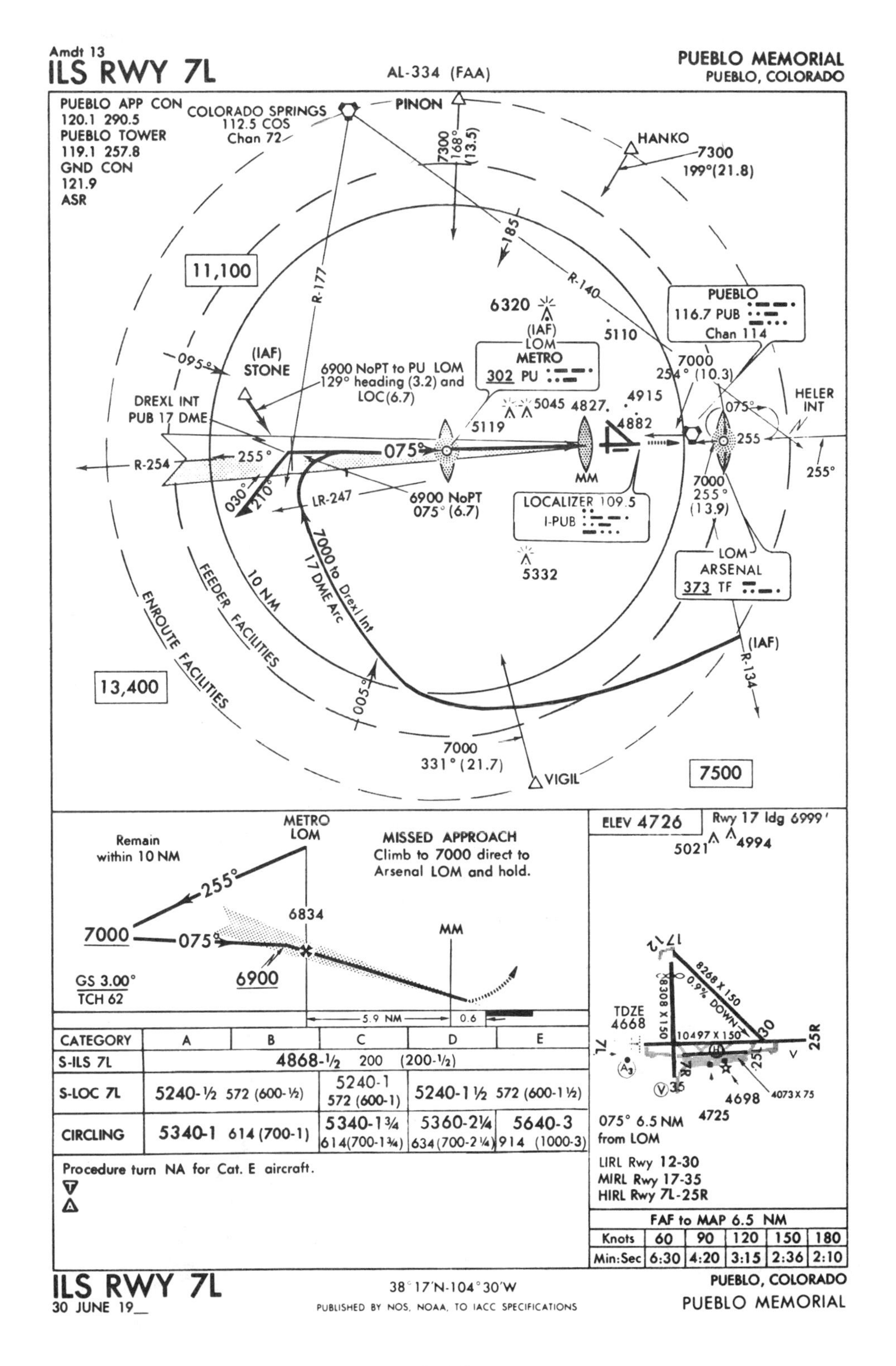

CATEGORY	A	B	C	D	E
S-ILS 7L	4868-½ 200 (200-½)				
S-LOC 7L	5240-½ 572 (600-½)		5240-1 572 (600-1)	5240-1½ 572 (600-1½)	
CIRCLING	5340-1 614 (700-1)		5340-1¾ 614(700-1¾)	5360-2¼ 634(700-2¼)	5640-3 914 (1000-3)

FAF to MAP 6.5 NM					
Knots	60	90	120	150	180
Min:Sec	6:30	4:20	3:15	2:36	2:10

Fig. A-9

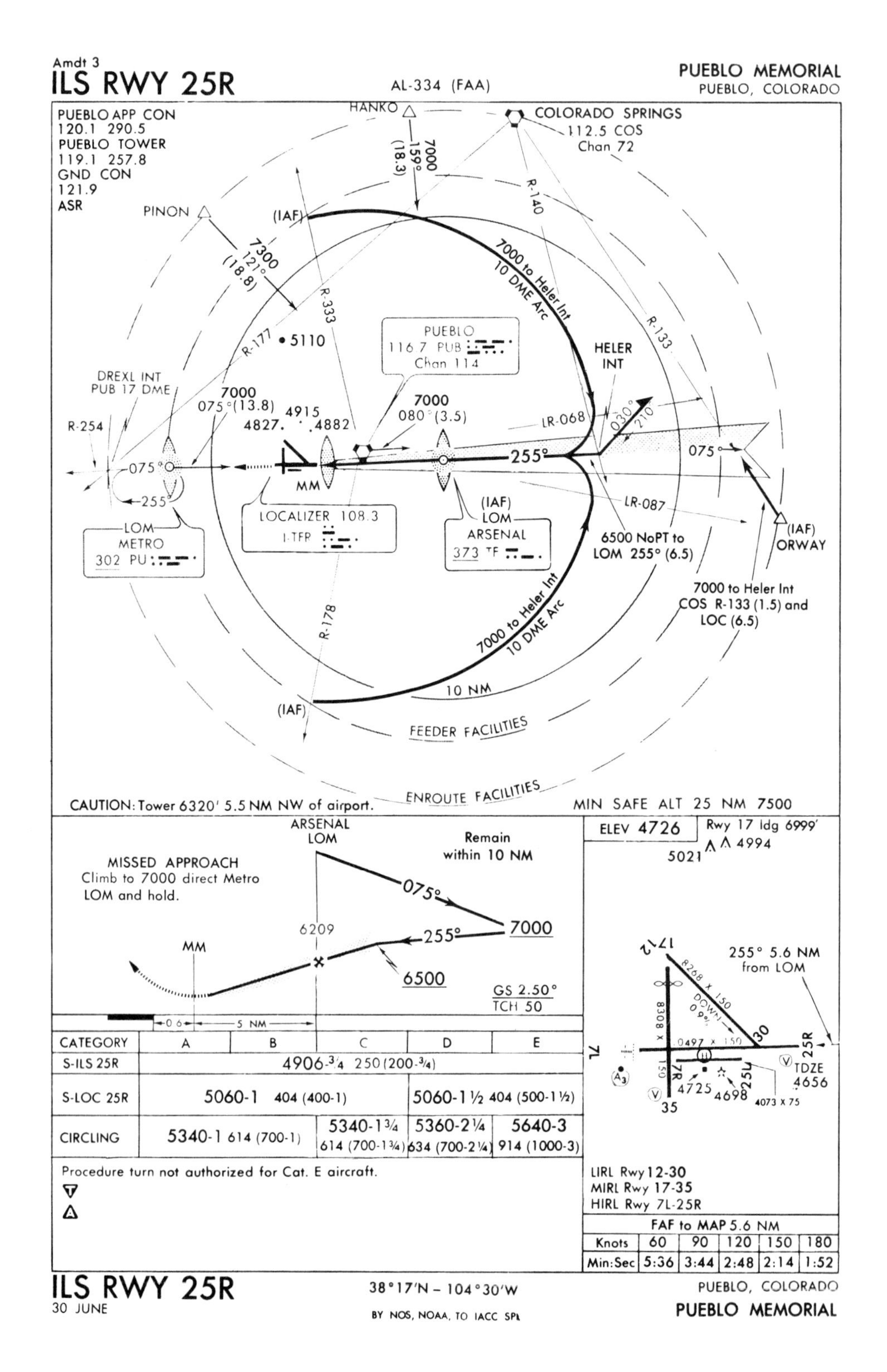

CATEGORY	A	B	C	D	E
S-ILS 25R	4906-¾ 250 (200-¾)				
S-LOC 25R	5060-1 404 (400-1)			5060-1½ 404 (500-1½)	
CIRCLING	5340-1 614 (700-1)		5340-1¾ 614 (700-1¾)	5360-2¼ 634 (700-2¼)	5640-3 914 (1000-3)

FAF to MAP 5.6 NM					
Knots	60	90	120	150	180
Min:Sec	5:36	3:44	2:48	2:14	1:52

Fig. A-10

appendix 2

FEDERAL AVIATION REGULATIONS

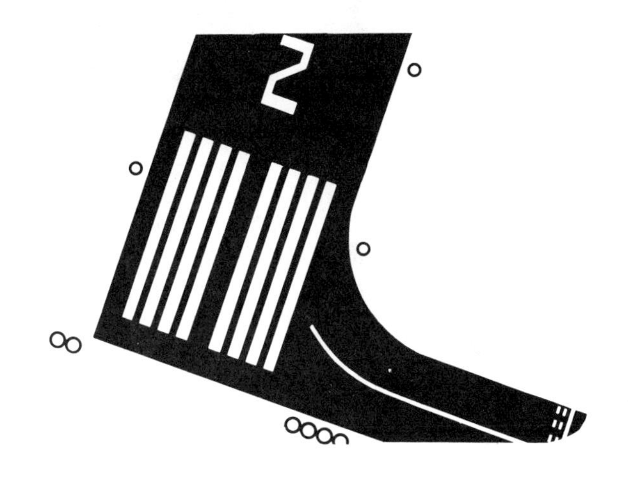

Just as the automobile driver must learn the laws concerning the operation of his vehicle, the pilot must know the Federal Aviation Regulations that pertain to the operation of aircraft. By understanding and observing these regulations, he will be able to fly confidently and safely with the knowledge that all aircraft in the sky are operating under the same rules.

The Federal Aviation Regulations are divided into over a hundred parts, each dealing with a particular aspect of aviation. In this chapter, excerpts from Parts 61, 91, and NTSB Part 830 have been included. It should be noted that these excerpts are reproduced *verbatim* as received from the U.S. Government Printing Office. The excerpts included are those considered most pertinent to the instrument pilot.

PART 61—CERTIFICATION: PILOTS AND FLIGHT INSTRUCTORS

SUBPART A—GENERAL

61.1 APPLICABILITY.

(a) This part prescribes the requirements for issuing pilot and flight instructor certificates and ratings, the conditions under which those certificates and ratings are necessary, and the privileges and limitations of those certificates and ratings.

(b) Except as provided in 61.71 of this part, an applicant for a certificate or rating may, until November 1, 1974, meet either the requirements of this part, or the requirements in effect immediately before November 1, 1973. However, the applicant for a private pilot certificate with a free balloon class rating must meet the requirements of this part.

61.3 REQUIREMENT FOR CERTIFICATES, RATING, AND AUTHORIZATIONS.

(a) *Pilot certificate.* No person may act as pilot in command or in any other capacity as a required pilot flight crewmember of a civil aircraft of United States registry unless he has in his personal possession a current pilot certificate issued to him under this part. However, when the aircraft is operated within a foreign country a current pilot license issued by the country in which the aircraft is operated may be used.

(c) *Medical certificate.* Except for free balloon pilots piloting balloons and glider pilots piloting gliders, no person may act as pilot in command or in any other capacity as a required pilot flight crewmember of an aircraft under a certificate issued to him under this part, unless he has in his personal possession an appropriate current medical certificate issued under Part 67.

(e) *Instrument rating.* No person may act as pilot in command of a civil aircraft under instrument flight rules, or in weather conditions less than the minimums prescribed for VFR flight unless—

(1) In the case of an airplane, he holds an instrument rating or an airline transport pilct certificate with an airplane category rating on it;

(f) *Category II Pilot Authorization.*

(1) No person may act as pilot in command of a civil aircraft in a Category II operation unless he holds a current Category II pilot authorization for that type aircraft or, in the case of a civil aircraft of foreign registry, he is authorized by the country of registry to act as pilot in command of that aircraft in Category II operations.

(2) No person may act as second in command of a civil aircraft in a Category II operation unless he holds a current appropriate instrument rating or an airline transport pilot certificate (airplane) or, in the case of a civil aircraft of foreign registry, he is authorized by the country of registry to act as second in command of that aircraft in Category II operations.

This paragraph does not apply to operations conducted by the holder of a certificate issued under Part 121 of this chapter.

(h) *Inspection of certificate.* Each person who holds a pilot certificate, flight instructor certificate, medical certificate, authorization or license required by this part shall present it for inspection upon the request of the Administrator, an authorized representative of the National Transportation Safety Board, or any Federal, State, or local law enforcement officer.

61.5 CERTIFICATES AND RATINGS ISSUED UNDER THIS PART.

(a) The following certificates are issued under this part:

(1) Pilot certificates:
- (i) Student pilot.
- (ii) Private pilot.
- (iii) Commercial pilot.
- (iv) Airline transport pilot.

(2) Flight instructor certificates.

(b) The following ratings are placed on pilot certificates (other than student pilot) where applicable:

(1) Aircraft category ratings:
- (i) Airplane.
- (ii) Rotorcraft.
- (iii) Glider.
- (iv) Lighter-than-air.

(2) Airplane class ratings:
- (i) Single-engine land.
- (ii) Multiengine land.
- (iii) Single-engine sea.
- (iv) Multiengine sea.

(3) Rotorcraft class ratings:
- (i) Helicopter.
- (ii) Gyroplane.

(4) Lighter-than-air class ratings:
- (i) Airship.
- (ii) Free balloon.

(5) Aircraft type ratings are listed in Advisory Circular 61-1 entitled "Aircraft Type Ratings." This list includes ratings for the following:
- (i) Large aircraft, other than lighter-than-air.
- (ii) Small turbojet-powered airplanes.
- (iii) Small helicopters for operations requiring an airline transport pilot certificate.
- (iv) Other aircraft type ratings specified by the Administrator through aircraft type certificate procedures.

(6) Instrument ratings (on private and commercial pilot certificates only):
- (i) Instrument—airplanes.
- (ii) Instrument—helicopter.

61.13 APPLICATION AND QUALIFICATION.

(a) Application for a certificate and rating, or for an additional rating under this part is made on a form and in a manner prescribed by the Administrator.

(b) An applicant who meets the requirements of this part is entitled to an appropriate pilot certificate with aircraft ratings. Additional aircraft category, class, type and other ratings, for which the applicant is qualified, are added to his certificate. However, the Administrator may refuse to issue certificates to persons who are not citizens of the United States and who do not reside in the United States.

(c) An applicant who cannot comply with all of the flight proficiency requirements prescribed by this part because the aircraft used by him for his flight training or flight test is characteristically incapable of performing a required pilot operation, but who meets all other requirements for the certificate or rating sought, is issued the certificate or rating with appropriate limitations.

(d) An applicant for a pilot certificate who holds a medical certificate under 67.19

with special limitations on it, but who meets all other requirements for that pilot certificate, is issued a pilot certificate containing such operating limitations as the Administrator determines are necessary because of the applicant's medical deficiency.

(f) Unless authorized by the Administrator—
  (1) A person whose pilot certificate is suspended may not apply for any pilot or flight instructor certificate or rating during the period of suspension;

(g) Unless the order of revocation provides otherwise—
  (1) A person whose pilot certificate is revoked may not apply for any pilot or flight instructor certificate or rating for 1 year after the date of revocation;

61.15 OFFENSES INVOLVING NARCOTIC DRUGS, MARIHUANA, AND DEPRESSANT OR STIMULANT DRUGS OR SUBSTANCES

(a) No person who is convicted of violating any Federal or State statute relating to the growing, processing, manufacture, sale, disposition, possession, transportation, or importation of narcotic drugs, marihuana, and depressant or stimulant drugs or substances, is eligible for any certificate or rating issued under this part for a period of 1 year after the date of final conviction.

(b) No person who commits an act prohibited by 91.12(a) of this chapter is eligible for any certificate or rating issued under this part for a period of 1 year after the date of that act.

(c) Any conviction specified in paragraph (a) of this section or the commission of the act referenced in paragraph (b) of this section, is grounds for suspending or revoking any certificate or rating issued under this part.

61.17 TEMPORARY CERTIFICATE.

(a) A temporary pilot or flight instructor certificate, or a rating, effective for a period of not more than 120 days, is issued to a qualified applicant pending a review of his qualifications and the issuance of a permanent certificate or rating by the Administrator. The permanent certificate or rating is issued to an applicant found qualified and a denial thereof issued to an applicant found not qualified.

(b) A temporary certificate issued under paragraph (a) of this section expires—
  (1) At the end of the expiration date stated thereon: or
  (2) Upon receipt by the applicant, of—
    (i) The certificate or rating sought; or
    (ii) Notice that the certificate or rating sought is denied.

61.19 DURATION OF PILOT AND FLIGHT INSTRUCTOR CERTIFICATES

(a) *General.* The holder of a certificate with an expiration date may not, after that date, exercise the privileges of that certificate.

(c) *Other pilot certificates.* Any pilot certificate (other than a student pilot certificate) issued under this part is issued without a specific expiration date. However, the holder of a pilot certificate issued on the basis of a foreign pilot license may exercise the privileges of that certificate only while the foreign pilot license on which that certificate is based is effective.

(e) *Surrender, suspension, or revocation.* Any pilot certificate or flight instructor certificate issued under this part ceases to be effective if it is surrendered, suspended, or revoked.

(f) *Return of certificate.* The holder of any certificate issued under this part that is suspended or revoked shall, upon the Administrator's request, return it to the Administrator.

61.23 DURATION OF MEDICAL CERTIFICATES.

(a) A first-class medical certificate expires at the end of the last day of—
  (1) The sixth month after the month of the date of examination shown on the certificate, for operations requiring an airline transport pilot certificate;
  (2) The 12th month after the month of the date of examination shown on the certificate, for operations requiring only a commercial pilot certificate; and
  (3) The 24th month after the month of the date of examination shown on the certificate, for operations requiring only a private or student pilot certificate.

(b) A second-class medical certificate expires at the end of the last day of—
  (1) The 12th month after the month of the date of examination shown on the certificate, for operations requiring a commercial pilot certificate; and
  (2) The 24th month after the month of the date of examination shown on the certificate, for operations requiring only a private or student pilot certificate.

(c) A third-class medical certificate expires at the end of the last day of the 24th month after the month of the date of examination shown on the certificate, for operations requiring a private or student pilot certificate.

61.25 CHANGE OF NAME.

An application for the change of a name on a certificate issued under this part must be accompanied by the applicant's current certificate and a copy of the marriage license, court order, or other document verifying the change. The documents are returned to the applicant after inspection.

61.27 VOLUNTARY SURRENDER OR EXCHANGE OF CERTIFICATE.

The holder of a certificate issued under this part may voluntarily surrender it for cancellation, or for the issue of a certificate of lower grade, or another certificate with specific ratings deleted. If he so requests, he must include the following signed statement or its equivalent:

This request is made for my own reasons, with full knowledge that my (insert name of certificate or rating, as appropriate) may not be reissued to me unless I again pass the tests prescribed for its issue.

61.29 REPLACEMENT OF LOST OR DESTROYED CERTIFICATE.

(a) An application for the replacement of a lost or destroyed airman certificate issued under this part is made by letter to the Department of Transportation, Federal Aviation Administration, Airman Certification Branch, Post Office Box 25082, Oklahoma City, OK 73125. The letter must—
  (1) State the name of the person to whom the certificate was issued, the permanent mailing address (including zip code), social security number (if any), date and place of birth of the certificate holder, and any available information regarding the grade, number, and date of issue of the certificate, and the ratings on it; and
  (2) Be accompanied by a check or money order for $2, payable to the Federal Aviation Administration.

(b) An application for the replacement of a lost or destroyed medical certificate is made by letter to the Department of Transportation, Federal Aviation Administration, Aeromedical Certification Branch, Post Office Box 25082, Oklahoma City, OK 73125, accompanied by a check or money order for $2.

(c) A person who has lost a certificate issued under this part, or a medical certificate issued under Part 67 or both, may obtain a telegram from the FAA confirming that it was issued. The telegram may be carried as a certificate for a period not to exceed 60 days pending his receipt of a duplicate certificate under paragraph (a) or (b) of this section, unless he has been notified that the certificate has been suspended or revoked. The request for such a telegram may be made by letter or prepaid telegram, including the date upon which a duplicate certificate was previously requested, if a request had been made, and a money order for the cost of the duplicate certificate. The request for a telegraphic certificate is sent to the office listed in paragraph (a) or (b) of this section, as appropriate. However, a request for both airman and medical certificates at the same time must be sent to the office prescribed in paragraph (a) of this section.

61.31 GENERAL LIMITATIONS.

(a) *Type ratings required.* A person may not act as pilot in command of any of the following aircraft unless he holds a type rating for that aircraft:
  (1) A large aircraft (except lighter-than-air).
  (2) A helicopter, for operations requiring an airline transport pilot certificate.
  (3) A turbojet powered airplane.
  (4) Other aircraft specified by the Administrator through aircraft type certificate procedures.

(b) *Authorization in lieu of a type rating.*
   (1) In lieu of a type rating required under paragraphs (a)(1), (3), and (4) of this section, an aircraft may be operated under an authorization issued by the Administrator, for a flight or series of flights within the United States, if—
      (i) The particular operation for which the authorization is requested involves a ferry flight, a practice or training flight, a flight test for a pilot type rating, or a test flight of an aircraft, for a period that does not exceed 60 days;
      (ii) The applicant shows that compliance with paragraph (a) of this section is impracticable for the particular operations; and
      (iii) The Administrator finds that an equivalent level of safety may be achieved through operating limitations on the authorization.
   (2) Aircraft operated under an authorization issued under this paragraph—
      (i) May not be operated for compensation or hire; and
      (ii) May carry only flight crewmembers necessary for the flight.
   (3) An authorization issued under this paragraph may be reissued for an additional 60-day period for the same operation if the applicant shows that he was prevented from carrying out the purpose of the particular operation before his authorization expired.

The prohibition of paragraph (b)(2)(i) of this section does not prohibit compensation for the use of an aircraft by a pilot solely to prepare for or take a flight test for a type rating.

(c) *Category and class rating: Carrying another person or operating for compensation or hire.* Unless he holds a category and class rating for that aircraft, a person may not act as pilot in command of an aircraft that is carrying another person or is operated for compensation or hire. In addition, he may not act as pilot in command of that aircraft for compensation or hire.

(d) *Category and class rating: Other operations.* No person may act as pilot in command of an aircraft in solo flight in operations not subject to paragraph (c) of this section, unless he meets at least one of the following:
   (1) He holds a category and class rating appropriate to that aircraft.
   (2) He has received flight instruction in the pilot operations required by this part, appropriate to the category and class of aircraft for first solo, given to him by a certificated flight instructor who found him competent to solo that category and class of aircraft and has so endorsed his pilot logbook.
   (3) He has soloed and logged pilot-in-command time in that category and class of aircraft before November 1, 1973.

(e) *High performance airplanes.* A person holding a private or commercial pilot certificate may not act as pilot in command of an airplane that has more than 200 horsepower, or that has a retractable landing gear, flaps, and a controllable propeller, unless he has received flight instruction from an authorized flight instructor who has certified in his logbook that he is competent to pilot an airplane that has more than 200 horsepower, or that has a retractable landing gear, flaps, and a controllable propeller, as the case may be. However, this instruction is not required if he has logged flight time as pilot in command in high performance airplanes before November 1, 1973.

(f) *Exception.* This section does not require a class rating for gliders, or category and class ratings for aircraft that are not type certificated as airplanes, rotorcraft, or lighter-than-air aircraft. In addition, the rating limitations of this section do not apply to—
   (1) The holder of a student pilot certificate;
   (2) The holder of a pilot certificate when operating an aircraft under the authority of an experimental or provisional type certificate;
   (3) An applicant when taking a flight test given by the Administrator; or
   (4) The holder of a pilot certificate with a lighter-than-air category rating when operating a hot air balloon without an airborne heater.

61.33 TESTS: GENERAL PROCEDURE.

Tests prescribed by or under this part are given at times and places, and by persons, designated by the Administrator.

61.35 WRITTEN TEST: PREREQUISITES AND PASSING GRADES.

(a) An applicant for a written test must—
  (1) Show that he has satisfactorily completed the ground instruction or home study course required by this part for the certificate or rating sought:
  (2) Present as personal identification an airman certificate, driver's license, or other official document; and
  (3) Present a birth certificate or other official document showing that he meets the age requirement prescribed in this part for the certificate sought not later than 2 years from the date of application for the test.

(b) The minimum passing grade is specified by the Administrator on each written test sheet or booklet furnished to the applicant.

This section does not apply to the written test for an airline transport pilot certificate or a rating associated with that certificate.

61.37 WRITTEN TESTS: CHEATING OR OTHER UNAUTHORIZED CONDUCT.

(a) Except as authorized by the Administrator, no person may—
  (1) Copy, or intentionally remove, a written test under this part;
  (2) Give to another, or receive from another, any part or copy of that test;
  (3) Give help on that test to, or receive help on that test from, any person during the period that test is being given;
  (4) Take any part of that test in behalf of another person;
  (5) Use any material or aid during the period that test is being given; or
  (6) Intentionally cause, assist, or participate in any act prohibited by this paragraph.

(b) No person whom the Administrator finds to have committed an act prohibited by paragraph (a) of this section is eligible for any airman or ground instructor certificate or rating, or to take any test therefor, under this chapter for a period of 1 year after the date of that act. In addition, the commission of that act is a basis for suspending or revoking any airman or ground instructor certificate or rating held by that person.

61.39 PREREQUISITES FOR FLIGHT TESTS.

(a) To be eligible for a flight test for a certificate, or an aircraft or instrument rating issued under this part, the applicant must—
  (1) Have passed any required written test since the beginning of the 24th month before the month in which he takes the flight test;
  (2) Have the applicable instruction and aeronautical experience prescribed in this part;
  (3) Hold a current medical certificate appropriate to the certificate he seeks or, in the case of a rating to be added to his pilot certificate, at least a third-class medical certificate issued since the beginning of the 24th month before the month in which he takes the flight test;
  (4) Except for a flight test for an airline transport pilot certificate, meet the age requirement for the issuance of the certificate or rating he seeks; and
  (5) Have a written statement from an appropriately certificated flight instructor certifying that he has given the applicant flight instruction in preparation for the flight test within 60 days preceding the date of application, and finds him competent to pass the test and to have satisfactory knowledge of the subject areas in which he is shown to be deficient by his FAA airman written test report. However, an applicant need not have this written statement if he—
    (i) Holds a foreign pilot license issued by a contracting State to the Convention on International Civil Aviation that authorizes at least the pilot privileges of the airman certificate sought by him;
    (ii) Is applying for a type rating only, or a class rating with an associated type rating; or
    (iii) Is applying for an airline transport pilot certificate or an additional aircraft rating on that certificate.

61.41 FLIGHT INSTRUCTION RECEIVED FROM FLIGHT INSTRUCTORS NOT CERTIFICATED BY FAA.

Flight instruction may be credited toward the requirements for a pilot certificate or rating issued under this part if it is received from—

(a) An Armed Force of either the United States or a foreign contracting State to the Convention on International Civil Aviation in a program for training military pilots; or

(b) A flight instructor who is authorized to give that flight instruction by the licensing authority of a foreign contracting State to the Convention on International Civil Aviation and the flight instruction is given outside the United States.

61.43 FLIGHT TESTS: GENERAL PROCEDURES.

(a) The ability of an applicant for a private or commercial pilot certificate, or for an aircraft or instrument rating on that certificate to perform the required pilot operations is based on the following:

 (1) Executing procedures and maneuvers within the aircraft's performance capabilities and limitations, including use of the aircraft's systems.

 (2) Executing emergency procedures and maneuvers appropriate to the aircraft.

 (3) Piloting the aircraft with smoothness and accuracy.

 (4) Exercising judgment.

 (5) Applying his aeronautical knowledge.

 (6) Showing that he is the master of the aircraft, with the successful outcome of a procedure or maneuver never seriously in doubt.

(b) If the applicant fails any of the required pilot operations in accordance with the applicable provisions of paragraph (a) of this section, the applicant fails the flight test. The applicant is not eligible for the certificate or rating sought until he passes any pilot operations he has failed.

(c) The examiner or the applicant may discontinue the test at any time when the failure of a required pilot operation makes the applicant ineligible for the certificate or rating sought. If the test is discontinued the applicant is entitled to credit for only those entire pilot operations that he has successfully performed.

61.45 FLIGHT TESTS: REQUIRED AIRCRAFT AND EQUIPMENT.

(a) *General.* An applicant for a certificate or rating under this part must furnish, for each flight test that he is required to take, an appropriate aircraft of United States registry that has a current standard or limited airworthiness certificate. However, the applicant may, at the discretion of the inspector or examiner conducting the test, furnish an aircraft of U.S. registry that has a current airworthiness certificate other than standard or limited, an aircraft of foreign registry that is properly certificated by the country or registry, or a military aircraft in an operational status if its use is allowed by an appropriate military authority.

(b) *Required equipment (other than controls).* Aircraft furnished for a flight test must have—

 (1) The equipment for each pilot operation required for the flight test;

 (2) No prescribed operating limitations that prohibit its use in any pilot operation required on the test;

 (3) Pilot seats with adequate visibility for each pilot to operate the aircraft safely, except as provided in paragraph (d) of this section; and

 (4) Cockpit and outside visibility adequate to evaluate the performance of the applicant, where an additional jump seat is provided for the examiner.

(c) *Required controls.* An aircraft (other than lighter-than-air) furnished under paragraph (a) of this section for any pilot flight test must have engine power controls and flight controls that are easily reached and operable in a normal manner by both pilots, unless after considering all the factors, the examiner determines that the flight test can be conducted safely without them. However, an aircraft having other controls such as nose-wheel steering, brakes, switches, fuel selectors, and engine air flow controls that are not easily reached and operable in a normal manner by both pilots may be used, if more than one pilot is required under its airworthiness certificate, or if the examiner determines that the flight can be conducted safely.

(d) *Simulated instrument flight equipment.* An applicant for any flight test involving flight maneuvers solely by reference to instruments must furnish equipment satisfactory to the examiner that excludes the visual reference of the applicant outside of the aircraft.

(e) *Aircraft with single controls.* At the discretion of the examiner, an aircraft furnished under paragraph (a) of this section for a flight test may, in the cases listed herein, have a single set of controls. In such case, the examiner determines the competence of the applicant by observation from the ground or from another aircraft.

(1) A flight test for addition of a class or type rating, not involving demonstration of instrument skills, to a private or commercial pilot certificate.

61.47 FLIGHT TESTS: STATUS OF FAA INSPECTORS AND OTHER AUTHORIZED FLIGHT EXAMINERS.

An FAA inspector or other authorized flight examiner conducts the flight test of an applicant for a pilot certificate or rating for the purpose of observing the applicant's ability to perform satisfactorily the procedures and maneuvers on the flight test. The inspector or other examiner is not pilot in command of the aircraft during the flight test unless he acts in that capacity for the flight, or portion of the flight, by prior arrangement with the applicant or other person who would otherwise act as pilot in command of the flight, or portion of the flight. Notwithstanding the type of aircraft used during a flight test, the applicant and the inspector or other examiner are not, with respect to each other (or other occupants authorized by the inspector or other examiner), subject to the requirements or limitations for the carriage of passengers specified in this chapter.

61.49 RETESTING AFTER FAILURE.

An applicant for a written or flight test who fails that test may not apply for retesting until after 30 days after the date he failed the test. However, in the case of his first failure he may apply for retesting before the 30 days have expired upon presenting a written statement from an authorized instructor certifying that he has given flight or ground instruction as appropriate to the applicant and finds him competent to pass the test.

61.51 PILOT LOGBOOKS.

(a) The aeronautical training and experience used to meet the requirements for a certificate or rating, or the recent flight experience requirements of this part must be shown by a reliable record. The logging of other flight time is not required.

(b) *Logbook entries.* Each pilot shall enter the following information for each flight or lesson logged:

(1) *General.*

(i) Date.

(ii) Total time of flight.

(iii) Place, or points of departure and arrival.

(iv) Type and identification of aircraft.

(2) *Type of pilot experience or training.*

(i) Pilot in command or solo.

(ii) Second in command.

(iii) Flight instruction received from an authorized flight instructor.

(iv) Instrument flight instruction from an authorized flight instructor.

(v) Pilot ground trainer instruction.

(vi) Participating crew (lighter-than-air).

(vii) Other pilot time.

(3) *Conditions of flight.*

(i) Day or night.

(ii) Actual instrument.

(iii) Simulated instrument conditions.

(c) *Logging of pilot time —*

(1) *Solo flight time.* A pilot may log as solo flight time only that flight time when he is the sole occupant of the aircraft. However, a student pilot may also log as solo flight time that time during which he acts as the pilot in command of an airship requiring more than one flight crewmember.

(2) *Pilot-in-command flight time.*

(i) A private or commercial pilot may log as pilot in command time only that flight time during which he is the sole manipulator of the controls of an aircraft for which he is rated, or when he is the sole occupant of the aircraft, or when he acts as pilot in command of an aircraft on which more than one pilot is required under the type certification of the aircraft, or the regulations under which the flight is conducted.

(ii) An airline transport pilot may log as pilot in command time all of the flight time during which he acts as pilot in command.

(iii) A certificated flight instructor may log as pilot in command time all flight time during which he acts as a flight instructor.

(3) *Second-in-command flight time.* A pilot may log as second in command time all flight time during which he acts as second in command of an aircraft on which more than one pilot is required under the type certification of the aircraft, or the regulations under which the flight is conducted.

(4) *Instrument flight time.* A pilot may log as instrument flight time only that time during which he operates the aircraft solely by reference to instruments, under actual or simulated instrument flight conditions. Each entry must include the place and type of each instrument approach completed, and the name of the safety pilot for each simulated instrument flight. An instrument flight instructor may log as instrument time that time during which he acts as instrument flight instructor in actual instrument weather conditions.

(5) *Instruction time.* All time logged as flight instruction, instrument flight instruction, pilot ground trainer instruction, or ground instruction time must be certified by the appropriately rated and certificated instructor from whom it was received.

(d) *Presentation of logbook.*

(1) A pilot must present his logbook (or other record required by this section) for inspection upon reasonable request by the Administrator, an authorized representative of the National Transportation Safety Board, or any State or local law enforcement officer.

(2) A student pilot must carry his logbook (or other record required by this section) with him on all solo cross-country flights, as evidence of the required instructor clearances and endorsements.

61.53 OPERATIONS DURING MEDICAL DEFICIENCY.

No person may act as pilot in command, or in any other capacity as a required pilot flight crewmember while he has a known medical deficiency, or increase of a known medical deficiency, that would make him unable to meet the requirements for his current medical certificate.

61.57 RECENT FLIGHT EXPERIENCE: PILOT IN COMMAND.

(a) *Flight review.* After November 1, 1974, no person may act as pilot in command of an aircraft unless, within the preceding 24 months, he has—

(1) Accomplished a flight review given to him, in an aircraft for which he is rated, by an appropriately certificated instructor or other person designated by the Administrator; and

(2) Had his logbook endorsed by the person who gave him the review certifying that he has satisfactorily accomplished the flight review.

However, a person who has, within the preceding 24 months, satisfactorily completed a pilot proficiency check conducted by the FAA, an approved pilot check airman or a U.S. armed force for a pilot certificate, rating or operating privilege, need not accomplish the flight review required by this section.

(b) *Meaning of flight review.* As used in this section, a flight review consists of—

(1) A review of the current general operating and flight rules of Part 91; and

(2) A review of those maneuvers and procedures which in the discretion of the person giving the review are necessary for the pilot to demonstrate that he can safely exercise the privileges of his pilot certificate.

(c) *General experience.* No person may act as pilot in command of an aircraft carrying passengers, nor of an aircraft certificated for more than one required pilot flight crewmember, unless within the preceding 90 days, he has made three takeoffs and three landings as the sole manipulator of the flight controls in an aircraft of the same category and class and, if a type rating is required, of the same type. If the aircraft is a tailwheel airplane, the landings must have been made to a full stop in a tailwheel airplane. For the purpose of meeting the requirements of the paragraph a person may act as pilot in command of a flight under day VFR or day IFR if no persons or property other than as necessary for his compliance thereunder, are carried. This paragraph does not apply to operations requiring an airline transport pilot certificate, or to operations conducted under Part 135.

(d) *Night experience.* No person may act as pilot in command of an aircraft carrying passengers during the period beginning 1 hour after sunset and ending 1 hour before sunrise (as published in the American Air Almanac) unless, within the preceding 90 days, he has made at least three takeoffs and three landings to a full stop during that period in the category and class of aircraft to be used. This paragraph does not apply to operations requiring an airline transport pilot certificate.

(e) *Instrument* —

(1) *Recent IFR experience.* No pilot may act as pilot in command under IFR, nor in weather conditions less than the minimums prescribed for VFR, unless he has, within the past 6 months—

(i) In the case of an aircraft other than a glider, logged at least 6 hours of instrument time under actual or simulated IFR conditions, at least 3 of which were in flight in the category of aircraft involved, including at least six instrument approaches, or passed an instrument competency check in the category of aircraft involved.

(2) *Instrument competency check.* A pilot who does not meet the recent instrument experience requirements of paragraph (e)(1) of this section during the prescribed time or 6 months thereafter may not serve as pilot in command under IFR, nor in weather conditions less than the minimums prescribed for VFR, until he passes an instrument competency check in the category of aircraft involved, given by an FAA inspector, a member of an armed force of the United States authorized to conduct flight tests, an FAA-approved check pilot, or a certificated instrument flight instructor. The Administrator may authorize the conduct of part or all of this check in a pilot ground trainer equipped for instruments or an aircraft simulator.

61.59 FALSIFICATION, REPRODUCTION, OR ALTERATION OF APPLICATIONS, CERTIFICATES, LOGBOOKS, REPORTS, OR RECORDS.

(a) No person may make or cause to be made—

(1) Any fraudulent or intentionally false statement on any application for a certificate, rating, or duplicate thereof, issued under this part;

(2) Any fraudulent or intentionally false entry in any logbook, record, or report that is required to be kept, made, or used, to show compliance with any requirement for the issuance, or exercise of the privileges, or any certificate or rating under this part;

(3) Any reproduction, for fraudulent purpose, of any certificate or rating under this part; or

(4) Any alteration of any certificate or rating under this part.

(b) The commission by any person of an act prohibited under paragraph (a) of this section is a basis for suspending or revoking any airman or ground instructor certificate or rating held by that person.

61.60 CHANGE OF ADDRESS.

The holder of a pilot or flight instructor certificate who has made a change in his permanent mailing address may not after 30 days from the date he moved, exercise the privileges of his certificate unless he has notified in writing the Department of Transportation, Federal Aviation Administration, Airman Certification Branch, Box 25082, Oklahoma City, OK 73125, of his new address.

SUBPART B—AIRCRAFT RATINGS AND SPECIAL CERTIFICATES

61.61 APPLICABILITY.

This subpart prescribes the requirements for the issuance of additional aircraft ratings after a pilot or instructor certificate is issued, and the requirements and limitations for special pilot certificates and ratings issued by the Administrator.

61.63 ADDITIONAL AIRCRAFT RATINGS (OTHER THAN AIRLINE TRANSPORT PILOT).

(a) *General.* To be eligible for an aircraft rating after his certificate is issued to him an applicant must meet the requirements of paragraphs (b) through (d) of this section, as appropriate to the rating sought.

(b) *Category rating.* An applicant for a category rating to be added on his pilot certificate must meet the requirements of this Part for the issue of the pilot certificate appropriate to the privileges for which the category rating is sought. However, the holder of a category rating for powered aircraft is not required to take a written test for the addition of a category rating on his pilot certificate.

(c) *Class rating.* An applicant for an aircraft class rating to be added on his pilot certificate must—
  (1) Present a logbook record certified by an authorized flight instructor showing that the applicant has received flight instruction in the class of aircraft for which a rating is sought and has been found competent in the pilot operations appropriate to the pilot certificate to which his category rating applies; and
  (2) Pass a flight test appropriate to his pilot certificate and applicable to the aircraft category and class rating sought

(d) *Type rating.* An applicant for a type rating to be added on his pilot certificate must meet the following requirements:
  (1) He must hold, or concurrently obtain, an instrument rating appropriate to the aircraft for which a type rating is sought.
  (2) He must pass a flight test showing competence in pilot operations appropriate to the pilot certificate he holds and to the type rating sought.
  (3) He must pass a flight test showing competence in pilot operations under instrument flight rules in an aircraft of the type for which the type rating is sought or, in the case of a single pilot station airplane, meet the requirements of paragraph (d)(3)(i) or (ii) of this section, whichever is applicable.
    (i) The applicant must have met the requirements of this subparagraph in a multiengine airplane for which the type rating is required.
    (ii) If he does not meet the requirements of paragraph (d)(3)(i) of this section and he seeks a type rating for a single-engine airplane, he must meet the requirements of this subparagraph in either a single or multiengine airplane, and have the recent instrument experience set forth in 61.57(e), when he applies for the flight test under paragraph (d)(2) of this section.
  (4) An applicant who does not meet the requirements of paragraphs (d)(1) and (3) of this section may obtain a type rating limited to "VFR only." Upon meeting these instrument requirements or the requirements of 61.73 (e)(2), the "VFR only" limitation may be removed for the particular type of aircraft in which competence is shown.
  (5) When an instrument rating is issued to the holder of one or more type ratings, the type ratings on the amended certificate bear the limitation described in paragraph (d)(4) of this section for each airplane type rating for which he has not shown his instrument competency under this paragraph.

61.65 INSTRUMENT RATING REQUIREMENT

(a) *General.* To be eligible for an instrument rating (airplane) or an instrument rating (helicopter), an applicant must—
  (1) Hold a current private or commercial pilot certificate with an aircraft rating appropriate to the instrument rating sought;
  (2) Be able to read, speak, and understand the English language; and
  (3) Comply with the applicable requirements of this section.

(b) *Ground instruction.* An applicant for the written test for an instrument rating must have received ground instruction, or have logged home study in at least the following areas of aeronautical knowledge appropriate to the rating sought.
  (1) The regulations of this chapter that apply to flight under IFR conditions, the Airman's Information Manual, and the IFR air traffic system and procedures;
  (2) Dead reckoning appropriate to IFR navigation, IFR navigation by radio aids using the VOR, ADF, and ILS systems, and the use of IFR charts and instrument approach plates;
  (3) The procurement and use of aviation weather reports and forecasts, and the elements of forecasting weather trends on the basis of that informa-

tion and personal observation of weather conditions; and

(4) The safe and efficient operation of airplanes or helicopters, as appropriate, under instrument weather conditions.

(c) *Flight instruction and skill—airplanes.* An applicant for the flight test for an instrument rating (airplane) must present a logbook record certified by an authorized flight instructor showing that he has received instrument flight instruction in an airplane in the following pilot operations, and has been found competent in each of them:

(1) Control and accurate maneuvering of an airplane solely by reference to instruments.

(2) IFR navigation by the use of the VOR and ADF systems, including compliance with air traffic control instructions and procedures.

(3) Instrument approaches to published minimums using the VOR, ADF, and ILS systems (instruction in the use of the ADF and ILS may be received in an instrument ground trainer and instruction in the use of the ILS glide slope may be received in an airborne ILS simulator).

(4) Cross-country flying in simulated or actual IFR conditions, on Federal airways or as routed by ATC, including one such trip of at least 250 nautical miles, including VOR, ADF, and ILS approaches at different airports.

(5) Simulated emergencies, including the recovery from unusual attitudes, equipment or instrument malfunctions, loss of communications, and engine-out emergencies if a multiengine airplane is used, and missed approach procedure.

(e) *Flight experience.* An applicant for an instrument rating must have at least the following flight time as a pilot:

(1) A total of 200 hours of pilot flight time, including 100 hours as pilot in command, of which 50 hours are cross-country in the category of aircraft for which an instrument rating is sought.

(2) 40 hours of simulated or actual instrument time, of which not more than 20 hours may be instrument instruction by an authorized instructor in an instrument ground trainer acceptable to the Administrator.

(3) 15 hours of instrument flight instruction by an authorized flight instructor, including at least 5 hours in an airplane or a helicopter, as appropriate.

(f) *Written test.* An applicant for an instrument rating must pass a written test appropriate to the instrument rating sought on the subjects in which ground instruction is required by paragraph (b) of this section.

(g) *Practical test.* An applicant for an instrument rating must pass a flight test in an airplane or a helicopter, as appropriate. The test must include instrument flight procedures selected by the inspector or examiner conducting the test to determine the applicant's ability to perform competently the IFR operations on which instruction is required by paragraph (c) or (d) of this section.

61.71 GRADUATES OF CERTIFICATED FLYING SCHOOLS: SPECIAL RULES.

(a) A graduate of a flying school that is certificated under Part 141 of this chapter is considered to meet the applicable aeronautical experience requirements of this part if he presents an appropriate graduation certificate within 60 days after the date he is graduated. However, if he applies for a flight test for an instrument rating he must hold a commercial pilot certificate, or hold a private pilot certificate and meet the requirements of 61.65(e)(1) and 61.123 (except paragraphs (d) and (e) thereof). In addition, if he applies for a flight instructor certificate he must hold a commercial pilot certificate.

(b) An applicant for a certificate or rating under this part is considered to meet the aeronautical knowledge and skill requirements, or both, applicable to that certificate or rating, if he applies within 90 days after graduation from an appropriate course given by a flying school that is certificated under Part 141 of this chapter and is authorized to test applicants on aeronautical knowledge or skill or both.

61.73 MILITARY PILOTS OR FORMER MILITARY PILOTS: SPECIAL RULES.

(a) *General.* A rated military pilot or former rated military pilot who applies for a private or commercial pilot certificate, or an aircraft or instrument rating, is entitled to that certificate with appropriate ratings or to the addition of a rating on the pilot certificate he holds, if he meets the applicable requirements of this section. This section does not apply to a military pilot or former military pilot who has been removed from flying status for lack of proficiency or because of disciplinary action involving aircraft operations.

(b) *Military pilots on active flying status within 12 months.* A rated military pilot or former rated military pilot who has been on active flying status within the 12 months before he applies must pass a written test on the parts of this chapter relating to pilot privileges and limitations, air traffic and general operating rules, and accident reporting rules. In addition, he must present documents showing that he meets the requirements of paragraph (d) of this section for at least one aircraft rating, and that he is, or was at any time since the beginning of the twelfth month before the month in which he applies—

(1) A rated military pilot on active flying status in an armed force of the United States; or

(2) A rated military pilot of an armed force of a foreign contracting State to the Convention on International Civil Aviation, assigned to pilot duties (other than flight training) with an armed force of the United States who holds, at the time he applies, a current civil pilot license issued by that foreign State authorizing at least the privileges of the pilot certificate he seeks.

(c) *Military pilots not on active flying status within previous 12 months.* A rated military pilot or former military pilot who has not been on active flying status within the 12 months before he applies must pass the appropriate written and flight tests prescribed in this part for the certificate or rating he seeks. In addition, he must show that he holds an FAA medical certificate appropriate to the pilot certificate he seeks and present documents showing that he was, before the beginning of the twelfth month before the month in which he applies, a rated military pilot as prescribed by either paragraph (b)(1) or (2) of this section.

(d) *Aircraft ratings: Other than airplane category and type.* An applicant for a category, class, or type rating (other than airplane category and type rating) to be added on the pilot certificate he holds, or for which he has applied, is issued that rating if he presents documentary evidence showing one of the following:

(1) That he has passed an official United States military checkout as pilot in command of aircraft of the category, class, or type for which he seeks a rating since the beginning of the twelfth month before the month in which he applies.

(2) That he has had at least 10 hours of flight time serving as pilot in command of aircraft of the category, class, or type for which he seeks a rating since the beginning of the twelfth month before the month in which he applies and previously has had an official United States military checkout as pilot in command of that aircraft.

(3) That he has met the requirements of paragraph (b)(1) or (2) of this section, has had an official United States military checkout in the category of aircraft for which he seeks a rating, and that he passes an FAA flight test appropriate to that category and the class or type rating he seeks. To be eligible for that flight test, he must have a written statement from an authorized flight instructor, made not more than 60 days before he applies for the flight test, certifying that he is competent to pass the test.

A type rating is issued only for aircraft types that the Administrator has certificated for civil operations. Any rating placed on an airline transport pilot certificate is limited to commercial pilot privileges.

PART 91—GENERAL OPERATING AND FLIGHT RULES

SUBPART A—GENERAL

91.1 APPLICABILITY

(a) Except as provided in paragraph (b) of this section, this part describes rules governing the operation of aircraft (other than moored balloons, kites, unmanned rockets, and unmanned free balloons) within the United States.

91.3 RESPONSIBILITY AND AUTHORITY OF THE PILOT IN COMMAND

(a) The pilot in command of an aircraft is directly responsible for, and is the final authority as to, the operation of that aircraft.

(b) In an emergency requiring immediate action, the pilot in command may deviate from any rule of this Subpart or of Subpart B to the extent required to meet that emergency.

(c) Each pilot in command who deviates from a rule under paragraph (b) of this section shall, upon the request of the Administrator, send a written report of that deviation to the Administrator.

91.5 PREFLIGHT ACTION

Each pilot in command shall, before beginning a flight, familiarize himself with all available information concerning that flight. This information must include:

(a) For a flight under IFR or a flight not in the vicinity of an airport, weather reports and forecasts, fuel requirements, alternatives available if the planned flight cannot be completed, and any known traffic delays of which he has been advised by ATC.

(b) For any flight, runway lengths at airports of intended use, and the following takeoff and landing distance information:
   (1) For civil aircraft for which an approved airplane or rotorcraft flight manual containing takeoff and landing distance data is required, and takeoff and landing distance data contained therein; and
   (2) For civil aircraft other than those specified in subparagraph (1) of this paragraph, other reliable information appropriate to the aircraft, relating to aircraft performance under expected values of airport elevation and runway slope, aircraft gross weight, and wind and temperature.

91.7 FLIGHT CREWMEMBERS AT STATIONS

(a) During takeoff and landing, and while en route, each required flight crewmember shall —
   (1) Be at his station unless his absence is necessary in the performance of his duties in connection with the operation of the aircraft or in connection with his physiological needs; and
   (2) Keep his seat belt fastened while at his station.

(b) After July 18, 1978, each required flight crewmember of a U.S. registered civil airplane shall, during takeoff and landing, keep the shoulder harness fastened while at his station. This paragraph does not apply if —
   (1) The seat at the crewmember's station is not equipped with a shoulder harness; or
   (2) The crewmember would be unable to perform his required duties with the shoulder harness fastened.

91.9 CARELESS OR RECKLESS OPERATION

No person may operate an aircraft in a careless or reckless manner so as to endanger the life or property of another.

91.10 CARELESS OR RECKLESS OPERATION OTHER THAN FOR THE PURPOSE OF AIR NAVIGATION

No person may operate an aircraft other than for the purpose of air navigation, on any part of the surface of an airport used by aircraft for air commerce (including areas used by those aircraft for receiving or discharging persons or cargo), in a careless or reckless manner so as to endanger the life or property of another.

91.11 LIQUOR AND DRUGS

(a) No person may act as a crewmember of a civil aircraft:
   (1) Within 8 hours after the consumption of any alcoholic beverage;
   (2) While under the influence of alcohol; or
   (3) While using any drug that affects his faculties in any way contrary to safety.

(b) Except in an emergency, no pilot of a civil aircraft may allow a person who is obviously under the influence of intoxicating liquors or drugs (except a medical patient under proper care) to be carried in that aircraft.

91.14 USE OF SAFETY BELTS

(a) Unless otherwise authorized by the Administrator —
 (1) No pilot may take off a U.S. registered civil aircraft (except a free balloon that incorporates a basket or gondola and an airship) unless the pilot in command of that aircraft ensures that each person on board is briefed on how to fasten and unfasten that person's safety belt.
 (2) No pilot may take off or land a U.S. registered civil aircraft (except free balloons that incorporate baskets or gondolas and airships) unless the pilot in command of that aircraft ensures that each person on board has been notified to fasten his safety belt.
 (3) During the takeoff and landing of U.S. registered civil aircraft (except free balloons that incorporate baskets or gondolas and airships), each person on board that aircraft must occupy a seat or berth with a safety belt properly secured about him. However, a person who has not reached his second birthday may be held by an adult who is occupying a seat or berth, and a person on board for the purpose of engaging in sport parachuting may use the floor of the aircraft as a seat.

(b) This section does not apply to operations conducted under Part 121, 123, or 127 of this chapter. Paragraph (a)(3) of this section does not apply to persons subject to 91.7.

91.19 PORTABLE ELECTRONIC DEVICES

(a) Except as provided in paragraph (b) of this section, no person may operate, nor may any operator or pilot in command of an aircraft allow the operation of, any portable electronic devide on any of the following U.S. registered civil aircraft:
 (1) Aircraft operated by an air carrier or commercial operator; or
 (2) Any other aircraft while it is operated under IFR.

(b) Paragraph (a) of this section does not apply to:
 (1) Portable voice recorders;
 (2) Hearing aids;
 (3) Heart pacemakers;
 (4) Electric shavers; or
 (5) Any other portable electronic device that the operator of the aircraft has determined will not cause interference with the navigation or communication system of the aircraft on which it is to be used.

(c) In the case of an aircraft operated by an air carrier or commercial operator, the determination required by paragraph (b) (5) of this section shall be made by the air carrier or commercial operator of the aircraft on which the particular device is to be used. In the case of other aircraft, the determination may be made by the pilot in command or other operator of the aircraft.

91.21 FLIGHT INSTRUCTION: SIMULATED INSTRUMENT FLIGHT AND CERTAIN FLIGHT TESTS

(a) No person may operate a civil aircraft (except a manned free balloon) that is being used for flight instruction unless that aircraft has fully functioning, dual controls. However, instrument flight instruction may be given in a single-engine airplane equipped with a single, functioning throwover control wheel, in place of fixed, dual controls of the elevator and ailerons, when;
 (1) The instructor has determined that the flight can be conducted safely; and
 (2) The person manipulating the controls has at least a private pilot certificate with appropriate category and class ratings.

(b) No person may operate a civil aircraft in simulated instrument flight unless —
 (1) An appropriately rated pilot occupies the other control seat as a safety pilot;
 (2) The safety pilot has adequate vision forward and to each side of the aircraft, or a competent observer in the aircraft adequately supplements the vision of the safety pilot; and
 (3) Except in the case of lighter-than-air aircraft, that aircraft is equipped with fully functioning dual controls. However, simulated instrument flight may

be conducted in a single-engine airplane, equipped with a single, functioning, throwover control wheel, in place of fixed, dual controls of the elevator and ailerons, when —

(i) The safety pilot has determined that the flight can be conducted safely; and

(ii) The person manipulating the control has at least a private pilot with appropriate category and class ratings.

(c) No person may operate a civil aircraft that is being used for a flight test for an airline transport pilot certificate or a class or type rating on that certificate, or for a Federal Aviation Regulation Part 121 proficiency flight test, unless the pilot seated at the controls, other than the pilot being checked, is fully qualified to act as pilot in command of the aircraft.

91.23 FUEL REQUIREMENTS FOR FLIGHT IN IFR CONDITIONS

(a) Except as provided in paragraph (b) of this section, no person may operate a civil aircraft in IFR conditions unless it carries enough fuel (considering weather reports and forecasts, and weather conditions) to —

(1) Complete the flight to the first airport of intended landing;

(2) Fly from that airport to the alternate airport; and

(3) Fly after that for 45 minutes at normal cruising speed.

(b) Paragraph (a)(2) of this section does not apply if —

(1) Part 97 of this subchapter prescribes a standard instrument approach procedure for the first airport of intended landing; and

(2) For at least 1 hour before and 1 hour after the estimated time of arrival at the airport, the weather reports or forecasts or any combination of them, indicate —

(i) The ceiling will be at least 2,000 feet above the airport elevation; and

(ii) Visibility will be at least 3 miles.

91.24 ATC TRANSPONDER AND ALTITUDE REPORTING EQUIPMENT

(a) *All airspace: U.S. registered civil aircraft.* — For operations not conducted under Parts 121, 123, 127, or 135 of this chapter, ATC transponder equipment installed after January 1, 1974, in U.S. registered civil aircraft not previously equipped with an ATC transponder, and all ATC transponder equipment used in U.S. registered civil aircraft after July 1, 1975, must meet the performance and environmental requirements of any class of TSO-C74b or any class of TSO-C74c as appropriate, except that the Administrator may approve the use of TSO-C74 or TSO-C74a equipment after July 1, 1975, if the applicant submits data showing that such equipment meets the minimum performance standards of the appropriate class of TSO-C74c and environmental conditions of the TSO under which it was manufactured.

(b) *Controlled airspace: all aircraft.* — Except for persons operating helicopters in terminal control areas at or below 1,000 feet AGL under the terms of a letter of agreement, and except for persons operating gliders above 12,500 feet m.s.l. but below the floor of the positive control area, no person may operate an aircraft in the controlled airspace prescribed in paragraphs (b)(1) through (b)(4) of this paragraph unless that aircraft is equipped with an operable coded radar beacon transponder having a mode 3/A 4096 code capability, replying to mode 3/A interrogation with the code specified by ATC, and is equipped with automatic pressure altitude reporting equipment having a mode C capability that automatically replies to mode C interrogations by transmitting pressure altitude information in 100 foot increments. This requirement applies —

(1) In Group I Terminal Control Areas governed by 91.90(a);

(2) In Group II Terminal Control Areas governed by 91.90(b), except as provided therein;

(3) In Group III Terminal Control Areas governed by 91.90(c), except as provided therein; and

(4) In all controlled airspace of the 48 contiguous States and the District of Columbia, above 12,500 feet MSL, excluding the airspace at and below 2,500 feet AGL.

(c) *ATC authorized deviations.* ATC may authorize deviations from paragraph (b) of this section —

(1) Immediately, to allow an aircraft with an inoperative transponder to continue

to the airport of ultimate destination, incuding any intermediate stops, or to proceed to a place where suitable repairs can be made, or both;

(2) Immediately, for operations of aircraft with an operating transponder but without operating automatic pressure altitude reporting equipment having a Mode C capability; and

(3) On a continuing basis, or for individual flights, for operations of aircraft without a transponder, in which case the request for a deviation must be submitted to the ATC facility having jurisdiction over the airspace concerned at least four hours before the proposed operation.

91.25 VOR EQUIPMENT CHECK FOR IFR OPERATIONS

(a) No person may operate a civil aircraft under IFR using the VOR system of radio navigation unless the VOR equipment of that aircraft —

(1) Is maintained, checked, and inspected under an approved procedure; or

(2) Has been operationally checked within the preceding 30 days and was found to be within the limits of the permissible indicated bearing error set forth in paragraph (b) or (c) of this section.

(b) Except as provided in paragraph (c) of this section, each person conducting a VOR check under paragraph (a)(2) of this section shall —

(1) Use, at the airport of intended departure, an FAA operated or approved test signal or a test signal radiated by a certificated and appropriately rated radio repair station or, outside the United States, a test signal operated or approved by appropriate authority, to check the VOR equipment (the maximum permissible indicated bearing error is plus or minus 4 degrees);

(2) If a test signal is not available at the airport of intended departure, use a point on an airport surface designated as a VOR system checkpoint by the Administrator or, outside the United States, by appropriate authority (the maximum permissible bearing error is plus or minus 4 degrees);

(3) If neither a test signal nor a designated checkpoint on the surface is available, use an airborne checkpoint designated by the Administrator or, outside the United States, by appropriate authority (the maximum permissible bearing error is plus or minus 6 degrees); or

(4) If no check signal or point is available, while in flight —

(i) Select a VOR radial that lies along the centerline of an established VOR airway;

(ii) Select a prominent ground point along the selected radial preferably more than 20 miles from the VOR ground facility and maneuver the aircraft directly over the point at a reasonably low altitude; and

(iii) Note the VOR bearing indicated by the receiver when over the ground point (the maximum permissible variation between the published radial and the indicated bearing is 6 degrees).

(c) If dual system VOR (units independent of each other except for the antenna) is installed in the aircraft, the person checking the equipment may check one system against the other in place of the check procedures specified in paragraph (b) of this section. He shall tune both systems to the same VOR ground facility and note the indicated bearings to that station. The maximum permissible variation between the two indicated bearings is 4 degrees.

(d) Each person making the VOR operational check as specified in paragraph (b) or (c) of this section shall enter the date, place, bearing error, and sign the aircraft log or other record.

91.29 CIVIL AIRCRAFT AIRWORTHINESS

(a) No person may operate a civil aircraft unless it is in an airworthy condition.

(b) The pilot in command of a civil aircraft is responsible for determining whether that aircraft is in condition for safe flight. He shall discontinue the flight when unairworthy mechanical or structural conditions occur.

91.31 CIVIL AIRCRAFT OPERATING LIMITATIONS AND MARKING REQUIREMENTS

(a) Except as provided in paragraph (d) of this section, no person may operate a civil aircraft without compliance with the operating limitations for that aircraft prescribed by the certificating authority of the country of registry.

(b) No person may operate a U.S. registered civil aircraft —
   (1) For which an Airplane or Rotorcraft Flight Manual is required by 21.5 unless there is available in the aircraft a current approved Airplane or Rotorcraft Flight Manual or the manual provided for in 121.141(b); and
   (2) For which an Airplane or Rotorcraft Flight Manual is not required by 21.5, unless there is available in the aircraft a current approved Airplane or Rotorcraft Flight Manual, approved manual material, markings, and placards, or any combination thereof.

(c) No person may operate a U.S. registered civil aircraft unless that aircraft is identified in accordance with Part 45 of this chapter.

(e) The Airplane or Rotorcraft Flight Manual, or manual material, markings and placards required by paragraph (b) of this section must contain each operating limitation prescribed for that aircraft by the Administrator, including the following:
   (1) Powerplant (e.g., r.p.m., manifold pressure, gas temperature, etc.).
   (2) Airspeeds (e.g., normal operating speed, flaps extended speed, etc.).
   (3) Aircraft weight, center of gravity, and weight distribution, including the composition of the useful load in those combinations and ranges intended to ensure that the weight and center of gravity position will remain within approved limits (e.g., combinations and ranges of crew, oil, fuel, and baggage).
   (4) Minimum flight crew.
   (5) Kinds of operation.
   (6) Maximum operating altitude.
   (7) Maneuvering flight load factors.
   (8) Rotor speed (for rotorcraft).
   (9) Limiting height-speed envelope (for rotorcraft).

91.32 SUPPLEMENTAL OXYGEN

(a) *General.* No person may operate a civil aircraft of U.S. registry:
   (1) At cabin pressure altitudes above 12,500 feet (MSL) up to and including 14,000 feet (MSL), unless the required minimum flight crew is provided with and uses supplemental oxygen for that part of the flight at those altitudes that is of more than 30 minutes duration;
   (2) At cabin pressure altitudes above 14,000 feet (MSL), unless the required minimum flight crew is provided with and uses supplemental oxygen during the entire flight time at those altitudes; and
   (3) At cabin pressure altitudes above 15,000 feet (MSL), unless each occupant of the aircraft is provided with supplemental oxygen.

91.33 POWERED CIVIL AIRCRAFT WITH STANDARD CATEGORY U.S. AIRWORTHINESS CERTIFICATES; INSTRUMENT AND EQUIPMENT REQUIREMENTS

(a) *General.* Except as provided in paragraphs (c)(3) and (e) of this section, no person may operate a powered civil aircraft with a standard category U.S. airworthiness certificate in any operation described in paragraphs (b) through (f) of this section unless that aircraft contains the instruments and equipment specified in those paragraphs (of FAA-approved equivalents) for that type of operation, and those instruments and items of equipment are in operable condition.

(a) Visual flight rules (day). For VFR flight during the day, the following instruments and equipment are required:
   (1) Airspeed indicator.
   (2) Altimeter.
   (3) Magnetic direction indicator.
   (4) Tachometer for each engine.
   (5) Oil pressure gauge for each engine using pressure system.
   (6) Temperature gauge for each liquid-cooled engine.
   (7) Oil temperature gauge for each air-cooled engine.
   (8) Manifold pressure gauge for each altitude engine.
   (9) Fuel gauge indicating the quantity of fuel in each tank.
   (10) Landing gear position indicator, if the aircraft has a retractable landing gear.
   (11) If the aircraft is operated for hire over water and beyond power-off glid ing distance from shore, approved flotation gear readily available to each occupant, and at least one pyrotechnic signaling device.

(12) Except as to airships, an approved safety belt for all occupants who have reached their second birthday. After December 4, 1980, each safety belt must be equipped with an approved metal to metal latching device. The rated strength of each safety belt shall not be less than that corresponding with the ultimate load factors specified in the current applicable aircraft airworthiness requirements considering the dimensional characteristics of the safety belt installation for the specific seat or berth arrangement. The webbing of each safety belt shall be replaced as required by the Administrator.

(13) For small civil airplanes manufactured after July 18, 1978, an approved shoulder harness for each front seat. The shoulder harness must be designed to protect the occupant from serious head injury when the occupant experiences the ultimate inertia forces specified in 23.561(b)(2) of this chapter. Each shoulder harness installed at a flight crewmember station must permit the crewmember, when seated and with his safety belt and shoulder harness fastened, to perform all functions necessary for flight operations. For purposes of this paragraph —

(i) The date of manufacture of an airplane is the date the inspection acceptance records reflect that the airplane is complete and meets the FAA Approved Type Design Data; and

(ii) A front seat is a seat located at a flight crewmember station or any seat located alongside such a seat.

(c) *Visual flight rules (night).* For VFR flight at night the following instruments and equipment are required:

(1) Instruments and equipment specified in paragraph (b) of this section.

(2) Approved position lights.

(3) An approved aviation red or aviation white anti-collision light system on all U.S. registered civil aircraft. Anti-collision light systems initially installed after August 11, 1971, on aircraft for which a type certificate was issued or applied for before August 11, 1971, must at least meet the anti-collision light standards of Parts 23, 25, 27, or 29, as applicable, that were in effect on August 10, 1971, except that the color may be either aviation red or aviation white. In the event of failure of any light of the anti-collision light system, operations with the aircraft may be continued to a stop where repairs or replacement can be made.

(4) If the aircraft is operated for hire, one electric landing light.

(5) An adequate source of electrical energy for all installed electric and radio equipment.

(6) One spare set of fuses, or three spare fuses of each kind required.

(d) *Instrument flight rules.* For IFR flight the following instruments and equipment are required.

(1) Instruments and equipment specified in paragraph (b) of this section and for night flight, instruments and equipment specified in paragraph (c) of this section.

(2) Two-way radio communications system and navigational equipment appropriate to the ground facilities to be used.

(3) Gyroscopic rate-of-turn indicator, except on the following aircraft:

(i) Large airplanes with a third attitude instrument system usable through flight attitudes of 360 degrees of pitch and roll and installed in accordance with 121.305(j) of this chapter; and

(4) Slip-skid indicator.

(5) Sensitive altimeter adjustable for barometric pressure.

(6) A clock displaying hours, minutes, and seconds with a sweep-second pointer or digital presentation.

(7) Generator of adequate capacity.

(8) Gyroscopic bank and pitch indicator (artificial horizon).

(9) Gyroscopic direction indicator (directional gyro or equivalent.)

(e) *Flight at and above 24,000 feet MSL.* If VOR Navigational equipment is required under paragraph (d)(2) of this section, no person may operate a U.S. registered civil aircraft within the 50 states, and the District of Columbia, at or above 24,000 feet MSL unless that aircraft is equipped with approved distance measuring equipment (DME). When DME required by this paragraph fails at and above 24,000 feet MSL, the pilot in command of the aircraft shall notify ATC immediately, and may then continue operations at and above 24,000 feet MSL to the next airport of intended landing at which repairs or replacement of the equipment can be made.

SUBPART B — FLIGHT RULES: GENERAL

91.70 AIRCRAFT SPEED

(a) Unless otherwise authorized by the Administrator, no person may operate an aircraft below 10,000 feet MSL at an indicated airspeed of more than 250 knots (288 m.p.h.).

(b) Unless otherwise authorized or required by ATC, no person may operate an aircraft within an airport traffic area at an indicated airspeed of more than —

(1) In the case of a reciprocating engine aircraft, 156 knots (180 m.p.h.); or

(2) In the case of a turbine-powered aircraft, 200 knots (230 m.p.h.)

Paragraph (b) does not apply to any operations within a terminal control area. Such operations shall comply with paragraph (a) of this section.

(c) No person may operate an aircraft in the airspace underlying a terminal control area, or in a VFR corridor designated through a terminal control area, at an indicated airspeed of more than 200 knots (230 m.p.h.).

However, if the minimum safe airspeed for any particular operation is greater than the maximum speed prescribed in this section, the aircraft may be operated at that minimum speed.

91.73 AIRCRAFT LIGHTS

No person may, during the period from sunset to sunrise (or, in Alaska, during the period a prominent unlighted object cannot be seen from a distance of three statute miles or the sun is more than six degrees below the horizon) —

(a) Operate an aircraft unless it has lighted position lights;

(b) Park or move an aircraft in, or in dangerous proximity to, a night flight operations area of an airport unless the aircraft —

(1) Is clearly illuminated;

(2) Has lighted position lights; or

(3) Is in an area which is marked by obstruction lights.

(c) Anchor an aircraft unless the aircraft —

(1) Has lighted anchor lights; or

(2) Is in an area where anchor lights are not required on vessels; or

(d) Operate an aircraft, required by 91.33(c)(3) to be equipped with an anticollision light system, unless it has approved and lighted aviation red or aviation white anticollision lights. However, the anticollision lights need not be lighted when the pilot in command determines that, because of operating conditions, it would be in the interest of safety to turn the lights off.

91.75 COMPLIANCE WITH ATC CLEARANCES AND INSTRUCTIONS

(a) When an ATC clearance has been obtained, no pilot in command may deviate from that clearance, except in an emergency, unless he obtains an amended clearance. However, except in positive controlled airspace, this paragraph does not prohibit him from cancelling an IFR flight plan if he is operating in VFR weather conditions.

(b) Except in an emergency, no person may, in an area in which air traffic control is exercised, operate an aircraft contrary to an ATC instruction.

(c) Each pilot in command who deviates, in an emergency, from an ATC clearance or instruction shall notify ATC of that deviation as soon as possible.

(d) Each pilot in command who (though not deviating from a rule of this subpart) is given priority by ATC in an emergency, shall, if requested by ATC, submit a detailed report of that emergency within 48 hours to the chief of that ATC facility.

91.77 ATC LIGHT SIGNALS

ATC light signals have the meaning shown in the following table:

Color and type of signal	Meaning with respect to aircraft on the surface	Meaning with respect to aircraft in flight.
Steady green	Cleared for takeoff.	Cleared to land.
Flashing green	Cleared to taxi	Return for landing (to be followed by steady green at proper time).
Steady red	Stop	Give way to other aircraft and continue circling.
Flashing red	Taxi clear of runway in use.	Airport unsafe — do not land.
Flashing white	Return to starting point on airport.	Not applicable.
Alternating red and green	Exercise extreme caution.	Exercise extreme caution.

91.79 MINIMUM SAFE ALTITUDES: GENERAL

Except when necessary for takeoff or landing, no person may operate an aircraft below the following altitudes:

(a) *Anywhere.* An altitude allowing, if a power unit fails, an emergency landing without undue hazard to persons or property on the surface.

(b) *Over congested areas.* Over any congested area of a city, town, or settlement, or over any open air assembly of persons, an altitude of 1,000 feet above the highest obstacle within a horizontal radius of 2,000 feet of the aircraft.

(c) *Over other than congested areas.* An altitude of 500 feet above the surface, except over open water or sparsely populated areas. In that case, the aircraft may not be operated closer than 500 feet to any person, vessel, vehicle, or structure.

91.81 ALTIMETER SETTINGS

(a) Each person operating an aircraft shall maintain the cruising altitude or flight level of that aircraft, as the case may be, by reference to an altimeter that is set, when operating:

(1) Below 18,000 feet MSL, to:

(i) The current reported altimeter setting of a station along the route and within 100 nautical miles of the aircraft;

(ii) If there is no station within the area prescribed in subdivision (i) of this subparagraph, the current reported altimeter setting of an appropriate available station; or

(iii) In the case of an aircraft not equipped with a radio, the elevation of the departure airport or an appropriate altimeter setting available before departure; or

(2) at or above 18,000 feet MSL, to 29.92" Hg.

(b) The lowest usable flight level is determined by the atmospheric pressure in the area of operation, as shown in the following table:

Current altimeter setting	Lowest usable flight level
29.92 or higher	180
29.91 through 29.42	185
29.41 through 28.92	190
28.91 through 28.42	195
28.41 through 27.92	200
27.91 through 27.42	205
27.41 through 26.92	210

(c) To convert minimum altitude prescribed under 91.79 and 91.119 to the minimum flight level, the pilot shall take the flight-level equivalent of the minimum altitude in feet and add the appropriate number of feet specified below, according to the current reported altimeter setting:

Current altimeter setting	Adjustment factor
29.92 or higher	None
29.91 through 29.42	500 feet
29.41 through 28.92	1,000 feet
28.91 through 28.42	1,500 feet
28.41 through 27.92	2,000 feet
27.91 through 27.42	2,500 feet
27.41 through 26.92	3,000 feet

91.83 FLIGHT PLAN: INFORMATION REQUIRED

(a) Unless otherwise authorized by ATC each person filing an IFR or VFR flight plan shall include in it the following information:

(1) The aircraft identification number and, if necessary, its radio call sign.

(2) The type of the aircraft or, in the case of a formation flight, the type of each aircraft and the number of aircraft, in the formation.

(3) The full name and address of the pilot in command or, in the case of a formation flight, the formation commander.

(4) The point and proposed time of departure.

(5) The proposed route, cruising altitude (or flight level), and true airspeed at that altitude.

(6) The point of first intended landing and the estimated elapsed time until over that point.

(7) The radio frequencies to be used.

(8) The amount of fuel on board (in hours).

(9) In the case of an IFR flight plan, an alternate airport, except as provided in paragraph (b) of this section.

(10) The number of persons in the aircraft, except where that information is otherwise readily available to the FAA.

(11) Any other information the pilot in command or ATC believes is necessary for ATC purposes.

When a flight plan has been filed, the pilot in command, upon canceling or completing the flight under the flight plan, shall notify the nearest FAA Flight Service Station or ATC facility.

(b) Paragraph (a)(9) of this section does not apply if Part 97 of this subchapter prescribes a standard instrument approach procedure for the first airport of intended landing and, for at least one hour before and one hour after the estimated time of arrival, the weather reports or forecasts or any combination of them, indicate —
   (1) The ceiling will be at least 2,000 feet above the airport elevation; and
   (2) Visibility will be at least 3 miles.

(c) *IFR alternate airport weather minimums.* Unless otherwise authorized by the Administrator, no person may include an alternate airport in an IFR flight plan unless current weather forecasts indicate that, at the estimated time of arrival at the alternate airport, the ceiling and visibility at that airport will be at or above the following alternate airport weather minimums:
   (1) If an instrument approach procedure has been published in Part 97 of this chapter for that airport, the alternate airport minimums specified in that procedure, or, if none are so specified, the following minimums:
      (i) Precision approach procedure: Ceiling 600 feet and visibility 2 statute miles.
      (ii) Nonprecision approach procedure: Ceiling 800 feet and visibility 2 statute miles.
   (2) If no instrument approach procedure has been published in Part 97 of this chapter for that airport, the ceiling and visibility minimums are those allowing descent from the MEA, approach, and landing, under basic VFR.

91.85 OPERATING ON OR IN THE VICINITY OF AN AIRPORT: GENERAL RULES

(a) Unless otherwise required by Part 93 of this chapter, each person operating an aircraft on or in the vicinity of an airport shall comply with the requirements of this section and of 91.87 and 91.89.

(b) Unless otherwise authorized or required by ATC, no person may operate an aircraft within an airport traffic area except for the purpose of landing at, or taking off from, an airport within that area. ATC authorizations may be given as individual approval of specific operations or may be contained in written agreements between airport users and the tower concerned.

91.87 OPERATION AT AIRPORTS WITH OPERATING CONTROL TOWERS

(a) *General.* Unless otherwise authorized or required by ATC, each person operating an aircraft to, from, or on an airport with an operating control tower shall comply with the applicable provisions of this section.

(b) *Communications with control towers operated by the United States.* No person may, within an airport traffic area, operate an aircraft to, from, or on an airport having a control tower operated by the United States unless two-way radio communications are maintained between that aircraft and the control tower. However, if the aircraft radio fails in flight, he may operate that aircraft and land if weather conditions are at or above basic VFR weather minimums, he maintains visual contact with the tower, and he receives a clearance to land. If the aircraft radio fails while in flight under IFR, he must comply with 91.127.

(c) *Communications with other control towers.* No person may, within an airport traffic area, operate an aircraft to, from, or on an airport having a control tower that is operated by any person other than the United States unless:
   (1) If that aircraft's radio equipment so allows, two-way radio communications are maintained between the aircraft and the tower; or
   (2) If that aircraft's radio equipment allows only reception from the tower, the pilot has the tower's frequency monitored.

(d) *Minimum Altitudes.* When operating to an airport with an operating control tower. each pilot of—
   (1) A turbine-powered airplane or a large airplane shall, unless otherwise required by the applicable distance from cloud criteria, enter the airport traffic area at an altitude of at least 1,500 feet above the surface of the airport and maintain at least 1,500 feet within the airport traffic area, including the traffic pattern, until further descent is required for a safe landing;

(2) A turbine-powered airplane or a large airplane approching to land on a runway being served by an ILS, shall, if the airplane is ILS equipped, fly that airplane at an altitude at or above the glide slope between the outer marker (or the point of interception with the glide slope, if compliance with the applicable distance from clouds criteria requires interception closer in) and the middle marker; and

(3) An airplane approching to land on a runway served by a visual approach slope indicator, shall maintain an altitude at or above the glide slope until a lower altitude is necessary for a safe landing.

However, subparagraphs (2) and (3) of this paragraph do not prohibit normal bracketing maneuvers above or below the glide slope that are conducted for the purpose of remaining on the glide slope.

(e) *Approaches.* When approaching to land at an airport with an operating control tower, each pilot of:

(1) An airplane, shall circle the airport to the left; and

(2) A helicopter, shall avoid the flow of fixed-wing aircraft.

(f) *Departures.* No person may operate an aircraft taking off from an airport with an operating control tower except in compliance with the following:

(1) Each pilot shall comply with any departure procedures established for that airport by the FAA.

(2) Unless otherwise required by the departure procedure or the applicable distance from clouds criteria, each pilot of a turbine-powered airplane and each pilot of a large airplane shall climb to an altitude of 1,500 feet above the surface as rapidly as practicable.

(g) *Noise abatement.* — Omitted

(h) *Clearances required.* No person may, at any airport with an operating control tower, operate an aircraft on a runway or taxiway, or takeoff or land an aircraft, unless an appropriate clearance is received from ATC. A clearance to "taxi to" the takeoff runway assigned to the aircraft is not a clearance to cross that assigned takeoff runway, or to taxi on that runway at any point, but is a clearance to cross other runways that intersect the taxi route to that assigned takeoff runway. A clearance to "taxi to" any point other than an assigned takeoff runway is a clearance to cross all runways that intersect the taxi route to that point.

91.89 OPERATION AT AIRPORTS WITHOUT CONTROL TOWERS

Each person operating an aircraft to or from an airport without an operating control tower shall:

(a) In the case of an airplane approaching to land, make all turns of that airplane to the left unless the airport displays approved light signals or visual markings indicating that turns should be made to the right, in which case the pilot shall make all turns to the right.

(c) In the case of an aircraft departing the airport, comply with any FAA traffic pattern for that airport.

91.90 FLIGHT IN TERMINAL CONTROL AREAS: OPERATING RULES AND PILOT AND EQUIPMENT REQUIREMENTS

(a) *Group I terminal control areas:*

(1) *Operating rules.* No person may operate an aircraft within a Group I terminal control area designated in Part 71 of this chapter except in compliance with the following rules:

(i) No person may operate an aircraft within a Group I terminal control area unless he has received an appropriate authorization from ATC prior to the operation of that aircraft in that area.

(ii) Unless otherwise authorized by ATC, each person operating a large turbine engine powered airplane to or from a primary airport shall operate at or above the designated floors while within the lateral limits of the terminal control area.

(2) *Pilot requirements.* The pilot in command of a civil aircraft may not land or take off that aircraft from an airport within a Group I terminal control area unless he holds at least a private pilot certificate.

(3) *Equipment requirements.* Unless otherwise authorized by ATC in the case of in-flight VOR, TACAN, or two-way authorized by ATC in the case of a tran-

sponder failure occurring at any time, no person may operate an aircraft within a Group I terminal control area unless that aircraft is equipped with —

(i) An operable VOR or TACAN receiver (except in the case of helicopters);

(ii) An operable two-way radio capable of communicating with ATC on appropriate frequencies for that terminal control area; and

(iii) The applicable equipment specified in 91.24.

(b) *Group II terminal control areas:*

(1) *Operating rules.* No person may operate an aircraft within a Group II terminal control area designated in Part 71 of this chapter except in compliance with the following rules:

(i) No person may operate an aircraft within a Group II terminal control area unless he has received in appropriate authorization from ATC prior to operation of that aircraft in that area, and unless two-way radio communications are maintained, within that area, between that aircraft and the ATC facility.

(ii) Unless otherwise authorized by ATC, each person operating a large turbine engine powered airplane to or from a primary airport shall operate at or above the designated floors while within the lateral limits of the terminal control area.

(2) *Equipment requirements.* Unless otherwise authorized by ATC in the case of in-flight VOR, TACAN, or two-way radio failure; or unless otherwise authorized by ATC in the case of a transponder failure occurring at any time, to person may operate an aircraft within a Group II terminal control area unless that aircraft is equipped with —

(i) An operable VOR or TACAN receiver (except in the case of helicopters);

(ii) An operable two-way radio capable of communicating with ATC on appropriate frequencies for that terminal control area; and

(iii) The applicable equipment specified in 91.24, except that automatic pressure altitude reporting equipment is not required for any operation within the terminal control area, and a transponder is not required for IFR flights operating to or from an airport outside of but in close proximity to the terminal control area, when the commonly used transition, approach, or departure procedures to such airport require flight within the terminal control area.

(c) *Group III Terminal Control Areas.* No person may operate an aircraft within a Group III terminal control area designated in part 71 unless the applicable provisions of 91.24(b) are complied with, except that such compliance is not required if two-way radio communications are maintained.

91.97 POSITIVE CONTROL AREAS AND ROUTE SEGMENTS

(a) Except as provided in paragraph (b) of this section, no person may operate an aircraft within a positive control area, or positive control route segment designated in Part 71 of this chapter, unless that aircraft is —

(1) Operated under IFR at a specific flight level assigned by ATC;

(2) Equipped with instruments and equipment required for IFR operations;

(3) Flown by a pilot rated for instrument flight; and

(4) Equipped, when in a positive control area; with

(i) The applicable equipment specified in 91.24; and

(ii) A radio providing direct pilot/controller communication on the frequency specified by ATC for the area concerned.

(b) ATC may authorize deviations from the requirements of paragraph (a) of this section. In the case of an inoperative transponder, ATC may immediately approve an operation within a positive control area allowing flight to continue, if desired, to the airport of ultimate destination, including any intermediate stops, or to proceed to a place where suitable repairs can be made, or both. A request for authorization to deviate from a requirement of paragraph (a) of this section, other than for operation with an inoperative transponder as outlined above, must be submitted at least 4 days before the proposed operation, in writing, to the ATC center having jurisdiction over the positive control area concerned. ATC may authorize a deviation on a continuing basis or for an individual flight, as appropriate.

VISUAL FLIGHT RULES

91.105 BASIC VFR WEATHER MINIMUMS

(a) Except as provided in 91.107, no person may operate an aircraft under VFR when the flight visibility is less, or at a distance from clouds that is less, than that prescribed for the corresponding altitude in the following tables:

Altitude	Flight Visibility	Distance from clouds
1,200 feet or less above the surface (regardless of MSL altitude) —		
Within controlled airspace	3 statute miles	500 feet below 1,000 feet above 2,000 feet horizontal
Outside controlled airspace	1 statute mile except as provided in 91.105 (b);	Clear of clouds.
More than 1,200 feet above the surface but less than 10,000 feet MSL		
Within controlled airspace	3 statute miles	500 feet below. 1,000 feet above. 2,000 feet horizontal
Outside controlled airspace	1 statute mile	500 feet below. 1,000 feet above 2,000 feet horizontal.
More than 1,200 feet above the surface and at or above 10,000 feet MSL	5 statute miles	1,000 feet below. 1,000 feet above. 1 mile horizontal.

(b) When the visiblity is less than one mile, a helicopter may be operated outside controlled airspace at 1,200 feet or less above the surface if operated at a speed that allows the pilot adequate opportunity to see any air traffic or other obstructions in time to avoid a collision.

(c) Except as provided in 91.107, no person may operate an aircraft, under VFR, within a control zone beneath the ceiling when the ceiling is less than 1,000 feet.

(d) Except as provided in 91.107, no person may take off or land an aircraft, or enter the traffic pattern of an airport, under VFR, within a control zone:

(1) Unless ground visibility at that airport is at least 3 statute miles; and

(2) If ground visibility is not reported at that airport, unless flight visibility during landing or takeoff, or while operating in the traffic pattern, is at least 3 statute miles.

(e) For the purposes of this section, an aircraft operating at the base altitude of a transition area or control area is considered to be within the airspace directly below that area.

91.107 SPECIAL VFR WEATHER MINIMUMS

(a) Except as provided in 93.113 of this chapter, when a person has received an appropriate ATC clearance, the special weather minimums of this section instead of those contained in 91.105 apply to the operation of an aircraft by that person in a control zone under VFR.

(b) No person may operate an aircraft in a control zone under VFR except clear of clouds.

(c) No person may operate an aircraft (other than a helicopter) in a control zone under VFR unless flight visibility is at least one statute mile.

(d) No person may take off or land an aircraft (other than a helicopter) at any airport in a control zone under VFR:

(1) Unless ground visibility at that airport is at least 1 statute mile; or

(2) If ground visibility is not reported at that airport, unless flight visibility during landing or takeoff is at least 1 statute mile.

(e) No person may operate an aircraft (other than a helicopter) in a control zone under the special weather minimums of this section, between sunset and sunrise (or in

Alaska, when the sun is more than 6 degrees below the horizon) unless:

(1) That person meets the applicable requirements for instrument flight under Part 61 of this chapter; and

(2) The aircraft is equipped as required in 91.33(d).

91.109 VFR CRUISING ALTITUDE OR FLIGHT LEVEL

Except while holding in a holding pattern of 2 minutes or less, or while turning, each person operating an aircraft under VFR in level cruising flight more than 3,000 feet above the surface shall maintain the appropriate altitude or flight level prescribed below, unless otherwise authorized by ATC.

(a) When operating below 18,000 feet MSL and —

(1) On a magnetic course of zero degrees through 179 degrees, any odd thousand foot MSL altitude plus 500 feet (such as 3,500, 5,500, or 7,500); or

(2) On a magnetic course of 180 degrees through 359 degrees, any even thousand foot MSL altitude plus 500 feet (such as 4,500, 6,500 and 8,500).

(b) When operating above 18,000 feet MSL to flight level 290 (inclusive), and —

(1) On a magnetic course of zero degrees through 179 degrees, any odd flight level plus 500 feet (such as 195, 215 or 235); or

(2) On a magnetic course of 180 degrees through 359 degrees, any even flight level plus 500 feet (such as 185, 205, or 235).

(c) When operating above flight level 290 and—

(1) On a magnetic course of zero degrees through 179 degrees, any flight level, at 4,000-foot intervals, beginning at and including flight level 300 (such as flight level 300, 340, or 380); or

(2) On a magnetic course of 180 degrees through 359 degrees, any flight level at 4,000 foot intervals, beginning at and including flight level 320 (such as flight level 320, 360, or 400).

INSTRUMENT FLIGHT RULES

91.115 ATC CLEARANCE AND FLIGHT PLAN REQUIRED

No person may operate an aircraft in controlled airspace under IFR unless—

(a) He has filed an IFR flight plan; and

(b) He has received an appropriate ATC clearance.

91.116 TAKEOFF AND LANDING UNDER IFR: GENERAL

(a) *Instrument approaches to civil airports.* Unless otherwise authorized by the Administrator (including ATC), each person operating an aircraft shall, when an instrument letdown to an airport is necessary, use a standard instrument approach procedure prescribed for that airport in Part 97 of this chapter.

(b) *Landing minimums.* Unless otherwise authorized by the Administrator, no person operating an aircraft (except a military aircraft of the United States) may land that aircraft using a standard instrument approach procedure prescribed in Part 97 of this chapter unless the visibility is at or above the landing minimum prescribed in that part for the procedure used. If the landing minimum in a standard instrument approach procedure prescribed in Part 97 of this chapter is stated in terms of ceiling and visibility, the visibility minimum applies. However, the ceiling minimum shall be added to the field elevation and that value observed as the MDA or DH, as appropriate to the procedure being executed.

(c) *Civil airport takeoff minimums.* Unless otherwise authorized by the Administrator, no person operating an aircraft under Part 121, 123, 129 or 135 of this chapter may take off from a civil airport under IFR unless weather conditions are at or above the weather minimums for IFR takeoff prescribed for that airport in Part 97 of this chapter. If takeoff minimums are not prescribed in Part 97 of this chapter, for a particular airport, the following minimums apply to take-offs under IFR for aircraft operating under those parts:

(1) Aircraft having two engines or less: 1 statute mile visibility.

(2) Aircraft having more than two engines: One-half statute mile visibility.

(d) *Military airports.* Unless otherwise prescribed by the Administrator, each person operating a civil aircraft under IFR into, or out of, a military airport shall comply with the instrument approach procedures and the takeoff and landing minimums prescribed by the military authority having jurisdiction on that airport.

(e) *Comparable values of RVR and ground visibility.*

(1) If RVR minimums for takeoff or landing are prescribed in an instru-

ment approach procedure, but RVR is not reported for the runway of intended operation, the RVR minimum shall be converted to ground visibility in accordance with the table in subparagraph (2) of this paragraph and observed as the applicable visibility minimum for takeoff or landing on that runway.

(2)

RVR (feet)	Visibility (statute miles)
1,600	¼
2,400	½
3,200	5/8
4,000	3/4
4,500	7/8
5,000	1
6,000	1¼

(f) *Operation on unpublished routes and use of radar in instrument approach procedures.* When radar is approved at certain locations for ATC purposes, it may be used not only for surveillance and precision radar approaches, as applicable, but also may be used in conjunction with instrument approach procedures predicated on other types of radio navigational aids. Radar vectors may be authorized to provide course guidance through the segments of an approach procedure to the final approach fix or position. When operating on an unpublished route or while being radar vectored, the pilot, when an approach clearance is received, shall, in addition to complying with 91.119, maintain his last assigned altitude— (1) unless a different altitude is assigned by ATC, or (2) until the aircraft is established on a segment of a published route or instrument approach procedure. After the aircraft is so established, published altitudes apply to descent within each succeeding route or approach segment unless a different altitude is assigned by ATC. Upon reaching the final approach fix or position, the pilot may either complete his instrument approach in accordance with the procedure approved for the facility, or may continue a surveillance or precision radar approach to a landing.

(g) *Use of low or medium frequency simultaneous radio ranges for ADF procedures.* Low frequency or medium frequency simultaneous radio ranges may be used as an ADF instrument approach aid if an ADF procedure for the airport concerned is prescribed by the Administrator, or if an approach is conducted using the same courses and altitudes for the ADF approach as those specified in the approved range procedure.

(h) *Limitations on procedure turns.* In the case of a radar initial approach to a final approach fix or position, or a timed approach from a holding fix, or where the procedure specifies "NOPT" or "FINAL," no pilot may make a procedure turn unless, when he receives his final approach clearance, he so advises ATC.

91.117 LIMITATIONS ON USE OF INSTRUMENT APPROACH PROCEDURES (other than Category II)

(a) *General.* Unless otherwise authorized by the Administrator, each person operating an aircraft using an instrument approach procedure prescribed in Part 97 of this chapter shall comply with the requirements of this section. This section does not apply to the use of Category II approach procedures.

(b) *Descent below MDA or DH.* No person may operate an aircraft below the prescribed minimum descent altitude or continue an approach below the decision height unless—

(1) The aircraft is in a position from which a normal approach to the runway of intended landing can be made; and

(2) The approach threshold of that runway, or approach lights or other markings identifiable with the approach end of that runway, are clearly visible to the pilot.

If, upon arrival at the missed approach point or decision height, or at any time thereafter, any of the above requirements are not met, the pilot shall immediately execute the appropriate missed approach procedure.

(c) *Inoperative or unusable components and visual aids.* The basic ground components of an ILS are the localizer, glide slope, outer marker, and middle marker. The approach lights are visual aids normally associated with the ILS. In addition, if an ILS approach procedure in Part 97 of this chapter prescribes a visibility minimum of 1,800 feet or 2,000 feet RVR, high-intensity runway lights, touchdown zone lights, centerline lighting and marking and RVR are aids associated with the ILS for those minimums. Compass locator or precision radar may be substituted for the outer or middle marker. Surveillance radar may be substituted for the outer marker. Except as provided in subparagraph (5) of this paragraph or unless otherwise specified by the Administrator, if a ground component, visual aid, or RVR is inoperative, or unusable or not utilized, the straight-in minimums prescribed in any approach procedure in Part 97 of this chapter are raised in accordance with the following tables. If the related airborne equipment for a ground component is inoperative or not utilized, the increased minimums applicable to the related ground component shall be used. If more than one component or aid is inoperative, or unusable, or not utilized, each minimum is raised to the highest minimum required by any one of the components or aids which is inoperative, or unusable, or not utilized.

(1) ILS and PAR.

Component or aid	Increase decision height	Increase visibility (statute miles)	Approach category
LOC[1]	ILS approach not authorized.	.....	All.
GS	As specified in the procedure.	.....	All.
OM,[1] MM[1]	50 feet	None ...	ABC.
OM,[1] MM[1]	50 feet	1/4	D.
ALS	50 feet	1/4	All.
SSALSR	50 feet	1/4	ABC.
MALSR ..	50 feet	1/4	ABC.

[1]Not applicable to PAR.

(2) ILS with visibility minimum of 1,800 or 2,000 feet RVR.

Component or aid	Increase decision height	Increase visibility (statute miles)	Approach category
LOC	ILS approach not authorized.	.....	All.
GS	As specified in the procedure.	.....	All.
OM, MM ..	50 feet	To 1/2 mi	ABC.
OM, MM	50 feet	To 3/4 mi	D.
ALS	50 feet	To 3/4 mi	All.
HIRL, TDZL, RCLS.	None	To 1/2 mi	All.
RCLM	As specified in the procedure.		All.
RVR	None	To 1/2 mi	All.

(3) VOR, LOC, LDA, and ASR.

Component or aid	Increase MDA	Increase visibility (statute miles)	Approach category
ALS, SSALSR, MALSR.	None ...	1/2 mile ..	ABC.
SSALS, MALS, HIRL, REIL.	None ...	1/4 mile ..	ABC.

(4) NDB (ADF) and LFR.

Component or aid	Increase MDA	Increase visibility (statute miles)	Approach category
ALS, SSALSR, MALSR.	None ..	1/4 mile ..	ABC

(5) The inoperative component tables in subparagraphs (1) through (4) of this paragraph do not apply to helicopter procedures. Helicopter procedure minimums are specified on each procedure for inoperative components.

91.119 MINIMUM ALTITUDES FOR IFR OPERATIONS

(a) Except when necessary for takeoff or landing, or unless otherwise authorized by the Administrator, no person may operate an aircraft under IFR below—
   (1) The applicable minimum altitudes prescribed in Parts 95 (New) and 97 (New) of this chapter; or
   (2) If no applicable minimum altitude is prescribed in those parts—
      (i) In the case of operations over an area designated as a mountainous area in Part 95 (New), an altitude of 2,000 feet above the highest obstacle within a horizontal distance of five statute miles from the course to be flown; or
      (ii) In any other case, an altitude of 1,000 feet above the highest obstacle within a horizontal distance of five statute miles from the course to be flown.

However, if both a MEA and a MOCA are prescribed for a particular route or route segment, a person may operate an aircraft below the MEA down to, but not below, the MOCA, when within 25 statute miles of the VOR concerned (based on the pilot's reasonable estimate of that distance).

(b) *Climb.* Climb to a higher minimum IFR altitude shall begin immediately after passing the point beyond which that minimum altitude applies, except that, when ground obstructions intervene, the point beyond which the higher minimum altitude applies shall be crossed at or above the applicable MCA.

91.121 IFR CRUISING ALTITUDE OR FLIGHT LEVEL

(a) *In controlled airspace.* Each person operating an aircraft under IFR in level cruising flight in controlled airspace shall maintain the altitude or flight level assigned that aircraft by ATC. However, if the ATC clearance assigns "VFR conditions on-top," he shall maintain an altitude or flight level as prescribed by 91.109.

(b) *In uncontrolled airspace.* Except while holding in a holding pattern of two minutes or less, or while turning, each person operating an aircraft under IFR in level cruising flight, in uncontrolled airspace, shall maintain an appropriate altitude as follows:

(1) When operating below 18,000 feet MSL and—

(i) On a magnetic course of zero degrees through 179 degrees, any odd thousand foot MSL altitude (such as 3,000, 5,000, or 7,000); or

(ii) On a magnetic course of 180 degrees through 359 degrees, any even thousand foot MSL altitude (such as 2,000, 4,000, or 6,000).

(2) When operating at or above 18,000 feet MSL but below flight level 290, and—

(i) On a magnetic course of zero degrees through 179 degrees, any odd flight level (such as 190, 210, or 230); or

(ii) On a magnetic course of 180 degrees through 359 degrees, any even flight level (such as 180, 200, or 220).

(3) When operating at flight level 290 and above, and—

(i) On a magnetic course of zero degrees through 179 degrees, any flight level, at 4,000 foot intervals, beginning at and including flight level 290 (such as flight level 290, 330, or 370); or

(ii) On a magnetic course of 180 degrees through 359 degrees, any flight level, at 4,000 foot intervals, beginning at and including flight level 310 (such as flight level 310, 350, or 390).

91.123 COURSE TO BE FLOWN

Unless otherwise authorized by ATC, no person may operate an aircraft within controlled airspace, under IFR, except as follows:

(a) On a Federal airway, along the centerline of that airway.

(b) On any other route, along the direct course between the navigational aids or fixes defining that route.

However, this section does not prohibit maneuvering the aircraft to pass well clear of other air traffic or the maneuvering of the aircraft, in VFR conditions, to clear the intended flight path both before and during climb or descent.

91.125 IFR, RADIO COMMUNICATIONS

The pilot in command of each aircraft operated under IFR in controlled airspace shall have a continuous watch maintained on the appropriate frequency and shall report by radio as soon as possible—

(a) The time and altitude of passing each designated reporting point, or the reporting points specified by ATC, except that while the aircraft is under radar control, only the passing of those reporting points specifically requested by ATC need be reported;

(b) Any unforecast weather conditions encountered; and

(c) Any other information relating to the safety of flight.

91.127 IFR OPERATIONS; TWO-WAY RADIO COMMUNICATIONS FAILURE

(a) *General.* Unless otherwise authorized by ATC, each pilot who has two-way radio communications failure when operating under IFR shall comply with the rules of this section.

(b) *VFR conditions.* If the failure occurs in VFR conditions, or if VFR conditions are encountered after the failure, each pilot shall continue the flight under VFR and land as soon as practicable.

(c) *IFR conditions.* If the failure occurs in IFR conditions, or if paragraph (b) of this section cannot be complied with, each pilot shall continue the flight according to the following:

(1) *Route.*

(i) By the route assigned in the last ATC clearance received;

(ii) If being radar vectored, by the direct route from the point of radio failure to the fix, route, or airway specified in the vector clearance;

(iii) In the absence of an assigned route, by the route that ATC has advised may be expected in a further clearance; or

(iv) In the absence of an assigned route or a route that ATC has advised may be expected in a further clearance, by the route filed in the flight plan.

(2) *Altitude.* At the highest of the following altitudes or flight levels for the route segment being flown;

(i) The altitude or flight level assigned in the last ATC clearance received;

(ii) The minimum altitude (converted, if appropriate, to minimum flight level as prescribed in 91.81 (c) for IFR operations; or

(iii) The altitude or flight level ATC has advised may be expected in a further clearance.

(4) *Leave holding fix.* If holding instructions have been received, leave the holding fix at the expect-further-clearance time received, or, if an expected approach clearance time has been received, leave the holding fix in order to arrive over the fix from which the approach begins as close as possible to the expected approach clearance time.

(5) *Descent for approach.* Begin descent from the en route altitude or flight level upon reaching the fix from which the approach begins, but not before--

(i) The expect-approach-clearance time (if received); or

(ii) If no expect-approach-clearance time has been received, at the estimated time of arrival, shown on the flight plan, as amended with ATC.

91.129 OPERATION UNDER IFR IN CONTROLLED AIRSPACE; MALFUNCTION REPORTS

(a) The pilot in command of each aircraft operated in controlled airspace under IFR, shall report immediately to ATC any of the following malfunctions of equipment occuring in flight:

(1) Loss of VOR, TACAN, ADF, or low frequency navigation receiver capability.

(2) Complete or partial loss of ILS receiver capability.

(3) Impairment of air/ground communications capability.

(b) In each report required by paragraph (a) of this section, the pilot in command shall include the—

(1) Aircraft Identification;

(2) Equipment affected;

(3) Degree to which the capability of the pilot to operate under IFR in the ATC system is impaired; and

(4) Nature and extent of assistance he desires from ATC.

SUBPART C—MAINTENANCE, PREVENTIVE MAINTENANCE, AND ALTERATIONS

91.161 APPLICABILITY

(a) This subpart prescribes rules governing the maintenance, preventive maintenance, and alteration of U.S. registered civil aircraft operating within or without the United States.

91.163 GENERAL

(a) The owner or operator of an aircraft is primarily responsible for maintaining that aircraft in an airworthy condition, including compliance with Part 39 of this chapter.

(b) No person may perform maintenance, preventive maintenance, or alterations on an aircraft other than as prescribed in this subpart and other applicable regulations, including Part 43 of this chapter.

91.165 MAINTENANCE REQUIRED

Each owner or operator of an aircraft shall have that aircraft inspected as prescribed in Subpart D or 91.169 of this part, as appropriate, and 91.170 of this part and shall, between required inspections, have defects repaired as prescribed in Part 43 of this chapter. In addition, he shall ensure that maintenance personnel make appropriate entries in the aircraft and maintenance records indicating the aircraft has been released to service.

91.169 INSPECTIONS

(a) Except as provided in paragraph (c) of this section, no person may operate an aircraft, unless, within the preceding 12 calendar months, it has had:
   (1) An annual inspection in accordance with Part 43 of this chapter and has been approved for return to service by a person authorized by 43.7 of this chapter; or
   (2) An inspection for the issue of an airworthiness certificate.

No inspection performed under paragraph (b) of this section may be substitutued for any inspection required by this paragraph unless it is performed by a person authorized to perform annual inspections, and is entered as an "annual" inspection in the required maintenance records.

(b) Except as provided in paragraph (c) of this section, no person may operate an aircraft carrying any person (other than a crewmember) for hire, and no person may give flight instruction for hire in an aircraft which that person provides, unless within the preceding 100 hours of time in service it has received an annual or 100-hour inspection and been approved for return to service in accordance with Part 43 of this chapter, or received an inspection for the issuance of an airworthiness certificate in accordance with Part 21 of this chapter. The 100-hour limitation may be exceeded by not more than 10 hours if necessary to reach a place at which the inspection can be done. The excess time, however, is included in computing the next 100 hours of time in service.

(c) Paragraphs (a) and (b) of this section do not apply to:
   (1) Any aircraft for which its registered owner or operator complies with the progressive inspection requirements of 91.171 and Part 43 of this chapter;
   (2) An aircraft that carries a special flight permit or a current experimental or provisional certificate;
   (3) Any airplane operated by an airtravel club that is inspected in accordance with Part 123 of this chapter and the operator's manual and operations specifications; or
   (4) Any aircraft inspected in accordance with an approved aircraft inspection program under Part 135 of this chapter and is identified, by registration number, in the operations specifications of the certificate holder having the approved inspection program.

91.170 ALTIMETER SYSTEM TESTS AND INSPECTIONS

(a) No person may operate an airplane in controlled airspace under IFR unless, within the preceding 24 calendar months, each static pressure system and each altimeter instrument has been tested and inspected and found to comply with Appendix E of Part 43 of this chapter. The static pressure system and altimeter instrument tests and inspections may be conducted by—
   (1) The manufacturer of the airplane on which the tests and inspections are to be performed;
   (2) A certificated repair station properly equipped to perform these functions and holding—
      (i) An instrument rating, Class I;
      (ii) A limited instrument rating appropriate to the make and model altimeter to be tested;
      (iii) A limited rating appropriate to the test to be performed;
      (iv) An airframe rating appropriate to the airplane to be tested; or
      (v) A limited rating for a manufacturer issued for the altimeter in accordance with 145.101 (b) (4) of this chapter; or
   (3) A certificated mechanic with an airframe rating (static pressure system tests and inspections only).

(b) (Revoked)

(c) No person may operate an airplane in controlled airspace under IFR at an altitude above the maximum altitude to which an altimeter of that airplane has been tested.

91.177 ATC TRANSPONDER TESTS AND INSPECTIONS

(a) After January 1, 1976, no person may use an ATC transponder that is specified in 91.24(a), 121.345(c), 127.123(b), or 135.143(c) of this chapter, unless, within the preceding 24 calendar months, that ATC transponder has been tested and inspected and found to comply with Appendix F of Part 43 of this chapter.

(b) The tests and inspections specified in paragraph (a) of this section may be conducted by--

(1) A certified repair station properly equipped to perform those functions and holding—

(i) A radio rating, class III;

(ii) A limited radio rating appropriate to the make and model transponder to be tested;

(iii) A limited rating appropriate to the test to be performed; or

(iv) A limited rating for a manufacturer issued for the transponder in accordance with 145.101(b)(4) of this chapter; or

(2) A certificate holder authorized to perform maintenance in accordance with 121.379 or 127.140 of this chapter; or

(3) The manufacturer of the aircraft on which the transponder to be tested is installed, if the transponder was installed by that manufacturer.

NTSB PART 830—RULES PERTAINING TO THE NOTIFICATION AND REPORTING OF AIRCRAFT ACCIDENTS OR INCIDENTS AND OVERDUE AIRCRAFT

SUBPART A—GENERAL

830.1 APPLICABILITY

This part contains rules pertaining to:

(a) Giving notice of and reporting, aircraft accidents and incidents and certain other occurrences in the operation of aircraft when they involve civil aircraft of the United States wherever they occur, or foreign civil aircraft when such events occur in the United States, its territories or possessions.

(b) Preservation of aircraft wreckage, mail, cargo, and records involving all civil aircraft in the United States, its territories or possessions.

830.2 DEFINITIONS

As used in this part the following words or phrases are defined as follows:

AIRCRAFT ACCIDENT means an occurrence associated with the operation of an aircraft which takes place between the time any person boards the aircraft with the intention of flight until such time as all such persons have disembarked, in which any person suffers death or serious injury as a result of being in or upon the aircraft or by direct contact with the aircraft or anything attached thereto, or the aircraft receives substantial damage.

FATAL INJURY means any injury which results in death within 7 days.

OPERATOR means any person who causes or authorizes the operation of an aircraft, such as the owner, lessee, or bailee of an aircraft.

SERIOUS INJURY means any injury which:

(1) Requires hospitalization for more than 48 hours, commencing within 7 days from the date the injury was received;

(2) Results in a fracture of any bone (except simple fractures of fingers, toes, or nose);

(3) Involves lacerations which cause severe hemorrhages, nerve, muscle, or tendon damage;

(4) Involves injury to any internal organ; or

(5) Involves second or third degree burns; or burns affecting more than 5 percent of the body surface.

SUBSTANTIAL DAMAGE:

(1) Except as provided in subparagraph (2) of this paragraph, substantial damage means damage or structural failure which adversely affects the structural strength, performance, or flight characteristics of the aircraft, and which would normally require major repair or replacement of the affected component.

(2) Engine failure, damage limited to an engine, bent fairings or cowling, dented skin, small punctured holes in the skin or fabric, ground damage to rotor or propeller blades, damage to landing gear, wheels, tires, flaps, engine accessories, brakes, or wingtips are not considered "substantial damage" for the purpose of this part.

SUBPART B—INITIAL NOTIFICATION OF AIRCRAFT ACCIDENTS , INCIDENTS, AND OVERDUE AIRCRAFT

830.5 IMMEDIATE NOTIFICATION

The operator of an aircraft shall immediately, and by the most expeditious means available, notify the nearest National Transportation Safety Board, Bureau of Aviation Safety Board, Bureau of Aviation Safety Field Office when:

(a) An aircraft accident or any of the following listed incidents occur:

(1) Flight control system malfunction or failure;

(2) Inability of any required flightcrew member to perform his normal flight duties as a result of injury or illness;

(3) Turbine engine rotor failures excluding compressor blades and turbine buckets;

(4) In-flight fire; or

(5) Aircraft collide in flight.

(b) An aircraft is overdue and is believed to have been involved in an accident.

830.6 INFORMATION TO BE GIVEN IN NOTIFICATION

The notification required in 830.5 shall contain the following information, if available:

(a) Type, nationality, and registration marks of the aircraft;

(b) Name of owner, and operator of the aircraft;

(c) Name of the pilot in command;

(d) Date and time of the accident,

(e) Last point of departure and point of intended landing of the aircraft;
(f) Position of the aircraft with reference to some easily defined geographical point;
(g) Number of persons aboard, number killed, and number seriously injured;
(h) Nature of the accident including weather and the extent of damage to the aircraft so far as is known; and
(i) A description of any explosives, radioactive materials, or other dangerous articles carried.

SUBPART C— PRESERVATION OF AIRCRAFT WRECKAGE, MAIL, CARGO, AND RECORDS

830.10 PRESERVATION OF AIRCRAFT WRECKAGE, MAIL, CARGO, AND RECORD

(a) The operator of an aircraft is responsible for preserving to the extent possible any aircraft wreckage, cargo, and mail aboard the aircraft, and all records, including tapes of flight recorders and voice recorders, pertaining to the operation and maintenance of the aircraft and to the airmen involved in an accident or incident for which notification must be given until the Board takes custody thereof or a release is granted pursuant to 831.17.

(b) Prior to the time the Board or its authorized representative takes custody of aircraft wreckage, mail, or cargo, such wreckage, mail and cargo may be disturbed or moved only to the extent necessary:
(1) To remove persons injured or trapped;
(2) To protect the wreckage from further damage; or
(3) To protect the public from injury.

(c) Where it is necessary to disturb or move aircraft wreckage, mail, or cargo, sketches, descriptive notes, and photographs shall be made, if possible, of the accident locale including original position and condition of the wreckage and any significant impact marks.

(d) The operator of an aircraft involved in an accident or incident as defined in this Part shall retain all records and reports, including all internal documents and memoranda dealing with the accident or incident, until authorized by the Board to the contrary.

SUBPART D—REPORTING OF AIRCRAFT ACCIDENTS,INCIDENTS, AND OVERDUE AIRCRAFT

830.15 REPORTS AND STATEMENTS TO BE FILED

(a) *Reports.* The operator of an aircraft shall file a report as provided in paragraph (c) of this section on Board Form 6120.1 or Board Form 6120.2 within 10 days after an accident, or after 7 days if an overdue aircraft is still missing. A report on an incident for which notification is required by 830.5(a) shall be filed only as requested by an authorized representative of the Board.

(b) *Crewmember statement.* Each crewmember, if physically able at the time the report is submitted, shall attach thereto a statement setting forth the facts, conditions, and circumstances relating to the accident or incident as they appear to him to the best of his knowledge and belief. If the crewmember is incapacitated, he shall submit the statement as soon as he is physically able.

(c) *Where to file the reports.* The operator of an aircraft shall file with the field office of the Board nearest the accident or incident any report required by this section.

NOTE:
The National Transportation Safety Board field offices are listed under U.S. Government in the telephone directories in the following cities; Anchorage, Alaska; Chicago, Ill.; Denver, Colo.; Fort Worth, Tex.; Kansas City, Mo.; Los Angeles, Calif.; Miami, Fla.; New York, N.Y.; Oakland.; Seattle, Wash.; Washington, D.C.

Forms are obtainable from the Board field offices (see footnote 1), the National Transportation Safety Board, Washington, D.C. 20594, and the Federal Aviation Administration, Flight Standards District Office.

INSTRUMENT RATING FINAL EXAMINATION

This examination will test your knowledge of the materials studied in the manual and will help prepare you for the FAA written examination. The test is derived from a group of questions which, like the FAA exams, are numbered consecutively from number 201. The references necessary for answering the questions generally are located near them. When you must turn a page to find the appropriate reference, the question will note the correct page.

The question selection sheet which informs you which questions to answer is located on the next page. Note that the selection sheet is arranged in three vertical columns. The left numerals in each column refer to the answer spaces on the answer sheet supplied with this manual. The numerals on the right refer to the appropriate examination questions located in the manual. When working the examination, you may wish to cut the question selection sheet from the manual to reduce the amount of "page turning" required. However, retain the selection sheet with your answer sheet for your instructor.

Note that the answer sheet items are arranged in vertical columns with question numbers from 1 to 80. When you have chosen the correct answer, use a pencil to blacken the space which corresponds numerically to the question number and answer choice you have selected. Be sure to select the correct question number as well as the correct answer choice. It may be helpful to write your answer choice next to the item number on the question selection sheet as you take the exam. Then, transfer the answer to the answer sheet after the exam is completed.

The example below shows that choice 2 has been selected as the correct answer to question 1, choice 4 for question 2, and choice 4 for question 3.

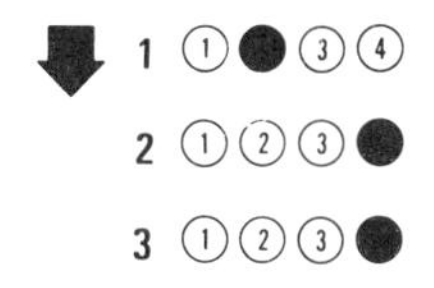

After you have completed the examination, give it to your instructor to grade and critique. After the exam has been graded, the question selection sheet and questions in the manual will be used to clarify any points of misunderstanding. Then, when your instructor is satisfied that you understand the material, you will be given a recommendation to take the FAA written exam.

INSTRUMENT RATING FINAL EXAMINATION

QUESTION SELECTION SHEET

NOTE:
The numbers on the left side of each column correspond to the numbers on the exam answer sheet. The numbers on the right side of each column correspond to the question numbers in the exam booklet.

ON ANSWER SHEET FOR ITEM NUMBER	ANSWER QUESTION NUMBER	ON ANSWER SHEET FOR ITEM NUMBER	ANSWER QUESTION NUMBER	ON ANSWER SHEET FOR ITEM NUMBER	ANSWER QUESTION NUMBER
1	201	28	234	55	268
2	202	29	235	56	269
3	203	30	237	57	271
4	205	31	238	58	272
5	206	32	239	59	273
6	207	33	240	60	274
7	208	34	242	61	276
8	210	35	243	62	277
9	211	36	244	63	278
10	212	37	245	64	279
11	213	38	247	65	281
12	215	39	248	66	282
13	216	40	249	67	283
14	217	41	251	68	284
15	218	42	252	69	286
16	219	43	253	70	287
17	221	44	254	71	288
18	222	45	256	72	289
19	223	46	257	73	290
20	224	47	258	74	292
21	226	48	259	75	293
22	227	49	261	76	294
23	228	50	262	77	295
24	229	51	263	78	297
25	230	52	264	79	299
26	232	53	266	80	300
27	233	54	267		

CUT HERE

INSTRUMENT RATING FINAL EXAMINATION

201. VOR equipment must be operationally checked before an IFR flight if the elapsed time since the last operational check is

    1. 10 days.
    2. 10 days or 10 flight hours, whichever occurs first.
    3. 20 days.
    4. 30 days.

202. What recent experience requirements must be met by the pilot in command in order to carry passengers on an instrument flight?

    1. Three takeoffs and landings to a full stop in an aircraft of the same category, class, and type within the preceding 60 days, and three hours of actual or simulated instrument time within the previous 30 days
    2. Three takeoffs and landings to a full stop in a twin-engine airplane within the preceding 90 days, and six hours of actual or simulated instrument time within the preceding 60 days
    3. Five takeoffs and landings to a full stop in an aircraft of the same category, class and type within 90 days, and three hours of simulated instrument time within the preceding 90 days
    4. Three takeoffs and landings in an aircraft of the same category and class within the preceding 90 days, and six hours of instrument flight within the past six months, including at least six instrument approaches

203. Assume a flight on August 4 requires a commercial pilot certificate with instrument and multi-engine ratings. What is the *lowest* class of medical certificate required, and what is the earliest date it could have been issued?

    1. First class, issued February 1 of the same year
    2. First class, issued July 31 of the previous year
    3. Second class, issued August 1 of the previous year
    4. Third class, issued August 31 two years previous

204. If an aircraft is operating in controlled airspace above 12,500 feet MSL and more than 2,500 above ground level, the transponder must be equipped with

    1. 64 codes.
    2. 4096 codes.
    3. 4096 codes and altitude encoder.
    4. 9999 codes and altitude encoder.

205. In addition to the gyroscopic instruments and those required for VFR flight, the instruments and equipment required for any IFR flight also include

    1. DME and transponder.
    2. a clock with sweep second hand or digital presentation and generator of adequate capacity.
    3. at least a two-axis autopilot and one landing light.
    4. a spare set of fuses and a vertical velocity indicator.

206. Except for the dual system VOR check, the maximum permissible bearing errors for ground and airborne VOR accuracy checks are

    1. plus or minus four degrees for both.
    2. plus or minus four degrees for ground and plus or minus six degrees for airborne.
    3. plus or minus six degrees for both.
    4. plus or minus six degrees for ground and plus or minus four degrees for airborne.

207. An airplane received an *annual* inspection and a *pitot-static system* inspection on January 1. The next annual and pitot-static system inspections must be completed before

1. January 1 next year.
2. January 1 next year for the annual and February 1 two years later for the pitot-static system.
3. February 1 next year.
4. February 1 next year for the annual and February 1 two years later for the pitot-static system.

208. During IFR operations under FAR Part 91, why should a pilot be concerned with nonstandard takeoff minimums, as indicated by the symbol ▼ on NOS approach charts?

1. Takeoff minimums govern all IFR departures.
2. The controller will deny takeoff clearance if the weather is below these minimums.
3. Although not mandatory for his flight, they assist the pilot in avoiding obstructions during departure.
4. They are mandatory for his flight and they provide instructions for avoiding obstructions in the departure path.

209. One of the circumstances which requires immediate notification of the National Transportation Safety Board is

1. an in-flight fire.
2. an in-flight engine failure.
3. ground damage to propeller blades.
4. damage to landing gear.

210. The chief advantage of an IFR clearance for VFR conditions on top is that

1. IFR position reporting is not required.
2. altitude changes can be made at the pilot's discretion.
3. it allows VFR flight with IFR traffic separation.
4. it provides a short IFR clearance, allowing the pilot to climb above local IFR conditions and proceed according to VFR to the destination.

211. What respective altimeter settings should be used by aircraft flying IFR along low altitude and high altitude routes?

1. The reported altimeter setting from the closest station within 100 nautical miles of the aircraft for both the low and high altitude sectors
2. The current reported altimeter setting from any station along the route and 29.92
3. The current reported altimeter setting from a station within 100 nautical miles of the aircraft's route of flight and 29.92, unless the lowest usable flight level requires a higher setting
4. The current reported altimeter setting from a station along the route of flight within 100 nautical miles of the aircraft and 29.92

212. To be acceptable for IFR navigation, what should the VOR bearing pointer of the RMI indicate when tuned to a VOT?

1. Between 174° and 186°
2. Between 176° and 184°
3. Between 354° and 006°
4. Between 356° and 004°

213. If an altimeter reads field elevation after it has been adjusted to the current altimeter setting, what type of altitude does it show?

1. Absolute
2. Pressure
3. Density
4. True

214. Vertigo is often caused by

1. acceleration forces which are misinterpreted as gravitational pull on the muscles and inner ear.
2. a sense of well-being resulting from oxygen deficiency.
3. pressure on the inner eardrum resulting from a head cold.
4. blood rushing to the brain and eyes during negative G loads.

215. A SID selected for an IFR departure contains the notation "This SID requires a minimum climb rate of 225 feet per n.m. to 5,000 feet." At a groundspeed of 78 knots, what minimum vertical velocity is required to make this departure?

1. 150 f.p.m.
2. 173 f.p.m.
3. 293 f.p.m.
4. 337 f.p.m.

216. The most current source of NOTAMs regarding the status of airport facilities is

1. AIM.
2. NOTAM summaries.
3. aviation weather reports.
4. terminal forecasts.

217. What is the best source of information about clear air turbulence (CAT)?

1. Area forecasts
2. Winds aloft forecasts
3. PIREPs
4. SIGMETs

218. Where can a pilot filing an IFR flight plan find information about substitute routes?

1. Flight advisories
2. AIM
3. Low altitude enroute charts
4. NOTAMs

219. Before flight, the switch for the emergency locator transmitter normally is set to

1. ARM.
2. OFF.
3. ON.
4. STANDBY.

220. During the final approach, if the upwind VASI lights are pink and the downwind lights are white, the aircraft is

1. on the glide slope.
2. slightly above the glide slope.
3. slightly below the glide slope.
4. well below the glide slope.

221. During an IFR departure, the pilot's attention is momentarily diverted to the enroute chart. When his scan returns to the instrument panel, the aircraft instruments show a 40° bank, even though he feels he is actually straight and level. What immediate action should he take?

1. The aircraft instruments probably have failed, so the indications should be disregarded and the climb continued to VFR conditions.
2. Control pressures should be relaxed so the aircraft will return to straight-and-level flight.
3. The autopilot should be disengaged until vertigo disappears.
4. The pilot should disregard body sensations, level the wings by instrument reference, and increase his instrument scan rate.

222. A pilot preparing to depart on an IFR flight receives the following information as part of his clearance, "*. . . maintain runway heading for departure vectors. Departure control will be 125.15. Squawk 1000 just before departure.*" To comply with this clearance, the earliest time the pilot should contact departure control is when

1. taking the active runway.
2. lifting off the departure runway.
3. passing 1,000 feet AGL.
4. the tower provides the handoff.

223. If the surface wind speed is omitted from a Terminal Forecast, it is assumed to be less than

1. 5 knots.
2. 10 knots.
3. 15 knots.
4. 20 knots.

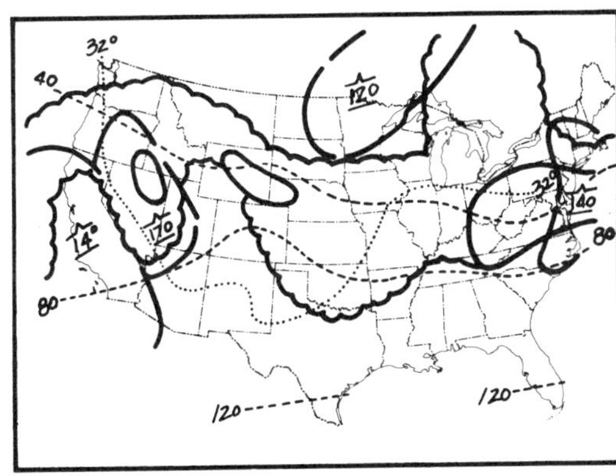

Question 224.

224. According to the depicted Sig Weather Prog Low Level Sfc—400 mb Chart on the preceding page, in central Utah there is

1. moderate turbulence from the surface to 17,000 feet.
2. moderate turbulence above 17,000 feet.
3. severe turbulence from the surface to 17,000 feet.
4. severe turbulence above 17,000 feet.

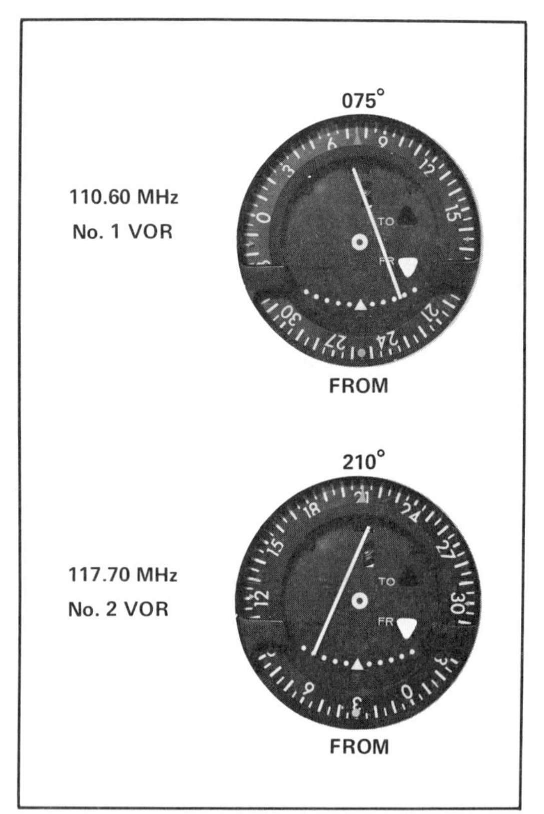

Question 225.

225. If the magnetic course is 075°, where is the airplane in relation to an intersection identified by the 075° radial on the No. 1 VOR and the 210° radial on the No. 2 VOR as shown in the accompanying illustration?

1. Left of course, past the intersection
2. Right of course, past the intersection
3. Left of course, intersection ahead
4. Right of course, intersection ahead

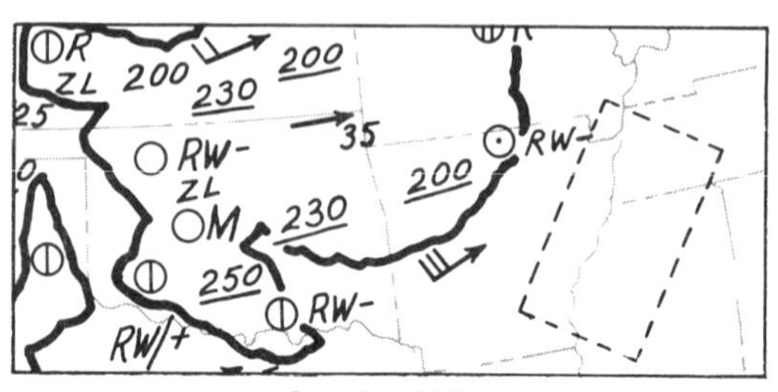

Question 226.

226. The Radar Summary Chart shows that the area over eastern Arkansas, northern Mississippi, and western Tennessee is

1. a restricted area.
2. a tornado watch area.
3. a severe thunderstorm watch area.
4. an area of severe turbulence.

```
SA 21ØØ

DEN SP M8 OVC 5R-F 187/5Ø/47/Ø812/ØØ9
```

Question 227.

227. The hourly aviation weather report indicates that the restriction to visibility at Denver is

1. light snow.
2. snow pellets.
3. thin fog.
4. light rain and fog.

```
HOU Ø71111 C2Ø BKN 6FK OCNL C1Ø OVC 4FK.
 17Z 15 SCT C1ØØ BKN 3125 OCNL C15 BKN.
 Ø5Z MVFR CIG..
```

Question 228.

228. According to the terminal forecast, after 1700Z the wind at Houston is forecast to be

1. 5 knots.
2. 10 knots.
3. 15 knots.
4. 25 knots.

ALABAMA

§ **BIRMINGHAM MUNI** (BHM) 4.3 NE GMT –6(–5DT) 33°33′50″N 86°45′16″W **ATLANTA**
643 B S4 **FUEL** 80, 100LL, JET A OX 1, 2 LRA CFR Index C **H-4G, L-14H**
RWY 05-23: H10000X150 (ASPH-GRVD) S-175, D-205, DT-350 HIRL CL **IAP**
RWY 05: ALSF1, TDZ, VASI—GA 2.08°, TCH 46′. Trée. Arrest device
RWY 23: MALSR. VASI—GA 3.0°, TCH 50′. Thld dsplcd 1780′. Trees. Arresting device.
RWY 18-36: H4856X150 (ASPH-CONC) S-55, D-75, DT-120 HIRL .58% up N.
RWY 18: Thld dsplcd 475′. Pole. **RWY 36**: Pole.
AIRPORT REMARKS: Attended continuously. MALSR Rwy 23 unmonitored but controlled by tower. Approach lights Rwy 23 controlled by Tower but operates unmonitored.
COMMUNICATIONS: UNICOM 122.95 **ATIS** 119.4
BIRMINGHAM FSS (BHM) on Arpt 123.65 122.2 122.1R 114.4T (205) 254-1387
Ⓡ **APP CON**: 124.5 (231°-049°) 124.9 (050°-230°)
TOWER: 119.9 **GND CON**: 121.7 **CLNC DEL**: 120.9 **PRE-TAXI CLNC**: 120.9
Ⓡ **DEP CON**: 124.5 (231°-049°) 124.9 (050°-230°)
STAGE III SVC ctc **APP CON**
RADIO AIDS TO NAVIGATION:
VULCAN (H) VORTAC 114.4 VUZ Chan 91 33°40′12″N 86°53′59″W 127° 9.6 NM to fld. 750/02E
McDEN NDB (H-SAB/LOM) 224 ■ BH 33°30′40″N 86°50′44″W 054° 4.5 NM to fld
ROEBY NDB (MHW/LOM) 201 RO 33°36′27″N 86°40′44″W 233° 4.0 NM to fld
ILS 110.3 I-BHM Rwy 05 LOM McDEN NDB
ILS/DME 109.5 I-ROE Chan 32 Rwy 23 LOM ROEBY NDB LOC only.
ASR ctc **APP CON**

Question 229.

229. According to the *Airport/Facility Directory* excerpt, the correct frequency for obtaining Transcribed Weather Broadcasts at Birmingham Municipal Airport is

1. 119.4.
2. 114.4.
3. 224.
4. not available at Birmingham Muni.

DAL E4 OVC 2F 53/5Ø/15Ø5/ØØ9/OVC 28

Question 230.

230. According to the surface aviation weather report for Dallas, Texas (DAL), which of the following conditions exist?

1. Base of overcast is at 280 feet.
2. Temperature is 50°F.
3. Ceiling is 400 feet MSL.
4. Top of overcast is at 2,800 feet.

231. To test for inadvertent activation of the emergency locator transmitter, the VHF receiver should be tuned to 121.50 MHz to check for a distinctive

1. series of dots and dashes.
2. variable tone.
3. steady, high pitched tone.
4. steady, low pitched tone.

232. The symbol on a constant pressure chart means the wind is from about

1. 300° at 15 knots.
2. 300° at 25 knots.
3. 330° at 15 knots.
4. 330° at 25 knots.

233. If two-way communications are lost, the first setting on the transponder should be

1. 0000.
2. 1200.
3. 7600.
4. 7700.

234. If a radar controller advised, *"Traffic at 8 o'clock,"* relative to the airplane's nose, the traffic would be located

1. 120° to the left.
2. 90° to the right.
3. 100° to the left.
4. 120° to the right.

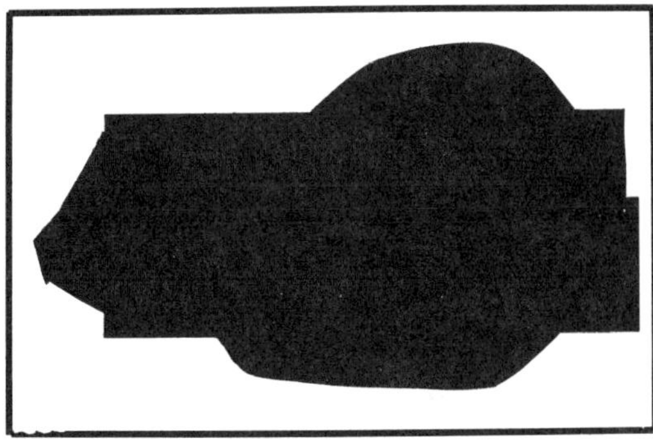

Question 235.

235. When flying outbound during the depicted NDB approach, a 10° correction is required to counteract a wind from the south. Following the procedure turn, positive interception of the inbound course is indicated if the fixed card ADF bearing pointer reads

1. 035°.
2. 045°.
3. 055°.
4. 082°.

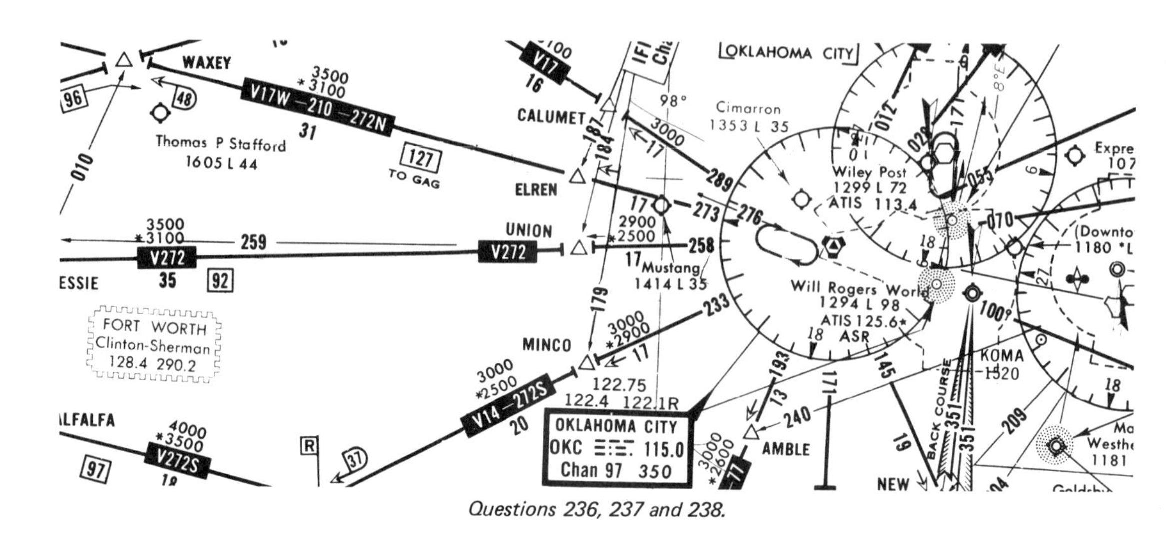

Questions 236, 237 and 238.

236. Using the Enroute Low Altitude Chart excerpt, the magnetic heading departing the Oklahoma City VORTAC (OKC) on V-272 should be approximately

   1. 289°.
   2. 273°.
   3. 258°.
   4. 233°.

237. According to the Enroute Low Altitude Chart, Union Intersection on V-272 is a

   1. turn point.
   2. non-compulsory reporting point.
   3. compulsory reporting point.
   4. changeover point.

238. Referring to the Enroute Low Altitude Chart, the symbol ←48 south of the Waxey Intersection on V17W denotes

   1. a minimum enroute altitude.
   2. a minimum obstruction clearance altitude.
   3. Waxey Intersection can be established as a DME fix.
   4. the distance to the next reporting point.

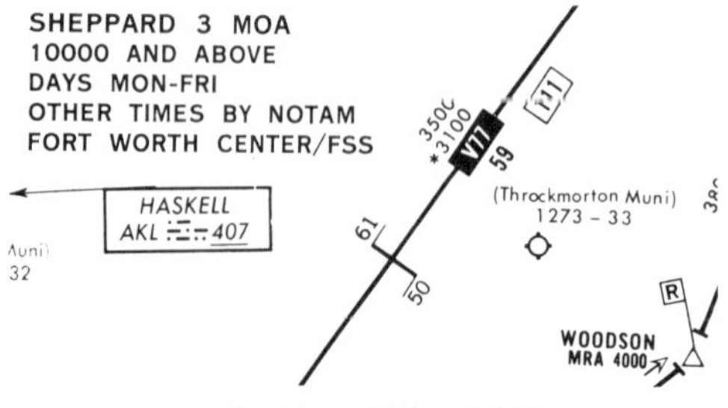

Questions 239 and 240.

239. The symbol R△ at the Woodson Intersection denotes a

   1. minimum crossing altitude.
   2. restricted intersection.
   3. mandatory reporting area.
   4. minimum reception altitude.

240. The controlling authority for the Sheppard 3 MOA is

   1. Sheppard Control Tower.
   2. Fort Worth Center.
   3. by NOTAM only.
   4. the local FSS.

TEXAS

Airborne—

Daisetta (Liberty Muni Arpt) : 195° ; 7.5 NM over hangar S end of arpt ; 1200′.

Question 241.

241. The illustrated indications are received during an airborne VOR receiver check over the checkpoint at Liberty Municipal Airport. These results indicate

   1. there is no error in the VOR receiver.
   2. the CDI is not within tolerances and should be corrected.

3. the CDI is within tolerances, but the TO/FROM indicator is faulty.
4. the VOR receiver is within tolerances for IFR flight.

242. If an airplane is 40 miles from a VOR with a FROM indication and the CDI needle is two dots to the left and remains that way for another 10 miles away from the VOR without a heading change, the airplane is

1. correcting to the radial.
2. on the radial.
3. moving away from the radial.
4. on course.

243. The predetermined geographical positions used for route definition or reporting points using a CLC RNAV system are called

1. TACAN stations.
2. VOR stations.
3. DME fixes.
4. waypoints.

244. When intercepting the 185° outbound bearing from an NDB at a 30° angle while on a magnetic heading of 155°, the ADF should read

1. 030°.
2. 225°.
3. 210°.
4. 015°.

245. In order to obtain a true airspeed of 120 knots at a pressure altitude of 9,000 feet at -10°C, what calibrated airspeed is required?

1. 106 knots
2. 111 knots
3. 118 knots
4. 125 knots

246. A true course can be converted to a magnetic course by

1. adding easterly variation to the true course.
2. subtracting easterly variation from the true course.
3. subtracting northerly variation from the true course.
4. adding westerly variation to the magnetic course.

247. The upper corners of an air/ground communications box are shaded if

1. an incoming call can be connected to a WSO by request.
2. the standard WSO frequency is not available.
3. enroute weather briefings are available on 122.00 MHz.
4. ATIS is broadcast over the navigational radio aid listed.

248. The frequency to use when declaring an emergency during an IFR flight in controlled airspace is

1. 121.50 MHz.
2. the currently assigned frequency.
3. the frequency of the closest FSS.
4. any frequency which will provide a response.

249. In order to fly a simulated instrument flight the other control seat must be occupied by

1. another pilot.
2. a flight instructor.
3. an appropriately rated pilot.
4. an instrument instructor.

250. If the CG is aft of limits, how will the aircraft react during the landing flare?

1. It will seem "heavy."
2. It will seem unusually stable.
3. The nose may pitch down uncontrollably.
4. The nose may pitch up uncontrollably.

251. After receiving a radar vector, the ATC controller advises, "*Resume normal navigation.*" Under these circumstances

1. contact the next controlling agency on the route.
2. squawk "standby" on the transponder.
3. the pilot is responsible for navigation.
4. radar service is terminated.

252. What is the purpose of "Preferred IFR Routes?"

1. For better radio navigation coverage
2. To provide positive control in controlled airspace
3. For more efficient departure, enroute, and arrival service
4. To provide adequate radar contact

253. During an IFR flight, ATC should be advised anytime the true airspeed changes by more than

1. 5 miles per hour.
2. 5 knots.
3. 10 miles per hour.
4. 10 knots.

254. What is the maximum allowable airspeed for a propeller-driven airplane beneath the lateral limits of a TCA?

1. 200 m.p.h.
2. 200 knots.
3. 250 m.p.h.
4. 250 knots.

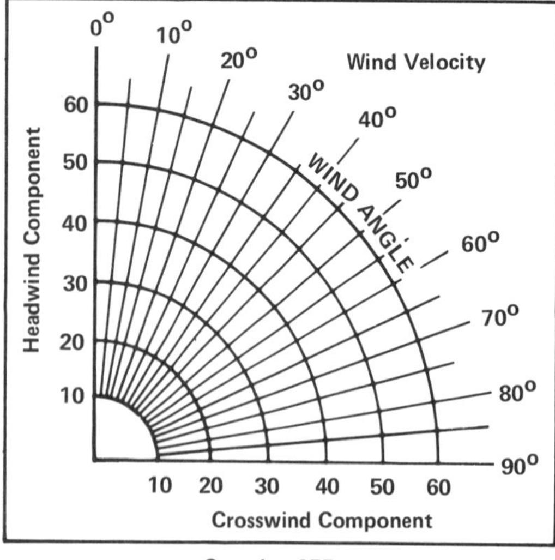

Question 255.

255. While taxiing for a takeoff on runway 22, the tower controller reports, "...*surface winds 180° at 26 knots.*" Under these conditions, the crosswind component will be

1. 13 m.p.h.
2. 15 m.p.h.
3. 19 m.p.h.
4. 20 knots

256. What is the maximum allowable airspeed for a propeller-driven airplane below 10,000 feet MSL within a TCA?

1. 200 m.p.h.
2. 200 knots.
3. 250 m.p.h.
4. 250 knots.

257. The alternate airport for an IFR flight has only a published LDA approach. What ceiling and visibility minimums must it be forecast to have at the ETA?

1. 600 feet, two miles
2. 800 feet, two miles
3. 2,100 feet, three miles
4. 2,500 feet, three miles

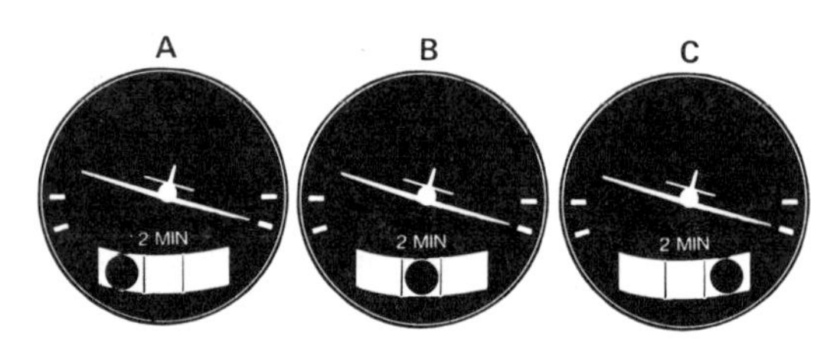

Question 258.

258. How does airplane A's turn performance compare with airplanes B and C?

1. Airplane A has the greatest angle of bank, radius of turn, and rate of turn.
2. Airplane A has the smallest angle of bank and radius of turn but the rate of turn is the same as airplanes B and C.
3. Airplane A's angle of bank is greater than B's, but less than C's, while the radius of turn and rate of turn are the same as B and C.
4. Airplane A's angle of bank is equal to B's or C's, while the radius of turn is smallest and the rate of turn is greatest.

259. If a new SIGMET is included with the scheduled weather broadcast at 1515Z, when will the next SIGMET broadcast occur?

1. 1530Z
2. 1545Z
3. 1600Z
4. 1615Z

260. As a turn is established, the attitude indicator provides information concerning the

    1. rate of turn.
    2. coordination of the turn.
    3. angle of bank.
    4. rate of climb or descent.

261. What traffic separation does ATC provide for IFR aircraft operating in accordance with a "VFR-on-top" clearance?

    1. Separation from other VFR traffic only
    2. Standard radar separation from other IFR traffic only
    3. Separation from VFR and IFR traffic equipped with transponders
    4. Separation is not normally provided.

262. Within what temperature range is structural ice accumulation most likely to occur in IFR conditions, and what are the most immediate and dangerous effects?

    1. Between minus 32° Fahrenheit and plus 32° Fahrenheit; the weight of the ice causes the airplane to exceed maximum gross weight
    2. Between minus 20° Celsius and plus 2° Celsius; an increase in stall speed and a deterioration of airplane performance
    3. Between minus 15° Fahrenheit and plus 15° Fahrenheit; inability to use flight controls and flaps
    4. Between minus 10° Celsius and plus 5° Celsius; propeller icing causes severe engine vibrations

263. After tuning the VOR and DME receivers to a VORTAC, the Morse code identification feature is received at approximately 40-second intervals. This is an indication that

    1. the VORTAC is operating normally.
    2. only the VOR portion of the VORTAC is operative.
    3. the DME receiver has malfunctioned.
    4. only the DME portion of the VORTAC is operative.

264. A pilot is cruising at 6,000 feet when he receives a clearance to "*. . . climb and maintain 9,000.*" To comply with this clearance, he should climb

    1. as fast as practical until 1,000 feet from the assigned altitude, then attempt to climb at the rate of 500 f.p.m. until reaching the assigned altitude.
    2. as fast as practical until within 500 feet of the assigned altitude.
    3. as fast as possible until within 1,000 feet of the assigned altitude and then maintain an absolute rate of 500 f.p.m. until reaching the assigned altitude.
    4. at the rate of 2,000 f.p.m., except for the last 1,000 feet of change where the rate of 1,000 f.p.m. applies.

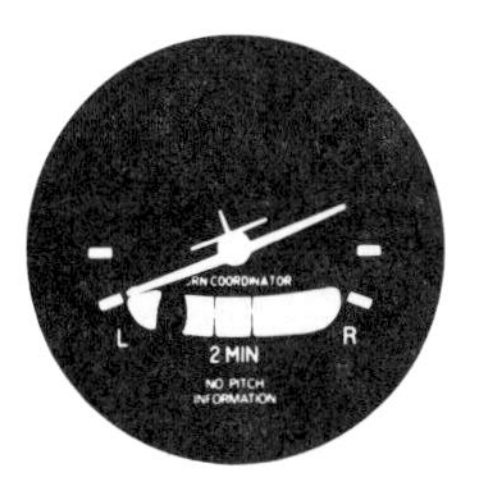

Question 265.

265. The illustrated turn coordinator indicates that the airplane is in a

    1. skid.
    2. slipping standard-rate turn.
    3. coordinated turn.
    4. four-minute turn.

266. In order to enter an assigned holding pattern at 10,000 feet MSL, the pilot of a propeller-driven aircraft with an airspeed of 210 knots is required to reduce the speed to

    1. 156 knots minimum TAS.
    2. 156 knots minimum IAS within three minutes prior to the ETA over the fix.
    3. 175 knots maximum IAS within three minutes prior to the ETA over the fix.
    4. 200 knots maximum TAS.

267. If the first inbound leg of a holding pattern at 10,000 feet is 45 seconds in duration, approximately how long should the next outbound leg be flown?

1. 60 seconds
2. 75 seconds
3. 100 seconds
4. 120 seconds

268. When a pilot is executing a circling approach, the visual portion of the approach is usually flown within what approximate distance from the runway?

1. .5 s.m.
2. 1.5 s.m.
3. 2.0 s.m.
4. 3.5 s.m.

269. The three authorized substitutes for the outer marker on an ILS approach are the compass locator, precision radar, and

1. RVR.
2. HIRL.
3. VOR.
4. ASR.

270. What flight instrument must be reset periodically in flight to correct for gyroscopic precession?

1. Vertical velocity indicator
2. Turn coordinator
3. Altimeter
4. Heading indicator

271. While taxiing out for a flight, the pilot notes that the vertical velocity indicator is showing a 380 f.p.m. rate of climb. What action, if any, must he take in order to comply with Federal Aviation Regulations?

1. The indicator must be repaired by an approved instrument repair station.
2. The pilot must adjust the indicator prior to flying in IFR conditions.
3. No action is required.
4. The pilot may file an IFR flight plan, but he must remain in VFR conditions.

272. A pilot under radar control who must make an instrument approach with only the magnetic compass for directional indications should

1. declare an emergency.
2. turn at minimum rates to minimize dip error.
3. request a no-gyro approach.
4. request a straight-in approach.

273. If a pilot arrives at DH on course and glide slope, but with only the approach lights clearly identifiable through rain showers, he should

1. execute a missed approach immediately.
2. continue the approach at DH until the runway threshold is visible.
3. continue the approach, but remain at DH until reaching the MAP.
4. continue the approach and descend toward the runway.

274. A DF instrument approach procedure will be authorized in IFR conditions only if the pilot

1. requests it.
2. declares an emergency.
3. squawks 7600 on his transponder.
4. desires a priority landing.

275. Which of the following combinations of conditions will result in the lowest true altitude if a constant indicated altitude is maintained with no adjustments to the altimeter setting window?

1. Flight from a cold airmass to a warm airmass and from a low pressure area to a high pressure area
2. Flight from a cold airmass to a warm airmass and from a high pressure area to a low pressure area
3. Flight from a warm airmass to a cold airmass and from a high pressure area to a low pressure area
4. Flight from a warm airmass to a cold airmass and from a low pressure area to a high pressure area

276. When landing with an inoperative radio receiver, a pilot sees a flashing red light from the tower as the landing roll is completed. What action is indicated?

1. Stop immediately
2. Taxi clear of runway
3. Expedite taxiing to the end of the runway
4. Make a 180° turn and taxi back on the runway

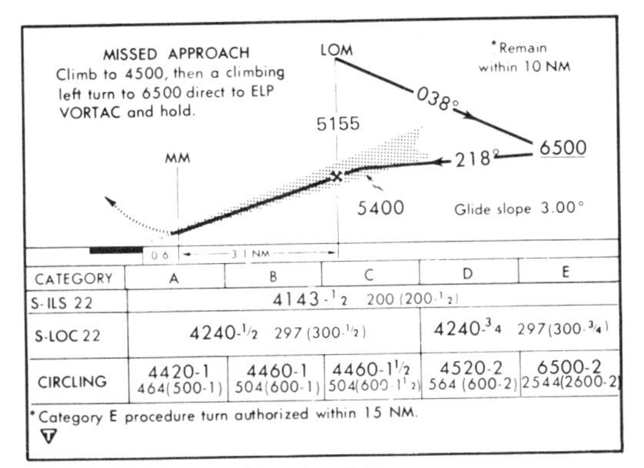

Question 277.

277. This profile view of an ILS to Runway 22 indicates that the glide slope will be intercepted

1. before reaching the LOM.
2. at the LOM.
3. after passing the LOM.
4. at the FAF.

Use the ILS Rwy 8 chart on the next page to answer questions 278 & 279.

278. Referring to the ILS RWY 8 approach chart for Houston International, what type of entry is made to enter the holding pattern at the Maxton Intersection? Assume the airplane's heading is 075°.

1. A direct entry
2. A teardrop entry
3. A parallel entry
4. Either a teardrop or direct

279. According to the ILS RWY 8 approach chart, when should a missed approach be performed?

1. At the decision height
2. At the middle marker
3. 2 minutes 26 seconds after passing the outer marker
4. .04 n.m. from the runway

280. Which instruments use the pitot-static system as a source of pressure for their operation?

1. Turn coordinator, attitude indicator, airspeed indicator, and suction gauge
2. Turn coordinator, airspeed indicator, and altimeter
3. Altimeter, vertical velocity indicator, and airspeed indicator
4. Attitude, heading, airspeed, and vertical velocity indicators

281. Upon reaching the MDA on a back course ILS approach, the VASI lights indicate red over red. The appropriate procedure in this situation is to

1. execute a missed approach.
2. continue descent at the same rate if the runway environment is clearly visible.
3. continue the approach and make a rapid descent to get on the proper visual glide slope.
4. continue the approach by leveling off momentarily in order to get on the proper visual glide slope.

282. What color changes will be seen on the standard two-bar VASI when using the back course localizer approach to runway 26 at Houston if an aircraft transitions from above the visual approach slope to below the slope?

1. Both bars will change from white through pink to red.
2. Both bars will change from red through pink to white.
3. The upwind bars will change from red to white.
4. The downwind bars will remain white, but the upwind bars will change from white to red.

283. The course width on an SDF approach is either

1. 3° or 6°.
2. 5° or 10°.
3. 6° or 12°.
4. 9° or 12°.

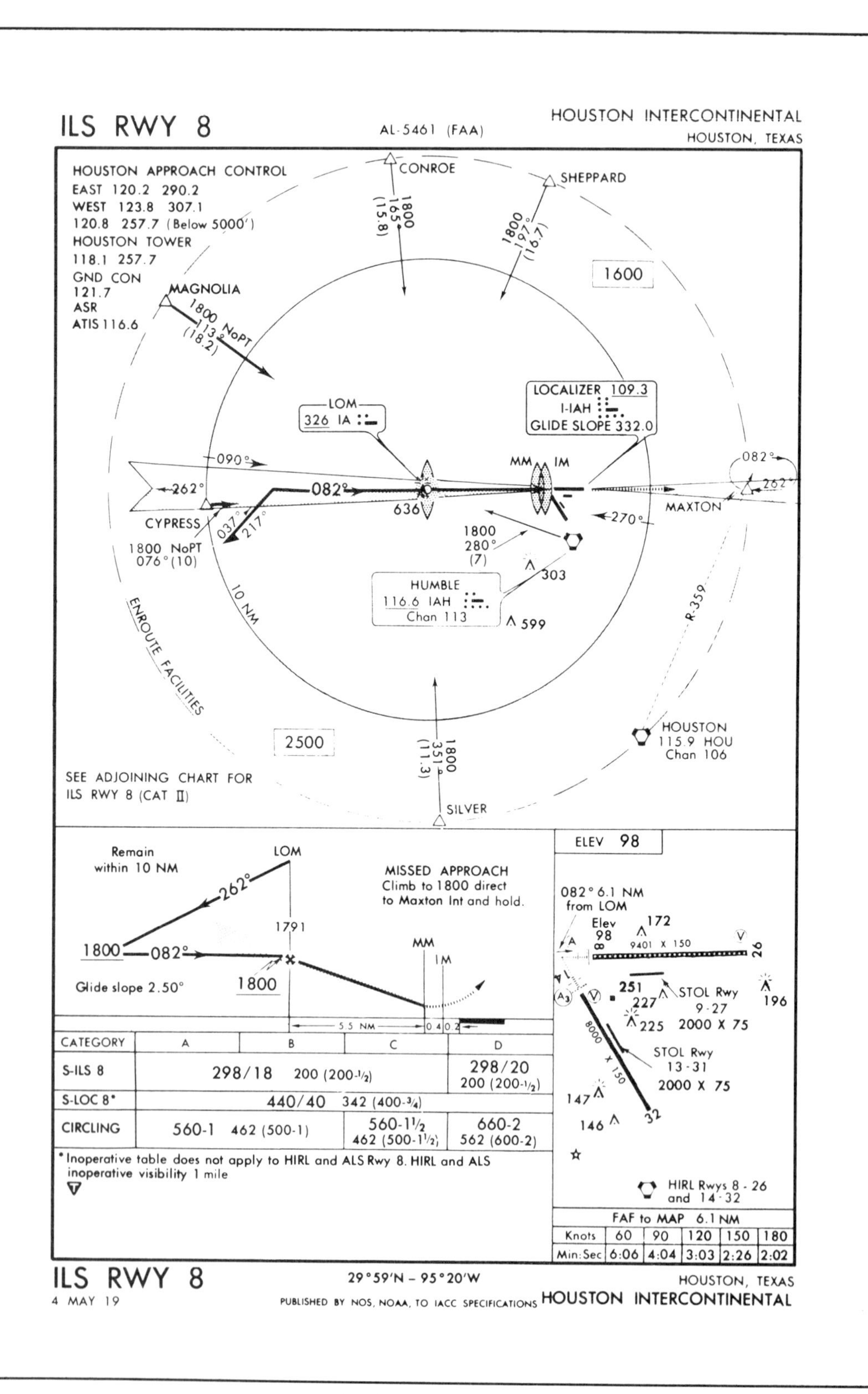

CATEGORY	A	B	C	D
S-ILS 8	298/18 200 (200-½)			298/20 200 (200-½)
S-LOC 8*	440/40 342 (400-¾)			
CIRCLING	560-1 462 (500-1)		560-1½ 462 (500-1½)	660-2 562 (600-2)

*Inoperative table does not apply to HIRL and ALS Rwy 8. HIRL and ALS inoperative visibility 1 mile

FAF to MAP 6.1 NM					
Knots	60	90	120	150	180
Min:Sec	6:06	4:04	3:03	2:26	2:02

Questions 278 and 279.

284. The major difference between the LDA and ILS approach courses is

  1. an LDA approach course is transmitted from a VOR type facility.
  2. an LDA approach requires a different type radio receiver.
  3. an LDA approach may not be aligned within 30° of the runway.
  4. an LDA localizer course is narrower than an ILS.

285. After takeoff from runway 35L, a shallow turn to the west is initiated. Which one of the compass indications applies as the airplane begins its bank to the left?

  1. A
  2. B
  3. C
  4. D

Use the LDA BC Rwy 8 chart on the following page to answer Questions 286, 287 & 288.

286. What is the distance between the final approach fix and the missed approach point?

  1. 2.1 n.m.
  2. 3.9 n.m.
  3. 6.0 n.m.
  4. 8.4 n.m.

287. The initial approach fix for the LDA BC RWY 8 approach to Pearson Airpark is the

  1. Portland VORTAC.
  2. Buxton Intersection.
  3. Beach Intersection.
  4. Morgan Intersection.

288. With a groundspeed of 90 knots, when will the missed approach be made after crossing the final approach fix?

  1. 2:00 minutes
  2. 2:24 minutes
  3. 3:00 minutes
  4. 4:00 minutes

289. "MALSR" means

  1. minimum altitude for surveillance radar.
  2. medium intensity approach light system.
  3. maximum intensity approach light system.
  4. medium intensity approach light system with runway alignment indicator lights.

A

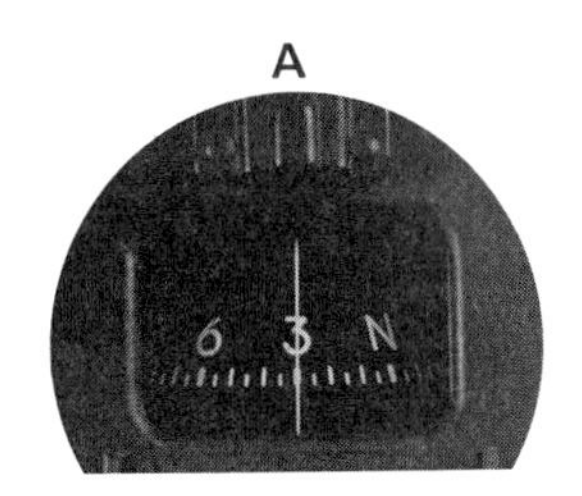

B

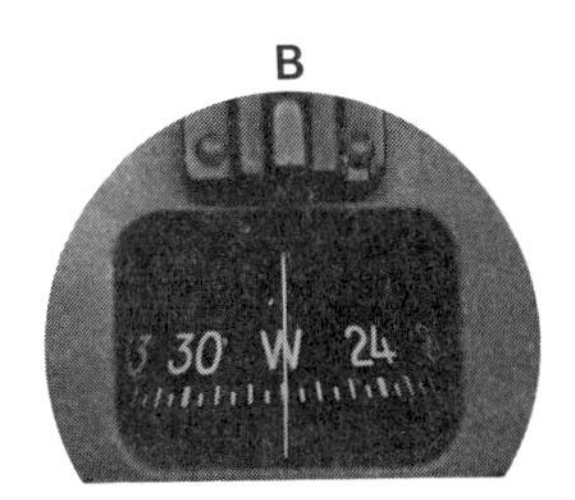

C

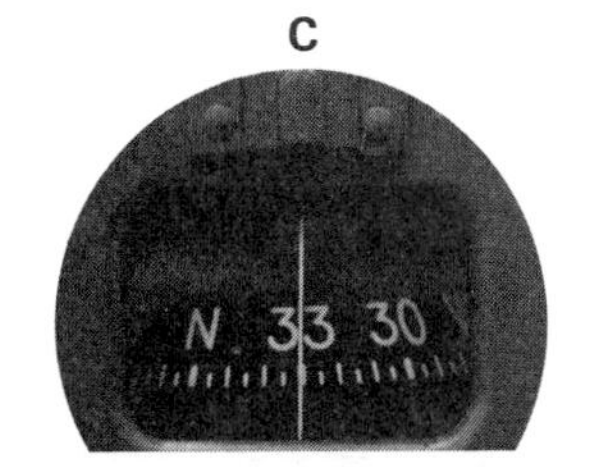

D

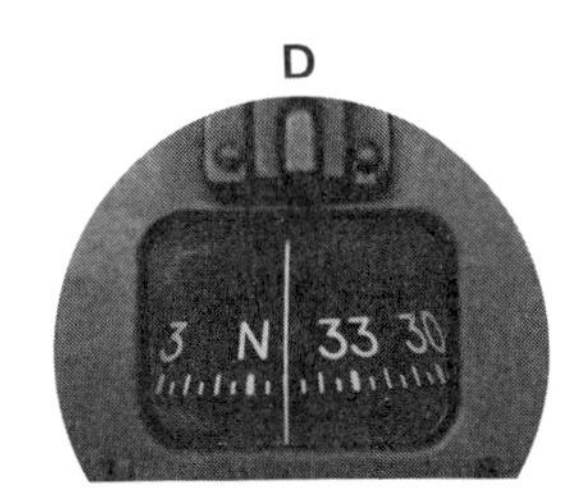

Question 285.

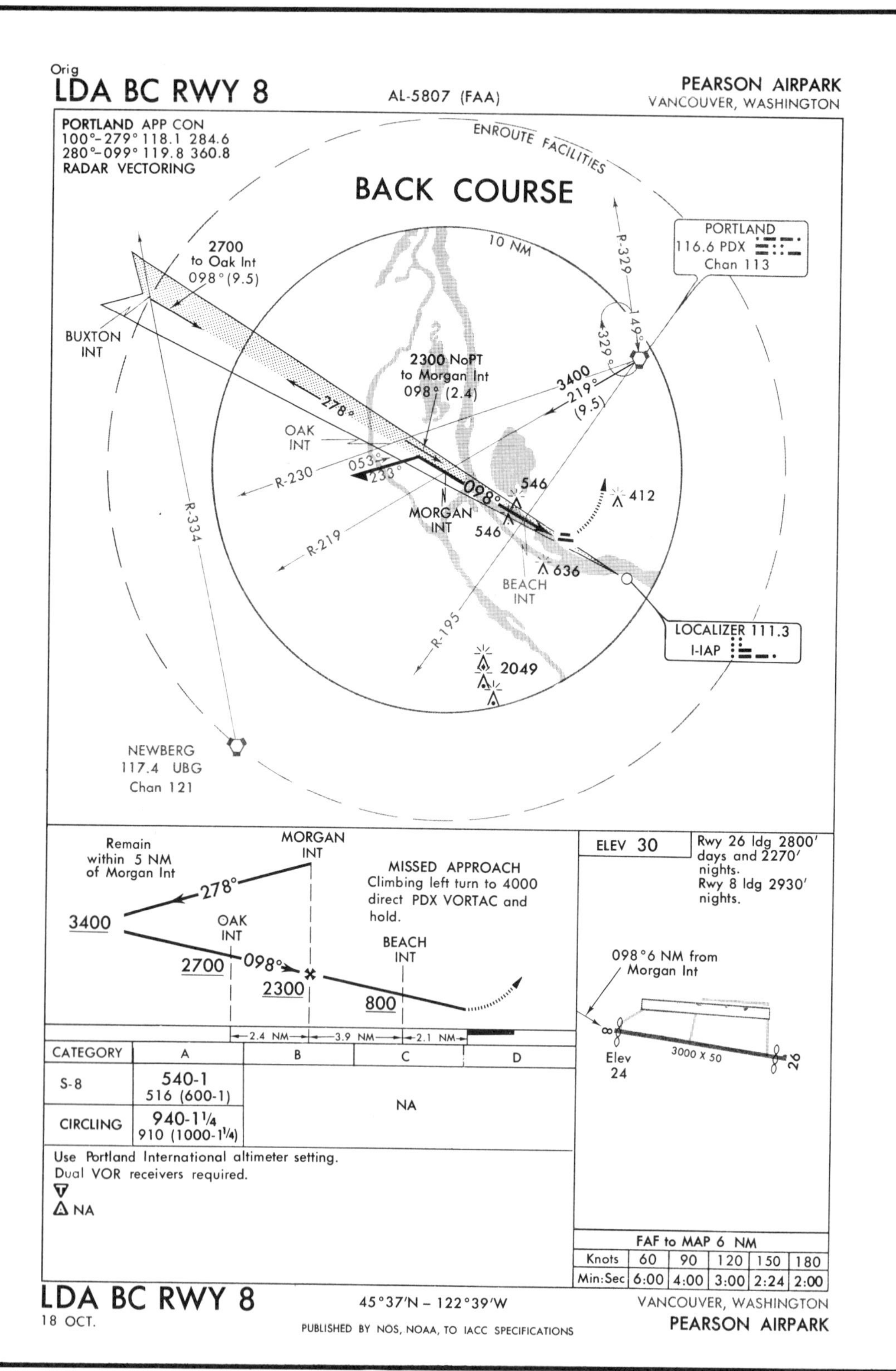

Questions 286, 287 and 288

290. Based on the following loading, are the CG and gross weight within limits for this flight? (Use average weights for pilots and passengers.)
   1. CG and gross weight are within limit.
   2. CG and gross weight are beyond limits.
   3. CG is aft of limits and gross weight is within limits.
   4. CG is forward of limits, gross weight is within limits.

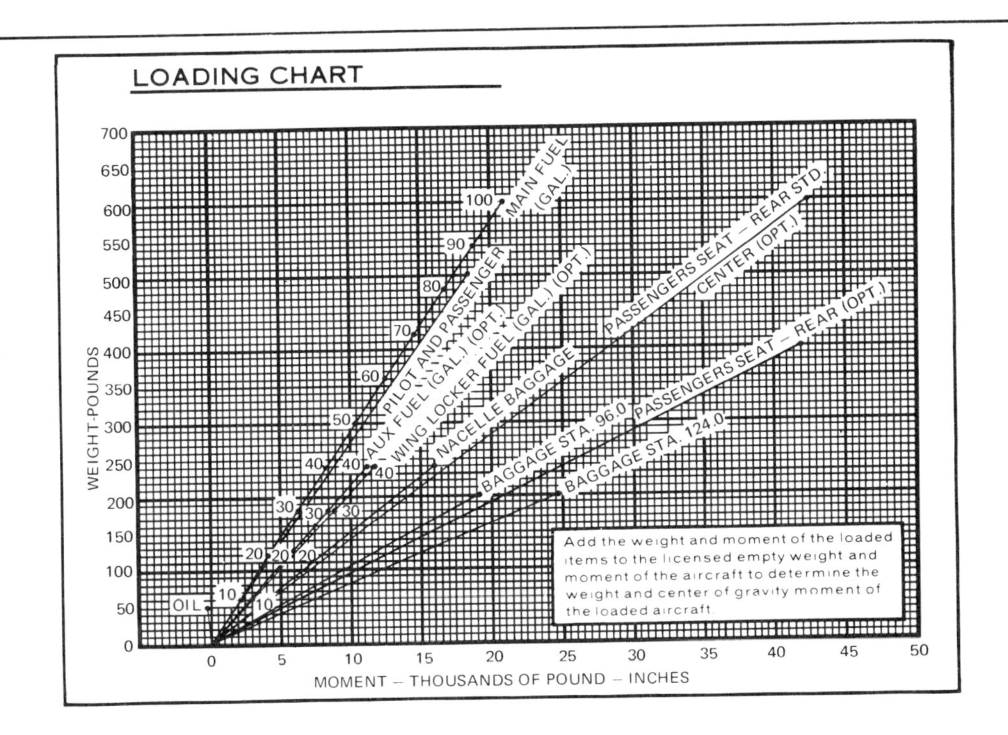

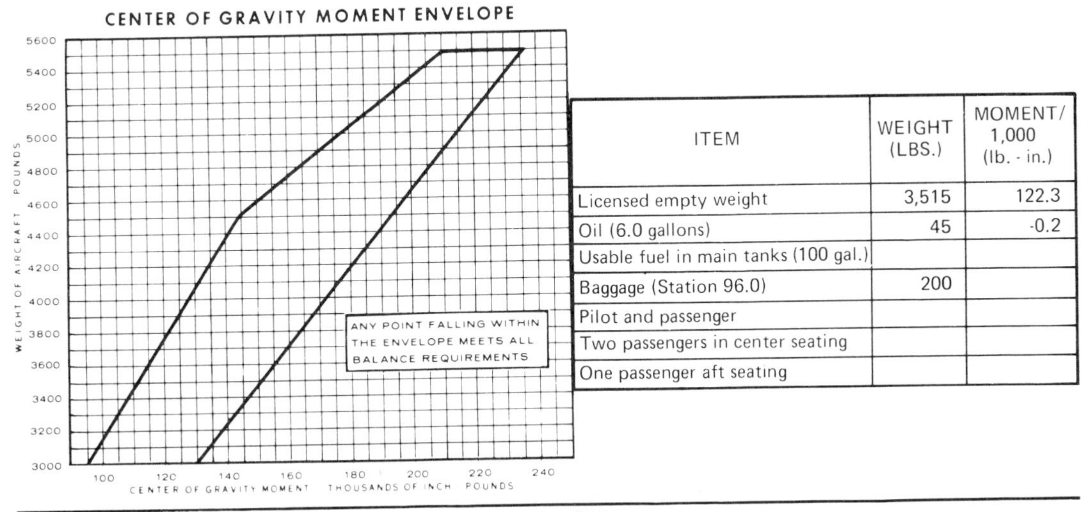

ITEM	WEIGHT (LBS.)	MOMENT/ 1,000 (lb. - in.)
Licensed empty weight	3,515	122.3
Oil (6.0 gallons)	45	-0.2
Usable fuel in main tanks (100 gal.)		
Baggage (Station 96.0)	200	
Pilot and passenger		
Two passengers in center seating		
One passenger aft seating		

Question 290.

LANDING PERFORMANCE									
		SEA LEVEL 59° F		2500 FT. 50° F		5000 FT. 41° F		7500 FT. 32° F	
Gross Weight Pounds	IAS at Obstacle MPH	Ground Run	Total Distance Over 50 Foot Obstacle	Ground Run	Total Distance Over 50 Foot Obstacle	Ground Run	Total Distance Over 50 Foot Obstacle	Ground Run	Total Distance Over 50 Foot Obstacle
5000	104	540	1690	582	1732	627	1777	677	1827
4600	100	450	1600	484	1634	522	1672	563	1713
4200	95	368	1518	396	1546	427	1577	461	1611

NOTE: WING FLAPS 35°, POWER OFF, HARD SURFACE RUNWAY, ZERO WIND, MAXIMUM BRAKING EFFORT. REDUCE LANDING DISTANCE 10% FOR EACH 10 MPH HEADWIND. INCREASE DISTANCE BY 25% OF THE GROUND RUN FOR OPERATION ON FIRM DRY SOD RUNWAYS.

Question 293.

291. A comparison of icing susceptibility between a fuel injection system and a standard, float-type carburetor system reveals that the fuel injection system is

    1. more susceptible to icing.
    2. less susceptible to icing.
    3. equally susceptible to icing.
    4. extremely susceptible to icing and requires 50 percent more heat to eliminate ice formation.

292. What is the danger of a night landing short of the green runway lights on runway 9 below?

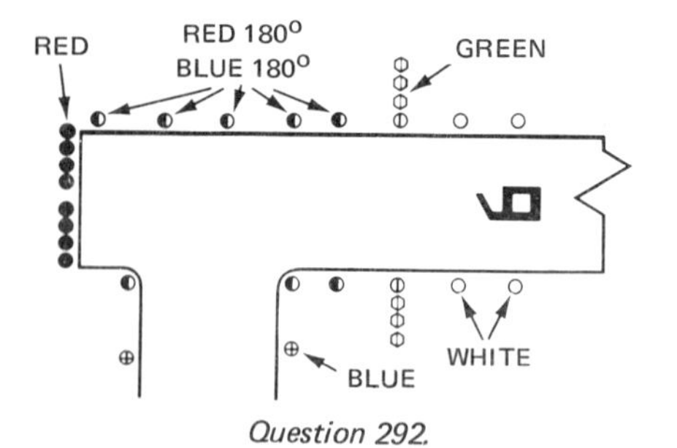

Question 292.

    1. The possibility of a collision with an obstruction exists during the final approach.
    2. The surface may not bear the weight of the aircraft.
    3. A violation may be filed, because no operations are permitted on that portion of the runway.
    4. Another aircraft may be on the taxiway short of the runway.

293. Based on the following conditions, what is the approximate minimum landing distance required to clear a 50-foot obstacle?

Runway	13L at Dallas Love Field
Gross weight	4,800 lb.
Elevation	487 ft.
Altimeter	29.92
Temperature	85°F
Tower winds	1510

    1. 1,750 feet
    2. 1,650 feet
    3. 1,515 feet
    4. 1,850 feet

294. The identifier for the outer marker of an ILS is a series of

    1. dashes.
    2. dots.
    3. dots and dashes.
    4. the identifier preceded by the letter "I."

295. During an ILS approach, which type of lighting system would require an increase in the decision height should it fail?

    1. VASI
    2. RAIL
    3. TDZL
    4. ALS

296. If an ATC transponder is required for any flight operation, it must have been tested inspected within the preceding

1. 6 months.
2. 12 months.
3. 24 months.
4. 36 months.

297. What are the main differences between a visual and a contact approach?

1. The pilot must have the field in sight in VFR conditions for a contact approach; the pilot must request a visual approach.
2. The pilot must request a contact approach; the pilot may be assigned a visual approach and higher weather minimums must exist.
3. Anytime the pilot reports the field in sight, ATC may clear the pilot for a contact approach; for a visual approach, the pilot must advise that the approach can be made under VFR conditions.
4. The pilot must request a visual approach and report having the field in sight; ATC may assign a contact approach if VFR conditions exist.

298. Which area in this runway diagram is 500 feet down the runway?

1. A
2. B
3. C
4. D

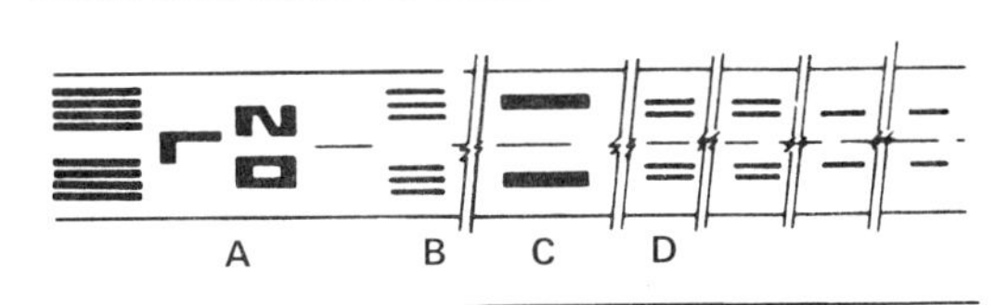

Questions 298 and 299.

299. The area used on this precision runway for landing out of a precision approach is point

1. A
2. B
3. C
4. D

300. What does the symbol Ⓡ denote when it precedes a departure control frequency in the *Airport/Facility Directory*?

1. Radar departure control
2. Receive-only frequency
3. Daylight hours only
4. TWEB frequency

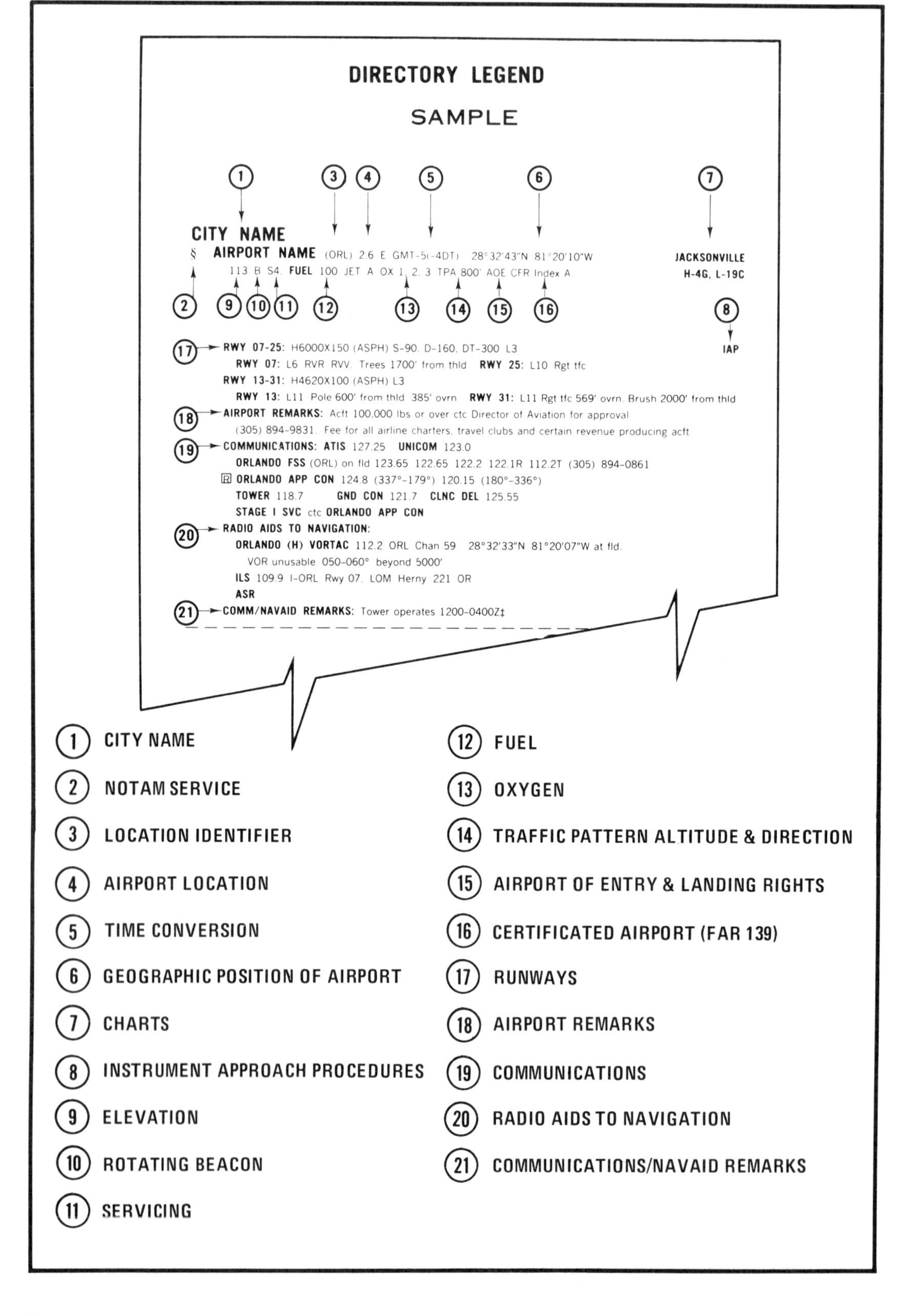
DIRECTORY LEGEND
SAMPLE
CITY NAME
§ AIRPORT NAME (ORL) 2.6 E GMT-5(-4DT) 28°32'43"N 81°20'10"W
113 B S4 FUEL 100 JET A OX 1, 2, 3 TPA 800' AOE CFR Index A
JACKSONVILLE
H-4G, L-19C
IAP
RWY 07-25: H6000X150 (ASPH) S-90, D-160, DT-300 L3
RWY 07: L6 RVR RVV Trees 1700' from thld RWY 25: L10 Rgt tfc
RWY 13-31: H4620X100 (ASPH) L3
RWY 13: L11 Pole 600' from thld 385' ovrn RWY 31: L11 Rgt tfc 569' ovrn. Brush 2000' from thld
AIRPORT REMARKS: Acft 100,000 lbs or over ctc Director of Aviation for approval
(305) 894-9831. Fee for all airline charters, travel clubs and certain revenue producing acft
COMMUNICATIONS: ATIS 127.25 UNICOM 123.0
ORLANDO FSS (ORL) on fld 123.65 122.65 122.2 122.1R 112.2T (305) 894-0861
Ⓡ ORLANDO APP CON 124.8 (337°-179°) 120.15 (180°-336°)
TOWER 118.7 GND CON 121.7 CLNC DEL 125.55
STAGE I SVC ctc ORLANDO APP CON
RADIO AIDS TO NAVIGATION:
ORLANDO (H) VORTAC 112.2 ORL Chan 59 28°32'33"N 81°20'07"W at fld.
VOR unusable 050-060° beyond 5000'
ILS 109.9 I-ORL Rwy 07. LOM Herny 221 OR
ASR
COMM/NAVAID REMARKS: Tower operates 1200-0400Z‡
1 CITY NAME
2 NOTAM SERVICE
3 LOCATION IDENTIFIER
4 AIRPORT LOCATION
5 TIME CONVERSION
6 GEOGRAPHIC POSITION OF AIRPORT
7 CHARTS
8 INSTRUMENT APPROACH PROCEDURES
9 ELEVATION
10 ROTATING BEACON
11 SERVICING
12 FUEL
13 OXYGEN
14 TRAFFIC PATTERN ALTITUDE & DIRECTION
15 AIRPORT OF ENTRY & LANDING RIGHTS
16 CERTIFICATED AIRPORT (FAR 139)
17 RUNWAYS
18 AIRPORT REMARKS
19 COMMUNICATIONS
20 RADIO AIDS TO NAVIGATION
21 COMMUNICATIONS/NAVAID REMARKS

UNITED STATES GOVERNMENT
FLIGHT INFORMATION PUBLICATION

ENROUTE LOW ALTITUDE – U. S.

For use up to but not including 18,000' MSL

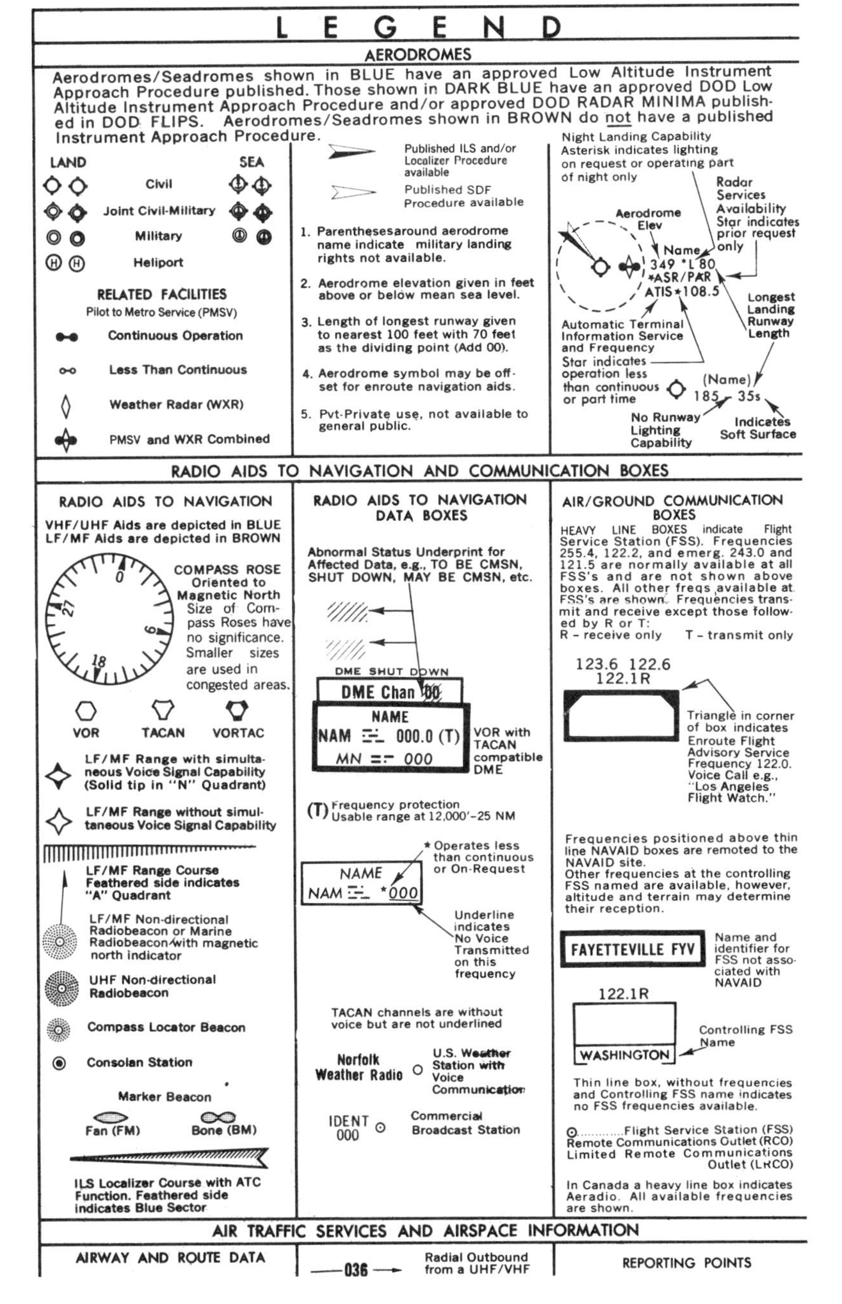

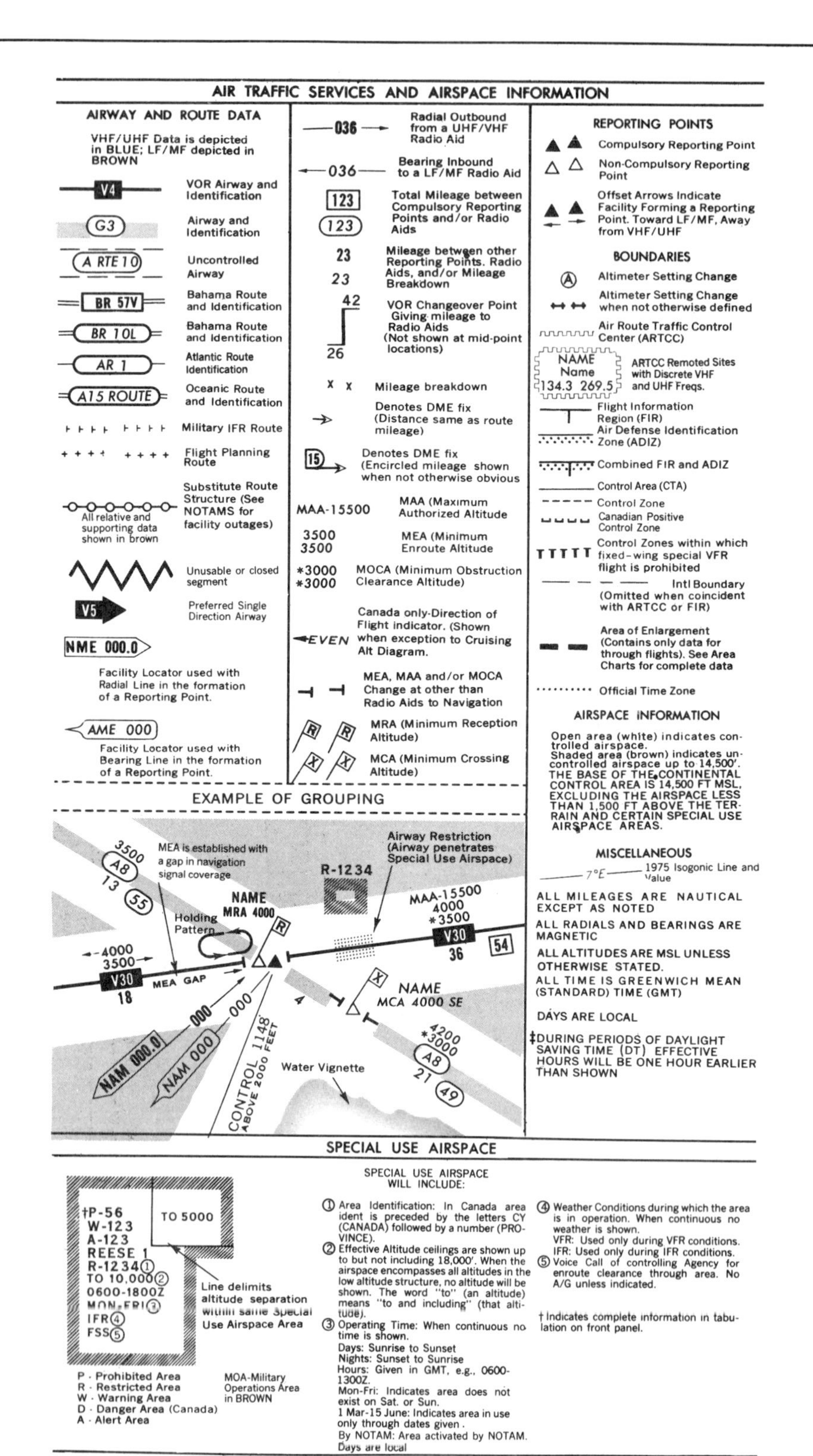
AIR TRAFFIC SERVICES AND AIRSPACE INFORMATION
AIRWAY AND ROUTE DATA
VHF/UHF Data is depicted in BLUE; LF/MF depicted in BROWN
V4
VOR Airway and Identification
G3
Airway and Identification
A RTE 10
Uncontrolled Airway
BR 57V
Bahama Route and Identification
BR 10L
Bahama Route and Identification
AR 1
Atlantic Route Identification
A15 ROUTE
Oceanic Route and Identification
Military IFR Route
Flight Planning Route
Substitute Route Structure (See NOTAMS for facility outages)
All relative and supporting data shown in brown
Unusable or closed segment
V5
Preferred Single Direction Airway
NME 000.0
Facility Locator used with Radial Line in the formation of a Reporting Point.
AME 000
Facility Locator used with Bearing Line in the formation of a Reporting Point.
036
Radial Outbound from a UHF/VHF Radio Aid
036
Bearing Inbound to a LF/MF Radio Aid
123
123
Total Mileage between Compulsory Reporting Points and/or Radio Aids
23
23
Mileage between other Reporting Points. Radio Aids, and/or Mileage Breakdown
42
26
VOR Changeover Point Giving mileage to Radio Aids (Not shown at mid-point locations)
X X
Mileage breakdown
Denotes DME fix (Distance same as route mileage)
15
Denotes DME fix (Encircled mileage shown when not otherwise obvious
MAA-15500
MAA (Maximum Authorized Altitude
3500
3500
MEA (Minimum Enroute Altitude
*3000
*3000
MOCA (Minimum Obstruction Clearance Altitude)
EVEN
Canada only-Direction of Flight indicator. (Shown when exception to Cruising Alt Diagram.
MEA, MAA and/or MOCA Change at other than Radio Aids to Navigation
R
MRA (Minimum Reception Altitude)
X
MCA (Minimum Crossing Altitude)
REPORTING POINTS
Compulsory Reporting Point
Non-Compulsory Reporting Point
Offset Arrows Indicate Facility Forming a Reporting Point. Toward LF/MF, Away from VHF/UHF
BOUNDARIES
A
Altimeter Setting Change
Altimeter Setting Change when not otherwise defined
Air Route Traffic Control Center (ARTCC)
NAME
Name
134.3 269.5
ARTCC Remoted Sites with Discrete VHF and UHF Freqs.
Flight Information Region (FIR)
Air Defense Identification Zone (ADIZ)
Combined FIR and ADIZ
Control Area (CTA)
Control Zone
Canadian Positive Control Zone
Control Zones within which fixed-wing special VFR flight is prohibited
Intl Boundary (Omitted when coincident with ARTCC or FIR)
Area of Enlargement (Contains only data for through flights). See Area Charts for complete data
Official Time Zone
AIRSPACE INFORMATION
Open area (white) indicates controlled airspace.
Shaded area (brown) indicates uncontrolled airspace up to 14,500'.
THE BASE OF THE CONTINENTAL CONTROL AREA IS 14,500 FT MSL, EXCLUDING THE AIRSPACE LESS THAN 1,500 FT ABOVE THE TERRAIN AND CERTAIN SPECIAL USE AIRSPACE AREAS.
MISCELLANEOUS
7°E
1975 Isogonic Line and Value
ALL MILEAGES ARE NAUTICAL EXCEPT AS NOTED
ALL RADIALS AND BEARINGS ARE MAGNETIC
ALL ALTITUDES ARE MSL UNLESS OTHERWISE STATED.
ALL TIME IS GREENWICH MEAN (STANDARD) TIME (GMT)
DAYS ARE LOCAL
‡DURING PERIODS OF DAYLIGHT SAVING TIME (DT) EFFECTIVE HOURS WILL BE ONE HOUR EARLIER THAN SHOWN
EXAMPLE OF GROUPING
3500
A8
13
55
MEA is established with a gap in navigation signal coverage
Holding Pattern
NAME
MRA 4000
R
R-1234
Airway Restriction (Airway penetrates Special Use Airspace)
MAA-15500
4000
*3500
V30
36
54
4000
3500
V30
18
MEA GAP
000
000
NAM 000.0
NAM 000
CONTROL 1148
ABOVE 2000 FEET
4
X
NAME
MCA 4000 SE
4200
*3000
A8
21
49
Water Vignette
SPECIAL USE AIRSPACE
†P-56
W-123
A-123
REESE 1
R-1234①
TO 10,000②
0600-1800Z
MON-FRI③
IFR④
FSS⑤
TO 5000
Line delimits altitude separation within same Special Use Airspace Area
P - Prohibited Area
R - Restricted Area
W - Warning Area
D - Danger Area (Canada)
A - Alert Area
MOA-Military Operations Area in BROWN
SPECIAL USE AIRSPACE WILL INCLUDE:
① Area Identification: In Canada area ident is preceded by the letters CY (CANADA) followed by a number (PROVINCE).
② Effective Altitude ceilings are shown up to but not including 18,000'. When the airspace encompasses all altitudes in the low altitude structure, no altitude will be shown. The word "to" (an altitude) means "to and including" (that altitude).
③ Operating Time: When continuous no time is shown.
Days: Sunrise to Sunset
Nights: Sunset to Sunrise
Hours: Given in GMT, e.g., 0600-1300Z.
Mon-Fri: Indicates area does not exist on Sat. or Sun.
1 Mar-15 June: Indicates area in use only through dates given.
By NOTAM: Area activated by NOTAM.
Days are local
④ Weather Conditions during which the area is in operation. When continuous no weather is shown.
VFR: Used only during VFR conditions.
IFR: Used only during IFR conditions.
⑤ Voice Call of controlling Agency for enroute clearance through area. No A/G unless indicated.
† Indicates complete information in tabulation on front panel.
CRUISING ALTITUDES - U. S.

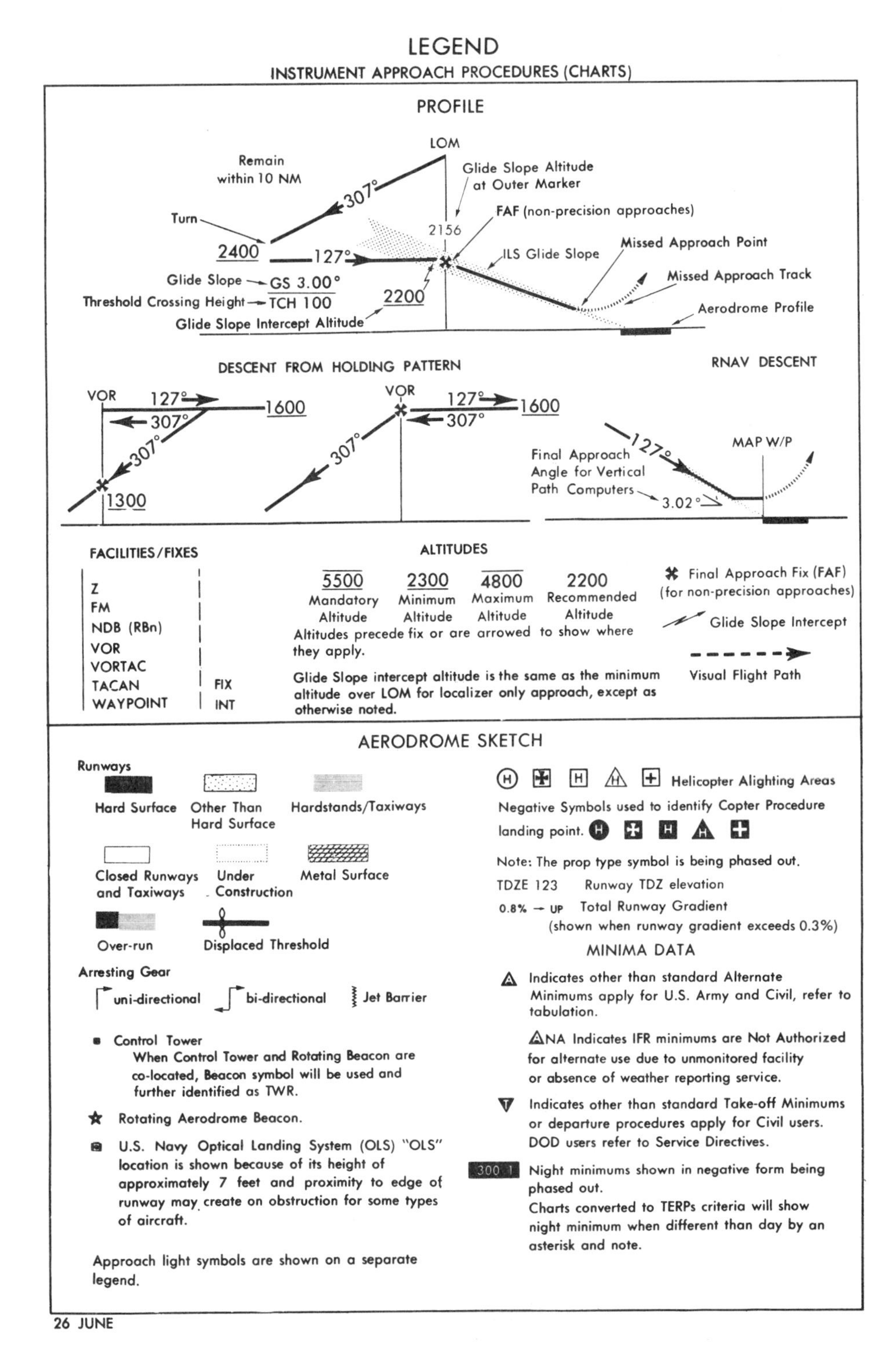

LEGEND
INSTRUMENT APPROACH PROCEDURES (CHARTS)
PROFILE
LOM
Remain within 10 NM
Glide Slope Altitude at Outer Marker
307°
Turn
2156
FAF (non-precision approaches)
2400
127°
ILS Glide Slope
Missed Approach Point
Glide Slope
GS 3.00°
Missed Approach Track
Threshold Crossing Height
TCH 100
2200
Aerodrome Profile
Glide Slope Intercept Altitude
DESCENT FROM HOLDING PATTERN
RNAV DESCENT
VOR
127°
1600
307°
1300
VOR
127°
1600
307°
127°
MAP W/P
Final Approach Angle for Vertical Path Computers
3.02°
FACILITIES/FIXES
Z
FM
NDB (RBn)
VOR
VORTAC
TACAN
WAYPOINT
FIX
INT
ALTITUDES
5500
Mandatory Altitude
2300
Minimum Altitude
4800
Maximum Altitude
2200
Recommended Altitude
Altitudes precede fix or are arrowed to show where they apply.
Glide Slope intercept altitude is the same as the minimum altitude over LOM for localizer only approach, except as otherwise noted.
Final Approach Fix (FAF) (for non-precision approaches)
Glide Slope Intercept
Visual Flight Path
AERODROME SKETCH
Runways
Hard Surface
Other Than Hard Surface
Hardstands/Taxiways
Closed Runways and Taxiways
Under Construction
Metal Surface
Over-run
Displaced Threshold
Arresting Gear
uni-directional
bi-directional
Jet Barrier
Control Tower
When Control Tower and Rotating Beacon are co-located, Beacon symbol will be used and further identified as TWR.
Rotating Aerodrome Beacon.
U.S. Navy Optical Landing System (OLS) "OLS" location is shown because of its height of approximately 7 feet and proximity to edge of runway may create on obstruction for some types of aircraft.
Approach light symbols are shown on a separate legend.
Helicopter Alighting Areas
Negative Symbols used to identify Copter Procedure landing point.
Note: The prop type symbol is being phased out.
TDZE 123 Runway TDZ elevation
0.8% → UP Total Runway Gradient
(shown when runway gradient exceeds 0.3%)
MINIMA DATA
Indicates other than standard Alternate Minimums apply for U.S. Army and Civil, refer to tabulation.
NA Indicates IFR minimums are Not Authorized for alternate use due to unmonitored facility or absence of weather reporting service.
Indicates other than standard Take-off Minimums or departure procedures apply for Civil users. DOD users refer to Service Directives.
300 1
Night minimums shown in negative form being phased out.
Charts converted to TERPs criteria will show night minimum when different than day by an asterisk and note.
26 JUNE

LEGEND

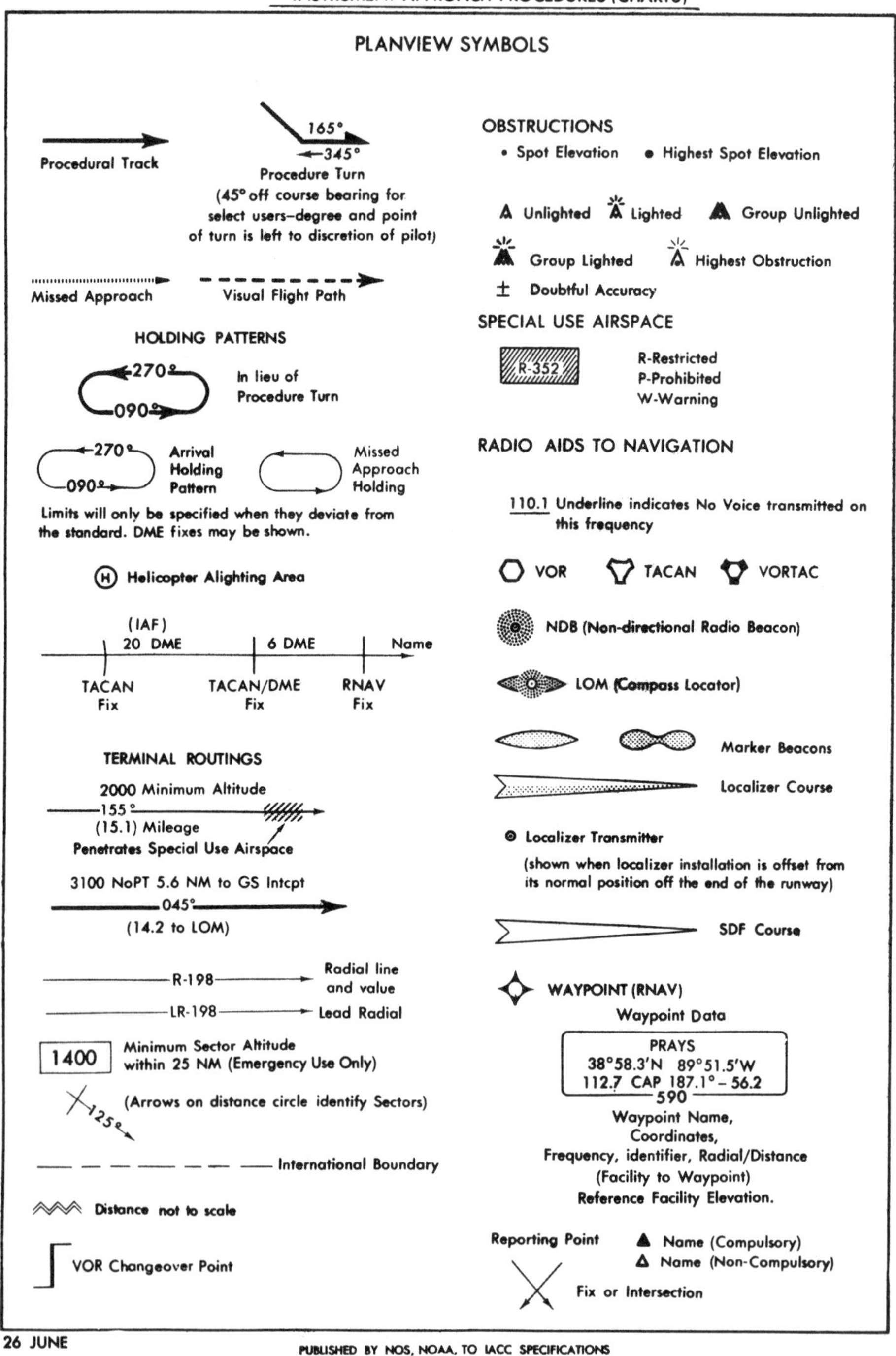
INSTRUMENT APPROACH PROCEDURES (CHARTS)
PLANVIEW SYMBOLS
Procedural Track
165°
345°
Procedure Turn
(45° off course bearing for select users–degree and point of turn is left to discretion of pilot)
Missed Approach
Visual Flight Path
HOLDING PATTERNS
270°
090°
In lieu of Procedure Turn
Arrival Holding Pattern
Missed Approach Holding
Limits will only be specified when they deviate from the standard. DME fixes may be shown.
H
Helicopter Alighting Area
(IAF)
20 DME
6 DME
Name
TACAN Fix
TACAN/DME Fix
RNAV Fix
TERMINAL ROUTINGS
2000 Minimum Altitude
155°
(15.1) Mileage
Penetrates Special Use Airspace
3100 NoPT 5.6 NM to GS Intcpt
045°
(14.2 to LOM)
R-198
Radial line and value
LR-198
Lead Radial
1400
Minimum Sector Altitude within 25 NM (Emergency Use Only)
125°
(Arrows on distance circle identify Sectors)
International Boundary
Distance not to scale
VOR Changeover Point
OBSTRUCTIONS
Spot Elevation
Highest Spot Elevation
Unlighted
Lighted
Group Unlighted
Group Lighted
Highest Obstruction
Doubtful Accuracy
SPECIAL USE AIRSPACE
R-352
R-Restricted
P-Prohibited
W-Warning
RADIO AIDS TO NAVIGATION
110.1 Underline indicates No Voice transmitted on this frequency
VOR
TACAN
VORTAC
NDB (Non-directional Radio Beacon)
LOM (Compass Locator)
Marker Beacons
Localizer Course
Localizer Transmitter
(shown when localizer installation is offset from its normal position off the end of the runway)
SDF Course
WAYPOINT (RNAV)
Waypoint Data
PRAYS
38°58.3'N 89°51.5'W
112.7 CAP 187.1°– 56.2
590
Waypoint Name,
Coordinates,
Frequency, identifier, Radial/Distance
(Facility to Waypoint)
Reference Facility Elevation.
Reporting Point
Name (Compulsory)
Name (Non-Compulsory)
Fix or Intersection
26 JUNE
PUBLISHED BY NOS, NOAA, TO IACC SPECIFICATIONS

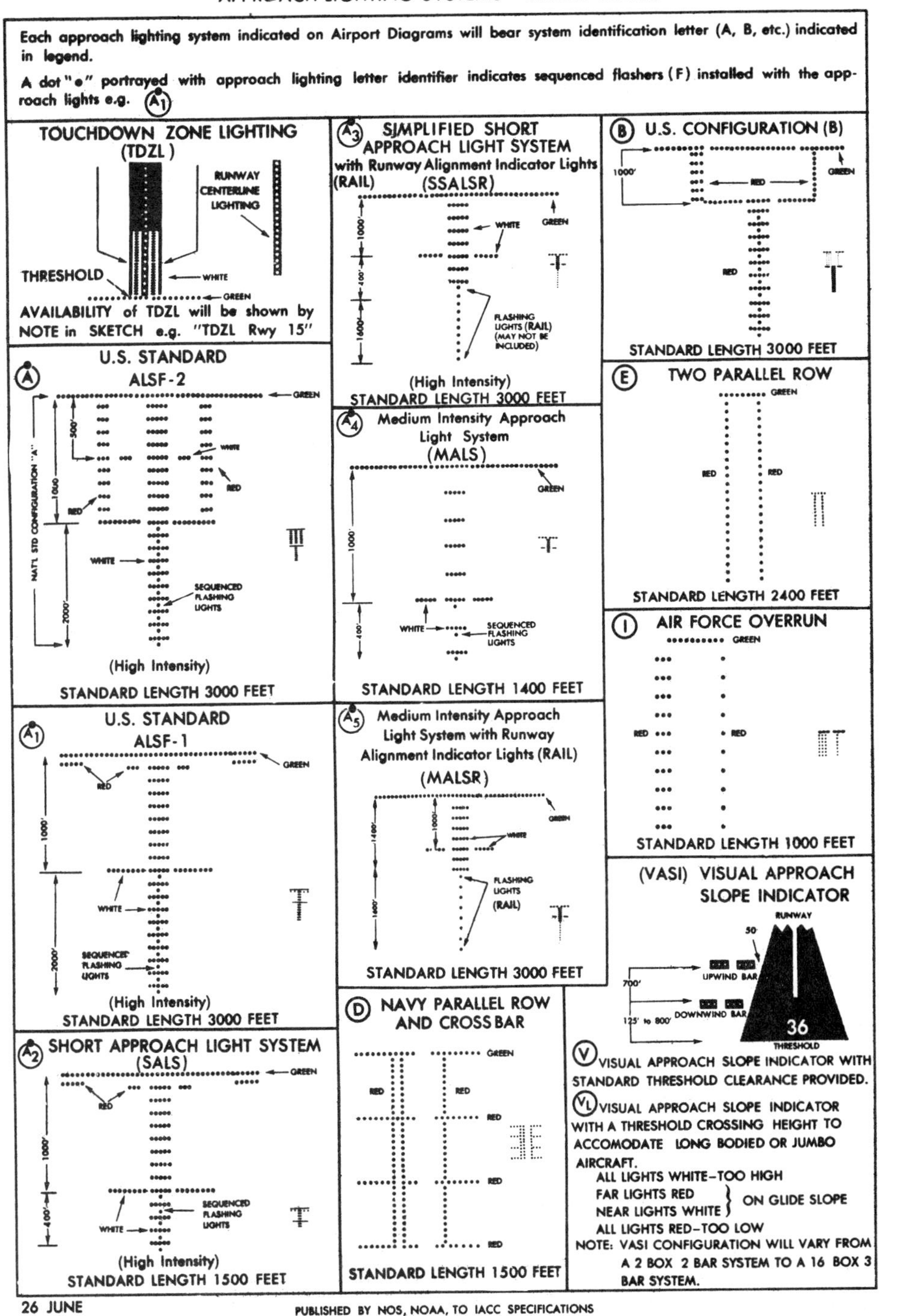

LEGEND
INSTRUMENT APPROACH PROCEDURES (CHARTS)
APPROACH LIGHTING SYSTEMS – UNITED STATES
Each approach lighting system indicated on Airport Diagrams will bear system identification letter (A, B, etc.) indicated in legend.
A dot "•" portrayed with approach lighting letter identifier indicates sequenced flashers (F) installed with the approach lights e.g. A1
TOUCHDOWN ZONE LIGHTING (TDZL)
RUNWAY CENTERLINE LIGHTING
THRESHOLD
WHITE
GREEN
AVAILABILITY of TDZL will be shown by NOTE in SKETCH e.g. "TDZL Rwy 15"
A3 SIMPLIFIED SHORT APPROACH LIGHT SYSTEM with Runway Alignment Indicator Lights (RAIL) (SSALSR)
FLASHING LIGHTS (RAIL) (MAY NOT BE INCLUDED)
(High Intensity)
STANDARD LENGTH 3000 FEET
B U.S. CONFIGURATION (B)
RED
GREEN
STANDARD LENGTH 3000 FEET
A U.S. STANDARD ALSF-2
NAT'L STD CONFIGURATION "A"
SEQUENCED FLASHING LIGHTS
(High Intensity)
STANDARD LENGTH 3000 FEET
A4 Medium Intensity Approach Light System (MALS)
STANDARD LENGTH 1400 FEET
E TWO PARALLEL ROW
STANDARD LENGTH 2400 FEET
I AIR FORCE OVERRUN
STANDARD LENGTH 1000 FEET
A1 U.S. STANDARD ALSF-1
(High Intensity)
STANDARD LENGTH 3000 FEET
A5 Medium Intensity Approach Light System with Runway Alignment Indicator Lights (RAIL) (MALSR)
FLASHING LIGHTS (RAIL)
STANDARD LENGTH 3000 FEET
(VASI) VISUAL APPROACH SLOPE INDICATOR
RUNWAY
UPWIND BAR
DOWNWIND BAR
THRESHOLD
36
A2 SHORT APPROACH LIGHT SYSTEM (SALS)
(High Intensity)
STANDARD LENGTH 1500 FEET
D NAVY PARALLEL ROW AND CROSS BAR
STANDARD LENGTH 1500 FEET
V VISUAL APPROACH SLOPE INDICATOR WITH STANDARD THRESHOLD CLEARANCE PROVIDED.
VL VISUAL APPROACH SLOPE INDICATOR WITH A THRESHOLD CROSSING HEIGHT TO ACCOMODATE LONG BODIED OR JUMBO AIRCRAFT.
ALL LIGHTS WHITE—TOO HIGH
FAR LIGHTS RED
NEAR LIGHTS WHITE
ON GLIDE SLOPE
ALL LIGHTS RED—TOO LOW
NOTE: VASI CONFIGURATION WILL VARY FROM A 2 BOX 2 BAR SYSTEM TO A 16 BOX 3 BAR SYSTEM.
26 JUNE
PUBLISHED BY NOS, NOAA, TO IACC SPECIFICATIONS

LEGEND

INSTRUMENT APPROACH PROCEDURES (CHARTS)

GENERAL INFORMATION & ABBREVIATIONS

★ Indicates control tower operates non-continuously.
All distances in nautical miles (except Visibility Data which is in statute miles and Runway Visual Range which is in hundreds of feet).
Runway dimensions in feet.
Elevations in feet Mean Sea Level.
All radials/bearings are Magnetic.

ADF . . . . Automatic Direction Finder
ALS . . . . Approach Light System
APP CON . . . . Approach Control
ARR . . . . Arrival
ASR/PAR . . . . Published Radar Minimums at this Aerodrome.
ATIS . . . . Automatic Terminal Information Service
BC . . . . Back Course
C . . . . Circling
CAT . . . . Category
CHAN . . . . Channel
CLNC DEL . . . . Clearance Delivery
DH . . . . Decision Height
DME . . . . Distance Measuring Equipment
DR . . . . Dead Reckoning
ELEV . . . . Airport Elevation
FAF . . . . Final Approach Fix
FM . . . . Fan Marker
GPI . . . . Ground Point of Interception
GS . . . . Glide Slope
HAA . . . . Height Above Aerodrome
HAL . . . . Height Above Landing
HAT . . . . Height Above Touchdown
HIRL . . . . High Intensity Runway Lights
IAF . . . . Initial Approach Fix
ICAO . . . . International Civil Aviation Organization
Intcp . . . . Intercept
INT, INTXN . . . . Intersection
LDA . . . . Localizer Type Directional Aid
Ldg . . . . Landing
LIRL . . . . Low Intensity Runway Lights
LDIN . . . . Lead in Light System
LOC . . . . Localizer
LR . . . . Lead Radial. Provides at least 2 NM (Copter 1 NM) of lead to assist in turning onto the intermediate/final course.
MALS . . . . Medium Intensity Approach Light System
MALS/R . . . . Medium Intensity Approach Light Systems/with RAIL
MAP . . . . Missed Approach Point
MDA . . . . Minimum Descent Altitude
MIRL . . . . Medium Intensity Runway Lights
NA . . . . Not Authorized
NDB . . . . Non-directional Radio Beacon
NoPT . . . . No Procedure Turn Required (Procedure Turn shall not be executed without ATC clearance)
RA . . . . Radio Altimeter Height
Radar Required . . . . Radar vectoring required for this approach
Radar Vectoring . . . . May be expected through any portion of the Nav Aid Approach, except final.
RAIL . . . . Runway Alignment Indicator Lights
RBn . . . . Radio Beacon
REIL . . . . Runway End Identifier Lights
RCLS . . . . Runway Centerline Light System
RNAV . . . . Area Navigation
RRL . . . . Runway Remaining Lights
RTB . . . . Return To Base
Runway Touchdown Zone . First 3000′ of Runway.
RVR . . . . Runway Visual Range
S . . . . Straight-in
SALS . . . . Short Approach Light System
(S) SALS/R . . . . (Simplified) Short Approach Light System/with RAIL
SDF . . . . Simplified Directional Facility
TA . . . . Transition Altitude
TAC . . . . TACAN
TCH . . . . Threshold Crossing Height (Height in feet Above Ground Level)
TDZ . . . . Touchdown Zone
TDZE . . . . Touchdown Zone Elevation
TDZL . . . . Touchdown Zone Lights
TLv . . . . Transition Level
W/P . . . . Waypoint (RNAV)

LANDING MINIMA FORMAT

In this example airport elevation is 1179, and runway touchdown zone elevation is 1152.

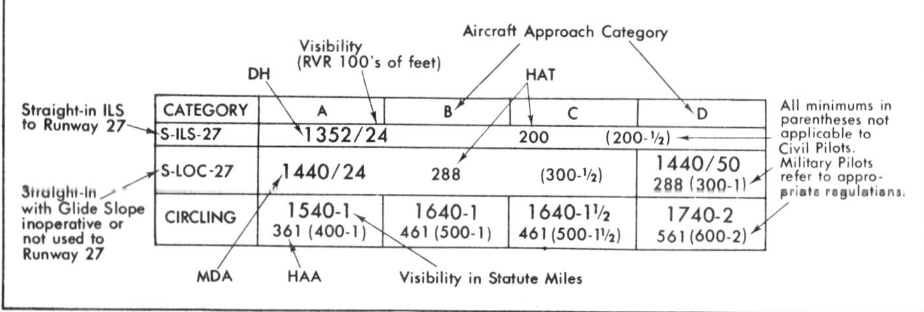

26 JUNE

Instrument Approach Procedures (Charts)
INOPERATIVE COMPONENTS OR VISUAL AIDS TABLE
Civil Pilots see FAR 91.117(c)

Landing minimums published on instrument approach procedure charts are based upon full operation of all components and visual aids associated with the particular instrument approach chart being used. Higher minimums are required with inoperative components or visual aids as indicated below. If more than one component is inoperative, each minimum is raised to the highest minimum required by any single component that is inoperative. ILS glide slope inoperative minimums are published on instrument approach charts as localizer minimums. This table may be amended by notes on the approach chart. Such notes apply only to the particular approach category(ies) as stated. See legend page for description of components indicated below.

(1) ILS, MLS, and PAR

Inoperative Component or Aid	Approach Category	Increase DH	Increase Visibility
MM*	ABC	50 feet	None
MM*	D	50 feet	¼ mile
ALSF 1 & 2, MALSR, & SSALR	ABCD	None	¼ mile

*Not applicable to PAR

(2) ILS with visibility minimum of 1,800 or 2,000 RVR.

MM	ABC	50 feet	To 2400 RVR
MM	D	50 feet	To 4000 RVR
ALSF 1 & 2, MALSR, & SSALR	ABCD	None	To 4000 RVR
TDZL, RCLS	ABCD	None	To 2400 RVR
RVR	ABCD	None	To ½ mile

(3) VOR, VOR/DME, VORTAC, VOR (TAC), VOR/DME (TAC), LOC, LOC/DME, LDA, LDA/DME, SDF, SDF/DME, RNAV, and ASR

Inoperative Visual Aid	Approach Category	Increase MDA	Increase Visibility
ALSF 1 & 2, MALSR, & SSALR	ABCD	None	½ mile
SSALS, MALS & ODALS	ABC	None	¼ mile

(4) NDB

ALSF 1 & 2, MALSR, & SSALR	C	None	½ mile
	ABD	None	¼ mile
MALS, SSALS, ODALS	ABC	None	¼ mile

PUBLISHED BY NOS, NOAA, TO IACC SPECIFICATIONS

AIRCRAFT APPROACH CATEGORY — A grouping of aircraft basd on a speed of 1.3 times the stall speed in the landing configuration at maximum gross landing weight. An aircraft shall fit in ony one category. If it is necessary to maneuver at speeds in excess of the upper limit of a speed range for a category, the minimums for the next higher category should be used. For example, an aircraft which falls in Category A, but is circling to land at a speed in excess of 91 knots, should use the approach Category B minimums when circling to land. The categories are as follows:

1. Category A — Speed less than 91 knots.
2. Category B — Speed 91 knots or more but less than 121 knots.
3. Category C — Speed 121 knots or more but less than 141 knots.
4. Category D — Speed 141 knots or more but less than 166 knots.
5. Category E — Speed 166 knots or more.

(Refer to FAR Parts 1 and 97)

RVR/Meteorological Visibility Comparable Values

The following table should be used for converting RVR to meteorological visibility when RVR is inoperative.

RVR (feet)	Visibility (statute miles)	RVR (feet)	Visibility (statute miles)
1600	¼	4000	¾
200	⅜	4500	⅞
2400	½	5000	1
3200	⅝	6000	1¼

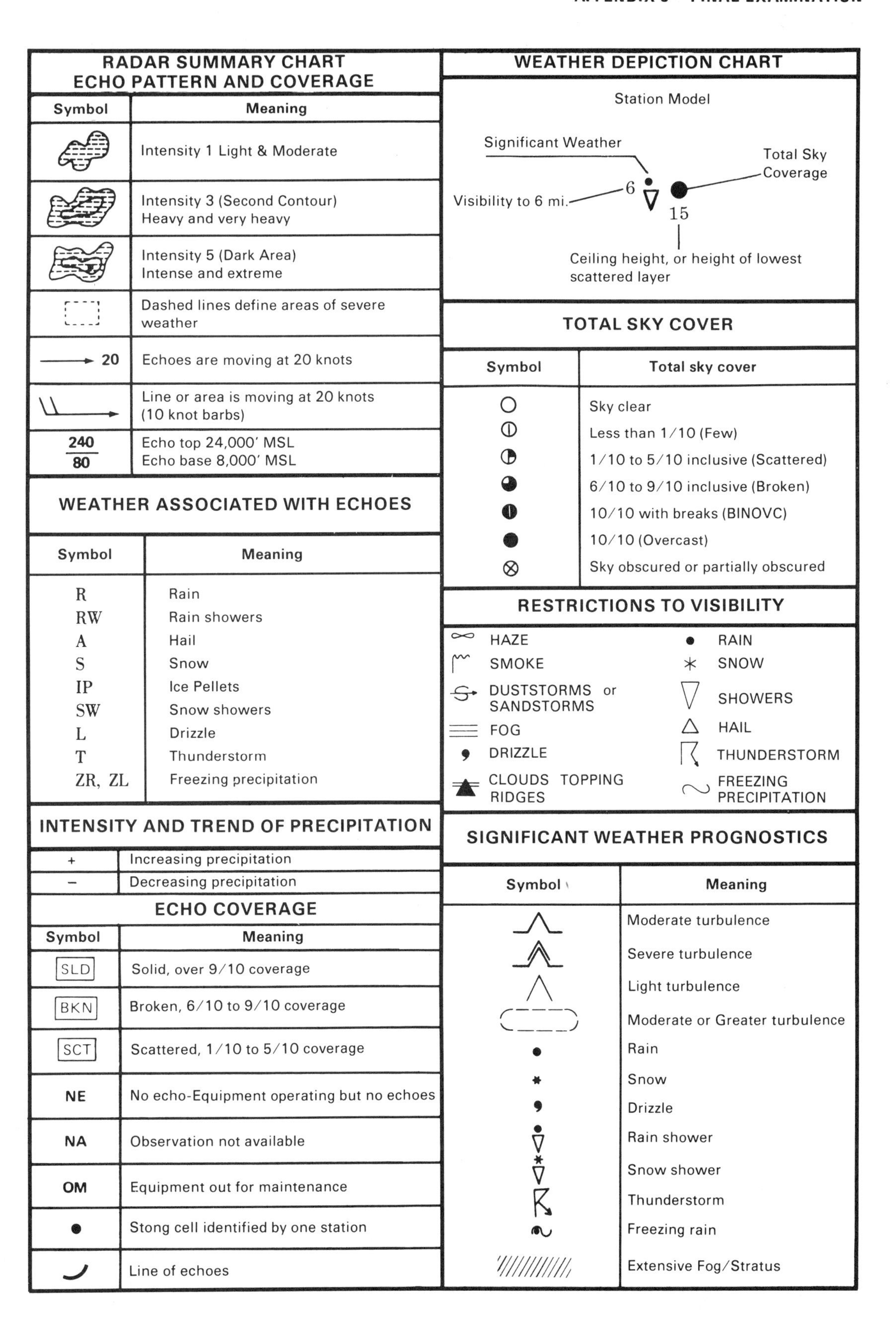

RADAR SUMMARY CHART ECHO PATTERN AND COVERAGE

Symbol	Meaning
	Intensity 1 Light & Moderate
	Intensity 3 (Second Contour) Heavy and very heavy
	Intensity 5 (Dark Area) Intense and extreme
	Dashed lines define areas of severe weather
20	Echoes are moving at 20 knots
	Line or area is moving at 20 knots (10 knot barbs)
240 / 80	Echo top 24,000′ MSL Echo base 8,000′ MSL

WEATHER ASSOCIATED WITH ECHOES

Symbol	Meaning
R	Rain
RW	Rain showers
A	Hail
S	Snow
IP	Ice Pellets
SW	Snow showers
L	Drizzle
T	Thunderstorm
ZR, ZL	Freezing precipitation

INTENSITY AND TREND OF PRECIPITATION

+	Increasing precipitation
–	Decreasing precipitation

ECHO COVERAGE

Symbol	Meaning
SLD	Solid, over 9/10 coverage
BKN	Broken, 6/10 to 9/10 coverage
SCT	Scattered, 1/10 to 5/10 coverage
NE	No echo-Equipment operating but no echoes
NA	Observation not available
OM	Equipment out for maintenance
●	Stong cell identified by one station
	Line of echoes

WEATHER DEPICTION CHART

TOTAL SKY COVER

Symbol	Total sky cover
○	Sky clear
	Less than 1/10 (Few)
	1/10 to 5/10 inclusive (Scattered)
	6/10 to 9/10 inclusive (Broken)
	10/10 with breaks (BINOVC)
●	10/10 (Overcast)
⊗	Sky obscured or partially obscured

RESTRICTIONS TO VISIBILITY

	HAZE	●	RAIN
	SMOKE	✳	SNOW
	DUSTSTORMS or SANDSTORMS	▽	SHOWERS
	FOG	△	HAIL
	DRIZZLE		THUNDERSTORM
	CLOUDS TOPPING RIDGES		FREEZING PRECIPITATION

SIGNIFICANT WEATHER PROGNOSTICS

Symbol	Meaning
	Moderate turbulence
	Severe turbulence
	Light turbulence
	Moderate or Greater turbulence
●	Rain
✱	Snow
	Drizzle
	Rain shower
	Snow shower
	Thunderstorm
	Freezing rain
	Extensive Fog/Stratus

INSTRUMENT RATING

FINAL EXAMINATION ANSWER KEY

1. 4
2. 4
3. 3
4. 2
5. 2
6. 4
7. 3
8. 2
9. 4
10. 2
11. 4
12. 3
13. 3
14. 3
15. 3
16. 1
17. 4
18. 4
19. 2
20. 1
21. 2
22. 4
23. 4
24. 3
25. 4
26. 3
27. 4
28. 1
29. 2
30. 2
31. 3
32. 4
33. 2
34. 3
35. 4
36. 3
37. 1
38. 3
39. 2
40. 3
41. 3
42. 3
43. 4
44. 2
45. 4
46. 2
47. 2
48. 1
49. 4
50. 2
51. 4
52. 1
53. 3
54. 2
55. 2
56. 4
57. 3
58. 3
59. 4
60. 2
61. 2
62. 1
63. 2
64. 1
65. 4
66. 1
67. 3
68. 3
69. 3
70. 4
71. 4
72. 4
73. 1
74. 1
75. 3
76. 1
77. 4
78. 2
79. 3
80. 1

ALPHABETICAL INDEX

A

B

C

D

E

F

G

I

J

K

L

M

N

O

P

R

W

Y